HOW TO USE EXCEL® IN ANALYTICAL C

AND IN GENERAL SCIENTIFIC DATA AN.

HOW TO USE

EXCEL®

IN ANALYTICAL

CHEMISTRY

AND IN GENERAL SCIENTIFIC

DATA ANALYSIS

Robert de Levie
Bowdoin College, Brunswick, ME

CAMBRIDGE
UNIVERSITY PRESS

PUBLISHED BY THE PRESS SYNDICATE OF THE UNIVERSITY OF CAMBRIDGE
The Pitt Building, Trumpington Street, Cambridge, United Kingdom

CAMBRIDGE UNIVERSITY PRESS
The Edinburgh Building, Cambridge CB2 2RU, UK
40 West 20th Street, New York, NY 10011–4211, USA
10 Stamford Road, Oakleigh, VIC 3166, Australia
Ruiz de Alarcón 13, 28014 Madrid, Spain
Dock House, The Waterfront, Cape Town 8001, South Africa

http://www.cambridge.org

First published 2001

Printed in the United Kingdom at the University Press, Cambridge

Typeface Utopia 9.25/13.5pt and Meta Plus   System QuarkXPress® [SE]

A catalogue record for this book is available from the British Library

Library of Congress Cataloguing in Publication data
de Levie, Robert.
How to use Excel in analytical chemistry and in general scientific
data analysis / Robert de Levie.
        p.   cm.
ISBN 0 521 64282 5 (hbk.). – ISBN 0 521 64484 4 (pbk.).
1. Chemistry, Analytic–Statistical methods–Data processing.
2. Science–Statistical methods–Data processing.   3. Microsoft
Excel (Computer file)   4. Electronic spreadsheets.   I. Title.
QD75.4.S8D4   1999
543′.00285′5369–dc21   98-50543   CIP

ISBN 0 521 64282 5 hardback
ISBN 0 521 64484 4 paperback

Disclaimer of warranty

CONTENTS

PREFACE

Chemistry is an experimental science, and primarily lives in the laboratory. No book on spreadsheets will change that. However, many aspects of chemical analysis have significant quantitative, mathematical components, and many of these can be illustrated effectively using spreadsheets. At the same time, the spreadsheet is a very accessible tool for data analysis, an activity common to all of the physical sciences. This book emphasizes the use of spreadsheets in data analysis, while at the same time illustrating some of the underlying principles. The basic strength of spreadsheets was summarized by the name of the very first spreadsheet, VisiCalc, in that it facilitates the *visualization of calculations*, and thereby can help to make theory and data analysis come to life.

Spreadsheets are well-recognized for their near-immediate response to changes in their input parameters, for their ease in making graphs, for their open format and intuitive layout, and for their forgiving error-handling. For these reasons they are usually considered to be the most easily learned computer tools for numerical data analysis. Moreover, they are widely available, as they are often bundled with standard word processors.

Spreadsheets used to be far inferior to the so-called higher-level computer languages in terms of the mathematical manipulations they would support. In particular, numerical methods requiring iterations used to be awkward on a spreadsheet. Fortunately, this has changed with the introduction, in version 5 of Excel, of a macro language (Visual BASIC for Applications, or VBA) that allows the inclusion of standard computer code. Now the immediacy of the spreadsheet and the convenience of its graphical representations can be combined with the wide availability in the literature of sophisticated higher-level programs to make the spreadsheet a powerful scientific as well as didactic tool.

Of course, spreadsheets cannot do everything. While they make quite competent graphs, they lack some of the stunning three-dimensional representations of more specialized, graphics-oriented packages. Moreover, spreadsheets cannot handle symbolic mathematics, and they are unsuitable for highly specialized, computation-intensive tasks such as molecular modeling. However, they are unmatched for ease of learning, and for general availability and price.

Spreadsheets can be used as glorified calculators. There is nothing wrong with that, but there is no need to write about such rather obvious applications here, since there are already a sufficient number of books devoted to this topic. Instead I have tried to illustrate some of the more subtle aspects of data analysis, some of the more specialized features of chemical equilibrium, some of the more abstract underpinnings of modern chemical instrumentation, and some of the finer points of numerical simulation. The choice and sequence of topics closely follows the order in which these are typically encountered in textbooks in analytical chemistry, so that this book can readily be used in courses in quantitative or instrumental chemical analysis. Since the choice of topics is rather wide, the reader is welcome to pick and choose from among these according to his or her own preference and need.

Most chapters start with a brief summary of the theory in order to put the spreadsheet exercises in perspective, and to define the nomenclature used. The standard versions of Excel 95 through Excel 2000 for Windows 95 or Windows 98 are used. Many exercises use the Solver and the Analysis ToolPak, both of which are available in the standard Excel packages but may have to be loaded separately, as add-ins, in case this was not done initially. When use of chapter 10 is contemplated, the VBA help file should also be loaded.

While the specific spreadsheet instructions in this book are for Excel 97 on IBM-compatible computers, they can all be implemented readily (i.e., with no or very minor modifications) in Excel 5 (for Windows 3.1), Excel 95 (for Windows 95), Excel 98 (for the Mac), or Excel 2000 (for Windows 98 or Windows 2000). Moreover, I have indicated where Excel 5 and Excel 95 require different procedures from those in Excel 97, 98, or 2000, namely in their handling of graphs and macros. There are some minor differences between the Excel versions for IBM-compatible and MacIntosh computers. The most important of these are listed in chapter 1; none of them are serious.

Many exercises also work in the earlier versions (1 through 4) of Excel. However, these earlier versions cannot handle VBA macros, so that those spreadsheet exercises that use macros for weighted least squares, fast Fourier transformation, and convolution, cannot be run with versions preceding Excel 5. (Specifically, these are exercises 3.4 and beyond in chapter 3, and all exercises in chapter 7.) Moreover, the macros described in chapter 10 cannot be used in these earlier versions.

Many of the exercises in this book can also be run on spreadsheets other than Excel. In that case, however, apart from the impossibility to import higher-level computer programs into the spreadsheet, the user may also lack the convenience of a powerful multi-parameter non-linear least squares routine such as Solver. Given the choice of writing a book to fit all spreadsheets, or one that exploits the extra power of modern Excel, I have opted for the latter.

The purpose of this book is not to provide its readers with a set of prepackaged routines, into which they merely enter some constants. Instead, the emphasis is on letting the readers gain enough familiarity and experience to enable them to use spreadsheets independently, and in other scientific contexts, while at the same time illustrating a number of interesting features of analytical chemistry. In most cases, no theory is derived, and the reader should consult standard texts on statistics and on quantitative and instrumental chemical analysis for the necessary background information, as well as for a perspective on the strengths and weaknesses of the various methods.

The reader may discover some unavoidable parallelism between the material in this book and that in my undergraduate textbook, *Principles of Quantitative Chemical Analysis*,

McGraw-Hill, 1997, and even some remnants of my *Spreadsheet Workbook for Quantitative Chemical Analysis*, McGraw-Hill, 1992. This is partially because I have retained some of the didactic innovations introduced in these earlier texts, such as an emphasis on the progress of a titration rather than on the traditional titration curve, the use of buffer strength rather than buffer value, and the use of the abbreviations h and k in the description of electrochemical equilibria. However, the present text exploits the power of Excel to go far beyond what was possible in those earlier books.

For a few problems that would require the reader to write some rather complex macros, these have been provided. They are fully documented and explained in chapter 10, and can be downloaded from http://uk.cambridge.org/chemistry/resources/delevie Note that their code is readily accessible, and that the reader is not only encouraged to modify them, but is given the tools to do so. Again, the idea is to empower the reader to incorporate existing higher-language code into macros, in order to increase the reach and usefulness of Excel.

The first chapter introduces the reader to the software; it can be speed-read or skipped by those already familiar with Windows- or Mac-based spreadsheets. The last chapter discusses macros, which can convert a spreadsheet into a powerful computing tool. Sandwiched between these are the four main parts of this book: statistics and related methods, chemical equilibrium, instrumental methods, and mathematical analysis. These parts can be used independently, although some aspects introduced in chapters 2 and 3 are used in subsequent chapters, and the spreadsheet instructions tend to become somewhat less detailed as the text progresses.

The treatment of statistics is focused on explicit applications of both linear and non-linear least-squares methods, rather than on the alphabet soup (F, Q, R, T, etc.) of available tests. However, within that rather narrow framework, many practical aspects of error analysis and curve fitting are considered. They are chosen to illustrate the now almost two centuries old dictum of de Laplace that the theory of probability is merely common sense confirmed by calculation.

Since the spreadsheet is eminently capable of doing tedious numerical work, exact mathematical expressions are used as much as possible in the examples involving chemical equilibria. Similarly, the treatment of titrations emphasizes the use of exact mathematical relations, which can then be fitted to experimental data. In some of the exercises, the student first computes, say, a make-believe titration curve, complete with simulated noise, and is then asked to extract from that curve the relevant parameters. The make-believe curve is clearly a stand-in for using experimental data, which can be subjected to the very same analysis.

For the more instrumental methods of quantitative chemical analysis, I have taken a rather eclectic approach, merely illustrating some aspects that are especially suitable for spreadsheet exploration, such as Beer's law and its applications to the analysis of multicomponent mixtures, chromatographic plate theory, polarography, and cyclic voltammetry.

Because of its important place in modern chemical instrumentation, an entire chapter is devoted to Fourier transformation and its applications, including convolution and deconvolution. The chapter on mathematical analysis illustrates several aspects of signal handling traditionally included in courses in instrumental analysis, such as signal averaging and synchronous detection, that deal with the relation between signal and noise. Its main focus, however, is on numerical analysis, and it covers such aspects as finding roots and fitting curves, integrating, differentiating, smoothing, and interpolating data. Numerical solution of differential equations is the focus of chapter 9, where we discuss a number of kinetic schemes, partially to counterbalance the earlier emphasis on equilibrium behavior.

The final chapter describes the nitty-gritty of macros, and illustrates how they can be used to make the spreadsheet do many amazing things in exchange for relatively little effort on the part of the user, who can simply incorporate pre-existing, well-documented, widely available algorithms.

The aim of this book, then, is to illustrate numerical applications rather than to explain fundamental concepts. Theory is mentioned only insofar as it is needed to define the nomenclature used, or to explain the approach taken. This book can therefore be used in conjunction with a regular textbook in analytical chemistry, in courses on quantitative or instrumental chemical analysis. It can also serve as a stand-alone introduction to modern spreadsheet use for students of chemistry and related scientific disciplines, provided they are already familiar with some of the underlying scientific concepts. Because of its emphasis on exercises, this book is also suitable for individual, home use.

I am grateful to Drs. T. Moisio and M. Heikonen of Valio Ltd, Helsinki, for permission to use their unpublished experimental data in chapter 4, to Professor Phillip Barak of the University of Minnesota for permission to include his adaptive-degree least-squares algorithm in chapter 10, and to Numerical Recipes Software of Cambridge Massachusetts for permission to use some subroutines from the Numerical Recipes.

I am indebted to Professors Nancy Gordon and Gale Rhodes of the University of Southern Maine, Professor Barry Lavine of Clarkson University, Professors Panos Nikitas and Nanna Papa-Louisi of Aristotle University, as well as to Mr. William H. Craig and Professors Andrew Vogt, George Benke, and Daniel E. Martire of Georgetown University, for their many helpful and constructive comments and suggestions. I am especially indebted to Professor Joseph T. Maloy of Scton Hall University for his extensive advice.

I am grateful to Georgetown University for a sabbatical leave of absence, which gave me the unbroken time to work on this book, and to Professor Nancy Gordon of the University of Southern Maine in Portland, Maine, and Professor Panos Nikitas of Aristotle University of Thessaloniki, Greece, for their gracious hospitality during the writing of it. Finally I thank my son, Mark, for his invaluable help in getting me started on this project, and my wife, Jolanda, for letting me finish it.

User comments, including corrections of errors, and suggestions for additional topics and/or exercises, are most welcome. I can be reached at RDELEVIE@BOWDOIN.EDU Corrections will be posted in the web site

http://uk.cambridge.org/chemistry/resources/ delevie

From this web site you can also download the data set used in section 4.11, and the macros of chapter 10.

CHAPTER 1

HOW TO USE EXCEL

First things first: this introductory chapter is intended for readers who have no prior experience with Excel, and only provides the minimum information necessary to use the rest of this book. Emphatically, this chapter is *not* meant to replace a spreadsheet manual; if it were, that part alone would occupy more space than that of this entire workbook. Instead, during and after using this workbook, you may be tempted to consult an Excel manual (of which there will be several in your local library and bookstore) to learn what else it can do for you – but that is up to you.

Second: this book is not intended to be read, but instead to be used while you sit at the computer keyboard, trying out whatever is described in the text. Learning to use a spreadsheet is somewhat like learning to swim, to ride a bicycle, or to paint: *you can only learn it by doing it.* So set aside a block of time (one or two hours should do for this chapter, unless you are really new to computers, in which case you might want to reserve several such sessions in order to get acquainted), make yourself comfortable, turn on the computer, and try things out as they are described in, say, the first three sections of this chapter. (If it confuses you on your first try, and there is nobody at hand to help you along, stop, do something else, and come back to it later, or the next day, but don't give up.) Then try the next sections.

In order to run Excel (or any other spreadsheet program), your computer will need an **operating system**. Here we will assume that you have Windows as the operating system on your personal computer, and that you have a compatible version of Excel. Although there are relatively minor differences between the various versions of Excel, they fall roughly into three categories. Excel versions 1 through 4 did not use VBA as their macro language, and the macros described and used in this book will therefore not run on them. The second category includes Excel 5 and Excel 95 (also called Excel version 7; there never was a version 6), which use VBA with readily accessible modules. Excel 97, Excel 98 (for the Mac), and Excel 2000 make up the third category, which has macro modules that are hidden from sight. The instructions given in this book are specifically for the second and third categories, starting with Excel 5. While they were mostly tested in Excel 97, all versions more recent than Excel 4 will do fine for most of the spreadsheet exercises in this book. Because Excel is *backward compatible*, you can run older software in a more recent version, but not necessarily the other way around.

When you have a Macintosh, your operating system will be different, but Excel will be very similar. After all, both IBM and Mac versions of Excel were written by Microsoft. With relatively minor modifications, mostly reflecting differences between the IBM and Mac keyboards, all exercises in this book will run on the Mac, provided you have Excel version 5 or later.

In either case, whether you use an IBM-compatible PC or a Macintosh, use at least Excel version 5, because earlier versions lacked some of the more useful features of Excel that will be exploited in this book. If you have Excel 4 or earlier, it is time to upgrade.

When you are already familiar with earlier versions of Windows and Excel, you may want to use this chapter as a refresher, or scan the text quickly and then go directly to the next chapter. When you are already familiar with Windows 95 or Windows 98, and with Excel 95 or 97, you may skip this chapter altogether.

1.1 Starting Windows

Windows is a so-called *graphical* user interface, in which many programs, files, and instructions are shown pictorially, and in which many operations can be performed by 'pointing and clicking', an approach pioneered in the early 1970s by the Xerox Corporation, and long familiar to Macintosh users. The pointing device is usually a **mouse** or a **trackball**; for many instructions, equivalent typed commands can be used as well. We will use 'mouse' as the generic term for whatever pointing device you may have. There are often several ways to let the computer know what you want it to do. Here we will usually emphasize how to do it with the mouse, because most users find that the easiest.

In what follows we will assume that Windows and Excel have been installed in their complete, standard forms. For some applications we will also use the Solver and the Analysis Toolpak. These come with Excel, but (depending on the initial installation) may have to be loaded as an add-in.

When you start Windows, your monitor will show a screen (the **desktop**) which typically displays, on its left side, a number of pictures (**icons**), each with its own explanatory label. The bottom icon is labeled 'Start', and acts as the *on* switch of Windows. (There is no simple *off* switch, since Windows requires a more elaborate turn-off routine, which rather illogically begins with the Start button, and via the Shut Down command leads you to the Shut Down Windows dialog box, where you can choose between several options.)

Icons, such as the start label, are also called **buttons**, as if you could actually push them. Move the mouse so that the sharp point of the arrow on the screen, the **pointer**, indeed 'points to' (i.e., is *inside*) the start button, and press the left mouse button once. (Left and right depend, of course, on the orientation of the mouse. By 'left' we mean the left button when the two or three mouse buttons are pointing away from you, so that you can hold the body of the mouse with your thumb and index finger, or with the palm of your hand, while your index finger, middle finger, and ring finger can play with the buttons.) To briefly depress the left mouse button we will call to **click** the mouse; when you need to do this twice in quick succession we will call it **double clicking**, whereas briefly depressing the right mouse button we will call **right clicking**.

As soon as you have clicked the start button, a dialog box will pop up above it, showing you a number of choices. Manipulate the mouse so that the arrow points to 'Programs', which will now be highlighted, and click. A second dialog box will pop up next to the first to show you the various programs available. One of these will be Excel; click on it to start the spreadsheet. Alternatively, click on the Excel icon if the desktop shows it.

1.2 A first look at the spreadsheet

After displaying the Excel logo, the monitor screen will show you a rather busy screen, as illustrated in Fig. 1.2-1. The actual screen you will see may have more bars, or fewer, depending on how the screen has been configured. Please ignore such details for the moment; few if any of the instructions to follow will depend on such local variations.

At the top of the screen is the **title bar**. In its right-hand corner are three icon buttons, to *minimize* the screen to near-zero size, to *restore* it to medium or full size, and to *close* it. To the left on the same bar you will find the Excel logo and the name of the **file** you use, where 'file' is the generic name for any unit in which you may want to store your work. Below the title bar is the **menu bar** (with such menu headings as File, Edit, View, Insert, etc.). This is usually followed by a **standard bar** with icons (pictograms showing an empty sheet, an opening file folder, a diskette, a printer, etc.) and a **formula bar**. At this point, the latter will show two windows, of which the larger one will be empty.

Starting from the bottom of the screen and moving upwards, we usually first encounter the **task bar**, which has the Start button in its left corner. Next to the start button you will find the name of the Workbook you are using. When you have not yet given it a name, Excel will just call it Book1, Book2, etc. Above the task bar is the **status bar**, which may be largely empty for now.

Title bar
Menu bar
Standard toolbar
Formula bar

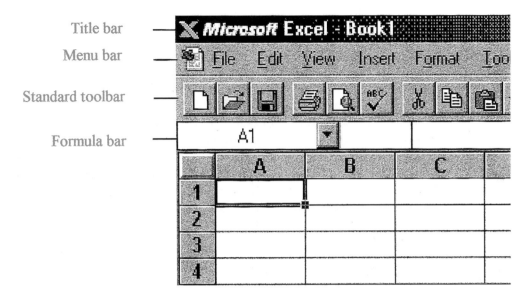

Fig. 1.2-1: The left top corner of the spreadsheet.

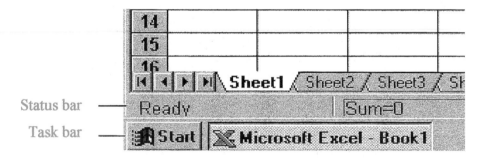

Status bar
Task bar

Fig. 1.2-2: The left bottom corner of the spreadsheet.

What we have described so far is the frame *around* the actual spreadsheet. Now we come to the spreadsheet itself, which is called a workbook, and is organized in different pages.

Above the status bar you will find a tab, in Fig. 1.2-2 labeled Sheet1, which identifies which *page* of the work book is open. Here, then, you see the general organization of individual spreadsheet pages into workbooks. You can have as many pages in your workbook as you wish (by adding or removing sheets), and again as many different workbooks as you desire. For the exercises in the present text, you may want to use a new sheet for each exercise, and a new workbook for each chapter, and label them accordingly.

In the region between the formula bar and the status bar you will find the actual working part of the spreadsheet page. It starts at the top with a sequence of rectangles, each containing one letter of the alphabet on a gray background. It ends, at the bottom, with a bar containing a series of tabs;

one such **tab**, such as the one labeled Sheet1 in Fig. 1.2-2, will have a white background, indicating the currently open (or 'active') sheet, while the others will be gray. In between these is a rectangular array of blank cells. Each such **cell** can be identified by its (vertical) **column** and its (horizontal) **row**. Columns are labeled by the letters shown just above row 1 of the spreadsheet, while rows are labeled by the numbers shown to the left of column A. The cell at the top left of the spreadsheet is labeled A1, the one below it A2, the one next to A2 is B2, etc. One cell will be singled out by a heavy black border; that is the highlighted, **active** cell in which the spreadsheet anticipates your next action. The **address** of the active cell is displayed in the left-most window of the formula bar; in Fig. 1.2-1 it is cell A1.

To **activate** another cell, move the mouse so that the pointer, which should now have the shape of a hollow **cross**, is within that cell, then click. The corresponding cell coordinates will show on the left-most window of the formula bar. When you move the mouse pointer to another cell and click again, that cell will now become the active one. Note that the left-most window in the formula bar will track the coordinates of the active cell. Play with moving the active cell around in order to get a feel for manipulating the mouse.

A cell can also be specified by typing its coordinates. The simplest way to do so is by using the **function key** labeled F5. (The function keys are usually located above the regular alphabet and number keys, and labeled F1 through F10 or F12. On some keyboards they are found to the left of the alphabet keys.) A **dialog box** will appear, and you just type the coordinates of the cell, say, D11, and deposit this by depressing the large 'enter' key (to the right of the regular alphabet keys). Another way, initially perhaps more convenient for those used to DOS-based spreadsheets, is to use the keystroke sequence Alt + e Alt + g. Here Alt + e denotes that you depress Alt and then, *while keeping Alt down*, also depress e; follow this by Alt + g. Alt specifies that you want to select an item from the menu bar, e selects the Edit command, and g the Go to command, where the underlining indicates the letter to be used: e in Edit, g in Go, o in Format, etc. As a gesture to prior users of Lotus 1-2-3 or QuattroPro, you can even use the slant instead of the Alternate key: / + e / + g . Any of the above methods will produce the dialog box in which to type the cell coordinates.

Below we will usually indicate how to accomplish something by using the mouse. For those more comfortable with using the keyboard rather than the mouse, keystrokes to accomplish the same goals are often available. There is no need to memorize these commands: just look for the underlined letters to find the corresponding letter code. Using keystrokes is often faster than pointing-and-shooting with a mouse, especially when you use a track ball.

Note that, inside the cell area of the spreadsheet, the mouse pointer usually shows as a cross. Select a cell, then move the pointer away from it and back again. You will see that, near the border of the active cell, the pointer changes its shape and becomes an **arrow**. When the pointer shows as an arrow, you can depress the left mouse button and, while *keeping it down*, move the pointer in the cell area. You will see that this will **drag** the cell *by its border*. By releasing the mouse button you can deposit the cell in a new location; the formula bar will then show its new coordinates.

Practice activating a set of neighboring, contiguous cells; such cell **blocks** or **arrays** are often needed in calculations. Move your mouse pointer to a particular cell, say cell F8, and click to activate it. You can now move the pointer away, the cell remains active as shown by its heavy border; also, the formula bar shows it as the active cell regardless of where you move the mouse pointer, as long as you don't click. Return the pointer to cell F8, and depress the left mouse button *without* releasing it, then (while still keeping the cell button down) move the mouse pointer away from cell F8 and slowly move it in a small circle around cell F8. You are now outlining a cell **block**; its size is clear from the reverse color used to highlight it (it will show as black on a white background, except for the cell with which you started, in this example F8, which will remain white, and which we will call the **anchor cell**). The size of the block will show in the formula bar in terms of rows and columns, e.g., $3R \times 2C$ will denote a block three rows high and two columns wide. By releasing the mouse button you activate the entire block, while the formula bar will return to showing the location of the anchor cell. You can then move away from it; the active block will remain. After you have selected the cell block, go back to it, grab its border (when the pointer is an arrow) and move the entire block around! To deposit the block in a new location, just release the mouse button. To abolish a block, release the mouse button to deposit it, then move the pointer to another cell and click on it.

To activate a block of cells from the keyboard, use F5 (or Alt + e Alt + g), then specify the block by the coordinates of its upper left cell and of its lower right cell, separated by a colon, as in D4:E9, and deposit it with the enter key.

There is yet another way to activate a block, starting from a single active cell. Again move the mouse pointer outside the active cell, but now approach the small square in the right bottom corner of the border around the active cell; this little square is the cell **handle**. The mouse pointer will change into a plus sign when it points to the cell handle; you can then drag the cell *by its handle* (rather than by its border) and make either columns or rows. Again, fix your choice by releasing the mouse button. You can drag it again to make a block out of a row or column. Practice these maneuvers to familiarize yourself with the mouse, and see how the pointer changes from a hollow cross (when you point at the middle of the cell) to an arrow (at its border) to a plus sign (at its handle). Below we will specifically indicate when to use the cell border or the cell handle; if nothing is specified, go to the center of the cell and use its standard pointer, the hollow cross.

1.3

A simple spreadsheet and graph

The spreadsheet is designed to facilitate making calculations, especially repeated calculations that would quickly become tedious and boring if you had to do them by hand, or even on a pocket calculator. (Unlike humans, computers do not tire of repetition.) The spreadsheet is also very useful when you have computations that would be too difficult for a pocket calculator. The tabular format, resembling an accountant's ledger, helps us to organize the calculations, while the so-called 'double precision' of the spreadsheet keeps round-off errors in check. When you change one number somewhere in a spreadsheet, the computer automatically recalculates all cells that depend on the one you have just changed. (There are a few exceptions to this statement: special functions and macros do not update automatically. We will alert you when we come to them.) The spreadsheet also makes it very easy to construct graphs. We will demonstrate this now that you know how to move around in the spreadsheet.

Among the things we can place inside a cell are a **number**, a **label** such as a column heading, or a **formula**. A cell can hold only one of these items at a time. Activate cell A1 by clicking on it, then type the letter x, followed by depressing the 'enter' key, or by moving the mouse pointer to a different cell and by then clicking on that other cell. Either method will deposit the typed letter.

Activate cell A3; to do this, either move the mouse pointer to cell A3 and click, or use the down arrow to get there. In cell A3 deposit the number 0. (As with the letter x, nothing will happen until you deposit it, using the Enter key. This lets the computer know that this is all you want to enter, rather than, say, 0.3 or 0.0670089.) Be careful to distinguish between the number 0 and the letter O; they are close neighbors on the keyboard but they are completely different symbols to the computer. Similarly, don't confuse the number 1, the lowercase letter L, and the capital I.

In cell A4 deposit the number 1. The letter x in A1 will usually show as left-justified (i.e., placed in the left corner of its cell), whereas the numbers 0 and 1 will usually be right-justified. (We hedge our bets with the 'will usually be' because all these features can easily be changed, as they may well have been on the computer you are using.) Return to cell A3, then activate both cells (by depressing the left mouse button while pointing to A3, keeping it down while moving to cell A4, then releasing the button). Both cells should now be active, as shown by their shared border.

Now comes a neat trick: grab both cells by their common handle (the little square at the right-hand bottom of their common border), drag the handle down to cell A11, and release the mouse button. With this simple procedure you have made a whole column of numbers, each one bigger by 1 than that in the cell above it!

	A	B	C
1	x	sine	
2			
3	0	0	
4	1	0.707107	
5	2	1	
6	3	0.707107	
7	4	1.23E-16	
8	5	-0.70711	
9	6	-1	
10	7	-0.70711	
11			

Fig. 1.3-1: Detail of the spreadsheet with its two columns and column headings.

Had you started with, say, the number 7 in cell A3, and 4.6 in cell A4, column A would have shown 7, 4.6, 2.2, -0.2, -2.6, and so on, each successive cell differing from its predecessor by $4.6 - 7 = -2.4$. In other words, this method of making a column generates constant increments or decrements, in arithmetic progression. Try this, with different values in A3 and A4. Then go back to deposit the series ranging from 0 to 7 with an increment of 1 or, in mathematical notation, the series 0 (1) 7. Incidentally, there are many other ways to fill a column, some of which we will encounter later.

In column B we will now calculate a sine wave. Activate cell B1 and deposit the heading 'sine'. Move to cell B3 and deposit the formula $= \sin(a3*pi()/4)$. The equal sign identifies this as a formula rather than as text; the asterisk indicates a multiplication. The spreadsheet uses the notation pi() to denote the value of π; the brackets alert the computer that this is a function. Excel instructions do not distinguish between lower case and capitals, but the formula bar always displays them as capitals, which are more clearly legible. By now your spreadsheet should look like that depicted in Fig. 1.3-1.

If you were to extend the columns to row 11, the value shown in cell B11 might baffle you, since it may not quite be 0 but a small number close to it, reflecting computer round-off error. But don't worry: the error will usually be below 1 part in 10^{15}.

There is a more convenient way to generate the second column. After you have entered the instruction $= \sin(a3*pi()/4)$ in cell B3, grab its handle (at which point the mouse arrow will show as a plus sign) and double-click. This will copy the instruction down as far as the column to its immediate right contains data! This is a very useful method, especially for long columns.

When there are no data to its immediate right, the column to its immediate left will do. When both are absent, the trick will not work.

Finally we will make a graph of this sine wave. Doing so is slightly different in Excel 97, Excel 98 for the Mac, or Excel 2000 on the one hand, and Excel 95 or Excel 5 on the other. We will here describe the procedure for each of these two versions.

Bring the mouse pointer to cell A3, click on it, drag the pointer (while keeping the mouse button depressed) to cell B11, then let go of the mouse button. This will activate (and highlight) the rectangular area from cell A3 through B11 (in spreadsheet parlance: A3:B11) containing the data to be graphed. Alternatively, you can highlight cell A3, then depress the Shift key, and while keeping this key down depress End, ↓, End, and finally →. (The sequence Shift + End, Shift + ↓ will highlight the column A3:A11, while Shift + End, Shift + → will include column B. As with double-clicking on the cell handle to copy an instruction, Shift + End looks for contiguous data.)

1.3a Making a graph in Excel 97 or a more recent version

If this is your first reading, and you use Excel 95 or Excel 5, skip the following, and continue with section 1.3b.

In Excel 97 or a more recent version, go with the mouse pointer to the menu bar, click on Insert, and in the resulting drop-down submenu click on Chart. Or achieve the same result with the keystrokes Alt + i, Alt + h. Either method will produce a dialog box labeled Chart Wizard – Step 1 of 4 – Chart Type.

In the list of Chart types, click on XY (Scatter); do *not* select the Line plot, which in Excel means something quite different from what a scientist might expect. The line plot can give you very misleading graphs because it presumes that the x-values are always equidistant.

As soon as you have selected the XY plot, the right-hand side of the dialog box will show five Chart sub-types: loose points, points connected by smooth or straight lines, or just smooth or straight lines. For now, pick the points connected by smooth lines – you can always change it later. (This is a general property of working with Windows Excel: you need not agonize over a choice, because there are almost always opportunities to change it later. So the best strategy is: when in doubt, pick *something*, move on, and worry about the details later.) Click on the Next > button.

Step 2 of the Chart Wizard shows the Data range selected. Also, under the Series tab, it shows which column will be used for X-values, and which for Y-values. The default (i.e., the assumption the spreadsheet makes in case you do not overrule it) is to use the left-most column of the selected block for X-values, so you need not take any action here, just press on with Next >. But it is handy to know that you can here, in step 2 of the Chart Wizard, change the assignments for X and Y.

Step 3 lets you enter a Chart title and axes labels. Click on the Chart title window, and enter Sine wave. Then click in the Value (X) Axis window, and enter angle. Finally, click in the Value (Y) Axis window, and enter sine. A picture will show you what your graph is going to look like.

There are other things you can specify at this point, such as the axes, grid-lines, legends, and data labels, but we will forgo them here in order to keep things simple for now, and to illustrate later how to modify the end product. So, on to the Next >.

Step 4 defines the chart location, either As a new sheet, or As object in a spreadsheet page. Select the latter, and Finish. This will place the graph on the spreadsheet.

Now click on the graph, preferably inside its outer frame near its left edge, where the computer cannot misinterpret your command. This will adorn the graph with eight black handles, which allow you to change its size and location. First, locate the mouse pointer on the graph, depress the mouse button, and while keeping it down move the graph to any place you like on the spreadsheet, preferably somewhere where it does not block data from view. To release, simply release the mouse button. Note that the graph as it were floats on the page, and does not obliterate the underlying information. To fit the graph in the cell grid, depress the Alt key, then (while keeping Alt depressed) bring the mouse pointer to a handle in the middle of the side of the graph, where the pointer should change into a two-sided arrow, and pull that pointer toward a cell boundary. Repeat with the other sides. For greater efficiency you can combine this for two adjacent sides by pulling or pushing on two opposing corners.

In the final result, click on the little rectangular box to the right of the graph, then press Delete.

If you want to remove the gray background (which seldom prints well) just click somewhere in the plot area (where the label shows Plot Area), right-click, highlight Format Plot Area, and under Area either select None or, in the choice of colors, click on white. Exit with OK.

If you want to get rid of the horizontal grid lines, point to them (the label will identify Value (Y) Axis Major Gridlines), right-click, and select Clear.

To change the range of the x-scale, point to the axis (the label will show Value (X) Axis), right-click, select Format Axis, and under the Scale tab pick the scale properties you want. And, while you're at it, please note that you can also change the font, size, color, position, and alignment of the numbers of the x-axis. Ditto for the numbers on the vertical axis.

To change the type of graph itself, point at the curve, right-click, and select Format Data Series. Then for the Line pick the Style, Color, and Weight you like, and for Marker the Style, Foreground and Background color, and Size.

And so it goes: you can point at virtually every detail of the graph, and modify it to your taste. Figure 1.3-2 shows you what you might have wrought.

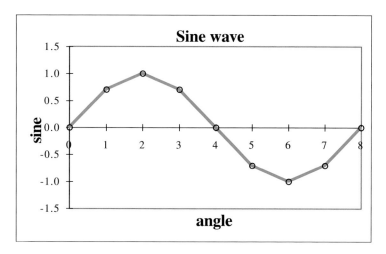

Fig. 1.3-2: The graph showing your sine wave.

1.3b Making a graph in Excel 5 or Excel 95

If this is your first reading, and you use Excel 97, Excel 98, or Excel 2000, skip to the last two paragraphs of this section.

Go with the mouse pointer to the menu bar, click on Insert, and in the resulting drop-down submenu click on Chart. A second box will appear, which lets you select a graph either On the spreadsheet, or As a separate sheet. Select the former by clicking on it. You will now see a succession of ChartWizard boxes that let you specify how the graph should look. You can achieve the same result with the keystrokes Alt + i, Alt + h, Alt + o, Enter, with i for Insert, h for Chart, etc. Either method will produce a dialog box labeled ChartWizard.

The first ChartWizard box, labeled Step 1 of 5, asks you what area of the spreadsheet you want to be graphed. Since you already selected that area, the window with the heading Range should show = $A$3:$B$11. If it does, move the pointer to the Next > button, and click. If it does not, first move the pointer to the Range window, click, if necessary replace its present contents by A3:B11 (use the Delete key located to the right of the enter key, type A3:B11, and click again to deposit this), then click on the Next > button and proceed to step 2.

The second ChartWizard box lets you specify the type of graph you want. Click on the XY (Scatter) plot; your choice will be highlighted. (Do *not* select the Line plot, because it will automatically assume that all X-values are equidistant. This is convenient when you want to plot, e.g., income or expense as a function of the month of the year, or the region of the country. In scientific applications, however, it makes no sense to treat the X-values merely as labels, and it can yield quite misleading graphs.) Click on Next > to move to the next ChartWizard.

The third box lets you define the data presentation. Let's just select 2, which will show the individual data points in a linear graph, connected by line segments. If you want to see what the other presentation styles look like, try them out, either now or, better yet, after you have made your first few charts. Excel has many options, and often several ways to achieve each of them. Here we describe only a few simple ways to get you started, without confusing you with many possible alternatives. After you have become familiar with the spreadsheet, by all means play to find out how to move around in Excel, what all is available, and what formats and shortcuts you like; then use those.

The fourth box shows you a sample chart. The top right-hand corner will let you specify whether you want to plot rows or columns; we will usually plot columns, and that will most probably already have been selected. On to the Next > step of the ChartWizard.

Step 5 allows you to add a legend, and to label the axes. If the question Add a Legend? is answered affirmatively, push the radio button to Yes. Point to the rectangular window under the heading Chart Title, click on it, then type a title of your choice, say, Sine wave, and deposit that title. Similarly, enter a legend for the X-axis (in the text box next to Category [X]:), and a legend for the Y-axis (in the box next to Value [Y]:). That is all for now: click on the Finish button in the lower right-hand corner of the ChartWizard. You should see the graph, properly scaled, with tick marks and associated numbers, and it should look more or less like Fig. 1.3-2 (although there will almost certainly be differences in the exact scaling, letter type used, and so on, details that will not concern us here). If you had made the graph As a separate sheet, click the mouse on the **tab** labeled Sheet1 at the bottom of the spreadsheet; to go back again to the graph, click on the tab labeled Chart1, etc.

We will now add a few finishing touches. The numbers for the horizontal scale in Fig. 1.3-2 are placed just below the horizontal axis, at $y = 0$. It is nice that Excel selects and labels the scales for you, automatically, but you may want to have the numbers outside rather than inside the graph area. In that case, point with your mouse to a number with the horizontal axis, and click on it. This will result in two black blocks, one on each end of the axis, showing that you have activated the axis. Right-click to produce a small pop-up menu, and click on Format Axis, then select the tab Patterns, click on Tick mark lables Low, and end with OK.

Figure 1.3-2 contains the few points you have calculated, with connecting line segments. In this case, where we deal with a continuous function, it will look much better when we use a 'French curve' to connect the points with a smooth line. There are two ways to do so. The obvious one is to calculate more points per cycle, so that the points get closer together, the linear segments are shorter, and therefore more closely approach a smooth curve. The easier one (OK as long as you do not use the curve for precise interpolation) is to let the computer draw a smooth curve through the points, which

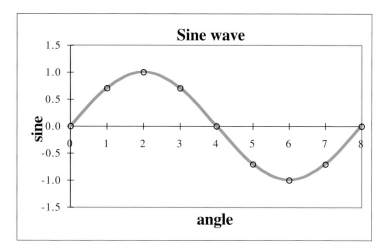

Fig. 1.3-3: The same graph after smoothing.

it will do with ease using what is called a **cubic spline**. You can do this as
follows: double-click on the graph, click on a connecting line segment,
right-click on it to get its properties, then click on Format Data Series. In the
Format Data Series dialog box, click on S_moothed Line, followed by OK.
That does it. The effect is shown in Fig. 1.3-3.

   Finally, we change the font of the legends and labels. First get the
Formatting toolbar with _View ⇒ _Toolbars ⇒ Formatting. Now click on the
axis numbers, then in the Formatting toolbar select Times New Roman and,
in the adjacent Font Size window, click on 12 (points). Do this for both axes.
Then click on the axis labels and the graph title and adjust them likewise. It
doesn't matter whether you prefer the cleaner-looking sans-serif fonts like
Ariel, or the more readable serif fonts such as Times New Roman; the
purpose of the present exercise is merely to show you how to change it to
your taste. Incidentally, instead of using the Formatting bar you can click on,
say, the axis numbers, and then use F_ormat ⇒ Se_lected Axis to get the
Format Axis dialog box, in which you can accomplish the same tasks as with
the Formatting toolbar.

Addressing a spreadsheet cell

It is useful to go back to the spreadsheet and see what you have done. Bring
the mouse pointer to cell B3, click on it, and observe the instruction shown
in the formula bar: it should read $= \text{SIN}(\text{A3*PI()/4})$. Now move the pointer to
cell B4 (again it should show a cross) and click on it. The formula bar will
show the instruction as $= \text{SIN}(\text{A4*PI()/4})$. Move to the cell below, and
examine its instruction: it will read $= \text{SIN}(\text{A5*PI()/4})$, and so on. Clearly, as
you copied the instruction from cell B3 down, the **address** of the cell to

which the instruction referred was also pulled down, from A3 to A4 to A5 etc. This is called **relative addressing**, and is a main feature of all spreadsheets. In other words, the instruction refers to a cell in a given position *relative* to that of the cell from which it is called. It is as if the instruction reads: take the sine of $\pi/4$ times the contents of the cell to my immediate left. In copying a formula in a spreadsheet from one cell to another, relative addressing is the norm, i.e., the **default**, the operation you get without specifying anything special. An example of relative addressing in a different context is the movement of a knight on a chess board. In fact, most chess moves are relative to the starting position of the moving piece.

Sometimes we need to refer to a particular cell, for instance when such a cell contains a constant. In that case we must specify that we want **absolute addressing**; we do this by preceding both components of the cell address (its column letter and its row number) by that symbol of stability, the dollar sign. (We already encountered this notation in the previous section, where the block A3:B11 showed in the first ChartWizard dialog box as the range $= \$A\$3:\$B\$11$.) We can also protect the column but not the row, by placing a dollar sign in front of the column letter, or vice versa; we will occasionally encounter such **mixed address** modes in subsequent chapters. To return to our earlier analogy: the movement of a chess pawn is relative, except at its first move, or when it reaches the opposite end of the board, at which points its absolute address counts.

Now go back to column A, and examine its cell contents. Here we find no specific formula, but only numbers. The way we generated that column of numbers, by dragging its top two cells by their common handle, was convenient and quick, but did not give us much flexibility to change it later. If we anticipate that wc might subsequently want to modify the contents of column A, here are two alternative ways to do so.

First, deposit the number 1 in cell F1. Then go to cell A4, and there deposit the instruction $+A3+\$F\$1$. (You can type it as shown or, faster, first type $+A3+F1$ followed by depressing the function key F4, which will insert the two dollar signs for you. Please don't get confused: F1 here means column F row 1, while 'function key F4' signifies the function key so labeled.)

Now copy this instruction down to cell A11; again, there are several ways to do this. They all start with cell A4 as the active cell; if cell A4 is not the active cell, make it so by clicking on it. Then try out the alternative methods described below:

(a) Depress the control key labeled Ctrl (there are two on the usual keyboard, one on each side of the 'space bar') and, with the Ctrl key down, also depress the letter c; this combination will from now on be denoted by Ctrl + c. (If you have been brought up with the DOS taboo never to use Ctrl + c, there are numerous other ways to do the same thing. For example, click on Edit in the menu bar, then on Copy, or use the keystrokes Alt + e Alt + c instead. You can also click on the copy icon in the icon bar, indicated by two sheets to the right of the icon showing scissors. In Excel 95 and subsequent versions, you can point to the icon if you are not sure of its meaning, and wait one or two seconds: an explanatory note will appear to tell you its function.

Ctrl + c makes a **copy** of the active cell, and stores it in a place in the computer memory called the **clipboard**. Drag the active cell down to generate a column from A4 through A11 (make sure that the mouse pointer is the cross, so that you make a column rather than just move a single cell around), then **paste** the contents of the clipboard in this column with the command Ctrl + v (or E̲dit ⇨ P̲aste on the menu bar, the Paste icon on the icon bar, or Alt + e, Alt + p from the keyboard).

(b) When you want to make a long column, from A4 all the way to, say, A1394, it is more convenient to use the PageDown key rather than to drag the active cell. In that case we again start with copying the active cell with Ctrl + c. Now depress the Shift key while depressing the PageDown key until you are roughly where you want to be, and fine-tune with the up or down keys to reach your destination, all the time keeping the Shift key down. Release the shift key only when your column has the required length, then press Ctrl + v to paste the instruction from the clipboard into the now activated column A4:A1394.

(c) Even faster (for such a long column) is the following method. Activate cell A4, copy it onto the clipboard (Ctrl + c), then select the Goto function key F5. This invokes the Go To dialog box; in its R̲eference window type A1397, click on OK, and you will now find yourself in cell 1397. While keeping down the shift key, now select End and the arrow up key, ↑, then paste with Ctrl + v. Bingo.

The above methods illustrate the use of relative and absolute addressing. Now let us look at the result. Go to cell F1 and deposit the value 2; immediately, column A will show the sequence 0, 2, 4, 6, etc. Play with it, and satisfy yourself that the constant value stored in cell F1 indeed determines the increment. The constant in F1 can be a fraction, a negative number, whatever. Then go to cell A3 and deposit a new starting value, say − 3. Again the data in column A adjust immediately, as do the values in column B that depend on it. You now have much more flexibility to modify the contents of column A, without having to reprogram the spreadsheet.

1.5 More on graphs

Graphs are such an important part of spreadsheets because most of us can take in the meaning of a figure much faster than that of formulas or of a column of numbers.

First we lengthen the columns in the spreadsheet to contain more data. Go back to the (left-hand) top of the spreadsheet; the fastest way to do so is with Ctrl + Home (i.e., by depressing Control while hitting the Home key, which you will usually find in the key cluster above the arrow keys). Using any of the methods described in section 1.4, you can now extend column A3:A11 to A83, then go to cell B11 and double-click on its handle. Alternatively you can extend columns A and B simultaneously: highlight the two adjacent cells A11:B11, copy these with Ctrl + c as if they were one cell, go down to cell A83, use Shift + End + Up to highlight A12:A83, and paste with Ctrl + p. This will copy both columns.

The spreadsheet should now contain several complete cycles of the sine wave. However, the graph does not yet reflect this, because you had earlier specifically instructed it to plot A3:B11. Check that this is, indeed, the case. We will now modify this.

With the mouse, point to the line in the graph, and press the Enter key. You will see some points in the graph highlighted, while the formula bar will contain the graph range, in a statement such as = SERIES (,Sheet1!$A$3:$A$11,Sheet1!$B$3:$B$11,1). Quite a mouthful, but let that be so. Simply move your mouse pointer to that statement, specifically go to the 11's in it, and change them into 83's. Then press Enter; the graph will now show the entire set, B3:B83 versus A3:A83.

Instead of modifying Chart1 we can also make a new graph. Because our earlier graph was embedded in the spreadsheet, now make a separate graph. Embedding a graph has the advantage that you can see it while you are working on the spreadsheet, and the disadvantage that it tends to clutter up your workspace, and that (in order to keep them visible on the screen) embedded graphs are usually quite small. On the other hand, graphs on the spreadsheet can be moved around easily, because they as it were float on the spreadsheet. Likewise, their size can be changed readily. (In Excel 97 etc., the two types of graph are treated as fully equivalent, and you can readily change them from one type to another. Activate the chart, then select Chart ⇨ Location and use the dialog box. Note that the Chart menu appears only after you have activated a chart, otherwise the same location hosts the Data menu label.)

The next two paragraphs are intended specifically for users of Excel 5 or Excel 95. If you use a more recent version of Excel, which treats embedded and separate charts the same way, you may want to speed-read (or skip) this part.

Highlight (activate) block A3:B83. (You can do this most conveniently as follows: go to cell A3 and, while keeping the Shift key down, press End →, then End ↓.) Click on Insert Chart, then select On this sheet. The mouse pointer will change into a cross with a small histogram attached, the histogram being Excel's idea of a graph. Bring the pointer to the left top corner of cell D1, and click. Reenter the ChartWizard, which will show the highlighted area as = $A$3:$B$83. Click on Next. In step 2, select the XY(Scatter) plot, then click on Next. In step 3, select 2, then Next. In step 4 use Data Series in

Columns, Use First 1 Column(s) for X Data, Use First 0 Row(s) for Legend Text, then Next >. In step 5, Add a Legend Yes, Chart Titles: Sine wave, Axis Titles Category (X): angle, Value (Y): sine, then press Finish.

If you are adventuresome, make alternative choices and see what they do. There is no penalty for experimenting; to the contrary, this is how you will quickly become familiar with the spreadsheet. If you don't like the choices you have made, select Back to back up in the ChartWizard steps, and change your choices; if you dislike the final result, just scrap it and start over again. To abolish the graph, bring the mouse pointer anywhere inside the graph area, click on it, then use the Delete key to abolish it. To modify it, highlight the curve and make your changes in the formula bar. *You* are in charge here, the spreadsheet is your willing servant.

Again, the graph you just made may need some adjusting. First let us do its positioning. Bring the mouse pointer to the graph (anywhere inside the figure or its edge will do) and click. The graph will now be identified by eight handles, one on each corner, and one in the middle of each side. These handles are there for you to grab if you want to move or resize the graph.

In order to move the graph as a whole rather than to resize it, click with the pointer anywhere inside the figure (but not on any handle), drag it to another place, then drop it there by releasing the mouse button. In order to move it again, click again on the graph, grab it, and this time move it right smack on top of the data in block A3:B83. As you will see, it does not matter: it really floats on top of the data, and you can pick up the chart again, and place it somewhere else on the spreadsheet, thereby freeing the A and B columns. These columns will emerge unscathed, since you did not erase them, but only placed an image over them. It is like the sun, which is not obliterated by a cloud moving in front of it, but is merely blocked from our view.

Now resize the graph. Activate the graph again, and go to the middle bottom handle. When you are on target, the pointer will change into a vertical double arrow. Now you can drag the handle, up or down. Likewise you can move the other borders. You can also grab a corner, which allows you to change the graph size simultaneously in two directions. If you like to nest the graph neatly inside the spreadsheet, you may want the borders to line up with cell boundaries. You can achieve this by depressing the Alt key while dragging the borders, in which case the graph boundaries will jump from line to line. Use this to make the graph fit the area D1:F9.

Place the label 'second sine' in cell C1. Go to cell C3, and deposit the formula = 0.7* sin(A3*pi()/16). (Note that you must use * to specify multiplication: = 0.7sin(A3*pi()/16) will *not* be accepted.) Copy this instruction all the way down to cell C83 by double-clicking on its handle. Now plot the second sine wave versus X, again embedding the graph in the spreadsheet. The more figures, the more fun!

Go to cell A3, and highlight the range A3:A83 (e.g., with Shift + End, Shift + ↓). Then release the shift key and, instead, depress the Ctrl key, and keep it down. With the mouse, move the pointer sideways to cell C83, release the Ctrl key, and use Shift + End, Shift + ↑, i.e., depress the Shift key, and press End ↑. You will now have marked two non-adjacent columns.

Click on Insert, Chart, On this sheet, place the new graph next (or below) the earlier one, and answer the ChartWizard; you already preselected the Range in step 1 as $A$1:$A$83,$C$1: $C$83. (When you prefer to type in the range rather than to point to it, this shows you the format to use, except that you can leave out the dollar signs: just type A1:A83,C1:C83.) Answer the other ChartWizard queries, look at the result, and if necessary reposition the graphs to resemble Fig. 1-5.

(In Excel 97 and later versions there is an even easier way: activate the plot, click on Chart ⇨ Add Data, then specify the Range in the Add Data dialog box.)

Do you want to change the markers indicating the individual points? Click on a graph. Then position the mouse to point to a marker, and click again (sometimes it requires a few clicks) until a few markers are highlighted. At that point double-click, and a Format Data Point or Format Data Series dialog box will appear. (The latter is actually a whole series of boxes, each selectable by clicking on its tab. The top dialog box is labeled Patterns, and is the one to play with here.)

Either dialog box allows you to select or modify the type of plot: whether you want to show the data as a line, as points only, or as their combination; what color and line thickness you want for the line, and/or what type and color of markers you wish to use. Either box shows you what the line and marker will look like; click on OK when you are done making your selection, or on Cancel when you do not want any changes.

At this point you get the idea: once you have learned to ride this horse, it will do most anything you want from it to make life easy for you. You want to change the axes: click on them, then double-click, and a magic box will appear to ask for your wishes. You want to change the legend, the font used, whatever – the possibilities are endless. Most changes beyond the simplest use dialog boxes: they allow you to order your graphs à la carte.

Back to serious business: these graphs represent your spreadsheet data. Even if you now modify those data, the graphs will reflect the numbers in your spreadsheet. For example, go to cell B36, there deposit the instruction = 0.2*cos(A36*pi()/8) + $G$1, then copy this down through cell B67. The top graph will immediately show the modification, because it plots column B. Now go to cell C3 and modify it (again using the edit keystroke, F2) by adding to the already existing instruction = SIN(A3*PI()/16) a second term, + 0.3*B3, and deposit it (with the Enter key). Copy the instruction down to C83, by double-clicking on its handle. Look at the second graph, which represents column C: it now shows the sum of a sine and cosine wave, including the

	A	B	C	D	E	F
1	x	sine	second sine			
2						
3		0	0	0		
4		1	0.70710678	0.13656323		
5		2	1	0.2678784		
6		3	0.70710678	0.38889916		
7		4	1.2251E-16	0.49497475		
8		5	-0.7071068	0.58202873		
9		6	-1	0.64671567		
10		7	-0.7071068	0.6865497		
11		8	-2.45E-16	0.7		
12		9	0.70710678	0.6865497		
13		10	1	0.64671567		
14		11	0.70710678	0.58202873		
15		12	3.6754E-16	0.49497475		
16		13	-0.7071068	0.38889916		
17		14	-1	0.2678784		
18		15	-0.7071068	0.13656323		
19		16	-4.901E-16	8.576E-17		
20		17	0.70710678	-0.1365632		
21		18	1	-0.2678784		
22		19	0.70710678	-0.3888992		

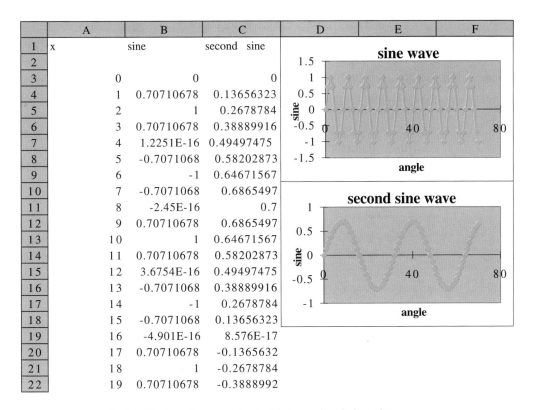

Fig. 1-5: The top of a spreadsheet with two embedded graphs.

modified section in column B, and the vertical scale has changed to accommodate the data. Then deposit a number in cell G1, and see what happens.

In the Chart Wizard we have encountered how to give the graph a title and how to label its axes; now we will see how to introduce annotating text anywhere in the actual graph. To activate the inner frame of the graph, locate the mouse pointer inside this inner frame but away from any specific feature such as a data point or curve, and click so that this inner frame becomes accentuated. Now click on the formula window (the larger window in the formula bar), type the text you want to introduce, and hit the enter key, whereupon the text will appear somewhere inside the figure, in a small box. As long as it is selected (as indicated by the surrounding box; which you can select again by clicking on it) and the mouse pointer shows as an arrow-tipped plus sign, you can move that box with its contents to any position in the graph. Moreover, you can change the properties of the lettering by moving the mouse pointer over it until it shows as a capital I, highlighting part or all of the text you want to be changed with the mouse key, and then, change its letter type, point size, color, etc. Try it out, and play with it. Again, in Excel 97 and subsequent versions, you can activate the graph, then use Chart ⇒ Chart Options to achieve the same result.

Once you have a graph, it is easy to *add* another curve to it, provided it has the same *x*-axis. Merely highlight the column containing the new *y*-values, press Ctr + c, activate the chart so that its inner (coordinate) frame is highlighted (this may require clicking twice inside that frame), then press Ctrl + v. This will convert a column of numbers into a new curve or set of points. The reverse process, *removing* a particular curve from a graph, is even easier: highlight the curve, erase its description in the formula bar, then press Enter.

How do we name the spreadsheet? Bring the mouse pointer to the tab at the bottom of the central area of the spreadsheet, which will show a generic name, such as Sheet1. Right-click on the mouse, which is the general method of gaining access to the *properties* of the item to which you point. Select Rename; in the resulting Rename Sheet dialog box, click on Sheet1 in the Name window, and replace it with your own choice of spreadsheet name.

Copying a graph embedded in a spreadsheet to another location on the *same sheet* merely requires that we activate the graph, copy it with Ctrl + c, then click on a new location and paste it there with Ctrl + v. Make sure that the spreadsheet shows a zoom value of 100%, otherwise the copy will differ in size from the original.

Copying a graph to *another sheet* is another matter. It is just as easy to do, but the graph you get will still refer to the original sheet, because the coordinates of the graphed columns or rows contain the name of the sheet. That may be just what you want, in which case everything is fine. However, when you copy a graph, or an entire sheet, including its graphs, to another sheet, and you want that graph to refer to the data on the *new* sheet, you must activate each curve and then, in the formula box, change the associated sheet name (just before the exclamation mark). The same, incidentally, applies to names. Regardless of how many worksheets you use, in one workbook a given name can only be assigned once.

When you copy a graph to *another workbook*, and want it to refer to its new environment, you must also change the workbook name.

How do we save the spreadsheet? When you are ready to stop, click on File, then on Save As. In the resulting Save As dialog box, a name of your choice should go in the window File name. The location where the spreadsheet will be saved is specified in the Save in window. If you don't want to save in the My Documents file, click on the arrow to the right of My Documents. A list of options appears; select one of them by double clicking.

To end this section on a playful note: let's move some of the embedded graphs around. Take one graph, and click on it while keeping the Ctrl button down. Now move the entire graph: you are moving a copy of it, the original remains in place! You can move it anywhere, deposit it by releasing the mouse button, pick it up again (or leave it, and only pick up a copy of it by using Ctrl) and move it all over the place. Drop it partway over another graph, or over data; it does not bother either of them, just temporarily blocks you from seeing them. Move it away again, or make a pleasing pattern with them. To erase a graph, highlight it (so that it shows its eight handles), then use the Delete button.

1.6 Mathematical operations

In this section we will summarize some of the most useful *mathematical* operations available in Excel. This section is merely for your information, just to give you an idea of what is available; it is certainly not meant to be memorized. There are many more functions, *not* listed here, that are mainly used in connection with statistics, with logic (Boolean algebra), with business and database applications, with the manipulation of text strings, and with conversions between binary, octal, decimal and hexadecimal notation.

The mathematical operations and functions are organized here in order of increasing complexity, and are listed in Tables 1.6-1 through 1.6-6. We often use only a few of them; the list is given here only to illustrate the wide range of available spreadsheet operations. More detailed information on any of these worksheet functions can be found under <u>H</u>elp, as described in section 1.11.

Table 1.6-1: *The basic calculator operations.*

Operator	Description	Notes	Precedence[1]
^	Exponentiation	** will *not* work; E sometimes works[2]	1
*	Multiplication	×, . or · will *not* work[3]	2
/	Division[4]		2
+	Addition		3
−	Subtraction or negation		3
&	Concatenation[5]		4
=	Equal to		5
<	Less than		5
>	Greater than		5

Notes:
[1] Precedence indicates the order in which operations will be performed *in the absence of brackets*. For example, $4*3\wedge2 = 4*(3\wedge2) = 4*9 = 45$; if you want to compute the square of $4*3$ you must use brackets, as in $(4*3)\wedge2 = 12\wedge2 = 144$. Likewise, $8\wedge2/3 = 64/3 = 21.333$; $8\wedge(2/3) = 4$. Note that Excel has one annoying quirk: *negation* is performed *before* exponentiation: $-2\wedge2 = 4$. *When in doubt, use brackets*: $(-2)\wedge2 = 4$, $-(2\wedge2) = -4$.
[2] E-notation works as long as the exponent is an *integer*, i.e., 3.4E–5 (for 3.4×10^{-5}) is OK but 3E–5.2 is not.
[3] The multiplication sign is always needed; instructions such as $A$1(A3 + 4)$ will be flagged as incorrect, and should instead read $A$1*(A3 + 4)$.
[4] Do not use repeated dividers, as in N3/D2/D2, which is ambiguous, but instead use N3/(D2*D2) or N3/(D2$\wedge$2).
[5] Concatenation is the joining together of two or more strings of text or digits. For example, when cell B3 contains the string R2D2 and cell C6 contains 4U then = B3&C6 will yield R2D24U.

Table 1.6-2: *Some of the most common mathematical operations.*

Function	Description and example
ABS(*x*)	Absolute value: ABS(− 3) = 3; SQRT(ABS(− 25)) = 5
AVERAGE(*range*)	Average of the range specified, as in AVERAGE(A3:A7)
COUNT(*range*)	Counts number of entries in given range, as in COUNT(A3:A7)
COUNTA(*range*)	Counts number of non-blank values in range, as in COUNTA(A3:A7)
COUNTBLANK(*range*)	Counts number of blank cells in specified range
COUNTIF(*range, criterion*)	Counts only those values in the range that satisfy a given criterion, as in COUNTIF(A3:A7,">0"), which counts only positive entries
DEGREES(*angle in radians*)	Converts radians to degrees: DEGREES(PI()) = 180
EVEN(*x*)	Rounds a number to the nearest even integer away from zero: EVEN(1.9) = 2; EVEN(2.1) = 4; EVEN(− 0.1) = − 2
EXP(*x*)	Exponentiates, i.e., raises e = 2.7182818284904 to the power *x*: EXP(2) = 7.389056; EXP(LN(3)) = 3
FACT(*n*)	Factorial of a non-negative integer, $n! = n(n − 1)(n − 2)\cdots 3 \times 2 \times 1$: FACT(4) = $4 \times 3 \times 2 \times 1$ = 24; FACT(5) = $5 \times 4 \times 3 \times 2 \times 1$ = 120, FACT(0) = 1. Non-integer positive arguments are first truncated
FACTDOUBLE(*n*)	Double factorial of a non-negative number: $n!! = n(n − 2)(n − 4)\cdots 4 \times 2$ for *n* even, $n!! = n(n − 2)(n − 4)\cdots 3 \times 1$ for *n* odd: FACTDOUBLE(4) = 4×2 = 8; FACTDOUBLE(5) = $5 \times 3 \times 1$ = 15
INT(*x*)	Rounds a number down to the nearest integer: INT(1.9) = 1; INT(− 1.9) = − 2
LN(*x*)	Natural logarithm of positive number: LN(2) = 0.693147; LN(EXP(3)) = 3
LOG(*x,n*)	Logarithm of base *n*, where *n* is optional, with a default value of 10: LOG(10) = 1; LOG(10,2) = 3.321928; LOG(8) = 0.90309; LOG(8,2) = 3
LOG10(*x*)	Ten-based logarithm, LOG10(*x*) = LOG(*x*): LOG10(100) = 2; LOG10(8) = 0.90309
MAX(*range*)	Finds the maximum value in a specified range or in up to 30 ranges, as in MAX(H3:H402) or MAX(H3:H402, P6:Q200)
MDETERM(*array*)	Yields the determinant of a square array of numbers; MDETERM(A1:B2) = A1*B2 − A2*B1; MDETERM(D3:F5) = D3*(E4*F5 − F4*E5) + E3*(F4*D5 − D4*F5) + F3*(D4*E5 − E4*D5)
MEDIAN(*range*)	The median of a set of numbers: the middle value for an odd number of values, the average of the two middle numbers for an even number of values
MIN(*range*)	Finds the smallest number in a range or number of ranges
MOD(*x,y*)	The remainder of the division *x/y*
ODD(*x*)	Rounds away from zero to the nearest odd number: ODD(1.5) = 3

Table 1.6-2: (*cont.*)

Function	Description and example
PI()	The number $\pi = 3.14159265358979$; note that PI() contains no argument, but still requires (empty) brackets: SIN(PI()/2) = 1
POWER(*x,y*)	= x^y, i.e., raises x to the power y: POWER(4.3,2.1) = 21.39359. Equivalent instruction: 4.3^2.1
PRODUCT(*range*)	Product of numbers in specified range, as in PRODUCT(A2:C2) = A2*B2*C2
RADIANS(*angle*)	Converts from degrees to radians: RADIANS(270) = 4.712389 (= $3\pi/2$)
RAND()	Generates a random number between 0 and 1. Note that the brackets remain empty, as with PI(). The random number will change every time a spreadsheet calculation is made. If you do not want that to happen, highlight the cell where you want the random number, then go to the formula bar, type = RAND(), and deposit that instruction with the function key F9 (for the MacIntosh use COMMAND + =)
ROUND(*x,n*)	Rounds x to n decimal places: ROUND(21.49,1) = 21.4; ROUND(− 21.49,− 1) = − 20
ROUNDDOWN(*x,n*)	As ROUND but always rounds away from zero
ROUNDUP(*x,n*)	As ROUND but always rounds towards zero
SIGN(*x*)	SIGN(x) = 1 for $x > 1$, 0 for $x = 0$, and −1 for $x < 0$
SQRT(*x*)	Square root of non-negative number: SQRT(9) = 3; SQRT(− 9) = #NUM!; SQRT(ABS(− 9)) = 3
SUM(*range*)	Sums values in specified range or ranges, as in SUM(C6:C65) or SUM(C6:C65, D80:F93)
SUMPRODUCT(*array1,array2, ...*)	Computes the sums of the products of two or more arrays of equal dimensions; SUMPRODUCT(A1:A3,C1:C3) = A1*C1 + A2*C2 + A3*C3
SUMSQ(*range*)	Sum of squares of specified range or ranges, as in SUMSQ(G7:G16); SUMSQ(3,4) = 25
SUMX2MY2(*xarray,yarray*)	Sum of x^2 minus y^2: SUMX2MY2 = $\Sigma(x^2 - y^2)$
SUMX2PY2(*xarray,yarray*)	Sum of x^2 plus y^2: SUMX2PY2 = $\Sigma(x^2 + y^2)$
SUMXMY2(*xarray,yarray*)	Sum of (x minus y)^2: SUMXMY2 = $\Sigma(x - y)^2$
TRUNC(*x*)	Truncates a number to an integer: TRUNC(PI()) = 3, TRUNC(2.9) = 2, TRUNC(− 2.9) = − 2

Table 1.6-3: *Trigonometric and related operations.*

Function	Description and example
ACOS(x)	The inverse cosine, arccos, in radians: ACOS(0.5) = 1.047198
ACOSH(x)	The inverse hyperbolic cosine, arcosh, in radians: ACOSH(1.5) = 0.962424
ASIN(x)	The inverse sine, arcsin, in radians: ASIN(0.5) = 0.523599
ASINH(x)	The inverse hyperbolic sine, arsinh, in radians: ASINH(0.5) = 0.481212
ATAN(x)	The inverse tangent, arctan, in radians: ATAN(0.5) = 0.463648
ATAN2(x,y)	= ATAN(y/x)
ATANH(x)	The inverse hyperbolic tangent, artanh, in radians: ATANH(0.5) = 0.549306
COS(x)	The cosine, in radians: COS(0.5) = 0.877583
COSH(x)	The hyperbolic cosine, cosh, in radians: COSH(0.5) = 1.127626
SIN(x)	The sine, in radians: SIN(0.5) = 0.479426
SINH(x)	The hyperbolic sine, in radians: SINH(0.5) = 0.521095
TAN(x)	The tangent, in radians: TAN(0.5) = 0.546302
TANH(x)	The hyperbolic tangent, in radians: TANH(0.5) = 0.462117

The functions listed in Tables 1.6-4 and 1.6-5 require that the Analysis Toolpak has been loaded.

Table 1.6-4: *Some engineering functions.*

Function	Description and example
BESSELI(x,n)	The modified Bessel function $I_n(x) = i^{-1}J_n(ix)$: BESSELI(1.5, 1) = $I_1(1.5)$ = 0.981666
BESSELJ(x,n)	The Bessel function $J_n(x)$: BESSELJ(1.9, 2) = $J_2(1.9)$ = 0.329926
BESSELK(x,n)	The modified Bessel function $K_n(x)$: BESSELK(1.5, 1) = $K_1(1.5)$ = 0.277388
BESSELY(x,n)	The Bessel function $Y_n(x)$: BESSELY(1.5, 1) = $Y_1(1.5)$ = 0.145918
CONVERT($n,fromUnit,toUnit$)	Converts a number from one measurement system to another: CONVERT(1.0, 'lbm', 'kg') = 0.453592; CONVERT(64, 'F', 'C') = 20
DELTA(n,m)	Kronecker delta, tests whether two values are equal: DELTA(4,5) = 0; DELTA(5,5) = 1
ERF(n)	The error function: ERF(1) = 0.8427
ERFC(n)	The complementary error function, erfc(x) = 1 − erf(x): ERF(1) = 0.1573
GESTEP($n,step$)	Tests whether a number n exceeds (is Greater than or Equal to) a threshold value *step*: GESTEP(4,5) = 0; GESTEP(5,5) = 1, GESTEP(6,5) = 1.
RANDBETWEEN(n,m)	Generates a random integer between the integer values n and m; it will change every time a spreadsheet calculation is performed.

Table 1.6-5: *Functions involving complex numbers.*

Function	Description and example
COMPLEX(a,b)	Converts a real and an imaginary number to a complex one: COMPLEX(3,4) $= 3 + 4i$
IMABS('$a + bi$')	Absolute value (modulus) of a complex number, $(a^2 + b^2)^{1/2}$: IMABS('3 $+ 4i$') $= 5$
IMAGINARY('$a + bi$')	The imaginary component of a complex number: IMAGINARY('3 $+ 4i$') $= 4$
IMARGUMENT('$a + bi$')	The argument of a complex number, in radians, $= \arctan(b/a)$: IMARGUMENT('3 $+ 4i$') $= 0.927295$
IMCONJUGATE('$a + bi$')	The complex conjugate of a complex number: IMCONJUGATE ('3 $+ 4i$') $= 3 - 4i$
IMCOS('$a + bi$')	The cosine of a complex number: IMCOS('3 $+ 4i$') $= -27.034946 - 3.851153i$
IMDIV('$a + bi$','$c + di$')	The quotient of two complex numbers: IMDIV('1 $+ 2i$', '3 $+ 4i$') $= 0.44 + 0.08i$
IMEXP('$a + bi$')	The exponential of a complex number: IMEXP('3 $+ 4i$') $= -13.128783 - 15.200784i$
IMLN('$a + bi$')	The natural logarithm of a complex number: IMLN('3 $+ 4i$') $= 1.609438 + 0.927295i$
IMLOG10('$a + bi$')	The base-10 logarithm of a complex number: IMLOG10('3 $+ 4i$') $= 0.698970 + 0.402719i$
IMLOG2('$a + bi$')	The base-2 logarithm of a complex number: IMLOG2('3 $+ 4i$') $= 2.321928 + 1.337804i$
IMPOWER('$a + bi$', n)	The complex number raised to an integer power: IMPOWER('3 $+ 4i$', 3) $= -17 + 44i$
IMPRODUCT('$a + bi$','$c + di$')	The product of two complex numbers('1 $+ 2i$', '3 $+ 4i$') $= -5 + 10i$
IMREAL('$a + bi$')	The real component of a complex number: IMREAL('3 $+ 4i$') $= 3$
IMSIN('$a + bi$')	The sine of a complex number: IMSIN('3 $+ 4i$') $= 3.853738 - 27.06813i$
IMSQRT('$a + bi$')	The square root of a complex number: IMSQRT('3 $+ 4i$') $= 2 + i$
IMSUB('$a + bi$','$c + di$')	The difference between two (or more) complex numbers: IMSUB ('1 $+ 2i$', '3 $+ 4i$') $= -2 - 2i$
IMSUM('$a + bi$','$c + di$')	The sum of two (or more) complex numbers: IMSUM('1 $+ 2i$', '3 $+ 4i$') $= 4 + 6i$

Note:
Operations on complex numbers all start with IM, and use text strings to squeeze the two components of a complex number into one cell. In order to use the results of complex number operations, you must therefore first *extract* its real and imaginary components, using IMREAL() and IMAGINARY(). Instead of i you can use j to denote the square root of minus one (which you must then *specify* as such), but you *cannot* use the corresponding capitals, I or J.

Table 1.6-6: *Matrix operations.*[1]

Function	Description and example
INDEX(*array,row#,column#*)	Looks up an individual matrix element in given array: INDEX({1,2,3;4,5,6},2,3) = 6
MINVERSE(*array*)[2]	The matrix inverse of a square array: MINVERSE({3,4;5,6}) = $\{-3,2;2.5,-1.5\} = \begin{vmatrix} -3 & 2 \\ 2.5 & -1.5 \end{vmatrix}$, MINVERSE(B3:C4) = $\begin{vmatrix} -3 & 2 \\ 2.5 & -1.5 \end{vmatrix}$ when B3:C4 contains the dat.a $\begin{smallmatrix} 3 & 5 \\ 4 & 6 \end{smallmatrix}$.
MMULT(*array*)[2]	The matrix product of two arrays (where the number of columns in the first array must be equal to the number of rows in the second array): MMULT({3,4;5,6},{-3,2;2.5,-1.5}) = $\begin{vmatrix} 1 & 0 \\ 0 & 1 \end{vmatrix}$ and, likewise, MMULT(B3:C4,E6:F7) = $\begin{vmatrix} 1 & 0 \\ 0 & 1 \end{vmatrix}$ when B3:C4 and E6:F7 contain the data $\begin{smallmatrix} 3 & 5 \\ 4 & 6 \end{smallmatrix}$ and $\begin{smallmatrix} -3 & 2 \\ 2.5 & -1.5 \end{smallmatrix}$ respectively

Notes:
[1] A fourth matrix operation, TRANSPOSE, is performed as part of the <u>E</u>dit ⇒ Paste <u>S</u>pecial operation.
[2] Matrix inversion and matrix multiplication work only on data *arrays*, i.e., rectangular blocks of cells, but not on single cells. To enter these instructions, enter the array with CRTL + SHIFT + ENTER (on the MacIntosh: COMMAND + RETURN).

In addition, several special data analysis tools are available through <u>T</u>ools ⇒ <u>D</u>ata Analysis. While most of these are for statistical and business use, we will use two of them, for Random Number Generation, and for Regression. Data Analysis also contains a Fourier Analysis tool, which we will not use because a simpler macro is provided, see chapters 7 and 9. Likewise, the Regression tool can be replaced by the Weighted Least Squares macro discussed in chapters 3 and 9, which is somewhat simpler to use but does not provide as much statistical information.

1.7　Error messages

Excel is very forgiving when you ask it to do something it does not know how to do. For instance, when you use SQRT( ) to take the square root of a series of numbers, and one or more of these numbers is negative, Excel does not come to a screeching halt, but simply prints the somewhat cryptic error message #NUM! to alert you of the problem, then goes on taking the other square roots. There are only seven different error messages, as listed in Table 1.7-1. While it is not absolutely necessary to take corrective action when an error message appears, it is usually wise to heed the warning and to correct the underlying problem.

Table 1.7-1: *Excel's error messages.*

Error message	Problem
#DIV/0!	Division by zero or by an empty cell
#NAME?	Excel does not recognize the name; perhaps it has been deleted
#N/A	Some needed data are *not a*vailable
#NULL!	The formula refers to a non-existing intersection of two ranges
#NUM!	The number is of incorrect type, e.g., it is negative when a positive number is expected
#REF!	The reference is not valid; it may have been deleted
#VALUE!	The argument or operand is of the wrong type

1.8 Naming and annotating cells

You can assign descriptive names to cells, such as Ka1 and Ka2 to refer to the two acid dissociation constants of a diprotic acid. It is usually much easier to write and read formulas that contain descriptive names rather than cell addresses such as $B$2 and $C$2. Names can be used only to refer to individual *absolute* addresses. Names must start with a letter, may contain only letters, numbers, periods and underscores, and cannot be R, C, or possible cell addresses. Consequently, C1 and Ca1 are not valid names, but Ca, caa3 and (in all current versions of Excel) Ka1 are, since the rightmost column label is IV; in future versions of Excel you might have to use Kaa1and Kaa2 instead.

   Assigning names to constants is easy: highlight the cell you want to be named, then click on the address box in the formula bar, type the name, and press enter. Alternatively, you can use Insert ⇨ Name ⇨ Define, then use the resulting Define Name dialog box. After you have named a cell, you can refer to it either by its regular address or by its given name. To delete a name, use the Define Name dialog box (via Insert ⇨ Name ⇨ Define). When you erase a name that is used in a formula, that formula will become invalid, and will show the error message #NAME?

   In recent versions of Excel it is possible to attach a descriptive or explanatory note to a cell, reminding you of its source, explaining its function, giving a literature reference, expressing doubts about the correctness of the answer, or whatever. In order to attach such a note, use Insert ⇨ Note, in the resulting Cell Note dialog box enter the text of the note, then deposit it by clicking on OK. The cell will have a small red square in its right top corner to indicate that a note is attached. The text of the note will show when you use the mouse to point to the cell. If you want to see what notes are attached, again use Insert ⇨ Note which lists them all in the column Notes in Sheet.

1.9

Viewing the spreadsheet

The monitor screen may display only a part of a spreadsheet, perhaps the first 10 columns and 20 or so rows. In order to get an overview of a much larger spreadsheet you may want to display a larger area, but then the cell contents may then become unreadable. Or you may want to zoom in on a smaller area so that you can examine a graph in detail. In all those cases you can use the Zoom Control, which allows you to enlarge all distances on the screen up to four times (to 400%), thereby showing only one-sixteenth (¼ × ¼) of the usual area, or to reduce the display to provide a larger (but less detailed) overview, reducing all linear distances to a minimum of 10%, in which case the screen displays a (10 × 10=) 100 times larger spreadsheet area. The Zoom Control is found as a window on the right-hand side of the Standard Toolbar. Click on the arrow to the right of that window, and click on any of the fixed percentages shown. Or enter any integer between 10 and 400, then deposit it with the Enter key. If the Standard Toolbar is not displayed, use View ⇨ Zoom instead.

In order to move the zoomed area use the scroll bars (to the right and at the bottom of the spreadsheet, for vertical and horizontal movement respectively) or the associated arrows. The arrows let you move one cell height (or cell width) at the time; clicking on the gray areas where the sliding bar moves lets you move in bigger steps, of about one screen height or screen width.

When you anticipate that a spreadsheet will be too large to keep on screen, it is best to organize it in such a way that all important information is in one area, say at its top. Still, it may sometimes be desirable to view different part of the spreadsheet simultaneously. This can be done in several ways. You can copy the worksheet with Edit ⇨ Move or Copy Sheet, then display both copies as multiple views with Windows ⇨ New Window. Each window can then be manipulated independently. Or you can divide the screen into two (and even four) separate parts that can be moved individually with the scroll bars, so that you can keep different parts of the spreadsheet on the screen, using the command Window ⇨ Split. The location of the pointer determines how many pieces will show: if the pointer is in the top row, it will cause a vertical split at that position; a horizontal split is made with the pointer in the first column; you will get a four-way split when the pointer is somewhere else. You can grab the dividing bars and drag them to change their positions. To undo the split, use Window ⇨ Remove Split. Alternatively, you can double-click on the tiny rectangular space just above the top arrow of the vertical scroll bar to generate a split of the screen into a top and bottom part. Again, the position of the split will be determined by the position of the active cell, except when the active cell sits in the top row, in which case the split is half-way down. Double-clicking on a dividing bar will undo the split.

Similarly, double-clicking on the small rectangular space to the right of the arrow in the right-hand corner of the horizontal scroll bar will generate a split screen with a right- and left-hand part.

When you need to work with long columns, it is often useful to keep the column headings visible. To do so, place the pointer in the cell below and to the right of the area you want to keep; make sure that this cell is empty. Then use Window ⇨ Freeze. When you now scroll through the column, the headings stay in place. To undo, use Window ⇨ Unfreeze.

1.10 Printing

An important aspect we have not yet described so far is how to print, assuming of course that you have a printer attached to the computer, and that it is turned on. The simplest way is to click on the printer icon on the toolbar, usually just under the View menu command. In our case, the spreadsheet is somewhat too complicated for that, so we will use the mouse to select, from the menu, the File Print Preview. This will show us the first page; by clicking on the Next button we can see the next page; to return to page 1 click on Previous. Clearly, the spreadsheet could use some cleaning up before we print it. Close the preview (this time the Close button is on the icon bar) to return to File, then click outside the menu to go back to the spreadsheet.

If the spreadsheet is just a little too large to fit on one page, you may want to scale it down to fit the paper. If the spreadsheet is too long, activate the row numbers (in the gray cells to the left of column A) which will highlight the corresponding rows. Then right-click, select Row Height, and reduce the row height appropriately. (You may have to reduce the font size to keep the cell contents from being decapitated.) Likewise you can change the column widths by highlighting the column letters, right-clicking to get Column Width, and changing that to suit your needs. Incidentally, this is also an easy way to scale figures displayed on the sheet, especially when you have several of them and want to fill the page with them as efficiently as possible.

Say that your spreadsheet only contains a few long columns, so that changing the column height would not be practical. In that case you may want to reorganize the spreadsheet layout. Note where the page break occurs, then go to the first cell past that page break in column A. (For the sake of convenience we will call that cell A52, even though, on your spreadsheet, the page break may not occur between rows 51 and 52.) Now go to cell E20, deposit the instruction $=$A52, and copy this instruction to block E20:G51. Now that the entire information is on page one, select Print from the File menu. A Print dialog box will pop up, in which, under the heading Page Range (near the bottom), you select Pages (rather than All) and then specify from 1 to 1. Finally, click on OK to start the printing process.

Please don't forget that you also have a graph stored as Chart1; move to that chart (by clicking on its tab), and make the graph more presentable now that you know how to do so, e.g., by using the French curve rather than straight-line segments to connect the few widely spaced points, with Format Data Series $\Rightarrow$ Patterns $\Rightarrow$ Smoothed Line. Then print the graph.

1.11 Help!

Excel is a rather complex program, that can be used at various levels of sophistication. In this book we will mostly stay with the basic operations common to most spreadsheets, although we will occasionally exploit some of Excel's specific capabilities, as when we use functions from the add-in menu or special macros. Because printing manuals is expensive, and many users do not consult them anyway, Excel has an extensive, *built-in* help library that can be called while you are working on a spreadsheet. Call it by clicking Help, then find your way to the topic you want. For example, let us assume that you want to use the Analysis ToolPak, but cannot find it under Tools. Apparently it was not installed. How would you find out how to install it?

In Excel 97 and later versions you would click on Help, then select Contents and Index. This will give you three tabbed options; first select Index, and type 'Analysis ToolPak' (without the quotes). A number of choices appear, from which you can select 'general information'. You are now offered three options: Install and use Analysis ToolPak, Supplemental information about statistical methods and algorithms, and Ways to analyze statistics. Click on the first choice, and find out how to install the Analysis ToolPak. The same information would be available from Ways to analyze statistics.

Say that you don't remember the name of the Analysis ToolPak, but instead look under 'data analysis'. You will find a choice labeled Data analysis tools in the Analysis ToolPak, and from there the path to the information is the same. Or assume that you have selected the Contents instead of the Index. Browse the options (and use the vertical scroll bar to see more of them than can be displayed in the window) till you come across something that might fit. In this case, you will find the item 'Analyzing Statistical Data', which will again lead you to 'Install and use the Analysis ToolPak'. In other words, with a little persistence you can find the required information almost no matter where you start: there are many roads that lead to Rome.

Similarly, in Excel 5 and 95, Help gets you to Microsoft Excel Help Topics and to the Answer Wizard. Click on either one, and you will see four tabs: Contents, Index, Find, and Answer Wizard. When you select the Answer Wizard, type 'data analysis' in the top window, then select an appropriate item from the list that appears in window 2. In this example, select 'Tell Me About Analyzing Statistics'. Or you type 'analysis toolpak', and find 'How Do I Enable the Analysis ToolPak'. Even 'How Do I Use the Analysis ToolPak' will get you to the installation instructions.

You could also have used Index, Find, or even Contents. Typing 'data analysis' in the Index would get you there via data analysis tools ⇨ Analysis ToolPak ⇨ Analyzing Statistics, which gives you a hypertext-like button to Enable the Analysis ToolPak. With Find your trail might also lead to Analyzing Statistics. And in Contents, a direct path to the required information might be Retrieving and Analyzing Data ⇨ Statistical Analysis of Data ⇨ Enable the Analysis ToolPak.

This is not to suggest that there are not an equal number of plausible options that won't get you there: when you navigate without a road map, some streets will turn out to be dead ends. But you get the idea: try a reasonable keyword, and if that does not work try another, until you succeed. You will usually find an answer faster than it would take to use the index in a book.

For mathematical functions use the button labeled f_x in the standard toolbar; in Excel 95 it is called the Function Wizard; in Excel 97 the Paste Function. Both will give you two side-by-side columns, one for Function Category, the other for Function Name. Pick a category, then in that category the name of the function you want to use. (Since there are so many functions listed under, say, Math & Trig, you most likely will need to use the vertical scroll bar to find your function.)

Once you have selected your function, you will not only find it described, but you will also get help in placing the arguments. For example, when you look under Math & Trig ⇨ SUMX2MY2, a function we will often use when computing least-squares fits, you will get two windows in which you can place the addresses of the two columns or rows you want to use for X and Y.

1.12 The case of the changing options

There is one aspect of Excel that initially may confuse you; it is the problem of changing options, i.e., of menu items that appear or disappear depending on prior action on the spreadsheet. While this can greatly enlarge the usefulness of the spreadsheet, it can be quite unsettling to the novice, hence this alert. Below we will illustrate it with Trendline, a very useful feature of Excel (to be described in more detail in chapter 2) that allows you to draw a number of least-squares lines or curves through graphed data.

The shorthand instruction for using Trendline in Excel 95 might read Insert ⇨ Trendline, but if you look in the pull-down menu under Insert you may not find Trendline. The problem is that the Trendline option appears only after you have activated the graph (by double-clicking on an embedded chart, or by clicking on the tab of a separate chart). Even then, it shows but cannot be used; for the latter, you must first select the particular data set in the graph to which you want it to apply. Only then can you select Trendline (or, for that matter, Error Bars).

A similar situation applies to Excel 97. Here, Trendline is part of the Chart menu, and the shorthand instruction might read Chart ⇨ Trendline. But you will usually not find Chart on the menu bar. It only appears there, instead of Data, after you have activated the chart, and again becomes accessible only after you have selected a particular data set.

In any case, consulting Help and asking for Trendline will give you a box with precise road instructions. As long as you know the proper name of the procedure, Help will get you there.

1.13 Importing macros and data

Spreadsheets are created to facilitate computation. Commonly used mathematical operations (such as SIN, LOG, SQRT, and MINVERSE) are built-in as **functions**, and some more complicated procedures (e.g., Solver, Random Number Generation, Regression) are provided as **macros**. However, no spreadsheet maker can anticipate the needs of all possible users, and Excel therefore allows the introduction of so-called user-defined functions and macros. In section 9.2d we will describe some user-defined functions, while chapter 10 deals extensively with user-defined macros. However, beyond the simple exercises of section 10.1, it makes no sense to enter long macros by hand, and they are therefore provided in a web site from which they can be downloaded and stored onto your own computer disk or diskette. The web site also contains a sample file that is, likewise, larger than you might want to enter manually.

Alternatively, you may have data or macros on a diskette, or receive them as e-mail attachments. In all such cases, the questions are (1) where and how to install the macros in Excel, and (2) how to enter the data in the spreadsheet.

The macros find their home in a **module** that becomes part of the spreadsheet. We therefore need to make the module first, then import the macros into that module. The procedures are slightly different for earlier versions (through Excel 95) and for more recent ones (starting with Excel 97), and are therefore described separately. For the sake of simplicity we will assume that the macros and data are stored in either a computer file (i.e., on a 'hard' disk) or a diskette.

1 To make a module in *Excel 5 or Excel 95*, move the pointer to the tab at the bottom of your spreadsheet, and right-click on it. A small menu will pop up. Click on the first item, Insert, which will give you several options. Highlight or double-click on Module, and click OK. You will now see a blank sheet, with a tab carrying the name Module1 (or, if the spreadsheet already contains modules, Module*N* where *N* is a sequence number Excel assigns automatically).

2   If your macro is in the form of a text file on diskette, insert it into the diskette drive. Go to the Open File icon (the second from the left on the standard toolbar, depicting an opening manila folder) or select File ⇨ Open. In either case you will see the Open dialog box. Click on the arrow on the right-hand side of the window labeled Look in:, select the file location, such as Systemdisk (C:) or 3½ Floppy (A:), or whatever suits the configuration of your computer).

3   Now highlight the name of the file, or go to the bottom of the dialog box, click on the arrow in the window for File name, and type that name. Then go to Files of type:, and select All Files (*.*). Push the Open button.

4   In *Excel 97 or later versions*, start from the spreadsheet with Alt + F11, then use Insert ⇨ Module, File ⇨ Import File. The resulting Import File dialog box is equivalent to the Open dialog box in Excel 95, as described above under points (2) and (3). Select the disk or diskette in Look in:, and proceed as indicated above. Switch back to the spreadsheet with Alt + F11.

5   In either case, the macros will now be stored in the module, where you may see that they have interesting colors: the macro labels and comment lines will show in dark green, the actual instructions in black and dark blue. Once you see those colors (which are sometimes hard to distinguish, depending on the monitor used) you can be sure that the spreadsheet has accepted the text as genuine macro instructions, and that they are available for your use.

6   For data, the approach is similar, except that these go directly into the spreadsheet rather than in a module. Therefore, in Excel 95, delete step (1) above, but proceed directly to steps (2) and (3), except that you now import the file called Data. Similarly, for Excel 97, there is no need to find the module, and in fact the procedure is now the same for Excel 97 and Excel 95. Make sure that the mouse pointer points to the top of an empty column or, better yet, the left-top corner of an empty spreadsheet, so that the imported data will not overwrite anything in the spreadsheet.

7   Usually (though not for our exercise data files), importing data into the spreadsheet will involve the Text Import Wizard. This asks you about the nature of the data file (e.g., whether and how the data are **delimited**, i.e., how the various data points are separated from each other) and then helps you along. But that is beyond what we need to learn now; it may become relevant if you want to import a long file with experimental data from some instrument.

1.14 Differences between the various versions of Excel

This book was originally written for Excel 95 and Excel 97, but can also be used with the subsequent Excel 98 and Excel 2000, and with the earlier Excel 5. Versions 1 through 4 are *not* recommended because they do not use VBA but, instead, use a different, much more restricted macro language. Even so, only programs in chapters 3, 7, and 10 use such macros; all other chapters should give no problems even with earlier versions of Excel. Similarly, users of Macintosh computers can use it, with some minor modifications, provided they have Excel version 5 (for System 7) or more recent versions.

The most important difference is that Excel version 5 runs in Windows 3.1, whereas Excel 95 and later versions require at least Windows 95. There are major differences between Windows 3.1 and Windows 95, but they hardly affect Excel, which provides its own environment. As far as Excel is concerned, the differences between Windows 95 and Windows 98 are rather insignificant.

The differences between Excel 5 and subsequent versions are mostly trivial and cosmetic. The look of several features and dialog boxes is somewhat different in the two versions. There are some minor changes in convenience; for example, in Excel 5, cell notes (mentioned in section 1.9) are not displayed automatically when you point to the cell. Excel 5 also has somewhat less extensive Help features, and provides less online VBA help. However, none of these will seriously affect the spreadsheet exercises in this book.

The differences between Excel 95, Excel 97, and Excel 2000 are even smaller, except that Excel 97 introduced an improved Chart Wizard, which is why we split the discussion in section 1.3. Starting with Excel 97, macros are also stored in a quite different way, to be described in chapter 10. Other major changes in Excel 97 and, especially, Excel 2000, include built-in facilities to address World Wide Web sites, and the use of the Office Assistant, both of which are of no consequence to the applications described in this book. Starting with Excel 97, the spreadsheet has a higher capacity, of 65 530 rows, and allows graphs to contain 32 000 rather than 4000 points. Furthermore, printing is made somewhat easier in Excel 97 and later versions, and creating dialog boxes has been simplified. Thanks to backward compatibility, you can import Excel 95 spreadsheets into Excel 97, but of course not the other way around.

For users of Macintosh computers the main differences (which are still rather minor) stem from differences in mouse and keyboard. The Macintosh mouse has only one button, so that the equivalent of right-clicking on a Windows machine is achieved by the combination CTRL + click. Many operations performed on a Windows machine with the CTRL button instead use the COMMAND button on the Macintosh keyboard, or sometimes the OPTION or Apple key. The Microsoft Excel User's Guide nicely juxtaposes the corresponding Windows and Macintosh keystrokes where these are different. But, again, these are only superficial, easily learned differences: the underlying spreadsheets appear to be identical.

1.15 Some often-used spreadsheet commands

To move around quickly in the spreadsheet (rather than with mouse or arrow keys):

PageDown	Goes down one 'page' (typically a screenful of about 20 lines)
PageUp	Goes up one 'page'
Ctrl + Home	Goes to the upper left corner of the sheet, i.e., to cell A1
Ctrl + End	Goes to the bottom of the sheet

To move quickly to the end of a *continuous* column or row of numbers or instructions:

End ↑	Goes to the top cell of the column
End →	Goes to the right-most cell of the row
End ↓	Goes to the bottom cell of the column
End ←	Goes to the left-most cell of the row

To enter something (labels, numbers, formulas) in a cell, move the pointer to that cell (using the arrow keys, or by moving the mouse and then clicking on the cell), type what you want to enter, then either use the Enter key or click to deposit the information in that cell.

All cell contents starting with a letter are considered to be labels, all cell contents starting with a number are treated as data, and all cell contents starting with an equal sign as formulas. Formulas often have a special syntax, such as SIN(), where the brackets must enclose an argument, or PI(), where the brackets should be left empty. Excel does not mind whether you use lower-case and capital letters, but always displays them in the formula window as capitals for better readability.

For copying data or formulas down short columns it is often convenient to grab a handle and pull them down. For longer columns it is usually faster to copy and paste.

To copy:	Ctrl + c	Places active area in Clipboard, leaves original in place.
To cut:	Ctrl + x	Places active area in Clipboard, but erases the original.
To paste:	Ctrl + v	Places contents of Clipboard in active area of the spreadsheet.

When you want to copy the *values* rather than the formulas of cells or blocks of cells, use special paste values instead of paste: following Ctrl + c click on Edit ⇨ Paste Special ⇨, then click on Values ⇨ OK.

To save:	Ctrl + s	Saves it in the same place from where it was opened.

To save a file in a different place, use Edit ⇨ Save As … and specify the new location before using OK or the Enter key, ↵.

1.16 Changing the default settings

In Excel, as in Windows, almost anything can be changed. It is useful to have **default settings**, so that you need not specify everything every time you start up Excel. Moreover, to the novice, it is also helpful to have fewer choices to confuse you. However, when you become familiar with Excel, you may want to make changes in the default settings to make the spreadsheet conform to your specific needs and taste. Here are some of the common defaults, and how to change them.

By default, Excel displays the standard and formatting **toolbars**. Excel has many other toolbars, which you can select with View ⇒ Toolbars. You can even make your own toolbar with View ⇒ Toolbars ⇒ Customize. An existing toolbar can be positioned anywhere on the spreadsheet simply by dragging the two vertical bars at its left edge (when it is docked in its standard place) or by its colored top (when not docked).

Many aspects of the spreadsheet proper can be changed with Format ⇒ Style ⇒ Modify, including the way **numbers** are represented, the **font** used, cell **borders**, **colors**, and **patterns**.

Many Excel settings can be personalized in Tools ⇒ Options ⇒ General. Here you can specify, e.g., the number of **files listed** upon clicking on File, and change the Standard **font** (e.g., from Arial to more easily readable serif **font** such as Times New Roman) to perhaps a different **font Size**. Here you can also set the **Default file location** (from C:\My Documents) and even define another **Alternate startup file location**.

Under the View tab (i.e., under Tools ⇒ Options ⇒ View) you can toggle the appearance of **spreadsheet Gridlines** on or off. Under the Edit tab (Tools ⇒ Options ⇒ Edit) you can (de)select to **Edit directly in the cell**, which allows you to edit in the cell (after double-clicking) rather than in the formula bar. Here you can also **Allow cell drag and drop** or disallow it, and **Move selection after enter** in case you prefer the cursor to stay put or move sideways rather than move down one cell after each data, text, or formula entry.

Excel does not make **back-up files** by default. If you wish to change this, use Files ⇒ Save As ⇒ Options and select Always create backup.

When you print with the Print button on the Standard Toolbar, you use the default printing settings. File ⇒ Page Setup provides many alternatives, including **paper size** and **orientation**, as well as **margins**.

In Excel 97 and later versions, browse in Tools ⇒ Customize to see (and select) your Toolbars and their Commands. You can click on a command and then use the button to get its Description.

Likewise, in Excel 97 and beyond, the default settings for graphs are accessible after you activate a chart to make the Chart menu available. Now select Chart ⇒ Chart Type, under Chart type pick your choice, such as XY(Scatter), select a Chart sub-type such as with all data points connected by smoothed lines, and **Set as default chart**. Even better, you can define the format of the default chart. Say that you make a number of logarithmic concentration diagrams, with pH from 0 to 14 as your horizontal axis, and pc from 10 (at the bottom) to 0 (on top) in the vertical direction. Make such a graph, with axis labels, then (while the chart is activated, so that the Chart button is accessible in the menu toolbar) click on Chart ⇒ Chart Type, select the Custom Types tab, click on the User-defined radio button, then on Add. In the next dialog box, specify its Name, give an (optional) Description, Set as default chart, and exit with OK. Then, the next time you highlight a block and invoke Insert ⇒ Chart you get the selected format just by pushing the Finish button on step 1 of the Chart Wizard. Or, faster, highlight the area involved, and type Alt + i, Alt + h, Alt + f, or / + i, / + h, / + f. The details of the graph may differ from those of the sample, but even so this can be a time saver.

1.17 Summary

In this introductory chapter you have encountered some of the basic manipulations of Excel. The first time around you may feel overwhelmed by it, but don't worry: as you practice, you will quickly become familiar with the rules of spreadsheets, and with their internal logic. A spreadsheet not only allows you to perform repeated calculations (such as computing values for a sine wave) and to print them as a graph, but to perform many much more sophisticated operations, such as data analysis and mathematical simulation. The main attributes of a spreadsheet such as Excel are:

1 Lay-out: the highly intuitive organization of a spreadsheet displays all initial, intermediary, and final results in tabular form, making it very easy to see precisely what is being done.

2 User-friendliness: when you perform an 'illegal' operation, such as dividing by zero or taking the logarithm of a negative number, the spreadsheet does *not* come to a punishing halt but, instead, flags the problem area. Then it goes on with its task, and performs the operation wherever it can do so.

3 Automatic updating: with the exception of macros and some functions, all computations are adjusted automatically any time that you enter a new number or instruction in the spreadsheet, thereby keeping it up-to-date.

4 Precision: all calculations are performed in so-called 'double precision', that is, to a precision of about 1 in 10^{15}, even when only a few digits are shown. Truncation and round-off errors are therefore seldom a problem in spreadsheet calculations.

5 Graphing: a picture is often much more informative than a table of numbers. The spreadsheet makes it very convenient to display data in graphical form. Graphs are readily made. By placing them directly on the spreadsheet the user can immediately see the results of the calculations. And you will not need any other software to make publication-quality graphs.

6 Competence: modern spreadsheets contain a large number of functions, thereby facilitating rather complicated calculations.

7 Data analysis: modern spreadsheets contain several convenient data-analysis aids, such as Excel's Trendline, a flexible linear least-squares tool, and Excel's Solver, a powerful multi-parameter non-linear least-squares fitting routine. Both of these are described in detail in chapter 3, and are used throughout the remainder of this book. Excel also contains a large number of tools for statistical data analysis.

8 Flexibility: repeated tasks can easily be customized, and default settings adjusted, to suit the user.

9 Expandability: starting with version 5, Excel has the added feature that the user can write or import entire BASIC programs to perform even more complicated tasks. Chapter 10 will discuss this capability in considerable detail and with many examples.

In the next chapters we will illustrate some applications of spreadsheets to common problems of analytical chemistry. Once you have become familiar with the spreadsheet, you may want to use it for many other tasks, such as for plotting your experimental data for lab reports as well as for publications, or in quite different areas, such as to visualize theoretical expressions in physical chemistry. As with many modern computer tools, ultimately your imagination is the limit.

CHAPTER 2

INTRODUCTION TO STATISTICS

2.1 Gaussian statistics

Analyzing a number of observations, each subject to some experimental error, in an effort to obtain a more reliable answer from a multitude of measurements than can presumably be obtained from a single observation, is part of statistics. For example, while the age at which you, my dear reader, will die, is usually not well known in advance, the commercial providers of life insurance need only know the *average* life expectancy of your *cohort* (the group of persons of comparable age, gender, socioeconomic group, etc.) in order to compute a profitable premium, on the assumption that they will insure a large enough group so that the effects of early and late deaths will cancel each other out.

The underlying assumption in statistical analysis is that the experimental error is not merely repeated in each measurement, otherwise there would be no gain in multiple observations. For example, when the 'pure' chemical we use as a standard is contaminated (say, with water of crystallization), so that its purity is less than 100%, no amount of chemical calibration with that standard will show the existence of such a **bias**, even though all conclusions drawn from the measurements will contain consequent, **determinate** or **systematic errors**. Systematic errors act uni-directionally, so that their effects do not 'average out' no matter how many repeat measurements are made. Statistics does not deal with systematic errors, but only with their counterparts, **indeterminate** or **random errors**. This important limitation of what statistics does, and what it does not, is often overlooked, but should be kept in mind. Unfortunately, the sum-total of all systematic errors is often larger than that of the random ones, in which case statistical error estimates can be very misleading if misinterpreted in terms of the presumed reliability of the answer. The insurance companies know it well, and use exclusion clauses for, say, preexisting illnesses, for war, or for unspecified 'acts of God', all of which act uni-directionally to increase the covered risk.

In this chapter we will illustrate some properties of **Gaussian** statistics. Such statistics are often (but by no means always) applicable to experimental data. Examples where Gaussian statistics do *not* apply are the throwing of dice (or, more generally, when we use a *discrete* number of trials and each trial has a *discrete* outcome), which is described by **binomial** statistics, and the disintegration of radioactive nuclei (for a *continuous* trial with *discrete* outcomes) which obeys **Poissonian** statistics. However, for a sufficiently large number of independent observations, all observations tend to approach Gaussian statistics, which is why the Gaussian distribution is often (including in Excel) but somewhat misleadingly called the **normal** one. We will come back to the uses and misuses of Gaussian statistics in sections 2.8 and 2.9.

In a Gaussian distribution, errors of any size are allowed. The probability that a measurement deviates from its 'true' value is assumed to depend on the *square* of that deviation; a positive or negative deviation of the same size is therefore assumed to be equally likely. Within Gaussian statistics, the usual procedure to calculate the 'best' answer from a multitude of measurements is called the method of least squares; it consists of minimizing the sum of the squares of these deviations. However, since the true answer is not known (if it were, we would not need statistics!) we usually substitute the measured average for the true value, and then minimize the sum of the squares of the differences between the individual observations and the average of these observations. In this context, in using the term **average**, we may simply mean the sum of all the measurements, divided by the number of those measurements, as defined in equation (2.2-1), or some more sophisticated quantity, such as a **weighted** average, in which some measurements are given more credence than others.

In this workbook we will often use Gaussian 'noise' to simulate the measurement imprecisions associated with experimental data, and it will therefore be useful to familiarize ourselves with such noise. This is the purpose of the first spreadsheet exercise of this chapter.

Instructions for exercise 2.1

1 Open an Excel spreadsheet.

2 Point with the mouse pointer to the tab labeled Sheet1, and right-click to get the *properties* of the label. Click on Rename. In Excel 97, just type a new name, say 'Gauss', then depress the enter key. In earlier versions of Excel, you get a Rename Sheet dialog box. Replace the generic name Sheet1 in the Name box by the new name, then click OK.

3 Now fill column A with data containing Gaussian noise. Click on Tools ⇨ Data Analysis. In the resulting Data Analysis dialog box, double-click on Random Number Generation. In order to find it, you may have to grab the scrollbar inside the dialog box with your mouse pointer, and move it downwards.

4 Place the mouse pointer over the down arrow inside the Distribution window, click on it, then select the Normal distribution by clicking on it.

5 Below it, under the heading 'Parameters', click inside the window labeled Mean (a vertical bar should appear in the window to show you that it is ready for your input), and enter the value 10. Do *not* use the Enter key, or depress the OK button; keep away from them until you are done with the entire dialog box.

6 Move the pointer to the window defining the Standard deviation, and set it at 1.

7 Click on the round radio button for the Output Range, then specify it in the window to the right of the label as A1:A10000. (If you wish, you may also fill the entire column A, which contains 16 384 cells. In that case you can specify the range as either A1:A16384 or as $A:$A, which Excel interprets as the entire column A.)

8 Leave all other windows blank, and click on OK or press the Enter key. The computer will now take some time to calculate 10 000 (or 16 384) data with an average value of 10 plus Gaussian noise of standard deviation 1. (It will show you that it is busy with the message Calculating Random Number Generation ... on the status bar, just above the Start button.)

9 Now that a data set has been generated, we will analyze it. To that end we will specify **sorting bins** that define a range of data values. Place the value 6.2 in cell B3, in B4 place the instruction $= B3 + 0.2$, then copy this instruction all the way down to B41. This will generate bins for counting how many data fall in the range <6.2, between 6.2 and 6.4, between 6.4 and 6.6, etc., with the last bin for data between 13.6 and 13.8. Although the average is 10 and the standard deviation is 1, we cast a much wider net, anticipating that there may be data well *outside* the range from $10 - 1 = 9$ to $10 + 1 = 11$.

10 Now call the Histogram tool, which will count how many of the data fit in each of the bins. To simplify matters we will first analyze the first ten data.

11 Select Tools $\Rightarrow$ Data Analysis, and in the Data Analysis dialog box double-click on Histogram.

12 In the Histogram dialog box specify the Input Range as A1:A10 (or of any other set of 10 adjacent data, such as A469:A478), and the Bin Range as B3:B41.

13 Click on the radio button for the Output Range, and in its window specify the top left corner of the histogram output as E2. Leave all other fields blank, and click on OK.

14 The results of the analysis of the first 10 data will now appear in columns E and F: column E repeats the bins, while column F lists how many of the analyzed data have a value within the interval specified by the various bins. Any values greater than 13.8 will be listed in cell F42 under 'More'.

15 Verify that the total data count is indeed 10, e.g., by depositing in cell F1 the instruction $= \text{SUM(F3:F42)}$.

16 Since we only consider 10 data, and sort them into 40 bins, the majority of these bins will be empty (and therefore show a 0). Most of the remaining bins will contain a 1, while one or a few may show a larger number.

In order to compare the resulting frequency distribution in column F with a Gaussian distribution, we recall that the latter is given by

$$P = \frac{1}{\sigma\sqrt{2\pi}} \exp\left[\frac{-(x-\bar{x})^2}{2\sigma^2}\right]$$ (2.1-1)

where P is the probability density, and $P(x)\,dx$ is the corresponding probability of finding a particular value in the range between x and dx when the average is $\bar{x}$ and the standard deviation is σ.

17 For the sake of comparing the data in column F with the prediction of eq. (2.1-1), we now deposit in cell C3 the value 6.1, in cell C4 the instruction $=C3+0.2$, and copy this down to cell C42. The resulting values are precisely 0.1 less than those in column B. (Alternatively you can use the instruction $=B3-0.1$ in C3, and deposit the value 13.9 in cell C42.) We do this in order to specify the *midpoints* of the bin ranges, whereas the Histogram routine uses the upper limits of their ranges.

18 In cells D3:D42 calculate the frequency of finding data in a given range as the product of the total number of data considered (here: 10), the bin width (0.2), and the probability P according to (2.1-1). For instance, the instruction in D3 should read $=$ (2/SQRT(2*PI()))*EXP(-0.5*(C3 − 10)^2) because $10 \times 0.2 = 2$, $\sigma = 1$ and $\bar{x} = 10$.

19 In order to make a graph of columns C, D and F, first highlight C3:D42 (use the Shift key), then release the Shift key and depress the Control key, use the mouse to move the pointer to cell F42, release Ctrl and re-engage Shift, and go up in column E with either End ↑, Page Up, or with the ↑ key to E3. The non-adjacent blocks C3:D42 and F3:F42 will now be highlighted, and therefore activated.

20 Select Insert ⇨ Chart, and complete the ChartWizard. You should get a result looking somewhat like Fig. 2.1-1, although the specific details will look different because every data set is different. Note the discrete nature of the count, with frequencies of 0, 1, 2, 3 etc. Not surprisingly, the agreement is quite poor: a highly discrete distribution such as obtained here cannot be represented very well by a continuous expression such as (2.1-1).

21 Now repeat the same procedure for a 100-data set, such as A1:A100. Use the same bins as before; the only changes you need to make are to specify in the Histogram dialog box the Input Range A1:A100 and, in the comparison with (2.1-1), to change the multiplier of P to $100 \times 0.2 = 20$ (instead of 2). Make these calculations in new columns. The result should look like Fig. 2.1-2.

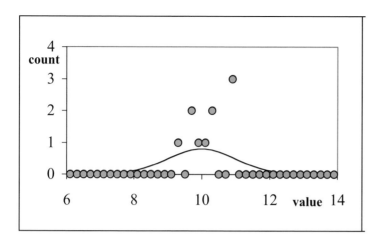

Fig. 2.1-1: Ten Gaussian data with mean 10 and standard deviation 1, sorted in 0.2-wide bins.

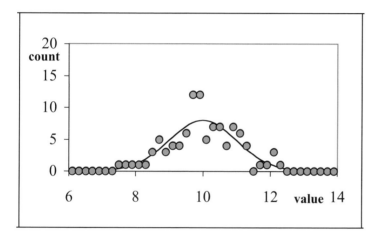

Fig. 2.1-2: 100 Gaussian data with mean 10 and standard deviation 1, sorted in 0.2-wide bins.

22 The agreement between the data and equation (2.1-1) is still quite poor. However, it clearly shows that quite a few data fall *outside* the average plus or minus one standard deviation: for a sufficiently large sample, approximately one-third of all data will do so when the fluctuations are Gaussian.

23 The larger the data set we examine, the better the agreement with (2.1-1) will be. Convince yourself of this by using a 1000-data set, then a 10 000-data set. Figures 2.1-3 and 2.1-4 illustrate the type of graphs you might obtain. The fit between your 'experimental' data and the theoretical expression becomes quite good when the sample is sufficiently large!

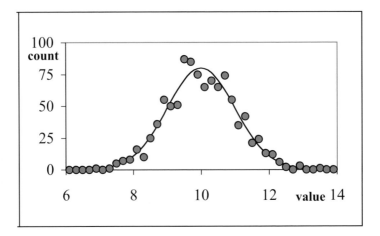

Fig. 2.1-3: 1000 Gaussian data with mean 10 and standard deviation 1, sorted in 0.2-wide bins.

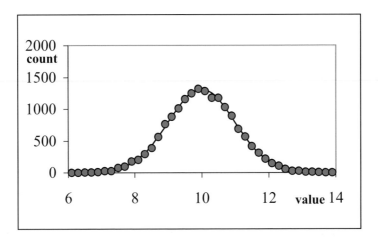

Fig. 2.1-4: 10 000 Gaussian data with mean 10 and standard deviation 1, sorted in 0.2-wide bins.

The above figures tell the story of statistics. When we consider a sufficiently large data set, as in Fig. 2.1-4, the distribution fits the theory rather closely. In the much smaller set of Fig. 2.1-1, the individual fluctuations dominate. As the sample size increases, the individual fluctuations become less visible, and equation (2.1-1) gradually becomes a better descriptor of its *aggregate* behavior. Statistics typically apply to large data sets, but are not meant to describe the behavior of individual data points, or small sets thereof. If they are nonetheless pushed to do so, as in Fig. 2.1-1, they usually fail miserably.

2.2

Replicate measurements

When a measurement y is repeated N times *under the same conditions*, we can calculate its **average** or **mean** value $\bar{y}$ as

$$\bar{y} = \frac{1}{N}\sum_{i=1}^{N} y_i \tag{2.2-1}$$

The **standard deviation** is a measure of the irreproducibility in the average, again assuming that the experiment is repeated under the same conditions. It is usually given the symbol σ, and is defined as

$$\sigma = \sqrt{\sum_{i=1}^{N} (y_i - \bar{y})^2 \Big/ (N-1)} \tag{2.2-2}$$

Its square σ^2 is called the **variance**. The next spreadsheet exercise will illustrate the meaning of the standard deviation.

Instructions for exercise 2.2

1 Open an Excel spreadsheet.

2 Point with the mouse pointer to the tab labeled Sheet1, and rename it Average. (For details, see instruction (2) of exercise 2.1.)

3 Deposit the label 'n' in cell A1, and the label 'data' in B1.

4 In cell A3 deposit the number 1, and in cell A4 the number 2.

5 Place the pointer in cell A3 (it should be a heavy cross), depress the mouse button, move the pointer to cell A4, then release the button. Both cells (A3 + A4) should now be activated.

6 Grab the common handle of cells A3 + A4 (when the mouse pointer has the shape of a plus sign) and drag the cells by this handle down to cell 302. This will establish N-values from 1 to 300 in column A. Or: in cell A4 use the instruction = A3 + 1, and copy this down to cell A302.

7 Click on Tools, then on Data Analysis, and in the Data Analysis dialog box select Random Number Generation.

8 In the Random Number Generation dialog box that now appears, select 'Normal' as the Distribution, 10 as the Mean, and 1 as the Standard Deviation. Furthermore, specify the Output Range as B3:B302. Click OK.

9 Select block A3:B302, e.g., by first using the mouse pointer to activate a small block such as A3:B6, and by then, while keeping the Shift depressed, keying in End followed by $\downarrow$.

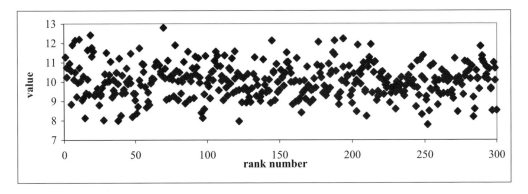

Fig. 2.2-1: 300 replicate data with Gaussian noise.

10 Select Insert ⇨ Chart, and use the ChartWizard to select an XY plot, showing individual data points without a connecting line. Complete the graph; Figure 2.2-1 suggests what it might look like.

11 We notice that the data all cluster around a Y-value of about 10, but that individual points can lie quite a bit farther from that average value: occasionally a point will lie more than 2 or 3 standard deviations from the average. That shows the true nature of such a distribution; if we consider a sufficiently large number of such data, we will find that about 68% of them lie within one standard deviation from the mean, but that the remainder, about one-third of all points, lie further away.

12 Click on the numbers with the X- or Y-axis, right-click, choose Format Axis, and select the Font and Scale to your liking. In our example we have used 16 point regular Times New Roman, and restricted the Y-scale to the range from 7 to 13, but please make your own decisions. We have also deleted the series marker (which usually appears in a sep-arate box to the right of the graph) by clicking on it to highlight it, and by then using the Delete key to remove it.

13 Activate cell C4, and deposit in it the instruction = AVERAGE(B3:B5) which is equiva-lent to the instruction = (B3 + B4 + B5)/3. Verify in an empty cell that the instruction indeed calculates the average, then erase your verification lest it will show as an odd point in one of the graphs you will make.

14 Activate cell D4, and make it carry the instruction = STDEV(B3:B5), which calculates the standard deviation according to eqn. (2.2-2). Again verify that STDEV indeed cal-culates correctly, then erase that test.

15 Highlight the area C3:D5, grab its common handle, and pull that handle all the way down to cell D302. Column C should now have 100 data, each the average of three suc-cessive data points in column B, while column D will now contain the corresponding standard deviations.

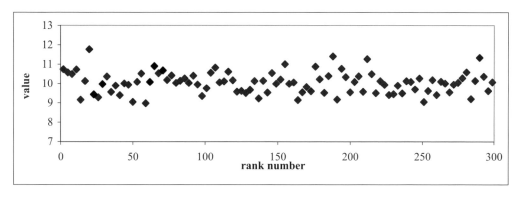

Fig. 2.2-2: Three-point averages of the data set of Fig. 2.2-1.

16 Take a look at the averages and the standard deviations: they will show considerable fluctuations, even though all numbers were calculated from Gaussian noise with a constant standard deviation.

17 The easiest way to 'take a look' at these averages and their standard deviations is, of course, to plot them. Activate cell A3, then drag the pointer to A302, depress the Ctrl key, move the mouse pointer over to cell C302, and drag the pointer to cell C3. This will highlight the data in columns A and C you want to graph.

18 Click on Insert ⇨ Chart, and answer the ChartWizard to make the graph showing the three-point averages. Plot it; it might look somewhat like Fig. 2.2-2.

19 Click on the chart, then on one of the data points in it, then right-click, and select Format Data Series. In the resulting dialog box go for the Y Error Bars tab, under Display select Both, and push the radio button for Custom. Then deposit in the two boxes labeled + and − the identical instruction: =AVERAGE!D3:D302, and use the OK button to enter these instructions. You should now obtain a graph resembling Fig. 2.2-3, in which all three-point averages are specifically labeled with their corresponding standard deviations.

Let's take a moment to consider what we have so far. Although all data in column B were generated with Gaussian noise, with a standard deviation of 1, we see that the three-point averages fluctuate rather wildly. Just look at the data in Fig. 2.2-3a (an enlargement of the first part of Fig. 2.2-3) around $n = 64$, where there are several sets of three consecutive data that lie close together and therefore have quite small standard deviations, of the order of 0.1, so that the error bars of successive data triplets do not overlap at all. To the right of these is a data triplet with a standard deviation of almost 2. And as the averages of points 22 through 24 and 28 through 30 show, this is not an isolated occurrence; similar (though somewhat less dramatic) cases are visible elsewhere in Fig. 2.2-3. You will, of course, have different data, but

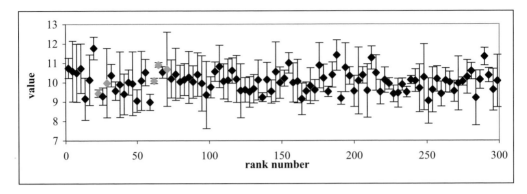

Fig. 2.2-3: Three-point averages with error bars of $\pm\sigma$ for the data set of Fig. 2.2-1.

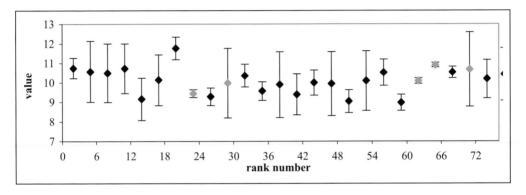

Fig. 2.2-3a: Detail of Fig. 2.2-3, with some points emphasized in color.

they will illustrate the same general phenomenon, viz. that *statistics don't apply to small data sets*, such as triplicates. This is the same point made in exercise 2.1.

20 Activate cell E8 and let it calculate the average of the 10 consecutive data in B3:B12.

21 Similarly, in cell F8, compute the corresponding standard deviation.

22 Highlight block E3:F12, grab it by its handle, and copy it down to F302. You should now have 30 averages of 10 points each, with their standard deviations.

23 Use columns A and E to make a graph, on a new sheet.

24 Add error bars to the averages plotted, using as before the Format Data Series dialog box, where you select the Y Error Bars tab, then specify the Error Amount under Custom as = AVERAGE!F3:F302 in both directions, and compare with Fig. 2.2-4.

25 In column G calculate the averages of 30 consecutive data points, and in column H the corresponding standard deviations. Plot the resulting thirty-point averages, with their individual error bars. Figure 2.2-5 shows an example.

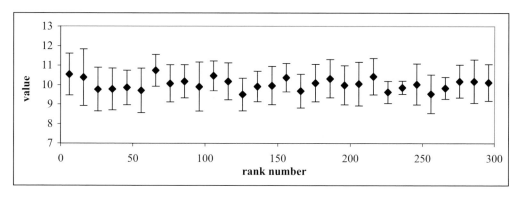

Fig. 2.2-4: Ten-point averages of the data set of Fig. 2.2-1.

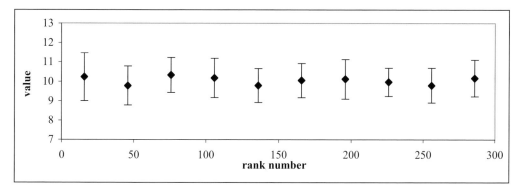

Fig. 2.2-5: Thirty-point averages of the data set of Fig. 2.2-1.

Again we step back for a moment. By taking the average over a sufficiently large number of data we see their statistics emerge: the ten-point averages have more uniform values for the averages and the standard deviations, while the thirty-point averages are a further improvement over the ten-point ones, but still exhibit some variability. None of these averages is exactly 10.000, nor do the standard deviations have the value 1.000, but the trend is clear: statistical averages do apply to these data provided the samples used are sufficiently large, and they are the more reliable the larger the data set that is used. A balance must be struck between the need for higher data precision and the time and effort needed to achieve it. In statistics, as in all other aspects of quantitative analysis, we encounter the law of diminishing returns: here, a twofold increase in precision usually requires a fourfold increase in the number of data (and hence in the experimental time) needed.

26 Finally, calculate the average and standard deviation of all 300 data points. For the data set shown in Fig. 2.2-1 the result is 10.04 ± 0.96, quite close to 10.00 ± 1.00 we might obtain for, say, a data set containing a million points. But there is no gain in taking that many data: thirty points was plenty for this (synthetic, and therefore highly idealized) data set to find the average as about 10 ± 1, while three was statistically inadequate.

At the risk of being repetitive (which this exercise is all about anyway), let us look once more at some of these data. (You will, of course, have different data, but you will most likely have comparable regions from which you can draw similar conclusions.) The data in B61:B63 just happen to be very close to each other, and to the true average. (In this case we know the true average to be 10.00 because we have used a synthetic data set; in a more realistic situation the true average is not known.) The data in the next triplet, B64:B66, lie even closer together, and therefore have an even smaller standard deviation. However, that high average value of about 11, with its very small standard deviation, would be quite misleading!

The triplet B70:B72 just happened to contain one fairly high reading, 12.79. Should we have thrown out that high point as an outlier? Absolutely not: the high point is a perfectly legitimate member of this Gaussian distribution. The conclusion we can draw from this or similar experiments is that, even for such an ideal case of a synthetic Gaussian distribution of errors, triplicate measurements are statistically inadequate to yield a standard deviation. The only benefit of a triplicate determination lies in the value of its *average*, which can be assumed to be somewhat more reliable than the value of an individual measurement.

Since the value of any standard deviation computed from just three replicas can vary wildly, such 'statistical estimates' have little scientific standing. A sufficiently large number of observations is required to justify the use of statistical analysis – otherwise we misuse statistics as an empty ritual, merely going through the motions, and giving our results a semblance of statistical respectability they do not deserve. In fact, one can estimate the relative standard deviation of the standard deviation, which comes to $1/\sqrt{(2N-2)}$ where N is the number of repeat measurements used, see J. R. Taylor, *An introduction to error analysis*, University Science Books, 2nd ed. 1997. Consequently the standard deviation of a triplicate measurement has, itself, a relative standard deviation of 50%! And it would require some 5000 repeat experiments to reduce that value to 1%. The moral of all this is: use statistics wherever their use is justified, and then use them well – but don't degrade them by applying them, inappropriately, to small data sets, or by over-specifying the precision of the standard deviation.

A picture is worth a thousand words, which is why we have used error bars to make our point. However, there is another, simpler way to demonstrate that, for our limited sample, the standard deviations we obtain are only *estimates* of the true standard deviation.

27 Determine the smallest value of the standard deviations in the three-point samples with the instruction = MIN(D3:D302), and for the corresponding largest value with = MAX(D3:D302). Likewise find the extreme values for the standard deviations of the ten-point samples in column F, and for the thirty-point samples in column H.

28 Record these values, then save the spreadsheet, and close it.

     In our particular example we find for the 100 three-point samples a minimum standard deviation of 0.12, and a maximum of 1.91, quite a spread around the theoretical value of 1.00! (The specific numerical values you will find will of course be different from the example given here, but the trends are likely to be similar.) For the 30 ten-point samples we obtain the extreme values 0.35 and 1.45, and for the 10 thirty-point samples 0.73 and 1.23. While taking more samples improves matters, even with thirty samples our estimate of the standard deviation can be off by more than 20%. And this is for by-the-book, synthetic Gaussian noise. From now on, take all standard deviations you calculate with an appropriate grain of salt: for any finite data set, the standard deviations are themselves estimates subject to chance.

2.3 The propagation of imprecision from a single parameter

Experimental data are often used to calculate some other, derived quantity *F*. When we have taken enough data to make a statistically significant estimate of its standard deviation, one can ask what will be the resulting standard deviation in the derived quantity. In other words, how does the standard deviation of the measured, experimental data *X* propagate through the calculation, to produce an estimate of the resulting standard deviation in the derived answer *F*? Such an estimate of experimental imprecision is often called the 'error' in the answer, and the method considered below is then called a 'propagation of errors'. However, *the true errors will almost always be much larger*, since they will also contain any systematic errors ('inaccuracy', 'determinate error' or 'bias'), including those due to inadvertent changes in experimental conditions when the experiment is reproduced at another time or place. In the example below, we will consider how the standard deviation, which measures *random* error (and is therefore a measure of the imprecision of the data rather than of their inaccuracy) propagates through a calculation.

    Simple rules suffice in a number of simple cases, such as addition and subtraction (where the standard deviations add), or multiplication and division (where the *relative* standard deviations add). Simple rules also apply to a few transforms, such as exponentiation or taking logs, where the nature of the variance changes from absolute to relative, or vice versa. However, those rules leave us stranded in many other cases, where we therefore must use a more general approach.

Say that we measure the cross-section of a sphere, and then calculate its volume. Given a standard deviation in the cross-section, what is the resulting standard deviation in the volume of the sphere? Or we might make a pH reading, then use it to compute the corresponding hydrogen ion concentration $[H^+]$. Again, based on the standard deviation of the pH measurement, what is the resulting standard deviation in $[H^+]$?

There are several ways to approach such problems. When you are uncomfortable with calculus, it may initially be the simplest to use algebraic expressions or series expansion. For example, the volume V of a sphere, expressed in terms of its diameter d, is $(4/3)\pi(d/2)^3 = \pi d^3/6$. When the measurement produces a diameter $d \pm \Delta d$, where Δd is an estimate of the experimental imprecision in d, then the volume follows as $V \pm \Delta V = \pi(d \pm \Delta d)^3/6 = (\pi/6) \times (d^3 \pm 3d^2\Delta d + 3d(\Delta d)^2 \pm (\Delta d)^3) \approx (\pi/6) \times (d^3 \pm 3d^2\Delta d) = (\pi d^3/6) \times (1 \pm 3\Delta d/d)$ when we make the usual assumption that $\Delta d \ll d$, so that all higher-order terms in Δd can be neglected. In other words, the relative standard deviation of the volume, $\Delta V/V$, is three times the relative standard deviation $\Delta d/d$ of the diameter, a result we could also have obtained from the above-quoted rules for multiplication, because $r^3 = r \times r \times r$.

Using calculus we can obtain the same result as follows: again assuming that the deviations Δy are small relative to the parameters y themselves, we have $\Delta V/\Delta d \approx dV/dd = d(\pi d^3/6)/dd = 3\pi d^2/6 = 3V/d$ or $\Delta V/V = 3\Delta d/d$. In general, when we have a function F of a single variable x, then the standard deviation σ_F of F is related to the standard deviation σ_x of x through

$$\sigma_F = \left|\frac{dF}{dx}\right|\sigma_x \qquad (2.3\text{-}1)$$

This is where the spreadsheet comes in, because we can use it to compute the numerical value of the derivative dF/dx even when you, my reader, may be uncomfortable (or even unfamiliar) with calculus. Here is the definition of the derivative:

$$\frac{dF}{dx} = \lim_{\Delta x \to 0}\frac{\Delta F}{\Delta x} = \lim_{\Delta x \to 0}\frac{F(x + \Delta x) - F(x)}{\Delta x} \qquad (2.3\text{-}2)$$

Consequently we can find the *numerical value* of the derivative of F with respect to x by calculating the function F twice, once with the original parameter x, and once with that parameter slightly changed (from x to $x + \Delta x$), and by then dividing their difference by the magnitude of that change, Δx. When Δx is sufficiently small, this will calculate the value of dF/dx without requiring formal differentiation. Here goes.

Instructions for exercise 2.3

1 Open an Excel spreadsheet, and name it (by renaming its label) Propagation.

2 Deposit the label 'd=' in cell A1, 'sd=' in A2, 'V=' in A3, 'dV/dd=' in A4 and 'sV=' in A5, where we use s instead of σ.

3 Optional: if you want to spend the extra effort on appearance, activate cell A2, use the function key F2 to select the editing mode, in the formula window highlight the letter s, then use the down-arrow next to the font listing (to the left of the formula window) to select the Symbol font, click on it, then click somewhere in the spreadsheet area. You can do the same for sV in A5, etc. Since such beautification of the text is more decorative than functional, we will not spend time on it, but instead use equivalents such as s for σ, and S for Σ.

4 In cells B1 and B2 deposit numerical values for d and sd respectively.

5 Activate cell B1, and click on Insert Name Define. In the resulting Define Name dialog box, the top window (under Names in Workbook:) will now show the d (otherwise type it in), while the bottom window (under Refers to:) will show =PROPAGATION!$B$1 (and, again, make it so if it is not so). Click on OK. Similarly, give the (future) contents of cell B2 the name sd. Note that Excel guesses the correct name (when already used in the cell to the left of that being named) and thereby reduces the amount of typing you need to do. Naming cells can only be done when we use absolute addressing of these cells, i.e., when they represent constants.

6 In cell B3 place the formula = PI()*(d^3)/6.

7 In cell B4 now calculate the derivative as = (PI()*((d + sd)^3)/6–B3)/sd.

8 In cell B5 calculate the final result, sV, as = ABS(B4)*sd.

9 In order to compare this with the theoretical result $\sigma_V = (dV/dd)\,\sigma_d$ we use cell B6 to calculate $\Delta V = (3V/d)\,\Delta d$. Convince yourself that the result in B6 is the same as that in B5 as long as sd $\ll$ d. Note that, in cell B4, you have performed a numerical differentiation without using calculus!

10 Save the spreadsheet, then close it.

As our second example, we will estimate the standard deviation in $[H^+]$ when the concentration of hydrogen ions is calculated from a pH reading p with a corresponding imprecision Δp. In order to see how the imprecision propagates, we now use the relation $[H^+] = 10^{-pH} = 10^{-(p\pm\Delta p)} = 10^{-p(1\pm\Delta p/p)} = 10^{-p} \times 10^{\pm\Delta p/p}$. There does not exist a closed-form expression for $10^{\pm\Delta p/p}$ analogous to that which we used for $(d + \Delta d)^3$ in our earlier example, but instead we can use the series expansion

$$a^\delta = 1 + \delta\ln(a) + (\delta\ln(a))^2/2! + (\delta\ln(a))^3/3! + (\delta\ln(a))^4/4! + \cdots \approx 1 + \delta\ln(a)$$
$$\text{when } \delta \ll 1$$

where we will again assume that $\Delta p/p$ is much smaller than 1, so that $10^{\pm\Delta p/p} \approx 1 \pm (\Delta p/p)$ ln(10) and $[H^+] = 10^{-p} \times 10^{\pm\Delta p/p} = 10^{-p} \times [1 \pm (\Delta p/p)$ ln(10)]. Consequently the hydrogen ion concentration $[H^+]$ has a relative standard deviation of $(\Delta p/p)$ ln(10), and an absolute standard deviation of $(\Delta p/p) \times 10^{-p} \times \ln 10 = [H^+] \times \ln(10) \times \Delta p/pH$.

Calculus again provides a much more general and direct approach to this problem. The calculation from pH to $[H^+]$ converts the measured parameter x (here: the pH) into the derived function F (here: $[H^+] = 10^{-pH}$). We have $d(a^{-x})/dx = -a^{-x} \ln(a)$ so that $d([H^+])/d(pH) = d(10^{-pH})/d(pH) = -10^{-pH} \times \ln(10) = -[H^+] \ln(10)$. Consequently, $\sigma_H = |d[H^+]/d(pH)|\sigma_{pH} = [H^+] \ln(10) \times \sigma_{pH}$, and the resulting standard deviation in $[H^+]$ is $\sigma_{pH} \times \ln(10)$.

Below we will again use the spreadsheet to bypass series expansion as well as differentiation when we only need a numerical (rather than a general, algebraic) result. When we deal with experimental imprecision, numerical rather than algebraic results are usually all that is required.

11 Reopen the spreadsheet Propagation, and deposit the label 'pH=' in cell D1, 'spH=' in D2, '[H]=' in D3, 'dH/dpH=' in D4 and 'sH=' in D5.

12 In cells E1 and E2 deposit numerical values for pH and spH respectively, and name their contents.

13 In cell E3 place the formula = 10^-pH.

14 In cell E4 calculate the derivative as = (10^-(pH + spH)–E3)/spH.

15 In cell E5 calculate the final result, sH, as = ABS(E4)*spH, and compare this with the theoretical result by calculating the latter, $[H^+] \ln 10 \times \sigma_{pH}$.

16 Save the spreadsheet again, and close.

2.4 The propagation of imprecision from multiple parameters

When the derived result is a function of several independent parameters, each with its own experimental imprecision, computing the propagation of such imprecision becomes more complicated when we attempt to do it with algebraic expressions and series expansions. However, the answer remains straightforward when we use *partial* derivatives, i.e., when we consider separately the effects of each of the input parameters. Given a function $F(x_1, x_2, x_3, \ldots, x_N)$ of N variables x_i, where each x_i has an associated standard deviation σ_i, the resulting standard deviation σ_F in the function F will be given by

$$\sigma_F = \sqrt{\sum_{i=1}^{N} \left(\frac{\partial F}{\partial x_i}\right)^2 \sigma_i^2} \qquad\qquad (2.4\text{-}1)$$

While the standard deviation σ_F is the desired final result, since it has the same dimension of the function F, the corresponding variances σ_F^2 yield somewhat simpler equations:

$$\sigma_F^2 = \sum_{i=1}^{N} \left(\frac{\partial F}{\partial x_i}\right)^2 \sigma_i^2 \qquad (2.4\text{-}2)$$

Again, in order to use these equations numerically, one need not know how to take partial derivatives (although that certainly would not hurt), but merely realize that the partial derivative of the function F with respect to x_i is defined as

$$\frac{\partial F}{\partial x_i} = \lim_{\Delta x_i \to 0} \frac{F(x_1, x_2, \ldots, (x_i + \Delta x_i), \ldots, x_N) - F(x_1, x_2, \ldots, x_i, \ldots, x_N)}{\Delta x_i} \qquad (2.4\text{-}3)$$

Therefore, one can calculate the partial derivative of F with respect to x_i by calculating the function F twice, once with the original parameters and once with just *one* of these parameters slightly changed (from x_i to $x_i + \Delta x_i$), and by then dividing their difference by the magnitude of that change, Δx_i. When Δx_i is sufficiently small, this will calculate $\partial F/\partial x_i$ without requiring formal partial differentiation. Below we will compare the calculus-based and numerical method, using one of the examples given by Andraos (*J. Chem. Educ.* 73 (1996) 150) as a test function, namely

$$F = \log(xy + z^2) - x/z^3$$

where

$$\partial F/\partial x = y/\left((xy + z^2)\ln(10)\right) - 1/z^3$$

$$\partial F/\partial y = x/\left((xy + z^2)\ln(10)\right)$$

$$\partial F/\partial z = 2z/\left((xy + z^2)\ln(10)\right) + 3x/z^4$$

Instructions for exercise 2.4

1 Again reopen the spreadsheet Propagation.

2 Deposit the label 'x = ' in cell A7, 'y =' in A8, and 'z =' in A9.

3 In cell B7 deposit a value for x, in B8 a value for y, and in B9 a value for z.

4 Likewise, in cells D7:D9 deposit the labels 'sx =', 'sy =', and 'sz =', and in cells E7:E9 the corresponding values for the standard deviations in x, y, and z respectively. The only requirements are that the value for sx should be much smaller than that for x, and the same applies for sy and sz. For example, you might use 4, 5, and 6 for x, y, and z, and 0.1, 0.2, and 0.3 for sx, sy and sz. Or use whatever other values suit your fancy, as long as the standard deviations in a parameter are much smaller than the parameter itself. Use the spreadsheet to find out what 'much smaller' means.

5 Name cells B7:B9 and E7:E9.

6 In cell A10 deposit the label 'F =', and in cell B10 the formula = LOG(x*y + z^2) − x/z^3, then name this cell F.

7 In cell A11 deposit the label 'dF/dx =', in cell A12 'dF/dy =', and in A13 'dF/dz =', in order to denote terms in (2.4-3) such as $\partial F/\partial x_i$.

   After these preliminaries we are now ready to calculate the standard deviation σ according to (2.4-3), using the given expressions for the partial derivatives.

8 In cell B11 calculate $\partial F/\partial x$ as = y/((x*y + z^2)*LN(10)) − 1/z^3. Note how much easier and more transparent it is to type the algebraic expressions rather than the corresponding absolute cell addresses, i.e., x instead of $B$7, y instead of $B$8 etc.

9 Likewise in B12 calculate = x/((x*y + z^2)*LN(10)) to compute $\partial F/\partial y$, and enter the corresponding expression for $\partial F/\partial z$ in cell B13.

10 In cell A14 place the label 'st dev =', and in B14 calculate the properly propagated estimate of the standard deviation in F as = SQRT((B11*sx)^2 + (B12*sy)^2 + (B13*sz)^2).

> Now we will make the equivalent calculation without using the results of calculus.

11 First, add a label (such as delx) and a corresponding value (say, 0.01) to the top of the spreadsheet, and assign it a name, say delta.

12 Copy the contents of cell B10 to cell C11.

13 In order to compute the term $\partial F/\partial x = [F(x + \Delta x, y, z) − F(x, y, z)]/\Delta x$, edit the contents of cell C11 as follows. Place a minus sign to the right of the contents of the cell, then copy the expression LOG(x*y + z^2) − x/z^3, and paste it back after the minus sign. Then replace x everywhere in the first half of the resulting expression by (x + delx). Finally, place the entire expression inside brackets, and divide it by delx. It should now read = (LOG((x + delx)*y + z^2) − (x + delx)/z^3 − LOG(x*y + z^2) − x/z^3)/delx.

14 Verify that you obtain the same result for $\partial F/\partial x$ as before.

15 Also verify that you obtain the same result for $\partial F/\partial x$ when you use different values for delx, such as 1E-6, as long as these values are very much smaller than x. Try out for yourself what values of delx are acceptable, and record your observations.

16 In a similar vein, compute $\partial F/\partial y$ as = (LOG(x*(y + dely) + z^2) − x/z^3 − LOG(x*y + z^2) − x/z^3)/dely in cell C12, and in cell C13 calculate $\partial F/\partial z$ as (LOG(x*y + (z + delz)^2)–x/(z + delz)^3–LOG(x*y + z^2)–x/z^3)/delz.

17 Finally, compute the resulting standard deviation of F in cell C14 as = SQRT((C11*sx)^2 + (C12*sy)^2 + (C13*sz)^2), an instruction you can obtain simply by copying B14.

18 Compare your results in C11:C14 with those in B11:B14; they should be the same. (If not, you either have made a mistake, or one or more of your values for delx, dely and delz are too large.)

19 Save the spreadsheet, print it, and close it.

Note that steps 11 through 19 do not require you to use calculus in order to compute the partial derivatives $\partial F/\partial x$, $\partial F/\partial y$, and $\partial F/\partial z$; the spreadsheet does that for you, based on (2.4-3). It can always do this as long as it deals with specific numerical values for x, delx, y, dely, z, delz, etc. The above approach illustrates a calculus-free yet perfectly legitimate way to compute the *general* propagation of standard deviations in *any* formula, no matter how complicated, provided that the parameters x, y, z, etc. are mutually independent. And it need not add much work because, especially for a complicated formula, you will most likely already use the spreadsheet to calculate the standard deviations anyway.

We will now carry the above to its logical spreadsheet conclusion. The spreadsheet is there to make life easy for us in terms of mathematical manipulations, and three-quarters of a page of instructions to describe how to do it may not quite be your idea of making life easy. Touché. But this was only the introduction: once we know how to make the spreadsheet propagate imprecision for us, we can encode this knowledge in a macro. That is what we have done, and have described in detail in chapter 10. The macro is called Propagation, and if you have downloaded the macros from the website (as described in section 1.13) you can now use that macro. Below we illustrate how to use Propagation.

20 Return to the spreadsheet Propagate.

21 Call the macro with Tools ⇨ Macro ⇨ Macros.

22 In the resulting Macro dialog box, highlight Propagation, then push Run.

23 A sequence of input boxes will appear. The first is labeled Input Parameters. Highlight the block B7:B9 which contains these parameters. The address will appear in the window of the input box. Push the OK button.

24 Similarly, highlight and enter E7:E9 as the Standard Deviations, and B10 as the Function.

25 After you have entered the function, and pushed the OK button, you will see the propagated imprecision appear in cell C10, in italics. Compare it with your earlier results. That's it, no mathematics, no manipulations, just enter the data and push the OK button; the macro does the rest. Figure 2.4-1 shows the result, and the entire region of the spreadsheet used.

	A	B	C	D	E
7	x= 4			sx= 0.1	
8	y= 5			sy= 0.2	
9	z= 6			sz= 0.3	
10	F= 1.729670	0.031503			

Fig. 2.4-1: The part of the spreadsheet used by the macro Propagate. The blocks B7:B9 and E7:E9, and the single cell B10, are used as input to the macro, while the result appears in C10.

If you want to test whether you can now propagate imprecision without further hand-holding, try the second example given by Andraos, $G = (v^2 \sin(2\theta))/g$, assuming numerical values for v, θ, and g, as well as for σ_v, σ_θ, and σ_g.

2.5 The weighted average

Say that the age of a wooden artifact from antiquity is determined by taking a few samples of it, and subjecting these samples to radiocarbon dating. For valuable artifacts, the number and size of the samples must usually be kept as small as possible, and such minimal samples will typically yield individual results with different standard deviations. The question is then how to combine the various answers from the individual samples to yield a single, most probable age, plus an estimate of the corresponding precision. For example, Arnold & Libby reported in *Science* 113 (1951) 111 that they had used radiocarbon dating (the method for which Libby earned the Nobel prize) to determine the age of wood from a single acacia beam in the tomb of Zoser in Sakkara, Egypt. Three different samples from the same beam were taken, and their analysis yielded the following ages (counted from 1951): 3699 ± 770 years, 4234 ± 600 years, and 3991 ± 500 years. How to combine these results into a single, most probable age?

Because their precisions are different, we will assign these three analyses different weights. More specifically, we will weigh the individual measurements according to the reciprocals of their variances, i.e., we replace (2.2-1) by

$$\bar{y} = \frac{\sum_{i=1}^{N} w_i y_i}{\sum_{i=1}^{N} w_i} \tag{2.5-1}$$

where we have introduced the **individual weights** $w_i = 1/\sigma_i^2$. Note that (2.2-1) is the special case of (2.5-1) for when all data have equal weights, in which case these equal w_i's can be taken out of the summations, and then cancel each other out in numerator and denominator, while (in the denominator) the remaining sum of N terms 1 is equal to N. For the standard deviation of the weighted average we have

$$\sigma = \sqrt{\frac{N\sum_{i=1}^{N}w_i(y_i - \bar{y})^2}{(N-1)\sum_{i=1}^{N}w_i}} \qquad (2.5\text{-}2)$$

which, again, reduces to (2.2-2) when all weights w_i are the same. In exercise 2.5 we will use the spreadsheet to calculate the best estimate of the age at which that acacia beam was cut, and thereby stopped exchanging CO_2 with the air (where the radioisotope ^{14}C is continually replenished by cosmic radiation).

It is often necessary to apply weights to statistical data. For example, in epidemiological studies, the sample sizes of various studies of the same phenomenon in different countries may differ widely, even if they are otherwise identical. When such studies are subsequently combined, they should then be accorded different weights, or the original data pooled and the statistics redone on the aggregate. Unweighted averages, such as discussed in section 2.2, should only be used in two cases:

1 when we have reason to assume that the individual weights are the same, or
2 when we have no good means of expressing individual weights quantitatively, and therefore use constancy of weights as the best we can do under the circumstances. The second reason seems to apply most frequently.

Instructions for exercise 2.5

1 Open a spreadsheet.

2 In the 4th row, enter the column labels y, s, w, wy, and wRR (for weighted Residual squared).

3 Leave a row blank, and starting in row 6 deposit the experimental data, i.e., the ages in column y, and the corresponding standard deviations in column s.

4 In cell C6 deposit the instruction = 1/B6^2, which will calculate the weight according to $w_i = 1/\sigma_i^2$. Copy this instruction down to cells C7 and C8.

5 In cell D6 calculate the product $w_i y_i$ with = A6*C6, and copy this down to row 8.

6 In cells C1 and D1 deposit the labels Sw and Swy for Sum of weights w and Sum of the products $w_i y_i$ respectively. Below these labels, i.e., in cells C2 and D2, compute these sums, e.g., in C2 enter the instruction = SUM(C6:C8).

7 In cell A1 place the label y(av), and in B1 the label s(av), or some other name indicating the standard deviation of the average.

8 Now calculate the average value of y according to (2.5-1) as = D2/C2. This is the best estimate of the age of that wooden beam.

	A	B	C	D	E
1	y(av)	s(av)	Sw	Swy	SwRR
2	4012.56	232.066	8.5E-06	0.03396	0.3039
3					
4	y	s	w	wy	wRR
5					
6	3699	770	1.7E-06	0.00624	0.16583
7	4234	600	2.8E-06	0.01176	0.13621
8	3911	500	4E-06	0.01596	0.00186

Fig. 2.5: The spreadsheet used to calculate the weighted average age of an ancient wooden beam dating from about 20.6 ± 2.3 centuries BC.

9   In column E compute the weighted residual squared, i.e., in E6, as $= C6*(A6 - \$A\$2)^2$.

10 In cell E1 put the label SwRR for the Sum of the weighted Residuals squared, and in E2 calculate that sum as $=$ SUM(E6:E8).

11 Finally, in B2, calculate the standard deviation of your answer according to (2.5-2) as $=$ SQRT(3*E2/(2*C2)). For a large data set, we would have used the spreadsheet to calculate N, but for just three data pairs that is more trouble than it is worth.

12 Your complete spreadsheet should now look like Fig. 2.5. The wood was cut some 4013 ± 232 years before 1951.

13 Save the spreadsheet as WoodAge.

2.6 Least-squares fitting to a proportionality

In this and subsequent spreadsheet exercises, we will use the method of least-squares to fit data to a *function* rather than to repeat measurements. This is based on several assumptions: (1) that, except for the effect of random fluctuations, the experimental data can indeed be described by a particular function (say, a straight line, a hyperbola, a circle, etc.), that (2) the random fluctuations are predominantly in the 'dependent' parameter, which we will here call y, so that random fluctuations in the 'independent' parameter x can be neglected, and (3) that those random fluctuations can be described by a single Gaussian distribution.

The assumption that the experimental 'noise' is restricted to a single, 'dependent' variable, greatly simplifies the mathematical problem, and can often (though certainly not always) be justified. For example, time measurements can often be made with such exquisite precision, even just using an inexpensive digital watch, that in most measurements of experimental parameters (such as absorbance or pH) versus time the fluctuations in the time measurements are negligible compared to those in the other measured parameter(s).

We will first consider the proportionality $y = ax$, where we can measure y as a function of x. In the absence of experimental or theoretical imprecision, a single measurement would suffice, from which a could then be determined simply as $a = y/x$. However, such a measurement might be affected strongly by any experimental 'error' in y, which is why it is usually preferred to take and analyze a large number of measurements, rather than a single one. Moreover, we will usually want to check whether the assumed proportionality is a reasonable assumption, and therefore make measurements at various x-values. The two requirements, of many data and of data at various x-values, can be satisfied simultaneously by measuring y at a large number of x-values. The assignment is then to calculate the most likely value (the 'best' estimate) of the proportionality constant a, within the context of the assumption $y = ax$, from a large set of data pairs y_i, x_i. It is here that the least-squares method can be used. The least-squares method per se does *not* address the question whether a proportionality is the correct assumption, or whether some other model (say, a straight line with arbitrary intercept, rather than one through the origin) would be better. To check whether the assumed proportionality is obeyed we usually rely (1) on theoretical models, which shape our expectations, (2) on direct (visual) observation of the fit, and (3) on any trends in the **residuals** $(y_i - ax_i)$, i.e., the differences between the experimental data and the model. (Because we assume that the terms x_i contain no experimental errors, the terms ax_i are supposed to contain no experimental error either, and can therefore serve as the 'model' for y_i.)

For the least-squares fitting of N experimental data pairs y_i, x_i under the above conditions (where the index i denotes the ith measurement pair) we have

$$a = \frac{\sum_{i=1}^{N} x_i y_i}{\sum_{i=1}^{N} x_i^2} \tag{2.6-1}$$

as can be derived readily by minimizing the sum of the squares of the residuals $(y_i - ax_i)$ with respect to a, i.e., by setting $d\Sigma (y_i - ax_i)^2/da = \Sigma d(y_i - ax_i)^2/da = \Sigma - 2x_i(y_i - ax_i) = -2\Sigma x_i y_i + 2a \Sigma x_i^2$ equal to 0, and solving the resulting equation for a. In order to provide a numerical estimate of the random fluctuations in y_i we can define a standard deviation σ_y for y, and a corresponding variance σ_y^2; the latter is given by

$$\sigma_y^2 = \frac{\sum_{i=1}^{N} (y_i - ax_i)^2}{N-1} \tag{2.6-2}$$

which can be compared with (2.2-2). For the proportionality constant a we then find the variance σ_a^2 as

$$\sigma_a^2 = \frac{\sum_{i=1}^{N}(y_i - ax_i)^2}{(N-1)\sum_{i=1}^{N}x_i^2} = \frac{\sigma_y^2}{\sum_{i=1}^{N}x_i^2} \tag{2.6-3}$$

and a corresponding standard deviation σ_a.

In the following exercise we will first generate a set of noisy test data, then calculate the necessary sums, and use these to compute a, σ_y and σ_a, and plot the data.

Instructions for exercise 2.6

1 Open a spreadsheet, and name it Proportionality.

2 In cell A1 deposit the label a, and in cell B1 the label na or noise ampl.

3 In cell A2 place a numerical value for a, and in cell B2 a value for the noise amplitude.

4 To name the contents of cell A2, first activate it, then click on the name box in the formula bar (the bar immediately above the cell column labels A, B, C, etc.), type 'a' (its name), and depress the enter key. Similarly name the contents of cell B2 as na.

5 Place the labels x, noise, and y in cells A4, B4, and C4 respectively.

6 Place the numbers 0, 1, 2, ... , 9 in cells A6:A15.

7 Fill cells B6:B15 with Gaussian noise with zero average (or 'mean') and unit standard deviation. (Reminder: such Gaussian noise can be found under Tools ⇨ Data Analysis ⇨ Random Number Generation ⇨ Distribution: Normal, Mean = 0, Standard Deviation = 1, Output Range: B6:B15 ⇨ OK.)

8 In cells C6:C15 calculate the product ax using the value named a (as stored in A2) and the values of x in column A, plus the product of the noise amplitude na (stored in B2) times the noise in column B. For example, the instruction in cell C6 might read = a*A6 + na*B6 (or, when you don't use names, = $A$2*A6 + $B$2*B6).

9 In cells D4 through G4 deposit the labels xy, xx, RR, and y(calc).

10 In cell D6 calculate the product xy (as = A6*C6), in E6 compute x^2, and copy both down to row 15.

11 In cell D1 through F1 place the labels Sxy, Sxx, and SRR respectively.

12 Now we calculate in cell D2 the sum Σxy as = SUM(D6:D15). Contrary to what an accountant might do, we usually keep these sums *at the top of the spreadsheet* so that they will remain in sight regardless of the length of the data columns.

13 Likewise, in cell E2, compute the sum Σx^2 as = SUM(E6:E15). Shortcut: activate cell D2, copy it with Ctrl + c, activate E2, and paste with Ctrl + v.

14 Enter some more labels: a(calc), sy, and sa in G1 through I1 respectively. For greater clarity we will here explicitly distinguish the recovered value a_{calc} from the initially assumed value a, just as we use y_{calc} to denote the reconstituted data $a_{calc}x$. And we again label sums with S (for Σ), and standard deviations with s (for σ).

15 In G2 compute a_{calc} according to equation (2.6-1), i.e., as = D2/E2.

16 Now that we have found the least-squares estimate a_{calc} we can compute the corresponding standard deviations. In cell F6 calculate $(y - y_{calc})^2 = (y - a_{calc}x)^2$ as = (C6–$G$2*A6)^2, and copy this down to F15.

17 In cell F2 calculate SRR as = SUM(F6:F15), and in cell H2 compute σ_y as = SQRT(F2/9) in accordance with (2.6-2), since $N = 10$ so that $N - 1 = 9$.

18 Vary the value of the noise amplitude na in cell B2, and verify that sy (i.e., σ_y) is a reasonably close estimate (usually within a factor of 2) of na. (On average it will track it more closely when we have a larger number of data pairs, say 100 or 1000.)

19 In cell I2 compute σ_a according to (2.6-3) as = G2/SQRT(E2).

20 Compare sa (i.e., σ_a) with (the absolute value of) the difference between a and a_{calc}; again, sa usually tracks $|a - a_{calc}|$ to within a factor of 2.

21 In rows 6 through 15 of column G calculate $y_{calc} = a_{calc}\,x$, then make a graph of columns A, C and G. Show the simulated 'experimental' points y of column C as markers, and the least-squares line y_{calc} in column G as a smooth line. You can do this by clicking in the graph on a data point of one of these series, which will highlight them.

Excel has three built-in facilities for least-squares calculations, which provide the same (and, if you wish, much more) information. The first, LINEST, is a simple *function*. The second is the Regression *macro* in the Analysis Toolpak, which is part of Excel but must be loaded if this was not already done at the time the software was installed. The third (and often simplest) method is to use the Trendline feature, which is only available once the data appear in a graph. Later we will encounter yet another option, by using the weighted least squares macro described in chapter 10. Truly an embarrassment of riches! Below we will illustrate how to use the first three of these tools. Table 2.6-1 lists their main attributes, so that you can make an informed choice of which one of them to use.

22 In cell F3 deposit the instruction = LINEST(C6:C15,A6:A15,FALSE). The result is immediate: cell F3 will contain the value of a_{calc}.

23 The syntax of this *line-esti*mating function is as follows: the first block (C6:C15) specifies the array of y-values, the second block (A6:A15) defines the x-array, and FALSE signifies that the function should not calculate an intercept (because the line is supposed to go through the origin). The three pieces of information are separated by commas. Note that LINEST is a **function**, i.e., it updates automatically whenever you change an input value. The input arrays of x- and y-values cannot include empty cells.

24 Click on the curve to activate it, then right-click, choose Format Data Series, Patterns, and select, say, Line $\underline{A}$utomatic and Marker N$\underline{o}$ne or vice versa. Your plot might now look like Fig. 2.6-1.

25 Vary the noise amplitude in cell B2, and see how the recovered value a_{calc} varies with the noise amplitude. Also verify that you recover $a_{\mathrm{calc}} = a$ and $\sigma_a = 0$ when the noise amplitude is set equal to zero.

26 It is always useful to look at the residuals, i.e., the differences between the data and the fitted function; in the present example, the residuals are the differences $y_i - a_{\mathrm{calc}} x_i$. The reason for this is that use of an incorrect model (such as fitting to, say, a linear or quadratic relation rather than to a proportionality) often leads to a discernible **trend** in the residuals, whereas random deviations do not. Therefore plot the residuals $y - y_{\mathrm{calc}} = y - a_{\mathrm{calc}} x$, as in Fig. 2.6-2.

27 A fourth, optional TRUE or FALSE statement in the LINEST instruction specifies whether you want to see the standard deviations and other statistical information. However, such auxiliary information requires additional space, which must be reserved in advance, and makes the instruction somewhat more cumbersome to use. We will illustrate such use in steps (23) through (27) of spreadsheet exercise 2.8.

28 Now for the Regression routine in the Analysis Toolpak. Select $\underline{T}$ools $\Rightarrow$ $\underline{D}$ata Analysis $\Rightarrow$ Regression, then specify Input $\underline{Y}$ Range: C6:C15, Input $\underline{X}$ Range: A6:A15, activate the window for Constant is $\underline{Z}$ero, and set the Output Range: to J1. Click OK.

29 You will now see 18 lines of text and statistical data. Of interest to us are the value of a_{calc}, which you will find near the bottom, under X Variable, *Coefficient*, and the value of σ_a, which is listed under X Variable, *Standard Error*. The value of σ_y is not given as such, but its square, the variance of y, can be found under Residual, *MS*. Verify that these numbers are the same as those you computed. Because you specified Constant is $\underline{Z}$ero, the Intercept is indeed 0. The remainder of the information shown we will leave to the statisticians. Incidentally, the Regression routine does not update automatically when you change input data, but must be invoked anew.

30 Finally, check that the standard deviation σ_a provides an estimate of the magnitude of the difference between a and a_{calc}.

31 Also check that you will get (slightly) different answers when you use the regression routine without specifying that the Constant is $\underline{Z}$ero, in which case you fit to a linear relation rather than to a proportionality.

32 In order to use the **Trendline** feature, you need to have the data in graphical form. Fortunately you already made such a graph for instruction (21).

33 Click on a data point in that graph. You may note that the menu item $\underline{D}$ata (in the menu bar, to the right of $\underline{T}$ools) has now been replaced by $\underline{C}$hart. Now either right-click on the data point, or click on $\underline{C}$hart in the menu bar, and in either case use the resulting pop-up menu to select Add $\underline{T}$rendline. This will show the first page (Type) of the Add Trendline dialog box.

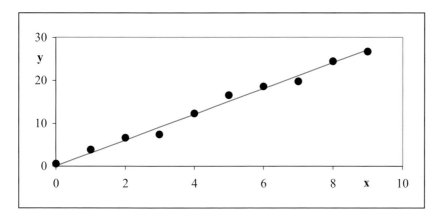

Fig. 2.6-1: A least-squares straight line through the origin.

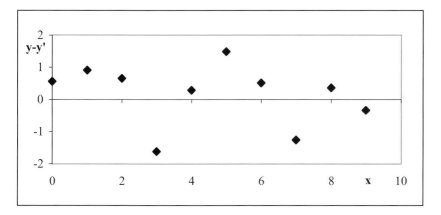

Fig. 2.6-2: The residuals of the plot of Fig. 2.6-1.

34 For fitting data to a line, click on the top left panel labeled L̲inear. Switch to the Options page, and select S̲et intercept = 0 for a line through the origin.

35 Also select Display e̲quation on chart, then click OK.

36 You will now see a heavy line drawn through the points, and the corresponding equation. By clicking and right-clicking on the line you can format the trendline, or clear it. Likewise you can click on the equation for the line, then drag it to another location, reformat it, or whatever.

37 Name and save the spreadsheet, then close it.

Table 2.6-1:   Comparison of the various least-squares methods available in Excel. The first three methods come with Excel, as does Solver. WLS is a weighted least-squares macro provided with this book, as is the macro SolverAid which yields the standard deviations for the parameter values provided by Solver.[1]

	linear, non-iterative	provides standard deviations	provides additional statistics	self-updating	fits data to a line	fits data to a power law	fits a multi-parameter equation	fits a general expression	accepts weights	user convenience
LINEST	+	+	+	+	+	+	+	−	−	low
Regression	+	+	+	−	+	+	+	−	−	medium
Trendline	+	−	−	+	+	+	−	−	−	high
WLS	+	+	−	−	+	+	+	−	+	medium
Solver	−	+[1]	−	−	+	+	+	+	+	medium

At this point it should be emphasized that the standard deviation σ_y provides an estimate of the goodness of the fit of the data *to a particular mathematical equation*, in this case to the proportionality $y = ax$. Note that this is quite different from an estimate of the (*model-independent*) *experimental reproducibility of a replicate measurement*. It is somewhat unfortunate that both measures are called standard deviations, and are denoted by the same symbol, σ.

There are many occasions where we can use the regression analysis tools incorporated in Excel. On the other hand, as we will see in chapter 3, there are also instances where we should *not* do so. Here we are merely getting acquainted with the mechanics of using a least-squares fit.

2.7 Least-squares fitting to a general straight line

The general equation for a straight line, $y = a_0 + a_1x$, does not require that the line through the data points go through the origin, as was the case with the proportionality $y = ax$. A general straight line has two adjustable parameters, where a_1 is the slope of the line, and a_0 its intercept with the vertical axis. In this case, the least-squares method minimizes the sum of the squares of the residuals with respect to both a_0 and a_1. This yields the following formulas for calculating the 'best estimates' a_0 and a_1 (where from now on we will delete the obvious subscript $_{calc}$ whenever there is little chance for confusion) and their respective standard deviations:

$$a_0 = \frac{\Sigma x^2 \Sigma y - \Sigma x \Sigma xy}{N \Sigma x^2 - (\Sigma x)^2} \tag{2.7-1}$$

$$a_1 = \frac{N \Sigma xy - \Sigma x \Sigma y}{N \Sigma x^2 - (\Sigma x)^2} \tag{2.7-2}$$

$$\sigma_y^2 = \frac{\Sigma(y - a_0 - a_1 x)^2}{N - 2} \tag{2.7-3}$$

$$\sigma_{a_0}^2 = \frac{\sigma_y^2 \Sigma x^2}{N \Sigma x^2 - (\Sigma x)^2} \tag{2.7-4}$$

$$\sigma_{a_1}^2 = \frac{N \sigma_y^2}{N \Sigma x^2 - (\Sigma x)^2} \tag{2.7-5}$$

where for the sake of notational simplicity we have left out the indices i, which always run from 1 to N, the number of data pairs entered. In this example, the quantity $N - 2$ is the **number of degrees of freedom**, and in general is equal to the value of N minus the number of constants derived from the data (here two: a_0 and a_1).

In the following spreadsheet we will use columns that extend beyond what can be seen on the screen. Moreover there are so many parameters that we will use two double rows of parameters above the actual columns of data. By organizing the spreadsheet in this fashion we can keep all important information within easy view on the monitor screen.

Instructions for exercise 2.7

1 Open a spreadsheet, and give it an appropriate name, such as Line.

2 In cells A1 through C1 deposit the labels a0, a1 and na (or noise ampl).

3 In cells A2 and B2 place assumed numerical values for a_0 and a_1, and in cell C2 a noise amplitude.

4 Place the labels x, y, and noise, in cells A7, B7, and C7 respectively.

5 Place the numbers 0, 1, 2, etc. in column A, starting with cell A9. Extend the column to some value N, say, 50.

The maximum column length Excel 95 can handle is $2^{14} = 16384$ entries; for Excel 97 it is $2^{16} = 65536$ entries. In either case, a spreadsheet with several such long columns may calculate very slowly, depending on the speed of the processor used and on the amount of available memory. For our purpose little is gained by using such long columns, while much time is lost. Therefore keep the column lengths reasonably short.

6 Fill cells C9:C58 (assuming here and in what follows that you use a column length of 50 data) with Gaussian noise with zero mean and unit standard deviation.

7 In cells B9:B58 calculate a simulated, noisy data set as $a_0 + a_1 x$ plus noise, using the values of a_0 and a_1 stored in A2 and B2 respectively, plus the product of the noise amplitude na stored in C2 times the Gaussian noise generated in column C. For example, the instruction in cell B9 might read = \$A\$2 + \$B\$2*A9 + \$C\$2*C9, or = a00 + a01*A9 + na*C9 when you use the names a00, a01 and na for the contents of A2, B2 and C2 respectively. (Names such as a1 or aa1 cannot be used because they are valid addresses.)

8 In cells D7 through H7 deposit the labels xy, xx, y(calc), R, and RR respectively.

9 In cells D9:D58 calculate the products xy, and in E9:E58 compute x^2.

10 In cells A4 through F4 place the labels Sx, Sy, N, Sxy, Sxx and denom, and in cell H4 the label SRR, for the Sum of the Residuals squared.

11 Calculate in cell A5 the sum Σx as = SUM(A9:A1000). Note that the empty cells below row 58 do not contribute to this sum.

12 Likewise, in cells B5, D5, E5 and H5 compute the sums Σy, Σxy, Σx^2 and $\Sigma(y - y_{calc})$^2 respectively, simply by copying cell A5 to cells B5:H5.

13 In C5 deposit the instruction = COUNT(A9:A1000), which counts all numerical values in the range specified. The advantage of using this instruction instead of specifying a fixed value for N is that it automatically adjusts when the size of the input data array is varied, as long as the range specified (here rows 9 through 1000) is not exceeded. In this way, the spreadsheet can be used over and over again. A disadvantage is that the instruction will count every filled cell in its range, so that one must clear the range before one can reuse it with a shorter data set, and must remember not to place any other data in it.

14 In cell F5 calculate the denominator $N\Sigma x^2 - (\Sigma x)^2$ common to equations (2.7-1), (2.7-2), (2.7-4) and (2.7-5), as = C5*E5 - A5^2.

15 Enter some more labels: a0(calc), a1(calc), sy, sa0, and sa1 in D1 through H1 respectively. Again, we distinguish between the initially assumed values of a_0 and a_1, and their recovered values $a_{0,calc}$ and $a_{1,calc}$.

16 In D2 calculate the least-squares estimate $a_{0,calc}$ of a_0 according to equation (2.7-1), e.g., as = (E5*B5 - A5*D5)/F5.

17 Similarly, in E2 calculate $a_{1,calc}$ using (2.7-2).

18 Now that we have found the least-squares estimates for a_0 and a_1 we can calculate the standard deviations for y, $a_{0,calc}$, and $a_{1,calc}$. In cell F9 calculate $a_{0,calc} + a_{1,calc} x$ as = \$D\$2 + \$E\$2*A9, in cell G9 $y - y_{calc}$ as = B9 - F9, and in H9 its square as = G9^2, then copy all of these down to row 58.

19 Use (2.6-3) to calculate σ_y in cell F2 as = SQRT(H5/(C5-2)).

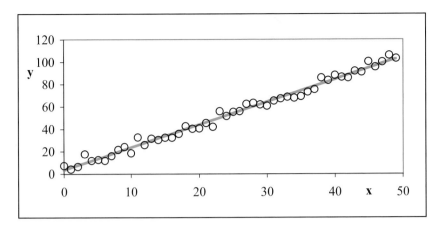

Fig. 2.7-1: The general line.

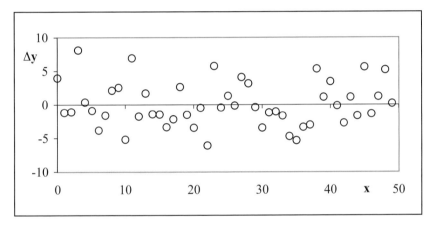

Fig. 2.7-2: The corresponding residuals.

20 In cell G2 compute σ_{a0} according to (2.7-4) as = F2*SQRT(E5/F5), and likewise calculate σ_{a1} in cell H2 based on (2.7-5).

21 Plot versus x the simulated data points y of B9:B58 together with the computed least-squares line y_{calc} in F9:F58, see Fig. 2.7-1.

22 Also plot the residuals $y - y_{calc}$ versus x, see Fig. 2.7-2.

23 Compare your results with those obtained with the LINEST function. We will here use it in its full form, which will require that you highlight an otherwise unused block of cells, 2 cells wide and 5 cells high, starting with the top-left cell.

24 In the formula window of the formula bar, type the instruction = LINEST(B9:B58,A9:A58,TRUE,TRUE). However, instead of depositing it with an Enter command, use Ctrl + Shift + Enter, i.e., hold down Ctrl and Shift while you press the Enter key. This is necessary to insert the instruction in the 5 × 2 block.

25 The top row of the block will now show the values of the slope and intercept, the second row the corresponding standard deviations, the third will contain the square of the correlation coefficient and the variance in y, then follow the value of the statistical function F and the number of degrees of freedom and, in the bottom row, the sums of squares of the residuals. The function LINEST provides a lot of information very quickly, albeit in a rather cryptic, unlabeled format.

26 To summarize the syntax of the argument of the LINEST( ) function, it is: (y-array, x-array, do you want the intercept?, do you want the statistical information?)

27 If you only want the values of a_0 and a_1, just highlight two adjacent cells in the same row, while still using Ctrl + Shift + Enter. If you specify a block two wide but fewer than five rows high, the values of a_0 and a_1 will be repeated in all rows. On the other hand, if you assign too large a block, the results will be fine, but the unused spaces will be filled with the error message #N/A.

28 Also compare your results with those from the Regression routine you can find under Tools ⇨ Data Analysis ⇨ Regression. This routine provides even more statistical information than the LINEST function, and labels it, but takes more time to execute and does not automatically update itself when the input data are changed. Regression also can make graphs.

29 This time, leave the box Constant is Zero blank. (For reasons that will become clear in exercise 2.8, place the output from the regression routine in cell J13.) Locate the places where Regression lists its estimates for $a_{0,calc}$, $a_{1,calc}$, σ_y^2, σ_{a0}, and σ_{a1}.

30 Verify that the difference between a_0 and $a_{0,calc}$ is of the order of magnitude (i.e., within a factor of two or three) of σ_{a0}, and that the difference between a_1 and $a_{1,calc}$ is likewise of the order of σ_{a1}. That is, of course, the significance of these standard deviations: they provide estimates of how close our 'best values' come to the true values, provided that all deviations are random and follow a single Gaussian distribution.

31 Save the spreadsheet, and close it.

In the above examples, we started from a precisely known expression such as $y = a_0 + a_1 x$, added Gaussian noise, and then extracted from the data the estimates $a_{0,calc}$ and $a_{1,calc}$. This allowed us to judge how closely we can reconstruct the true values of a_0 and a_1. In practice, however, the experimenter has no such luxury, since the true parameter values are generally not known, so that we will only have parameter estimates. In practice, then, there is little need to distinguish between the true parameters and their estimates, so that from now on the subscripts calc will be deleted whenever that can be done without introducing ambiguity.

Finally, a word of caution. In science, we usually have theoretical models to provide a basis for assuming a particular dependence of, say, y on x. Least-squares methods are designed to fit experimental data to such 'laws', and to give us some idea of the goodness of their fit. They are at their best when we have a large number of data points, justifying the statistical approach. However, they do not guarantee that the assumed model is the correct one. Always plot the experimental data together with the curve fitted through them, in order to make a visual judgment of whether the assumed model applies. And always plot the residuals, because such a plot, by removing the main trend of the data, is usually more revealing of systematic rather than random deviations than a direct comparison of experimental and reconstituted data. A plot of the residuals may show the presence of non-random trends, in which case the model chosen may have to be reconsidered.

Least-squares methods are usually favored over more subjective methods of fitting experimental data to a mathematical function, such as eyeballing or using French curves. However, the least-squares method is not entirely objective either: one still has to make the choice of model to which to fit the data. Least-squares fitting of data to a function gives the best fit *to that chosen function.* It is your responsibility to select the most appropriate function, preferably based on a theoretical model of the phenomenon studied. Absent theoretical guidance, one is most often led by Occam's parsimony rule according to which, all else being equal, the simplest of several satisfactory models is considered preferable.

2.8 Looking at the data

It is always useful to inspect the data visually, as plotted in a graph, rather than to just let the computer analyze them. The four data sets shown in Table 2.8-1 were carefully crafted by Anscombe (*Am. Statist.* 27#2 (1973) 17) to illustrate this point. Below we will fit all four data sets to a line $y = a_0 + a_1 x$, with the usual assumption that all errors reside in y.

Instructions for exercise 2.8

1 Recall the spreadsheet Line of the previous exercise. We will now make a copy of it to use here.

2 Right-click on the name tab Line, then click on Move or Copy. This will open the Move or Copy dialog box.

3 In this dialog box, click on Create a Copy, and in the window Before Sheet click on the spreadsheet you want to copy (here: Line). Click OK.

4 Automatically, the new copy will be called Line [2]. Rename it Anscombe.

5 Because we will not need any artificial noise, we simply put the noise amplitude in cell C2 to zero. The noise in C9:C58 then does not affect the analysis.

6 Enter the data from Table 2.8-1 in a block to the right of the region already used, e.g., in J1:Q11. Note that columns for X are the same for the first three data sets, so that you can copy them to save time and effort.

Table 2.8-1: *Four sets of x, y data pairs.*

Data set # 1		Data set # 2		Data set # 3		Data set # 4	
x	*y*	*x*	*y*	*x*	*y*	*x*	*y*
10	8.04	10	9.14	10	7.46	8	6.58
8	6.95	8	8.14	8	6.77	8	5.76
13	7.58	13	8.74	13	12.74	8	7.71
9	8.81	9	8.77	9	7.11	8	8.84
11	8.33	11	9.26	11	7.81	8	8.47
14	9.96	14	8.10	14	8.84	8	7.04
6	7.24	6	6.13	6	6.08	8	5.25
4	4.26	4	3.10	4	5.39	19	12.50
12	10.84	12	9.13	12	8.15	8	5.56
7	4.82	7	7.26	7	6.42	8	7.91
5	5.68	5	4.74	5	5.73	8	6.89

7 Import the first data set into the data analysis region, e.g., into block A9:B19. To do this, activate block J1:K11, copy it with Ctrl + c, activate cell A9, and paste using Ctrl + v.

8 Now erase all data in A20:H58, because these were not overwritten. Bingo! You will see the least-squares analysis of the newly reported data set in row 2. A quick check: make sure that N in cell C5 shows the proper value, 11.

9 Note down the values of the parameters $a_{0,calc}$, $a_{1,calc}$, σ_y, σ_{a_0}, and σ_{a_1}.

10 Verify your results with the Regression routine (which you can place, say, in cell J13), and also note down some of the other statistical parameters, such as the correlation coefficient ('Multiple R') and its square ('R Square').

11 Plot the data, and their residuals, On this Sheet. This will allow you to see all results (two plots each from four data sets) simultaneously, on the very same sheet. In order to accommodate eight graphs on one sheet, use the methods described in the second paragraph of section 1.10.

12 Now analyze the second data set, by copying block L1:M11 to A9. Row 2 will immediately provide the new values for the parameters $a_{1,calc}$, σ_y and σ_{a_1} (although you might hardly notice it, because they will be quite similar). On the other hand, the Regression routine does not automatically update, and must therefore be called in again.

　　Since you have noted down the results of the earlier regression analysis, just override it and write the new results over the old ones, in J13.

13 Note down the parameters, and plot the data and their residuals. Place the new graphs close to the earlier ones.

14 Repeat this process until you have analyzed all four data sets.

15 List the numerical results obtained.

The four data sets in this problem were selected by Anscombe (1973) to have the same values for their slopes a_1 $(= 0.50)$, their intercepts a_0 $(= 3.00)$, the sums of the squares of their residuals $\Sigma(y - y_{\mathrm{calc}})^2$ $(= 1.53)$, their standard deviations σ_y $(= 1.11)$, σ_{a_1} $(= 0.118)$, and σ_{a_0} $(= 1.12)$, as well as their correlation coefficients r $(= 0.816)$. By all these criteria, then, they fit the same equation of a straight line equally well. However, visual examination of the graphs, or of the residual plots, yields a quite different answer: only set #1 reasonably fits a straight line. In other words, the statistical analysis in terms of 'summary' statistics does not address the validity of the assumed model, and can produce results regardless of whether or not the model is appropriate. In the present case, the first data set reasonably fits the assumed model of a linear dependence, the second set should be fitted to a parabola instead of to a line, while the third and fourth sets both contain an intentional 'outlier'. Direct observation of the graphs before the analysis, and/or of the residuals afterwards, can often help us reject clearly inappropriate models.

The take-home message of this example is that a quick look at the original data (and, for more subtle differences, a quick look at the residuals) is often much more informative than the numerical values of the statistical analysis. A glance at the graphs can reveal a trend in the deviations from the model, and may suggest that the model used is inappropriate to the data, thereby sending you back to reconsidering the theory behind the phenomenon studied, or the method used to acquire the data. That theory may need to be modified or extended before you can benefit from a statistical data analysis, or a defect in the data acquisition method may have to be corrected. As is said in computer jargon, 'garbage in, garbage out', i.e., the quality of the results obtained by any computer analysis depends on the quality of the input data and, we might add here, on the appropriateness of the analysis model used. If the input data are of poor quality, no amount of statistical analysis can make them whole. Likewise, if the model used is inapplicable to the data, statistics cannot help. Just imagine trying to fit the coordinates of the numbers on a clock face to a straight line.

In short: it is seldom useful merely to analyze data without a thorough understanding of what the model used means or implies, and without a visual verification that the model is appropriate to the experimental data. Mindless application of statistical rules is no guarantee for valid conclusions. *Always graph your data and their residuals*; it may make you reconsider the experiment, or its interpretation.

2.9 What is 'normal'?

So far we have used the Gaussian distribution as our paradigm of experimental imprecision, because Excel makes it so readily available. The Gaussian distribution has indeed become the norm in much of science, as

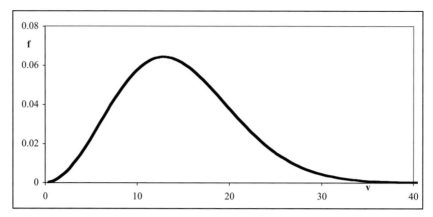

Fig. 2.9-1: The fraction f of molecules in an 'ideal' gas at a given speed v, in meters per second.

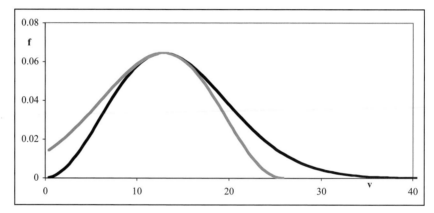

Fig. 2.9-2: The fraction f of molecules in an 'ideal' gas at a given speed v, as in Fig. 2.9-1, and (colored curve) its mirror image (mirrored around the maximum value of f) to emphasize the asymmetry of this distribution.

reflected in the fact that it is often called the **normal** distribution. The reason for its popularity, however, should be clearly understood: it is *not* that the Gaussian distribution is always, or even most often, followed by experimental data. We usually do not take the time to test the nature and distribution of the experimental imprecision, and the assignment is therefore more often based on hope or faith than on experimental evidence. There are several reasons that explain why the Gaussian distribution is so popular: it provides a *convenient mathematical model*, and it is usually *close enough* to approximate the actual distribution of errors or experimental imprecision. Moreover, a Gaussian distribution provides an *optimistic* estimate, because the assumption of a symmetrical distribution makes the imprecision look small. It also helps that a Gaussian distribution is quite compact: it has fewer far-out points than, e.g., a Lorentzian distribution would have. However, the following two examples should give us pause.

The kinetic theory of an ideal gas leads directly to an expression for the so-called root-mean-square velocity v of the gas molecules, viz.

$$v = \sqrt{3kT/m} = \sqrt{3RT/M} \qquad (2.9\text{-}1)$$

where k is the Boltzmann constant, T is the absolute temperature, m is the molecular mass, R is the gas constant and M is the molecular weight in Daltons. The Maxwell–Boltzmann distribution also provides an expression for the fraction of the gas molecules at any particular velocity,

$$f(v) = 4\pi v^2 \left(\frac{m}{2\pi RT}\right)^{2/3} \exp\left[\frac{-mv^2}{2RT}\right] \qquad (2.9\text{-}2)$$

Figure 2.9-1 shows that a plot of $f(v)$ versus v according to (2.9-2) has the general shape of a bell-shaped curve, and therefore resembles a Gaussian distribution. However, a more careful comparison indicates that the curve representing (2.9-2) is not quite Gaussian. Specifically, the distribution of molecular velocities is asymmetrical, with a 'tail' at high velocities. This asymmetry is more apparent in Fig. 2.9-2, where a colored line shows the same curve in reverse, mirrored around its maximum. For a symmetrical curve, such a mirror image would overlay the original curve; for an asymmetrical curve, it emphasizes the asymmetry. The parameters used in Figs. 2.9-1 and Fig. 2.9-2 are specified in the spreadsheet exercise.

Here, then, we have the simplest of *theoretical* situations (so that nobody can argue that we did not take a statistically valid sample) involving *random* thermal motion, and already we find that the velocities of the gas molecules only approximately follow a Gaussian distribution.

'Unfair,' I hear you mutter under your breath, 'velocities have a lower limit of zero but lack an upper limit, so it is no wonder that they exhibit an asymmetrical distribution.' True enough, but this also applies to many other quantities, such as absolute temperature, mass, concentration, and absorbance. The point here is not *why* a number of distributions are decidedly non-Gaussian, but *that* they are.

Instructions for exercise 2.9

1 The solid curve in Fig. 2.9-1 shows (2.9-2) for $R = 8.3143$ J mole^{-1} K^{-1} (1 J = 1 m^2 kg mole^{-1} sec^{-2} K^{-1}), $T = 300$ K and $M = 30$ Da (1 Da = 1 g mole^{-1} = 10^{-3} kg mole^{-1}) so that $m = M / N_{\text{Avogadro}} = 0.03/(6 \times 10^{23}) = 5 \times 10^{-26}$ kg. The value of M is appropriate for air molecules: $M = 28$ Da for N_2, $M = 32$ Da for O_2.

2 Open a spreadsheet.

3 In cell A1 deposit a label, such as v.

4 In cell A3 deposit the value 0, in cell A4 the instruction = A3 + 0.5, then copy this instruction and paste it in cells A4:A83. This will generate $v = 0$ (0.5) 40, i.e., numbers in the range from 0 to 40 with increments of 0.5.

5 In column B calculate the corresponding values for $f(v)$ as a function of v.

6 By inspection, find the approximate location of the maximum in the curve; it should lie close to $v = 13$.

7 Temporarily expand the v-scale around that maximum in order to get a better estimate. In the present example it is sufficient to expand the scale between $v = 12$ and $v = 14$ with increments of 0.1.

8 You will then see that the curve has a maximum close to $v = 12.9$.

9 Make the intervals symmetrical around that maximum, e.g., by using in A3 the value 0.4 instead of 0, so that column A now contains $v = 0.4 (0.5) 40.4$.

10 Copy the contents of column B for $f(v)$ and special paste its values in column C, but starting below where column B ends, i.e., in cell C85.

11 In A85:A136 compute 25.9 (-0.5) 0.4, by depositing the value 25.9 in A85, and the instruction = A85-0.5 in cell A86, and by copying this instruction down to row 136.

12 Highlight the block A3:C136, then plot the contents of columns B and C versus that of column A.

Our second example will show another reason why we should be careful with the assumption of a 'normal' distribution. Consider the weight of pennies. We usually assume that all pennies are born equal, but some may experience more wear than others (reducing their weights somewhat) while others may have been oxidized more (thereby increasing their weights). There is no a priori reason to assume that the weight loss by abrasion will be the same as the weight gain by oxidation (since abrasion and oxidation are rather independent processes), and therefore there is no reason to assume that the final distribution will be symmetrical, as in a Gaussian distribution. But that is not the main point we want to make here. When you actually weigh individual pennies, you will find that most of them weigh about 2.5 g, but once in a while you may encounter a penny that weighs well over 3 g. Say you weigh 10 pennies, and their weights are as follows: 2.5136 g, 2.5208 g, 2.5078 g, 2.4993 g, 2.5042 g, 2.5085 g, 2.5136 g, 3.1245 g, 2.5219 g, and 2.5084 g. What weight should you report?

You might just use (2.2-1) and (2.2-2) and calculate the weight as 2.57 ± 0.19 g. Or you might reason that the eighth measurement is so far off that it should be excluded as an 'outlier', in which case you would obtain 2.511 ± 0.007 g. The former result would seem to be the more honest one, because the heavy penny does not look much different from the other ones, and reweighing confirms the various weights. On the other hand, disregarding the heavy penny yields a result that certainly 'looks' much better, because it has a considerably smaller standard deviation. Which of these options should you choose?

Neither choice is correct. Look more carefully at the pennies, and you will find that the heavy one was minted in or before 1982, the lighter ones after that date. And if you were to dissolve the individual pennies in, say, concentrated nitric acid, and then were to analyze for their constituent metal ions, you would find that the heavy penny contains mostly copper, while the lighter ones are mostly zinc, which is a lighter metal, and therefore makes a lighter coin.

Indeed, the US government switched over from copper to copper-clad zinc when the value of a penny became less than the cost of the copper needed to make it. The assumption that all pennies are minted equal is therefore incorrect: pennies follow at least two different weight distributions, one for old, copper pennies, the other for the more recent, zinc ones. And, yes, there are still others, such as the steel pennies issued during World War II. But those you would have recognized immediately as different by their color.

Mixing the two distributions yields arbitrary results, because the average weight reflects what fraction of older pennies is included in the sample, and that fraction may depend on the source of the pennies: did they come from the bank (which usually issues new pennies), from your pocketbook, or from your older sister's penny collection? Arbitrarily throwing out the heavy ones is also incorrect. The only correct approach is (1) to recognize that there is a problem, (2) to identify its source (which in this case is relatively easy, because the year of minting is printed on each penny), (3) to report that there are two different types of pennies involved, and (4) to give the average weights and the corresponding standard deviations for both distributions. And if you don't have the time, resources and/or energy to collect enough old pennies to report a meaningful average weight for the heavy ones, at least mention that your result is valid for recent pennies, and that an older one was found to be much heavier.

Discussions of statistics often include a section on **outliers**. You have just read such a section, although it did not have that label, and certainly did not include a set of 'criteria' for outlier rejection. By definition, outliers are those results that do not seem to fit within the assumption that all experimental data obey a single, 'normal' distribution. Some outliers will result from outright errors, such as inadvertently exchanging the place of two numbers as you note down a weight, or experimental artifacts, such as the effect of a power glitch on the reading of an electronic instrument. The existence of such outliers may tempt you to reject all outliers. Please resist that temptation: many outliers reflect perfectly respectable measurements, of a phenomenon that just does not happen to follow a single, Gaussian distribution. We have just given two examples of such behavior. The distribution of molecular velocities in an ideal gas does not quite fit a Gaussian distribution, but instead exhibits an asymmetric distribution. And the penny weights show two distinct distributions rather than a single one. In general, then, there is *no justification* to reject outliers without good cause, and the mere fact that they are outliers, no matter how far off, is in itself insufficient cause: by that criterion, the heavy penny would have been rejected. As a chemist, you may sometimes have to cook your chemicals, but you should never cook your books. Do not let outliers make liars out of you.

After the above examples, the reader may well ask why it is that almost all natural scientists routinely use standard deviations and other measures based on a Gaussian distribution. And why these same assumptions are also used in most of the remainder of this book. For the answer we return to section 2.1, where we saw that the precise distribution of the experimental deviations can be observed only when we take a very large number of repeat measurements. We seldom take of the order of 10 000 repeat measurements; if we take only 100, we would not be able to tell from the data whether the underlying distribution is precisely Gaussian or only approximately so, as you will see by comparing Figs. 2.1-2 and 2.1-4. But this is an argument that can easily be inverted: even for a quite large number of repeat measurements, such as 100, the precise distribution is really immaterial. This is why the Gaussian distribution, with its well-established formalism, is commonly used, and justifiably so. As long as the actual distribution more or less resembles a bell-shaped curve, it is usually not worth the quite considerable effort required to establish and use a more appropriate distribution for each particular system being studied, because (except for very large data sets) the actual deviations do not yet adhere closely enough to such a distribution to make a perceivable difference. However, there are times when it is dangerous to assume a single Gaussian distribution, namely when evidence to the contrary stares us in the face, as it does in the case of outliers.

2.10 Poissonian statistics

In section 2.1 we already indicated that some types of measurements follow other than Gaussian types of statistics. Here we will briefly illustrate Poissonian statistics. These are in general called for whenever the experiment is a continuous one (e.g., it measures some parameter as a function of time t) yet its experimental result is quantized, as it is, e.g., in the measurement of radioactivity, in the opening and closing of ion-conducting channels in lipid bilayer membranes, or in single-photon counting. What all these have in common is that the outcome of the experiment is **discrete** rather than continuous: a radioactive nucleus is either in its original state or has decayed, an ion channel is either open or closed, a photon has either been counted or not.

The Poisson distribution describes the probability $P_N(t)$ that, in a time interval τ, N discrete events (such as radionuclide disintegrations, openings of a individual ion channels, or photon detections) will have taken place. That probability is

$$P_N(\tau) = \frac{(\lambda\tau)^N e^{-\lambda\tau}}{N!} \tag{2.10-1}$$

where λ is the likelihood of such an event occurring per unit time.

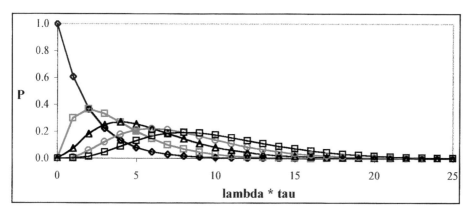

Fig. 2.10: The first five terms of the Poisson distribution (markers). The connecting line segments are drawn merely to indicate which points have the same N-values.

As can be seen in Fig. 2.10 this is a decidedly asymmetrical distribution. Moreover, since the outcome can only assume discrete values, the Poissonian distribution is a collection of points rather than a curve. Yet another difference between the Gaussian and Poissonian distribution is that (2.10-1) contains only one parameter, λ, whereas the Gaussian distribution (2.1-1) has two: the average value x and the standard deviation σ. The average value $\langle N \rangle$ of the Poissonian distribution is

$$\langle N \rangle = \lambda \tau \tag{2.10-2}$$

while its standard deviation is

$$\sigma = \sqrt{\lambda \tau} = \sqrt{\langle N \rangle} \tag{2.10-3}$$

Consequently, knowledge of N, the number of observed events, automatically implies the corresponding standard deviation. For instance, when 100 radioactive disintegrations have been counted, the standard deviation of the result is $\sqrt{100} = 10$, i.e., the result has a relative standard deviation of $10/100 = 0.10$ or 10%, whereas 40 000 events must be measured for the answer to have a relative standard deviation of 0.5%. These matters are mentioned here primarily in order to illustrate that the standard deviation, even of repeat measurements of the same basic phenomenon, is not always given by (2.2-2).

2.11 How likely is the improbable?

We will now briefly consider a question that is posed with increasing frequency in our society: how probable is the improbable? How likely is it that a spermicide or a drug used during pregnancy causes a birth defect, that power lines or portable phones cause cancer, or that working at a computer monitor causes a miscarriage? While this matter can be explained without benefit of a spreadsheet (as can almost any topic covered in this book) we will use the spreadsheet to illustrate combinatorics. To set the problem we start with a verbatim quote from a short review by K. R. Foster entitled 'Miscarriage and video display terminals: an update' (Chapter 6 in K. R. Foster, D. E. Bernstein, & P. W. Huber, *Phantom Risk*, MIT Press 1993):

> 'The link between miscarriages and use of video display terminals (VDTs) became a public issue around 1980 with the reports of clusters of reproductive mishaps in women users of VDTs.'
>
> 'All together, about a dozen clusters were reported. These included 7 adverse outcomes of 8 pregnancies at the offices of the solicitor general in Ottawa; 10 out of 19 at the offices of the attorney general in Toronto; 7 of 13 at the Air Canada offices at Dorval Airport, Montreal; 8 of 12 at Sears, Roebuck in Dallas, Texas; 10 of 15 at the Defense Logistics Agency in Atlanta; 3 of 5 at Pacific Northwest Bell in Renton, Washington; and 5 in 5 at Surrey Memorial Hospital in Vancouver. The problems included birth defects, spontaneous abortions, respiratory problems in the newborns, Down's syndrome, spina bifida, and premature birth.
>
> Despite attempts by health authorities to investigate the matter, the clusters were never adequately explained. I have been able to locate reports of a follow-up investigation by the US Army Environmental Hygiene Agency of the cluster at the Defense Logistics Agency (Tezak 1981), and by the Centers for Disease Control (1981) of the cluster at Sears, Roebuck. Both verified the existence of a cluster; neither established any apparent link to the women's use of VDTs.
>
> The interpretation of a cluster is problematic. Any unexpected grouping of problems (a cluster) may indicate some problem of public health significance. More commonly, investigation by health authorities of a reported cluster fails to identify a problem that can be remedied by public health measures. However tragic the outcomes may be to the people involved, the grouping of cases may have been a statistical event with no epidemiologic significance. Roughly one pregnancy in five ends in spontaneous abortion (the reported rates vary widely, depending on how early pregnancy is diagnosed); roughly 3 children in a hundred are born with a major birth defect. Simple calculation will show that many clusters will occur every year among the 10 million North American women who use VDTs. The issue, so easily raised, took a decade to resolve.'

Foster then goes on to describe the numerous studies aimed at proving or disproving a causal relation between use of computer monitors by pregnant women and birth defects in their offspring, especially the epidemiological evidence. He concludes that, while 'one can never achieve complete consistency in epidemiologic studies' … 'they certainly rule out the large increases in risk that some people inferred from the clusters.'

The question that will concern us here is the 'simple calculation'. In other words, are the observed clusters to be expected (on the basis of the statistical chances of spontaneous abortions and birth defects, and the number of women involved), or do they need an adequate explanation? In order to find out, we will, for the sake of the argument, assume that Foster's data are correct: that 20% of pregnancies end in spontaneous abortions, that 3% of children born (of the resulting 80% of pregnancies carried to completion) are born with a major birth defect (hence for a combined total of $20\% + 0.03 \times 80\% = 22.4\%$ of all pregnancies), and that the affected group consisted of 10 million North American women.

We first consider the simplest case: the five out of five women at Surrey Memorial Hospital. When a single woman has a chance of 22.4% or 0.224 of a problem pregnancy, the chance that two women will *both* have a problem pregnancy is $0.224^2 = 0.0502$ or just over 5%. Likewise, the chance that three, four, or five women will all have a problem pregnancy is 0.224^3, 0.224^4, and 0.224^5, respectively. We use a pocket calculator or a spreadsheet to find that $0.224^5 = 0.000564$ or 0.0564%. When we subdivide the 10 million women into 2 million groups of five, each group will have a chance of 0.00056 of having five out of five problem pregnancies. In two million possible groups of five women we therefore expect $2\,000\,000 \times 0.000564 = 1128$ of such clusters to occur. In this light, it is not very alarming to find that one such cluster has been reported, when one may expect many more to occur every year just on the basis of random chance.

The other examples are somewhat harder to calculate, because not all women in the cluster suffered problem pregnancies. It is here that we must use some combinatorics, and it is here that we will use the spreadsheet. For our example we will focus first on the three out of five women at Pacific Northwest Bell. We will call them Anne, Beth, Christine, Denise, and Elaine, or A, B, C, D, and E for short. Since all we know is that three out of five experienced problem pregnancies, but not which ones, we must count the various ways in which three of the five women can be involved. Here we go: the ten possible combinations of three specific women out of the group of five are

ABC, ABD, ABE, ACD, ACE, ADE, BCD, BCE, BDE, and CDE

The probability that three *specific* women out of five will have a problem pregnancy (with probability 0.224) and two will not (with a probability of $1 - 0.224 = 0.776$ for a problem-free pregnancy) will be $(0.224)^3 \times (0.776)^2 = 0.000677$ or 0.0677%. As we just saw, the probability that *any* three women of the group will experience problem pregnancies will be ten times larger, because there are ten different possible combinations of three in the group of five women. Consequently, the chance is $10 \times (0.224)^3 \times (0.776)^2 = 0.0677$ or 6.77%. Again assuming that we can make 2 million groups of five women out of the 10 million female workers exposed to VDTs, we have a probability of $2\,000\,000 \times 0.0677$, or more than one hundred thousand of such clusters

each year, *just by chance*. Of course, not all of the ten million women organize themselves in groups of five, but the point is still valid: given the rather
large prevalence of problem pregnancies, the results for the five women at
Surrey Memorial Hospital were almost certainly a chance occurrence, and
should not be used to imply that VDTs caused the problem.

Now to the combinatorics. The integers specifying how many combinations are possible, such as the number 10 above, can be expressed mathematically. Here we will use an alternative, more graphical approach, called
the Pascal triangle. (Incidentally, this same logic is used in determining the
multiplicity of proton NMR lines for nuclei with spin ½ such as ^{1}H and ^{13}C.)
In the Pascal triangle, each number is the sum of the two numbers diago-
nally above it; the triangle starts at its top with a single 1. It represents the
coefficients of the various terms in $(a+b)^n = a^n + na^{n-1}b + \cdots + b^n$, where
$b = 1 - a$. The mth coefficient can be expressed mathematically as $n! / \{m!
(n-m)!\}$, but the Pascal triangle will be easier to read for most non-
mathematicians. For $n = 5$, the spreadsheet gives the coefficients 1, 5, 10, 10,
5, and 1 for 5, 4, 3, 2, 1, and 0 problem pregnancies respectively.

Instructions for exercise 2.11

1 Open a spreadsheet.

2 In cell Z1 deposit the number 1.

3 In cell B2 deposit the instruction = A1 + C1.

4 Copy this instruction to cell C3, where it will read = B2 + D2.

5 Highlight block B2:C3, then grab its handle (at its right bottom) and drag it to cell C12.

6 Release the mouse, but keep the area B2:C22 highlighted. Now grab the handle again,
   and drag it to cell X12.

7 Release the mouse, and click somewhere outside the highlighted area. That is it: you
   have now computed all terms in the first 11 rows of the Pascal triangle!

8 If you want to compute more rows of the Pascal triangle, you need to use more than the
   top 12 rows and 24 columns of the spreadsheet, while the seed (the value '1' in cell W1)
   must be moved to a location further to the right in row 1. For example, move the seed to
   W1 and copy the instruction from B2:C3 to C22, then to AR22, to get the first 21 rows.

9 The special method of copying the instruction in B2 to the rest of the sheet is used here
   merely to keep the unused, interstitial spaces from filling up with zeroes, and thereby
   cluttering up the screen. Verify that you will indeed get the same result, but with zeroes
   in all the unfilled spaces, by deleting instruction (3), and by then simply copying the
   instruction of cell B2 to block B2:X12.

10 Even when you follow the above instructions (1) through (7), there will still be quite a
    few zeroes in the top of this table, which clutter it up. (Note that we are talking here
    only about the *appearance* of things; the actual computation is so simple and so fast

that we have nothing better to discuss!) Excel does not have an instruction to replace these zeroes by blanks.

11 Fortunately, Excel 97 allows you to make them invisible (which amounts to virtually the same thing) by selecting the command sequence Format ⇨ Conditional Formatting. In the resulting dialog box, select Cell Value is … equal to … 0, then press Format, under the 'Font' tab click on Color and select white (or whatever background color you use), then click OK twice to exit the dialog box. Now all zeroes will be displayed and printed in the background color, which will make them invisible. Sorry, this handy trick is not available in earlier versions of Excel.

Now that we have the coefficients, we can return to the problem posed earlier: how extraordinary are the reported clusters, or are they just what one might expect on the basis of pure chance, with or without video display terminals? The statement in the above quote that 'the clusters were never adequately explained' suggests that such an explanation is required, whereas pure chance neither requires nor has an explanation.

For seven out of eight we have the probability $8 \times (0.224)^7 \times (0.776) = 0.000176$, which must be multiplied by $10^7/8 = 1.25$ million for the number of possible groups of eight that can be formed from 10 million workers. The resulting probability of observing such a cluster of problem pregnancies is therefore 220 per annum.

For eight out of 12 we find, similarly, $495 \times (0.224)^8 \times (0.776)^4 \times 10^7/12 = 948$; for seven in a cluster of 13: $1716 \times (0.224)^7 \times (0.776)^6 \times 10^7/13 = 8156$; for 10 out of 15: $3003 \times (0.224)^{10} \times (0.776)^5 \times 10^7/15 = 179$; for 10 of 19: $92378 \times (0.224)^{10} \times (0.776)^9 \times 10^7/19 = 1578$. None of these are found to be rare events, and they therefore do not require a special explanation in terms of VDTs or other potential scapegoats. It is clearly the alarmist presentation of the data (or, to put it more charitably, our tendency to infer a causal relation even where none exists) that suggests that there is a problem. The combination of a high incidence of problem pregnancies (22.4%) and a very large group of women is the reason that these seemingly rare events are, actually, quite to be expected! Does his give VDTs a clean bill of health? Not necessarily, since they would have to have a quite significant effect before that could be measured above such a high background 'noise' of statistically expected problem pregnancies. But, perhaps, the efforts of society could be directed more profitably to bringing down the 'normal' rate of problem pregnancies, instead of spending scarce resources on highly speculative, unproven effects.

2.12 Summary

In this chapter we have encountered some of the principles of statistics. In the first spreadsheet exercise, we explored some properties of the Gaussian distribution, which is usually assumed to describe the distribution of random fluctuations of measurements around their mean values, as long as a sufficiently large number of such observations is considered. Likewise, we saw in the second spreadsheet exercise that random noise can be averaged out, but that doing so again requires a large data set, i.e., much redundancy. One may not always be willing or able to collect such a large set of observations, nor would it always be worth the time and effort spent.

In practice, then, we often take a much smaller sample; as a consequence, the calculated parameters, including their standard deviations, will themselves still be subject to random fluctuations, and therefore should be treated as such rather than as precise values. If you determine the mass of a precipitate from triplicate weighings, don't list the standard deviation of that determination to five significant figures: it is most likely that the first figure is already tentative.

In the next two sections we encountered the problem of propagation of experimental imprecision through a calculation. When the calculation involves only one parameter, taking its first derivative will provide the relation between the imprecision in the derived function and that in the measured parameter. In general, when the final result depends on more than one independent experimental parameter, use of partial derivatives is required, and the variance in the result is the sum of the variances of the individual parameters, each multiplied by the square of the corresponding partial derivative. In practice, the spreadsheet lets us find the required answers in a numerical way that does not require calculus, as illustrated in the exercises. While we still need to understand the principle of partial differentiation, i.e., *what* it does, at least in this case we need not know *how* to do it, because the spreadsheet (and, specifically, the macro PROPAGATION, see section 10.3) can simulate it numerically.

In section 2.5 we introduced the concept of weighting, i.e., of emphasizing certain data over others, by assigning individual weights inversely proportional to the variance of each point. In section 3.4 we will return to this subject, albeit with a somewhat different emphasis.

Section 2.6 illustrated the simplest example of least-squares fitting to a function, namely that of fitting data to the proportionality $y = ax$. This is the equation for a straight line through the origin, and has only one 'adjustable' parameter, the slope a. In section 2.7 we then considered the general straight line with arbitrary intercept, $y = a_0 + a_1x$, i.e., with two adjustable parameters, of which the earlier examples, $y = \bar{y} \ (= a_0)$, and $y = ax \ (= a_1x)$, are special cases. Again reflecting the statistical nature of a least-squares analysis, both of these methods work best when there is a large redundancy of input data, so that the experimental 'noise' is effectively averaged out as long as it is random. In that respect, our radiocarbon dating example was of marginal validity, and was used here only to illustrate the method.

We then emphasized the importance of looking at graphs of the data and their residuals, because such graphs can often show whether an inappropriate model is used. The moral of this exercise is that least-squares analysis, while very powerful in fitting data to a known relationship, cannot (and should not) be used to help select the type of relation to be fitted. That information must come from somewhere else, preferably from a sound understanding of the theory behind the phenomenon studied.

In section 2.9 we considered the usual assumption that random effects follow a single, Gaussian distribution. We took a theoretical example of a random distribution (and what better 'random' distribution could you get than that from the theory of randomly moving ideal gas molecules?), so that sampling error cannot be blamed for the result. We found what looks like a Gaussian distribution, but is not quite one. Then we looked at an example where an obvious 'outlier' is a perfectly legitimate member of another distribution. The take-home lesson of that section is: the assumption that imprecision follows a single Gaussian distribution is just that, an *assumption*. It is often a close approximation, but it is certainly no law of nature.

In section 2.10 we briefly considered another distribution, especially important for stochastic observations such as made in radiochemistry and electrophysiology, while in section 2.11 we took a quick look at the likelihood of seemingly unlikely events.

In connection with these later sections it might be well to realize that the role of statistics in chemistry is, usually, quite different from that in, say, epidemiology or sociology. In chemistry we typically start with a known relationship between a small (and typically known) number of parameters. We then minimize the role of experimental fluctuations by collecting an abundance of input data, and by using that large data set to determine the few underlying parameters. The resulting data reduction lessens the effect of the random fluctuations on the resulting parameters.

In the 'softer' sciences, the specific form of the relationship may not be known or, worse, it may not even be known whether a relationship exists at all. In that case, the question to be answered by statistics is not how to extract the best numerical parameters from the data, but how to establish whether or not a relationship exists in the first place. It is here that concepts such as correlation coefficients become relevant. In quantitative chemical analysis, there are few such ambiguities, since the causal relations are usually well-established and seldom at issue. On the other hand, further statistical measures such as confidence limits, based on a (seldom experimentally supported) presumption of a single Gaussian distribution, are more strongly favoring a particular, mathematically convenient model than seems to be realistic or prudent for the subject matter of this workbook, and thereby tend to provide an overly rosy picture of the data. For this reason, statistical measures beyond standard deviations will not be considered here.

We started this chapter by considering life insurance, and we will now return to this model. Life or death are, of course, binary options, while time is continuous. In principle, the appropriate statistics for life insurance are therefore based on the Poissonian distribution. Gauss was hesitant to publish his work on least squares, because he could find no fully satisfactory justification for it – consequently, he only published what he had found after Legendre had independently discovered and published it. In retrospect, there is indeed a much better theoretical foundation for Poissonian statistics, because we now know that mass is quantized, as are most forms of energy, while time is not.

Fortunately, for a sufficiently large cohort, the Poissonian distribution approaches the Gaussian one, a general limit more carefully described by the **central limit theorem** of statistics. Because of the large individual fluctuations in the human life span, insurance companies must operate with a large number of subscribers. Under those circumstances they can use statistics to set their premiums so as to provide a useful service to society while also making a profit.

Clearly, the life expectancies of different groups are different: women tend to outlive men, non-smokers on average live longer than smokers, etc. There are clearly genetic as well as behavioral factors involved here: gender is genetic, smoking is not. When the various subgroups are still sufficiently large, their subgroup statistics are still meaningful, and their distinct life expectancies can be established. Such statistical data are only valid *within the context of leaving all other variables constant.*

What such statistics cannot do is *predict* how the average life expectancy may change with changing circumstances (except retrospectively, which hardly qualifies as a prediction). For example, despite the fact that life expectancy is strongly linked to genetics (fruit flies on average have much shorter life spans that people, while bristlecone pines tend to outlive people), the life expectancies of people in the developed world have increased dramatically over the past century, as the result of improvements in the quality of drinking water, in hygienics, in the availability of sewers and antibiotics, etc. Such changes primarily affect the *bias* of the measurements, rather than their *spread.*

Statistics can only deal with effects that change the bias *after* they have occurred. This is so because statisticians are only able to draw their conclusions *by keeping all other factors constant.* When such other factors are *not* constant, statistics loses its predictive power. From the very beginning of statistics, this inherent limitation has confused some of its practitioners. For example, Francis Galton, an early statistician and the developer of the correlation coefficient, also coined the term eugenics, and believed that he could prove statistically that some races were superior to others. He couldn't, and he didn't, but similar, essentially self-serving arguments, dressed up in statistical clothes to give them a semblance of scientific objectivity, regularly reappear. For example, statistics showing a racial bias of IQ are sometimes offered as 'proof' of the superiority of one race (typically that of its authors) over another, implicitly assuming that societal race-dependent biases such as in education and in socioeconomic status are either absent or inconse-quential. There may well be such genetically linked IQ differences, but care-fully controlled experiments, in which the effects of environmental bias were minimized, such as in studies of the IQ of German children of black and white American GI's, have so far failed to demonstrate them. The difference in IQ between blacks and whites in the US is about 15%, similar to that between Sephardic and Ashkenazic Jews in Israel, or between (white) Catholics and (white) Protestants in Northern Ireland (T. Sowell, *Race and Culture*, 1994). If you are interested in such matters, read *The Bell Curve* (R. J. Herrnstein & C. Murray, Free Press, 1994) and its rebuttal in *The Bell Curve Debate* (R. Jacoby & N. Glauberman, eds, Times Books, 1995). The point here is not the complex relation between IQ, culture, and race, but the unwar-ranted over-extension of statistical inference.

In this chapter we have considered the random experimental fluctuations that can be described meaningfully by statistics, thereby yielding estimates of the *precision* of the experimental result, i.e., the repeatability of a particu-lar experiment under precisely the same conditions. The real question one usually would want to be answered is, of course, that of *accuracy*, i.e.: how reliable, how close to the truth, is our answer? Unfortunately, this is a ques-tion beyond the realm of statistics.

Little can be said in general about systematic error, and the consequent emphasis in this chapter on the effect of random error might suggest that the latter is the more important. However, comparison of results of the same experimental parameter as obtained by completely different methods usually indicates the opposite, namely that systematic errors are typically the more consequential ones. Statistics therefore should be applied, and interpreted, with a good deal of humility. *In no case should precision be mis-taken for an estimate of accuracy.*

The rate constant k of a first-order chemical reaction is the *characteristic parameter* of that reaction rate. Such a rate constant has the dimension of a reciprocal time, and one might therefore be tempted to assume that the reaction is complete in a time $1/k$. This is incorrect. The rate law for a first-order reaction, say A → products, is $[A] = [A]_{t=0} \exp[-kt]$, so that, far from being completely consumed, more than one-third of A is still *un*reacted at $t = 1/k$ or $kt = 1$: $[A]_{t=1/k} = [A]_{t=0} \exp[-1] = 0.37 [A]_{t=0}$. Now one could define a new time, say $3/k$ (after which the reaction is 95% complete, since $\exp[-3]$ $= 0.0498$), $4.6/k$ (after which it is 99% complete), or $7/k$ (after which it is 99.9% complete), but since the level of completeness (95%, 99%, 99.9%, etc.) is essentially arbitrary, and would apply only to first-order kinetics anyway, no such proposals have found favor in the chemical community. The char-acteristic parameter k contains the information concerning the reaction rate, and additional rate-related parameters are neither needed nor useful.

The characteristic parameter for experimental variability is the standard deviation. Just as $1/k$ does not indicate completion of the reaction, the standard deviation does not indicate the outer limits of the experimental variability. Alternative parameters can be defined in an attempt to more closely indicate the expected outer limits of that variability. One such measure is the so-called **confidence limit**, an unfortunate name that invokes associations with confidence artists and confidence games. Confidence limits are useful for indicating the imprecision in results calculated from a limited number of replicate measurements. As we saw in section 2.2, the *sample* standard deviation of a small set of samples is highly variable, and is a poor estimate of the *population* standard deviation of the underlying distribution.

The confidence limit starts from the estimated sample standard deviation, and multiplies it by a factor t that reflects both the number of measurements made, and the acceptable probability. For example, for a sufficiently large number of replicate measurements following a single Gaussian distribution, a 95% confidence limit will correspond with 1.96σ, and indicates that, for such a distribution, 19 out of 20 data can be expected to lie within those limits. In that case the confidence limit is $t\sigma$ where $t = 1.96$ and σ is the calculated sample standard deviation. But if the standard deviation is based on only triplicate measurements, $t = 4.3$, and for duplicate measurements, $t = 12.7$, representing the much larger imprecision in the result of such a small number of replicas. For 99% confidence limits one can expect to find 99 out of 100 data within those limits, again provided that we deal with a single Gaussian distribution and take a sufficiently large number of samples, in which case $t = 2.58$. For triplicate and duplicate measurements the corresponding values are $t = 9.92$ and $t = 63.7$ respectively.

The values of t for a particular percentage and number of replicate data are readily found in Excel with the function TINV (*probability, number of degrees of freedom*), where the probability is the complement of the percentage (0.05 for 95% confidence limits, 0.01 for 99%, 0.001 for 99.9%, etc.), and the number of degrees of freedom is one less than the number of replicate measurements. The above results for the 99% confidence limits of 1000, 3, and 2 replicate measurements are therefore obtained with the commands =TINV(0.01, 999), =TINV(0.01, 2), and =TINV(0.01, 1), respectively.

In this book we will not use confidence limits, primarily because they are of limited usefulness in fitting experimental data to *functions*, but also because there is no agreement on what percentage (95%, 99%, 99.9%, etc.) to use, and the term suggests a non-existing connection with accuracy rather than with mere experimental repeatability. Still, confidence limits do serve a useful purpose in emphasizing that statistics based on small numbers of measurements yield highly imprecise results.

As our final spreadsheet exercises of this chapter we will analyze experimental data published some 140 years ago, reported by J. D. Forbes in *Trans. Royal Soc. Edinburgh* 21 (1857) 135, and quoted in S. Weisberg, *Applied Linear Regression*, 2nd ed., Wiley 1985. Forbes suspected a relation between the logarithm of the barometric pressure, in those days used to determine altitude in the mountains, and the boiling point of water. The latter would be far easier to measure, since a (mercury) barometer was then a bulky, fragile, and generally mountaineer-unfriendly instrument. He therefore determined the barometric pressure (in inches of mercury) and the boiling point of water (in degrees Fahrenheit) at various high places in Scotland and in the Alps. Tables linking barometric pressure to altitude already existed.

Boiling point of water t_p / °F	Atmospheric pressure p / mm Hg	Boiling point of water t_p / °F	Atmospheric pressure p / mm Hg
194.5	20.79	201.3	24.01
194.3	20.79	203.6	25.14
197.9	22.40	204.6	26.57
198.4	22.67	209.5	28.49
199.4	23.15	208.6	27.76
199.9	23.35	210.7	29.04
200.9	23.89	211.9	29.88
201.1	23.99	212.2	30.06
201.4	24.02		

Forbes found that there is indeed a linear relation between the boiling point t_p of water and the logarithm of the barometric pressure p. Here are his data; use them to derive an equation to calculate the barometric pressure p from the boiling point t_p. For the resulting imprecision in p assume that t_p can be determined with a standard deviation of 0.1 °F.

Instructions for exercise 2.12

1 Open a spreadsheet.

2 Deposit appropriate column headings, and enter the experimental data.

3 Add a column in which you calculate log p.

4 Plot the experimental data points of log p versus t_p.

5 Use a least-squares fit to find the relation between t_p and log p.

6 In the graph made under (4) show the fitted line.

7 Also make a graph of the residuals.

8 Express p in terms of t_p including estimates of the resulting imprecision.

As you go through this exercise you will, of course, come upon the one point that does not seem to fit the line. Forbes agonized about this particular point, but he did report it, even though he considered it "evidently in error". The best thing you can do is to include it as well, or to make two calculations, one with and the other without the suspect point, and to list both results, in which case you leave the choice whether to include or reject the 'outlier' to the user.

CHAPTER 3

MORE ON LEAST SQUARES

In this chapter we will describe some of the more sophisticated uses of least squares, especially those for fitting experimental data to specific mathematical functions. First we will describe fitting data to a function of two or more independent parameters, or to a higher-order polynomial such as a quadratic. In section 3.3 we will see how to simplify least-squares analysis when the data are equidistant in the dependent variable (e.g., with data taken at fixed time intervals, or at equal wavelength increments), and how to exploit this for smoothing or differentiation of noisy data sets. In sections 3.4 and 3.5 we will use simple transformations to extend the reach of least-squares analysis to many functions other than polynomials. Finally, in section 3.6, we will encounter so-called non-linear least-squares methods, which can fit data to any computable function.

3.1 Multi-parameter fitting

We can expand least-squares fitting to encompass more than one dependent variable. Here we will fit data to an equation of the form $y = a_0 + a_1 x_1 + a_2 x_2$. Please note that the method to be used is not restricted to merely two dependent variables, but can in principle be applied to *any* number of x_i-values. In practice, the Regression routine of Excel can handle up to 16 different dependent variables. Typically, the different x_i represent independent parameters, although we will exploit the fact that this need not be so in section 3.2.

In order to find expressions for the least-squares coefficients for this situation, we can form the sum of the squares of the residuals, $SRR = \Sigma(y - a_0 - a_1 x_1 - a_2 x_2)^2$, and then derive the values of a_0, a_1, and a_2 by setting $\partial(SRR)/\partial a_0 = 0$, $\partial(SRR)/\partial a_1 = 0$, and $\partial(SRR)/\partial a_2 = 0$. This yields three simultaneous equations that can be solved to yield closed-form solutions for a_0,

a_1, and a_2. However, the resulting expressions are fairly complicated, even when written compactly in terms of three-by-three determinants:

$$a_0 = \begin{vmatrix} \sum x_2^2 & \sum x_1 x_2 & \sum x_2 y \\ \sum x_1 x_2 & \sum x_1^2 & \sum x_1 y \\ \sum x_2 & \sum x_1 & \sum y \end{vmatrix} \Big/ D \qquad\qquad (3.1\text{-}1)$$

$$a_1 = \begin{vmatrix} \sum x_2^2 & \sum x_2 y & \sum x_2 \\ \sum x_1 x_2 & \sum x_1 y & \sum x_1 \\ \sum x_2 & \sum y & N \end{vmatrix} \Big/ D \qquad\qquad (3.1\text{-}2)$$

$$a_2 = \begin{vmatrix} \sum x_2 y & \sum x_1 x_2 & \sum x_2 \\ \sum x_1 y & \sum x_1^2 & \sum x_1 \\ \sum y & \sum x_1 & N \end{vmatrix} \Big/ D \qquad\qquad (3.1\text{-}3)$$

$$D = \begin{vmatrix} \sum x_2^2 & \sum x_1 x_2 & \sum x_2 \\ \sum x_1 x_2 & \sum x_1^2 & \sum x_1 \\ \sum x_2 & \sum x_1 & N \end{vmatrix} \qquad\qquad (3.1\text{-}4)$$

and only somewhat simpler expressions for the corresponding standard deviations. Even on a spreadsheet, you would not want to evaluate these when you would not have to. Fortunately, that is precisely the case: *you don't have to*. This is so because the standard least-squares routine available in Excel will work as readily with multiple x_i-values as it does with one single x. We therefore need to give only a very simple example to demonstrate how this works.

Instructions for exercise 3.1

1 Open an Excel spreadsheet.

2 In row 1 enter the labels a0 = , a1 = , a2 = , and na = , and in row 2 some corresponding numbers, such as 2, 3, 4, and 0.

3 In cell A4 deposit the label y, in cells C4 and D4 the labels x1 and x2, and in cell F4 the label noise.

4 In cell A6 deposit the formula for $y = a_0 + a_1 x_1 + a_2 x_2 + na^* noise$, where *na* stands for the noise amplitude specified in row 2, and *noise* for the Gaussian noise (with zero mean and unit standard deviation) you will deposit in column F. Copy this instruction some distance down.

5 Starting with row 6, in columns C and D deposit some numbers, say, 1, 2, 3, ... and 1, 1, 2, 2, 3, 3, ... for x_1 and x_2 respectively. Make sure that the sequences x_1 and x_2 are not linearly related, otherwise the problem will not have a unique solution, in which case the program will fail and give you an error message.

6 In column F, starting with cell F6, deposit Gaussian noise. (Select Tools ⇨ Data Analysis, highlight Random Number Generation, click OK, select Distribution Normal, Mean = 0, Standard Deviation = 1, click on Output Range, then click on the adjacent window, specify the output range, and click OK.)

7 Make sure that columns A, C, D, and F have equal lengths.

8 Call the Regression: Tools ⇨ Data Analysis, select Regression, click OK, specify the Input Y Range, the Input X Range, and the Output Range (but leave four lines below the data columns empty for reasons that will soon become apparent), then click OK. You will find the coefficients of the Intercept, and of the two X Values, together with their standard deviations, here called Standard Errors.

9 Compare these results with your values of a_0, a_1, and a_2.

10 Change the noise amplitude to a non-zero value, say 0.1, and repeat the Regression analysis.

Later in this chapter you will encounter a weighted least-squares analysis program. For the latter, the usual, unweighted least-squares analysis is just a special case, with all the weights set equal to 1. Therefore you can also use this weighted least-squares program, which you will need later in this chapter anyway. It is organized differently, because it does not use an input dialog box, but instead requires a *fixed input format*: the first column must contain the *y*-values, the second the weights, the third (and subsequent) column(s) the *x*-value(s). You can leave the weight column empty, or fill it with 1's, but you cannot leave it out. This is why, so far, you have left the B column blank.

11 Highlight the block of data: the columns for *y* and *w*, as well as the two *x*-columns.

12 Select Tools ⇨ Macro ⇨ Macros, then activate WLS1 and press Run.

13 There you are: you should see the coefficients, and their standard deviations, right under the data columns.

14 Compare the results of the two routines.

15 Save the spreadsheet as MultiparameterFit.

3.2

Fitting data to a quadratic

We will now consider the special case of a multi-parameter least-squares fitting in which the various x_i's form a power series, i.e., $x_1 = x$, $x_2 = x^2$, etc. We will illustrate this with a simple example, viz. fitting data to a parabola, $y = a_0 + a_1 x + a_2 x^2$. Again, the algebraic expressions are cumbersome:

$$a_0 = \begin{vmatrix} \sum x^4 & \sum x^3 & \sum x^2 y \\ \sum x^3 & \sum x^2 & \sum xy \\ \sum x^2 & \sum x & \sum y \end{vmatrix} \Big/ D \tag{3.2-1}$$

$$a_1 = \begin{vmatrix} \sum x^4 & \sum x^2 y & \sum x^2 \\ \sum x^3 & \sum xy & \sum x \\ \sum x^2 & \sum y & N \end{vmatrix} \Big/ D \tag{3.2-2}$$

$$a_2 = \begin{vmatrix} \sum x^2 y & \sum x^3 & \sum x^2 \\ \sum xy & \sum x^2 & \sum x \\ \sum y & \sum x & N \end{vmatrix} \Big/ D \tag{3.2-3}$$

$$D = \begin{vmatrix} \sum x^4 & \sum x^3 & \sum x^2 \\ \sum x^3 & \sum x^2 & \sum x \\ \sum x^2 & \sum x & N \end{vmatrix} \tag{3.2-4}$$

and, again, the spreadsheet can handle this problem without any further ado, as demonstrated by the following exercise. The spreadsheet achieves this flexibility through the magic of matrices. If you are curious to see how that works, look at the even more general Weighted Least Squares program in chapter 10.

Instructions for exercise 3.2

1 Recall MultiparameterFit.

2 Compute the x_2-values in column D to be equal to x_1^2.

3 Use Regression as well as the Weighted Least Squares macro WLS1 to calculate the coefficients a_0, a_1, and a_2.

4 Use Trendline Polynomial Order 2, using the Option to Display Equation on Chart.

5 Verify that all three routines indeed yield the same results.

We note that this method can be used for any power law or sum thereof. It can even be mixed with a multiparameter fit. For instance, one could use it to fit y to a function such as $a_0 + a_1 x_1{}^4 + a_2 \log x_2 + a_3 \sqrt{x_3}$, or whatever suits your fancy. Keep in mind, though, that these various x-values are assumed to be free of experimental errors, and that a substantial data redundancy is needed when a large number of coefficients needs to be determined. Moreover, the parameters, though not necessarily mutually independent, should at least be distinguishable: if you try to fit $y = a\,e^{bx+c} = a\,e^c\,a^{bx}$, no computer in the world will be able to find unique values for a and c, because there are infinitely many combinations of a and c that yield the same value for $a\,e^c$. Finally, the larger the noise in the data, the less reliable the results will be. And, of course, there must be a good theoretical reason to use such a complicated model in the first place.

3.3 Least squares for equidistant data: smoothing and differentiation

When experimental data are obtained automatically, by an instrument, they are often equidistant in the dependent parameter, such as time, wavelength, or magnetic field strength. Many more data points may be collected than are really needed, but such data may need to be smoothed to remove instrumental noise; sometimes, they also need to be differentiated. In such cases, fitting the data with a **moving polynomial** is very useful.

In this method, a low-order polynomial, such as a parabola, is fitted to a small number of contiguous data. The (usually odd) number of data points included must be significantly larger than the number of parameters defining the polynomial, i.e., when one fits the data to a parabola, $y = a_0 + a_1 x + a_2 x_2$, at least five but preferably many more data points should be included. One then uses the fitted polynomial to calculate the smoothed y-value, or its derivative, at the midpoint of the polynomial. And it is here that you will appreciate using an odd number of data, because in that case the midpoint coincides with an already existing x-value.

Subsequently, the point at one extreme of the data set is dropped, a new point at the other end is added, and the process is repeated. In this way the polynomial slithers along the entire curve. Because a low-order polynomial is fitted to a still relatively small data set, the resulting distortion is often small even though, in this case, we do not utilize any knowledge regarding the shape of the underlying curve. The justifying assumption is that the data do not exhibit structural features on the scale of the length of the moving polynomial; if they do, then these features will be lost in the smoothing procedure.

As explained in chapter 8, the least-squares analysis for such an equidistant data set (i.e., with constant x-increments) can be simplified to a set of

multiplying integers. In a spreadsheet, least-squares smoothing or differentiation are then very easy, as the following exercise will demonstrate. The
method has been discovered and rediscovered repeatedly; in analytical
chemistry it is usually associated with the names of Savitzky and Golay.
Tables of the multiplying numbers, often called **convoluting integers**, are
available in the Savitzky–Golay paper, *Anal. Chem.* 36 (1964) 1627 and its
correcting complement, J. Steinier *et al.*, *Anal. Chem.* 44 (1972) 1906, as well
as in several textbooks (e.g., in Appendix B.6 of my *Principles of
Quantitative Chemical Analysis*). The principle of the method is explained
in section 8.5. Here we will use convoluting integers for 13-point smoothing
and 13-point differentiation, and will apply them to an artificially noisy
exponential of the form $y = y_0 e^{-kt}$.

Instructions for exercise 3.3-1

1 Open a new spreadsheet.

2 In row 1 deposit labels for y_0, k, and the noise amplitude *na*, and in row 2 some corresponding values, such as 1, 0.1, and 0.05 respectively.

3 In row 4, starting with cell A4, deposit the labels t, y, noise, y + noise, smooth, deriv, and
k.

4 Fill column A, starting with cell A6, with the numbers 0, 1, 2, ... , 50.

5 In cell B6 calculate *y* according to $y = y_0 e^{-kt}$, and copy this down to B56. This will generate the theoretical function $y_0 e^{-kt}$.

6 In cells C6:C56 deposit Gaussian noise with zero mean and unit standard deviation.

7 In cell D6 calculate $y = y_0 e^{-kt} + na^* noise$ to simulate a noisy exponential.

8 In column E we will compute a smoothed value of the noisy exponential, using a 13-
point moving parabola. Leave cells E6:E11 blank, because you cannot calculate a
smoothed value in the first (and last) 6 points of the data. In cell E12 calculate the
smoothed value as = (− 11*D6 + 9*D8 + 16*D9 + 21*D10 + 24*D11 + 25*D12 + 24*D13
+ 21*D14 + 16*D15 + 9*D16 − 11*D18)/143. The coefficients − 11, 0, 9, 16, 21, 24, 25,
24, 21, 16, 9, 0, and − 11 as well as the common divider, 143, are taken from a table of
convoluting integers. The above coefficients can also be found at the end of section
9.2b.

9 Copy this instruction down to cell E50, again leaving six cells (E51:E56) blank. This
should calculate smoothed values of the function.

10 Plot the original, noise-free curve in B6:B56, the noisy one in D6:D56, and the subsequently smoothed one in E6:E56, all versus the time in A6:A56. Make sure to include
the missing points, otherwise these points appear time-shifted in the graph. The result
should look like Fig. 3.3-1.

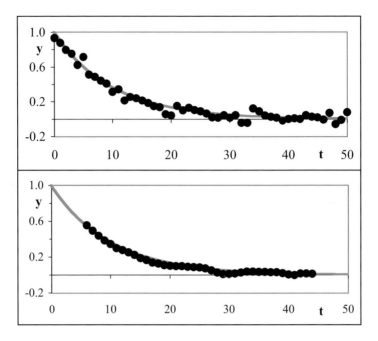

Fig. 3.3-1: Smoothing a noisy exponential curve. Top: to a noise-free curve (colored line) was added Gaussian noise, thereby generating the solid data points. Bottom: a moving 13-point polynomial was then used to smooth the latter.

11 In F12:F50 calculate the derivative, using the convoluting integers $-6, -5, -4, -3, -2, -1, 0, 1, 2, 3, 4, 5$, and 6, and the common divider 182 δ, where δ is the spacing between adjacent x-values; here, $\delta = 1$. In other words, the formula in F12 should read $= (-6*D6 - 5*D7 - 4*D8 - 3*D9 - 2*D10 - D11 + D13 + 2*D14 + 3*D15 + 4*D16 + 5*D17 + 6*D18)/182$.

12 You could compare this derivative with its correct value by calculating the latter in column G as $- k y_0 e^{-kt}$. Better yet, calculate k as the ratio of the derivative obtained and its correct value, obtainable in cell G12 as $= F12/B12$. Copy this instruction down to G50, also compute the average, $= \text{AVERAGE (G12: G50)}$, and compare these results with the value of k in row 2.

13 Save the spreadsheet as NoisyExponential.

The smoothing indeed removes some of the noise, but by no means all of it. Moreover, it drops the beginning and end of the curve. If we had used a longer polynomial of, say, 25 contiguous points, a smoother curve would have resulted, but more points would have been lost at the curve extremes. Moreover, use of a longer polynomial involves an enhanced risk of systematic distortion, since the method assumes that all those 25 contiguous data points fit a parabola, which they do not quite. In general, the more points we use for smoothing, the smoother the result will indeed look, but the more we risk distorting the underlying curve. With truly experimental data we cannot see that so clearly, because we do not know the underlying curve. This is why an experiment with artificial noise added to a known curve can be instructive.

A similar conclusion follows from a comparison of the calculated derivative with the correct one. The k-values obtained near the beginning of the curve are obviously more reliable than those from the tail end, where any theoretical decay is masked by the noise. Averaging all k-values is therefore a bad deal, since it indiscriminately mixes reasonably good data with lower-quality ones. In general, taking the derivative of noisy data is a quite demanding test, except when we know specifics about the function that we can exploit. In section 3.4 we will utilize the fact that the underlying curve is a single exponential to fit these same noisy data, using a least-squares criterion, and thereby extract a reliable value of k. Then you will be able to judge for yourself which is the better way.

The above example illustrates the quantitative limitations of so-called 'blind' smoothing or differentiating a curve with a moving polynomial. (The term 'blind' refers to the fact that no information regarding the nature of the underlying curve is used.) However, the method can be useful for *qualitative* applications, such as smoothing a noisy curve in a graph. We will demonstrate this in the next exercise, inspired by Press & Teukolsky (*Comp. Phys.* 4:6 (1990) 669), in which we simulate a spectrum containing Lorentzian peaks of various widths, add Gaussian noise, and then filter the resulting curve.

Instructions for exercise 3.3-2

1 Open a new spreadsheet.

2 In column A deposit a heading (such as #) and, in A3:A1002, the integers 1 (1) 1000 to simulate, say, equidistant wavelengths or wavenumbers.

3 In column B deposit Gaussian noise (Tools ⇨ Data Analysis ⇨ Random Number Generation ⇨ Distribution Normal, Mean 0, Standard Deviation 1).

4 In column C calculate a noise-free simulated spectrum. For this you may want to use Lorentzian line-shapes, which are of the form $y = 1/[a(x - b)^2 + c]$, where b defines the x-value at the center of the peak. The spectrum simulated in Fig. 3.3-2 is the sum of five such Lorentzians. For instance, cell C3 might contain the instruction = 1/($I$3*(A3 − $J$3) + $K$3) + 1/($I$4*(A3 − $J$4) + $K$4) + ⋯ + 1/($I$7*(A3 − $J$7) + $K$7), where the parameters a, b, and c are listed in a parameter table in block I3:K7.

5 In I9 deposit the label na, and in J9 its value.

6 In D3:D1002 calculate the sum of the noise-free spectrum and na times the noise of column B.

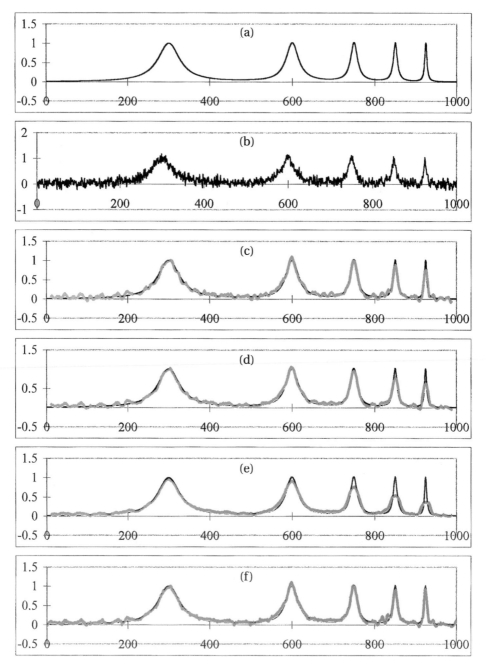

Fig. 3.3-2: A simulated 1000-point spectrum with five well-separated Lorentzian peaks of various widths, without (a) and with (b) Gaussian noise, and the noisy spectrum after filtering with a 25-point smoothing polynomial of (c) first-order, (d) third-order, and (e) fifth-order respectively. Parameters used for the simulated spectrum: $a_1 = 0.001$, $b_1 = 300$, $c_1 = 1$, $a_2 = 0.003$, $b_2 = 600$, $c_2 = 1$, $a_3 = 0.01$, $b_3 = 750$, $c_3 = 1$, $a_4 = 0.03$, $b_4 = 850$, $c_4 = 1$, $a_5 = 0.1$, $b_5 = 925$, $c_5 = 1$, na $= 0.1$. Curve (f) shows the fit of the same noisy data set (b) after using Barak's adaptive-degree polynomial filter for a 25-point moving polynomial of order between 0 and 10.

7 In H3:H7 deposit the numbers 1 through 5, and in I1:K1 the column headings a, b, and c.

8 Deposit numerical values in block I3:K7. You may want to start with those listed in the legend of Fig. 3.3-2, and then (after you have made the on-the-sheet graphs) vary those parameters to your taste.

9 Make on-the-sheet graphs of C3:C1002 and D3:D1002 vs. A3:A1002.

10 We start with the simplest possible filter, one that merely averages a number of adjacent data to reduce the noise. Go to cell E15, and there deposit the 25-point averaging formula = (D3 + D4 + D5 + ⋯ + D27)/25. Copy this instruction down to E991.

11 Add a graph of the so filtered spectrum. For reference you may want to include in that plot either the noise-free curve (as in Fig. 3.3-2) or the noisy one.

12 For the second filter, in cell F3, we use instead the formula = (− 253*D3 − 138*D4 − 33*D5 + 62*D6 + ⋯ − 253*D27)/5175, where we use the convoluting integers for a cubic 25-point smoothing filter: −253, −138, −33, 62, 147, 222, 287, 342, 387, 422, 447, 462, 467, 462, 447, 422, 387, 342, 287, 222, 147, 62, −33, − 138, and −253. Note that these numbers are symmetrical around the middle of the set. The corresponding normalizing factor is 5175. Again, copy this formula to F991, and plot the resulting curve.

13 Finally, in column G, calculate a third filtered curve, this time using the convoluting integers for a *fifth*-order 25-point smoothing filter: 1265, −345, −1122, −1255, −915, −255, 590, 1503, 2385, 3155, 3750, 4125, 4253, 4125, 3750, … , −345, 1265. The normalizing factor is 30 015. Again plot the data.

14 Save as NoisySpectrum.

Figure 3.3-2 shows typical results. The simple first-order averaging filter, panel (c) in Fig. 3.3-2, is most effective in reducing the noise, but also introduces the largest distortion, visible even on the broadest peaks. This is always a trade-off: noise reduction is gained at the cost of distortion. The same can be seen especially with the narrower peaks, where the higher-order filters distort less, but also filter out less noise. In section 10.9 we will describe a more sophisticated filter, due to Barak, which for each point determines the optimal polynomial order to be used, and thereby achieves a better compromise between noise reduction and distortion.

3.4 Weighted least squares

The least-squares analysis we have encountered so far works with a special class of functions, namely polynomials. (In chapter 2 we considered the simplest polynomials, the functions $y = a_0$, $y = a_1x$, and $y = a_0 + a_1x$; in sections 3.1 and 3.2 we used $y = a_0 + a_1 x_1 + a_2 x_2$ and $y = a_0 + a_1 x + a_2 x^2$ respectively.) Many types of experimental data can be described in terms of polynomials.

However, there are also many types of data that do not readily fit this mold. A prime example is a single exponential: we saw in the preceding section that fitting an exponential to a quadratic is not very satisfactory. Moreover, invoking higher-order terms does not really help.

Consider the concentration of a chemical species reacting according to a first-order irreversible reaction $A \rightarrow B$ with reaction rate constant k. In this case the concentration of A is described by $[A] = [A]_0 \exp(- kt)$ where $[A]$ is the concentration of species A, t is the time elapsed since the beginning of the experiment (at $t = 0$), and $[A]_0$ is the corresponding initial concentration, $[A]_0 = [A]_{t=0}$. Imagine that we follow the concentration of A spectrometrically, and want to extract from the resulting data the initial concentration $[A]_0$ and/or the rate constant k. In that case we must fit the exponential to a polynomial. Using a least-squares polynomial fit, this is no simple task.

On the other hand, it is easy to *transform* the exponential into a polynomial, by simply taking the (natural) logarithm. This yields the equation of a straight line, $\ln[A] = \ln[A]_0 - kt$, which is of the form $y = a_0 + a_1 x$ with $y = \ln[A]$, $a_0 = \ln[A]_0$, $a_1 = - k$, and $x = t$, the 'independent' variable. There is only one minor problem: in analyzing $\ln[A] = \ln[A]_0 - kt$ instead of $[A] = [A]_0 \exp(- kt)$ we minimize the sum of squares of the deviations in $\ln[A]$ rather than those in $[A]$. When the dominant 'noise' in $[A]$ is *proportional* to $[A]$, minimizing the sum of the squares of the deviations in $\ln[A]$ would indeed be appropriate. However, it will not be so when the magnitude of that noise is independent of $[A]$. In the latter case we can still use the transformation as long as we include with each input value y_i a **weight** w_i to transform the noise as well.

The weights depend on the particular transformation used. In general, when the experimental data y are transformed to some new function Y (as in the above example, where $Y = \ln y$), the corresponding **global weight** w will be given by

$$w = \frac{1}{(\mathrm{d}Y/\mathrm{d}y)^2} \tag{3.4-1}$$

Weighted least-squares analysis is also called for when we must average data of different precision. In section 2.5 we already encountered the need for weighting of the experimental data when their individual standard deviations are known. In that case the **individual weights** are simply the reciprocals of the variances of the individual measurements,

$$w = \frac{1}{\sigma^2} \tag{3.4-2}$$

In general, then, there are two quite different reasons for using weighted least squares: (a) *global* weights may be required when the data analysis involves a transformation of the dependent variable, while (b) *individual* weights are needed when we consider data of different (but known) precision. The variances of individual data are seldom known, which is why the use of global weights is the more common. However, when both (a) and (b) apply simultaneously, we have the general expression for the **total weights**

$$w = \frac{1}{\sigma^2 (dY/dy)^2} \tag{3.4-3}$$

of which (3.4-1) and (3.4-2) are the special cases for equal variances σ_i or constancy of dY/dy respectively.

We saw in sections 3.1 and 3.2 that the mathematical expressions to fit data to a multi-parameter function, or to a higher-order polynomial such as a quadratic, can be quite daunting, but that they are readily accessible when the spreadsheet has built-in facilities to do the least-squares analysis. Excel does not have comparable built-in capabilities for weighted least-squares analysis. However, it does have the facility to accept add-in programs coded in a language Excel understands, VBA (= Visual BASIC for Applications), and you may already have used the added Weighted Least Squares in sections 3.1 and 3.2. If you are curious how such an add-in program works, consult chapter 10, where it is described in detail, including its complete 'text'.

The example we will use below is again that of an exponential decay: $y = y_0 \exp(-kt)$. We already generated such a signal, in NoisyExponential, without and with added Gaussian noise. For the exponential $y = y_0 \exp(-kt)$ we have $Y = \ln y$ and $w = 1/(dY/dy) = y^2$.

Instructions for exercise 3.4

1 Recall NoisyExponential.

2 Add column headings in row 4 for Y, w, t, yw, and ynw.

3 In cell H6 deposit '= LN(D6)', in cell I6 '= D6^2', and in cell J6 '= A6', then copy these three instructions down to row 56. (We here repeat the column for t in order to fit the fixed format required by the Weighted Least Squares macro.)

4 Highlight these three columns, starting at H6 and extending it down as far as column H contains numbers. At a given moment, noise will make y negative, at which point Y cannot be calculated, which will show as #NUM!. That point, and the data following it, cannot be used, because they are clearly biased towards positive values of y, without the negative y-values contributing to Y.

5 Call the Weighted Least Squares routine (Tools ⇨ Macro ⇨ WLS1).

6 Also call the Regression analysis, and apply it to the same data set. (You could also use the Weighted Least Squares analysis for this, after copying the values in columns H and

J to, say, columns M and O, provided these are not used otherwise, and column N is empty. If you were to do the unweighted analysis in the block starting at H6, you would have to erase the data in column I, and you would overwrite the earlier results.)

7 In order to compare the results with the theoretical values in row 1, use the spreadsheet to take the antilogs (= 10^...) of the coefficient and standard deviation of the intercept, because both routines yield $Y_0 = \ln y_0$ rather than y_0.

8 In K6:K56 and L6:L56 calculate the curves y_w and y_{uw} for the *weighted* and *unweighted* curves, each reconstituted with the parameters provided by the weighted or unweighted least-squares fittings respectively, then plot these together with the data in B6:B56 and D6:D56. They might look like Fig. 3.4.

Because no two sets of noise are alike, your plot will be different, but most likely the conclusion will be similar: for this type of data, where the noise is constant, a weighted least-squares fit is far superior to an unweighted fit. The numerical data tell the same story: in this particular example, the weighted least-squares fit yielded a slope $- k$ of $- 0.0991 \pm 0.0036$, and an intercept ln $y_0 = -0.0214 \pm 0.0224$ so that $y_0 = 0.952 \pm 5.3\%$, whereas the unweighted fit gave $- k = -0.1181 \pm 0.0076$, ln $y_0 = 0.0800 \pm 0.1363$ hence $y_0 = 1.20 \pm 37\%$. However, the *opposite* conclusion would be drawn if the noise had been proportional to the magnitude of the signal y, and you had not taken that into account in the weights! You can use the spreadsheet to verify this yourself. The moral of this exercise is that you need to know something about the function to which you fit (in order to use the correct transform), *and* about the nature of the noise in your data, before you can get rid of most of it. And, as always, the less noise you have to start with, the better off you are.

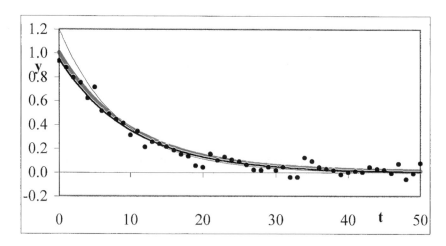

Fig. 3.4: Fitting a noisy exponential curve (points) with unweighted (thin black line) or weighted (solid black line) least-squares fit. The underlying, noise-free curve (heavy colored line) is shown for comparison.

3.5

Another example of weighted least squares: enzyme kinetics

Enzyme kinetics provide a prime example of the applicability of weighted least squares. The reason is twofold: (a) the experiments often have large experimental uncertainties, and are therefore in need of least-squares analysis to interpret them quantitatively, and (b) there are several methods to analyze the data, i.e., different ways of linearizing the theoretical expression. When unweighted least-squares analysis is used, these different analysis methods yield *different* results when operating on the *very same* experimental data! When weighted least squares are used, all analysis methods yield the *same* results.

The expression for the simplest form of enzyme kinetics was first given by Henri (*Compt. Rend.* 135 (1902) 916), but is often named after Michaelis & Menten (*Biochem. Z.* 49 (1913) 333), who investigated the same enzyme and used the same equation. The expression is

$$v = \frac{S v_m}{K + S} \tag{3.5-1}$$

where v is the *initial* rate of the enzyme-catalyzed reaction, S is the concentration of its substrate, v_m is the maximum rate, and K is a constant.

One popular way to rectify (3.5-1) is to convert it to the Lineweaver–Burk form

$$\frac{1}{v} = \frac{1}{v_m} + \frac{K}{S v_m} \tag{3.5-2}$$

so that a plot of $1/v$ versus $1/S$ yields a straight line with intercept $1/v_m$ and slope K/v_m. Another linearization, due to Hanes, is obtained by multiplying all terms with S:

$$\frac{S}{v} = \frac{S}{v_m} + \frac{K}{v_m} \tag{3.5-3}$$

where a plot of S/v versus S yields a straight line of intercept K/v_m and slope $1/v_m$. Typically, the determination of the initial rate v is the dominant source of experimental uncertainty, and below we will use that as our starting assumption. We will use a data set from M. R. Atkinson, J. F. Jackson, & R. K. Morton, *Biochem. J.* 80 (1961) 318, to illustrate the need to use weighted least-squares analysis in such a case. The experimental data are as follows:

Concentration S of nicotinamide mono-nucleotide, in mM	Initial rate v of nicotinamide-adenine dinucleotide formed, in micromoles
0.138	0.148
0.220	0.171
0.291	0.234
0.560	0.324
0.766	0.390
1.460	0.493

The weighting factors needed for the above two cases are different. For the Lineweaver–Burk plot we have $y = v$, $Y = 1/v$, and therefore, according to (3.4-1), $w = 1/(dY/dy)^2 = 1/(d(1/y)/dy)^2 = 1/(1/y^2)^2 = y^4 = v^4$ whereas for the Hanes plot we find $Y = S/v$ so that $w = v^4/S^2$.

Instructions for exercise 3.5

1 Open an Excel spreadsheet.

2 Enter the column headings S, v, 1/v, w, 1/S, S/v, w, and S.

3 Enter the data in the first two columns, labeled S and v.

4 In the next columns, calculate $1/v$, $w = v^4$, $1/S$, S/v, $w = v^4/S^2$, and copy S from the first column.

5 Highlight the data in the third through fifth column, and click on <u>T</u>ools ⇨ <u>M</u>acro ⇨ <u>M</u>acros ⇨ WLS1. The coefficients will appear below the data.

6 Similarly, highlight the data in the last three columns, and analyze them the same way.

7 The Lineweaver–Burk analysis yields slope K/v_m and intercept $1/v_m$. Use the next two rows to convert this information into the parameters of interest, K and v_m. Here, $K =$ slope / intercept, and $v_m = 1$ / intercept.

8 Calculate the corresponding standard deviations, either by hand (using the rules of error propagation through quotients, i.e., via relative errors) or (easier and less error-prone) with the macro Progression. Note that Progression handles only one parameter at a time, so that you must apply it separately to K and v_m. Also note that the weighted least-squares macro places the standard deviations *under* the coefficients, whereas Progression puts them *next* to the parameters.

9 Similarly, the Hanes analysis yields slope $1/v_m$ and intercept K/v_m, from which you again compute K and v_m, and the corresponding standard deviations. Note that the roles of slope and intercept are interchanged in the Lineweaver–Burk and Hanes plots. Compare the results of the two approaches.

10 Copy the entire block down.

11 Modify the just-copied block by replacing all weights by 1's.

12 Now repeat the analysis. Forcing all terms w to 1 will of course yield the equivalent, *unweighted* results.

13 Compare what you got; it should look like Fig. 3.5.

The *unweighted* Lineweaver–Burk and Hanes plots yield *different* answers even though they analyze the same data set. On the other hand, when we use appropriate global weighting factors, both methods yield identical results, as they should. Need we say more about the importance of proper weighting?

	A	B	C	D	E	F	G	H
1	WEIGHTED:			Lineweaver-Burk			Hanes	
2	S	v	1/v	w	1/S	S/v	w	S
3								
4	0.138	0.148	6.7568	0.0005	7.2464	0.9324	0.0252	0.1380
5	0.220	0.171	5.8480	0.0009	4.5455	1.2865	0.0177	0.2200
6	0.291	0.234	4.2735	0.0030	3.4364	1.2436	0.0354	0.2910
7	0.560	0.324	3.0864	0.0110	1.7857	1.7284	0.0351	0.5600
8	0.766	0.390	2.5641	0.0231	1.3055	1.9641	0.0394	0.7660
9	1.460	0.493	2.0284	0.0591	0.6849	2.9615	0.0277	1.4600
10								
11			Coeff.:	1.4709	0.8398	Coeff.:	0.8398	1.4709
12			St. Dev.:	0.0761	0.0559	St. Dev.:	0.0559	0.0761
13								
14			K=	0.5710	0.0481	K=	0.5710	0.0481
15			vm =	0.6799	0.0352	vm =	0.6799	0.0352
16								
17	UNWEIGHTED:			Lineweaver-Burk			Hanes	
18	S	v	1/v	w	1/S	S/v	w	S
19								
20	0.138	0.148	6.7568	1.0000	7.2464	0.9324	1.0000	0.1380
21	0.220	0.171	5.8480	1.0000	4.5455	1.2865	1.0000	0.2200
22	0.291	0.234	4.2735	1.0000	3.4364	1.2436	1.0000	0.2910
23	0.560	0.324	3.0864	1.0000	1.7857	1.7284	1.0000	0.5600
24	0.766	0.390	2.5641	1.0000	1.3055	1.9641	1.0000	0.7660
25	1.460	0.493	2.0284	1.0000	0.6849	2.9615	1.0000	1.4600
26								
27			Coeff.:	1.7085	0.7528	Coeff.:	0.8500	1.4603
28			St. Dev.:	0.3033	0.0782	St. Dev.:	0.0596	0.0818
29								
30			K=	0.4406	0.0906	K=	0.5821	0.0522
31			vm =	0.5853	0.1039	vm =	0.6848	0.0383

Fig. 3.5: Spreadsheet for the analysis of the data on the kinetics of nicotinamide mononucleotide adenyltransferase by Atkinson *et al.*, *Biochem. J.* 80 (1961) 318. The results should be read as, e.g., on line 30: $K = 0.4406 \pm 0.0906$ for the unweighted Lineweaver–Burk method, $K = 0.5821 \pm 0.0522$ for the unweighted Hanes plot, or, on line 31, $v_m = 0.5853 \pm 0.1039$ for Lineweaver–Burk, $v_m = 0.6848 \pm 0.0383$ for Hanes.

3.6 Non-linear data fitting

So far we have seen that experimental data can be fitted to many functions, such as a line, a polynomial, or (after transformation) an exponential. However, there are many more functions for which this does not seem possible. For example, no way is known to fit $y = a_1 \exp(-k_1 t) + a_2 \exp(-k_2 t)$ by using what is called a linear least-squares fit, where the term *linear* refers to the fact that the expression is linear *in the coefficients*, here a_1, a_2, k_1 and k_2. (As we have seen, the expressions for y can be quite non-linear in the independent parameter x; that is still considered a linear least-squares fit.)

In order to overcome this limitation of linear least-squares fitting, i.e., its restriction to a specific type of function, ingenious algorithms have been developed to fit data to *any* function that can be described analytically, using a single criterion, such as minimizing the sum of squares of the deviations between the data and the model. Such routines use a variety of techniques, often including a method of steepest descent as well as a Newton–Raphson algorithm, to as it were 'feel' their way towards that criterion. Excel has several such algorithms in its Solver. Below we will use several examples to demonstrate both the power, and some of the limitations, of using Solver.

3.6a Some kinetic data

First we will get acquainted with the method by applying it to a set of experimental multi-parameter data from a paper on the kinetics of the thermal isomerization of bicyclo[2.1.1]hexane by R. Srinivasan & A. A. Levi, *J. Am. Chem. Soc.* 85 (1963) 3363, as quoted in N. R. Draper & H. Smith, *Applied Regression Analysis*, 2nd ed., Wiley 1981). In order to reduce the tedium of having to enter 38 data sets, we have used a smaller subset for this example, leaving out all duplicate measurements as well as all data at 612 K and 631 K. The dependent variable y is the fraction of the parent compound remaining after a reaction time of t minutes, while T is the temperature of the experiment, in K. The data are as follows:

Temperature T, in K	Reaction time t, in min						
	15	30	45.1	60	90	120	150
600				0.949		0.900	
620	0.938	0.877	0.827	0.787	0.696		0.582
639	0.808	0.655		0.425	0.309		

You will notice that the temperature range investigated is rather limited, and that not all possible time–temperature combinations were measured. Still, the fractions range from 0.3 to 0.95, and there are enough data to work on.

Instructions for exercise 3.6-1

1 Open an Excel spreadsheet.

2 In cells A1 and A2 enter the labels a and b respectively.

3 In cells B1 and B2 deposit initial guess values; 0 and 0 will do for now.

4 In cell D1 deposit the label SRR.

5 In cells A4 through E4 place the labels time t, temp T, y(exp), y(calc), and RR.

6 In columns A through C, starting in row 6, deposit the data. For example, the first data set might read 60, 600, 0.949 in cells A6, B6 and C6 respectively, the next data set (in row 7) 120, 600, 0.900, and so on. Enter all 12 data sets.

7 In D6:D17 calculate values for the theoretical expression used by Srinivasan & Levi,
$y = \exp\{-a\,t\exp[-b\,(1/T-1/620)]\}$.

8 In column E compute the squares of the corresponding differences, e.g., in cell E6 use $= (C6 - D6)\wedge 2$.

9 In cell E1 deposit the instruction $= \text{SUM}(E6{:}E12)$. This sum of the squares of the differences between the actual data and the model will be the quantity you will want to be minimal.

10 Now call in the troops, with T̲ools ⇨ Sol̲ver. In the Solver Parameter dialog box, enter E1 in the top window, push the radio button with Mi̲n, and enter B1:B2 in the window below it, so that you can read the box as 'S̲et Target Cell E1 Equal to Mi̲n By Changing Cells B1:B2'. Then push the S̲olve button.

11 You will see the numbers in B1 and B2 change, as well as the value of SRR in E1, and those in columns D and E. After a while the program will come to a halt, announce that it has converged on a solution, and ask you whether it should keep that solution, or put the starting guess values back in B1 and B2. You should find the result $a = 0.00385$, $b = 2.66 \times 10^4$. The value of SRR is 0.00165, down from 1.34 when a and b were 0.

12 You were lucky in your choice of initial guess values for a_0 and b_0. Try again, with $a_0 = 1$ and $b_0 = 0$. Solver will again declare that it has converged on a solution, but it hasn't: the values of a and b have not changed, and SRR is 6.84. If you look at the data in columns C and D you will see why. For $a_0 = 1$, $b_0 = 0$, the values for y_{calc} are quite low, with all but two entries less than 10^{-10}, and those two only 3×10^{-7}. Even these two entries have very little influence on the difference $(y_{exp} - y_{calc})^2$ since the corresponding values for y_{exp} are at least 0.8, so that $(0.8 - 3 \times 10^{-7})^2 \approx 0.8^2 - 4.8 \times 10^{-7} = 0.6399995$, barely different from $0.8^2 = 0.64$. Apparently, when Solver varies either a or b, it finds insufficient change in the sum of squares of the differences, SRR, and interprets this as evidence for having found a minimum! The tricky part here is not so much that Solver fails, but that it sometimes (as in this example) announces its incorrect result with the same aplomb as when it has succeeded. Computers can lie without blushing.

13 Try again, this time with $a_0 = 0.6$ and $b_0 = 0$. You might obtain a different result, such as $a = 0.00467$, $b = 0.74$, and SRR $= 0.448$. Judging by the value of SRR, this is not a very close solution. The disturbing aspect of this observation is that $a_0 = 0.7$ or $a_0 = 0.8$ with $b_0 = 0$ will yield the earlier results, with the much lower value 0.00165 for SRR.

14 There are many other possible pitfalls in using Solver. For example, in the Solver Parameters dialog box you may have noticed the O̲ptions button. Push it, and you get the Solver Options dialog box, which lets you select such parameters as Precision and Tolerance, and also presents a number of choices. Change the Methods from Newton to Conjugate, and try Solver again with $a_0 = 0$ and $b_0 = 0$. You will get $a = 0.00469$, $b = 2.4 \times 10^{-9}$, which is clearly an incorrect result judging from the corresponding value of SRR, 0.45. And by selecting Quadratic Estimate, Central Derivative, and Conjugate Method, you will find $a = 0.00428$, $b = 2.24 \times 10^4$, and SRR $= 0.0098$, close but no cigar.

We now step back, and take a second look at the above problem. It should not have been treated this way, by just unleashing a non-linear least-squares routine as if we knew nothing about a and b. The experimental data are most numerous for 620 K, at which temperature the expression $y = \exp\{- a\,t\exp[- b\,(1/T - 1/620)]\}$ reduces to $y = \exp(- at)$. The data at 620 K can therefore be fitted with a weighted linear least squares to get the value of a. (Even b might be found, e.g., from the observations at 60 minutes, although that is more tenuous because there are so few data.) Only then, with a good first estimate for a (and perhaps also for b), should we have used Solver to refine the entire data set. In this way we minimize the risk of incorrect answers. In the end we will still use Solver, since it will allow us to include all experimental data in the analysis. In such an approach, non-linear least-squares fittings are not seen so much as primary tools for brute-force fitting, but more as aids in refining our answers. Try it out on the above data set.

15 Copy the data in C8:C13 to F8:F13, and the label from C4 to F6.

16 In G8:G13 compute $\ln(y_{\exp})$, and in H8:H13 the corresponding weights $(y_{\exp})^2$. Label these columns appropriately.

17 Copy the data in A8:A13 to I8:I13, and the label from A4 to I6. You now have all data at 620 K in the proper format for determining a.

18 Highlight block G8:I13, and call Tools ⇨ Data Analysis ⇨ Regression ⇨ Constant is Zero or, better yet, Tools ⇨ Macro ⇨ Macros ⇨ WLS0, which will yield $a = 0.0039 \pm 0.0001$.

19 You can now determine b as outlined above, or (since that is a very small data set, which might therefore not yield very reliable information anyway) return to Solver, either by fixing a at 0.00386 or, preferably, by using the value found for a as its initial estimate. (You can fix a by setting B1 equal to $= - \text{I}15$, and by deleting B1 from the list of cells to be changed in the Solver Parameter window By Changing Cells, which should now contain B2 only.) Either way you will find $a = 0.0039 \pm 0.0001$, $b = 2.7 \times 10^{-4}$, and SRR $= 0.0017$.

Since we are working in Excel, we should use its graphical facilities, and plot the data. Ideally we would want to use a three-dimensional plot for these data, but here Excel is not cooperative. Although Excel advertises that it has 3-D plots, none of these are true XYZ plots. Instead, at least one of the two independent axes is used to display categories rather than values. For example, if you assign the x-axis the values 0, 0.1, -0.3, 9 and 5, the plot will show them as such, properly labeled, but in the above order, at equidistant intervals! This is just what you may want when plotting profits as a function of the month, or of the geographic region, without having to assign a rank number to each month or region, but for scientific applications it makes no sense. (But then, if it were not for its business appeal, Excel would not be as powerful, as user-friendly, and as inexpensive as it is now.) Still, you can use them as long as you have equidistant data.

Back to the question: how do we graph the data? Below is a possible solution.

20 Click on the number 8 in the left-most column of your spreadsheet, to the left of column A. This will highlight the entire row 8. Right-click to get to the Properties, select Insert, and left-click, to insert a new, blank row.

21 Do the same in row 15. You have now separated the data into three separate sets, one for each temperature.

22 Now plot y_{exp} and y_{calc} versus time t. Display the experimental data as points, and the least-squares fit as a line. Note: make it a *smooth* line by clicking on that line, and in the resulting Format Data Series pick Patterns, then click on Smoothed Line.

23 In order to label the various data sets, click on the Formula Bar, and type T = 600 K. Pressing the Enter key will now produce a box with this text in the graph. Use the mouse to move it to the position you want for it. Make it bold, colored, change its letter type, whatever. Then copy and paste it to get a duplicate, modify it to read T = 620 K, move it in position, etc. Your finished product might look like Fig. 3.6-1.

24 Save the spreadsheet as Isomerization.

3.6b A double exponential

As our second example we will use the function $y = a_1 \exp(-k_1 t) + a_2 \exp(-k_2 t)$ to see more clearly, on a noise-free data set, some other limitations of non-linear fitting.

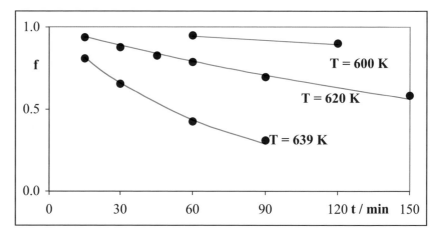

Fig. 3.6-1: The unreacted fraction f of bicyclo[2.1.1]hexane during its thermal isomerization, as a function of reaction time t and temperature T. Experimental data (solid circles)
selected from R. Srinivasan & A. A. Levi, *J. Am. Chem. Soc.* 85 (1963) 3363. The curves
(colored lines) show the relation $y = \exp\{-a\,t\exp[-b(1/T-1/620)]\}$ with $a = 0.0038_6$
s^{-1} and $b = 2.6_7 \times 10^4$ K as determined by least squares.

Instructions for exercise 3.6-2

1 Open an Excel spreadsheet.

2 Make a small parameter table, with the horizontal labels y(data) and y(fit), and the vertical labels a1, k1, a2, and k2.

3 Use some numbers in the y(data) column (in Figs. 3.6-2 through 3.6-4 we have used a_1
= 9, k_1 = 2, a_2 = 0.02, and k_2 = 0.5, but feel free to use other values), and copy those
same numbers into the y(fit) column.

4 Below this parameter table make space for the label SRR, and for a place to put the
associated number.

5 Below this, start the actual data table with four columns, labeled t, y(data), y(fit), and
RR respectively.

6 In column t place the numbers 0 (0.1) 10, i.e., 0, 0.1, 0.2, 0.3, ... , 10.

7 In the second column calculate $y = a_1 \exp(-k_1 t) + a_2 \exp(-k_2 t)$, using the constants
deposited in the table under y(data).

8 In the next column make the very same calculation, but this time based on constants
from the next column in the table, labeled y(fit). (Since you have entered identical constants in both columns of the table, you should also see identical numbers being calculated for the function.)

9 On the sheet, plot y(data) and y(fit) versus t. Since exponential functions are involved,
select the semi-logarithmic plot in the ChartWizard.

10 For the time being set the four constants in the y(fit) column of the parameter table to zero, which will make the corresponding data in the data table become zero as well.

11 Compute in the column labeled RR the square of the differences between corresponding values in the second and third columns.

12 In the place reserved for the value of SRR calculate the sum of the data in the RR column. This will be your least-squares fitting criterion, to be used in Solver's Target Cell.

13 Now call the Solver. Set Target Cell to the value of SRR, Equal to Min, and direct By Changing Cells to the four values in column y(fit) of the parameter table. Under Options again select Show Iteration Results.

14 See whether you can get a solution. Chances are you will get one, which will fit the beginning of the curve, but *not* its tail end, see Fig. 3.6-2. In that case, the constants found for both a's and for both k's may be identical, with $a_1 = a_2$ equal to half the a-value set for the dominant transient in your table.

Solver obviously found coefficients that fit the initial part of the curve, but has problems with its tail end. Now look at the graph, and keep in mind that its semi-logarithmic representation makes the tail end look much more important than it is. In the region Solver fails to fit, the signal is so much smaller than at the beginning of the curve that Solver is insensitive to the resulting differences in RR.

Data such as those used here may originate, e.g., from a radiochemical experiment involving two radioactive species, of which one decays with a much shorter half-life than the other. In that case the characteristic decay rate constant of the longer-lived species is often easier to find, by measuring at longer times, when the faster process has died out.

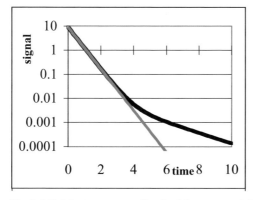

Fig. 3.6-2: A first attempt to fit a double exponential may fail to find the parameters of the second exponential when the amplitude of the latter is much smaller than that of the former. The colored parameters to the left of the graph are those adjusted by Solver.

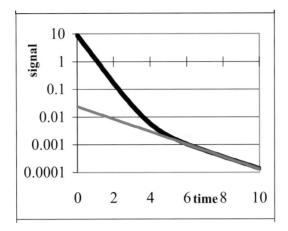

Fig. 3.6-3: Fitting the data in the restricted range $6 \le t \le 10$ yields a good fit to the tail end of the curve, while ignoring the initial part.

15 Since the data suggest that the second exponential is dominant at $t \ge 6$, change the instruction for the computation of SRR to restrict the sum of the squares of the residuals to the range $6 \le t \le 10$.

16 Try Solver again, with all four parameters (a1, k1, a2, and k2) in y(fit) set to zero, but with Solver only adjusting a2 and k2.

17 You will now obtain a single exponential (a straight line in the semilog plot) which yields a good fit to the tail end of the curve, while ignoring the initial part of that curve, as illustrated in Fig. 3.6-3.

18 Now you may be ready to fit the entire curve, as follows. First, change the computation of SRR back to encompass the entire curve, from $t = 0$ to $t = 10$. Then use Solver to fit a1 and k1, but let it *not* adjust a2 and k2, which should stay as you had found them under point 16. This makes it impossible for Solver to ignore the tail end, even though the larger (initial) data make by far the larger contribution to SRR.

19 Chances are that you will now see a good fit develop for the entire curve. Figure 3.6-4 illustrates some intermediate stages you may encounter with Show Iteration Results.

20 Once you have obtained a reasonably good fit to both ends of the curve, use Solver one more time, but only now let it adjust all four parameters simultaneously. Since it will start with close estimates, it should now produce a good answer. You may have to repeat Solver several times to get a value like that shown in Fig. 3.6-5.

21 The above clearly demonstrates that, in using Solver, starting with close initial estimates can make all the difference.

22 Save as DoubleExponential.

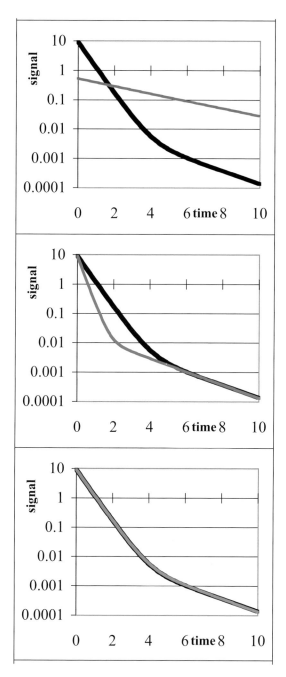

Fig. 3.6-4: Three stages in the fitting procedure for the double exponential. The colored data in the parameter table are those computed by Solver. The values of a_2 and k_2 were fixed at the values found in Fig. 3.6-3, and the adjustable initial values were $a_1 = k_1 = 0$.

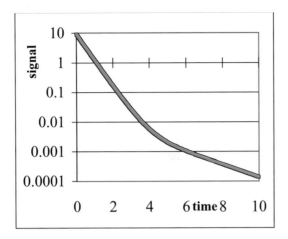

Fig. 3.6-5: The final fit of all data and all adjustable parameters. The resulting, extremely small value of SRR reflects the absence of noise in this simulation.

It may be useful to keep in mind that this is a rather unusual example, because very few other analytical measurements would have noise levels such that they would allow the determination of a second component with an amplitude of less than 0.5% of that of the major signal. In that respect, radiochemical measurements are rather unique, because quanta released in radioactive decay have such high energies that individual disintegrations can be counted, in counts that are often virtually free of electronic noise. Apart from the inherent randomness of radioactive disintegrations (see section 2.10), the only 'noise' in such measurements is background radiation, which usually can be kept extremely low by careful shielding.

The above examples clearly illustrate both the strengths and some of the inherent limitations of a non-linear search routine like Solver. First its strengths: it can often find a least-squares fit in situations where no linear least-squares algorithm will apply. And it may even do so for multi-parameter fits. Its main limitations are the following:

1 Solver needs reasonably close initial estimates of the parameter values, otherwise it can easily produce non-optimal results. If at all possible, subject a subset of the data to a linear least squares analysis to get an idea of at least some of the parameter values. And wherever possible, use a graph to see how close your initial estimate is, and follow the progress of the iterations visually, even though that slows down Solver. Where this is not feasible, as in multi-parameter fits, at least inspect the orders of magnitude of the values in the two columns Solver compares: y_{exp} and y_{calc}.

2 Even if Solver finds the dominant features of a curve without much trouble, some of its minor features may not make enough of a difference in the fitting criterion (here: SRR) to matter, in which case they may be fitted

poorly if at all. Noise (which was mercifully absent in the second example, to better illustrate this point) would only make matters worse: any noise in the initial data might easily be larger than the magnitude of the signal in the tail end of the curve in Fig. 3.4. In such a case you must provide Solver with separate, independent information, such as the value of k_2 in the last example.

Note that these limitations make eminent sense once you understand what Solver does, and are no different from those encountered when we extract information from ordinary experiments, without computer. It is just that we often come to expect too much of programs or of machines. Have you ever seen a horse win a race without a jockey? Solver is like that: a very competent program, that could be a champ – but only when you tell it where to go, and how to get there.

3.6c False minima

Solver travels down a multidimensional surface in search of a minimum value of SRR, just as water runs down a mountain under the influence of gravity. Often, the water finds its way to the ocean, but sometimes it collects in a lake without an outlet, and stays there. (Here, of course, the analogy stops, because the water can get back into the cycle by evaporation. And, of course, there are also lakes below sea-level.) The point is that a non-linear least-squares method can find a false minimum, and get stuck there, in which case you must help it to get out of that minimum. In fact, we already encountered an example of such a situation in Fig. 3.6-2, and we will now take a closer look at that case.

When we delete one of the four adjustable parameters from Solver, and instead give it a fixed value, we can observe how SRR changes. Say that we fix a_1 at a value that is slightly different from the one found, say at 5 instead of 4.5. If you apply Solver, not much will change, since a_2 can (and will) take up the slack by assuming a value close to 4. In that case, neither k_1, k_2, nor SRR changes. And when you fix the value of k_1 at, say, 2.5, the other three parameters will all change, and SRR will *in*crease. That is the characteristic of a false minimum: in its immediate neighborhood, it yields the lowest value of SRR. In order to find the good fit of Fig. 3.6-5 it was necessary to force one of the parameters quite a ways off, as we did in Fig. 3.6-4.

The possibility of ending up in a false minimum should put you on notice: don't just accept the answer of Solver as holy writ, and try whether other reasonable initial conditions yield the same answer. And, always, compare the y_{calc} data computed from the parameters found by Solver with the experimental data (directly or, better yet, by plotting the residuals) to see whether the parameters found make sense.

3.6d Enzyme kinetics revisited

In the context of the analysis of enzyme kinetics it is sometimes stated that one should always use a non-linear least-squares method for such data, because the usual, unweighted least-squares fits depend on the particular analysis method (Lineweaver–Burk, Hanes, etc.) used. We have seen in section 3.5 that the latter part of this statement is correct. But how about the former?

Instructions for exercise 3.6-3

1 Return to the spreadsheet shown in Fig. 3.5.

2 In the empty space below it copy (in columns A and B) the values of S and v, and relabel them S(exp) and v(exp).

3 Add columns C through F with the following column headings: v(calc), RR, S(calc), and RR.

4 Somewhere enter initial guess values for v_m and K, with their labels. In view of the data listed just before Fig. 3.5, enter guess values of 0.5 for both; those values should be close enough.

5 In column C compute v_{calc} from (3.5-1).

6 In column D calculate the square of the residuals, i.e., $(v_{exp} - v_{calc})^2$.

7 Find a place to calculate the sum of the squares of the residuals, SRR.

8 Unleash Solver, using SRR as its target, to be minimized, by changing the guess values for v_m and S. Note the result you obtain.

So far, so good. Now we rewrite (3.5-1) as $S = vK/(v_m - v)$. Same thing, isn't it? Mathematically fully equivalent.

9 In column E compute S_{calc} from v_{exp} and the initial guess values for v_m and S, which you should of course set back to 0.5.

10 In column F calculate RR $= (S_{exp} - S_{calc})^2$, and then compute SRR.

11 Again use Solver to find the best values for v_m and S by minimizing SRR. Note down your result. Is it the same as obtained under (8)? Not quite.

So here we are: with an unweighted non-linear least-squares program we encounter *precisely the same problems* as with unweighted linear least squares. The problem lies in defining which experimental parameter carries the dominant uncertainty, and selecting that parameter as the dependent one. In the unweighted least-squares analysis we used S/v in the

Hanes plot, and $1/v$ in the Lineweaver–Burk plot, and they did not give the same results. The weighting converted both to use v as the dependent variable, and then they agreed, of course, and we obtain the same result with the non-linear least-squares fit to (3.5-1). But when we use S as our dependent variable, we get a different result with the non-linear least squares, because we again compare apples and pears. And, of course, when we use non-linear least squares on (3.5-2) we get the same result as with the linear least-squares analysis of the Lineweaver–Burk plot, and non-linear least-squares of (3.5-2) would yield the same answers as linear least squares based on a Hanes plot.

The take-home message is the following. The spreadsheet makes it very easy to use least squares, either linear or non-linear, so that the *mechanics* of least-squares fitting are no longer a problem. But we will still get different results depending on which *assumption* we make regarding the dependent parameter. And *that* choice the spreadsheet cannot make for us. This is where the intelligent judgement of the experimentalist comes in. Least-squares analysis is non-trivial, not because its algebra is rather complicated (it is, but the spreadsheet can take care of that), but because it requires knowledge about the nature of the experimental uncertainties involved.

This brings us to a final comment: it is very easy to modify the non-linear least squares to include weighting, since the user determines the residuals. We can include any weighting factors we want in our column with residuals, thereby converting Solver to a *weighted* non-linear least-squares optimizer. The important part is to include the *proper* weighting. For that we need to know what is (are) the major source(s) of the experimental uncertainty. And therein lies the problem: we are often too much in a hurry to find out where the experimental uncertainty comes from. Unfortunately, without that knowledge, we cannot expect to get reliable answers, no matter how sophisticated the software used.

3.6e SolverAid

Solver does not provide estimates of the precision of its answers. Fortunately, this limitation is readily remedied, because it is relatively straightforward to write a macro that will compute the standard deviations of the parameters found by Solver. Such a macro is fully described in chapter 10, and is there called SolverAid. Here we will merely illustrate how to use it (assuming it has been installed), using as our example spreadsheet exercise 3.6-3 of the preceding section.

12 Call SolverAid. It should be findable in <u>T</u>ools ⇨ <u>M</u>acros.

13 SolverAid will sequentially present three input boxes, one each for the parameters Solver found, for SRR, and for the values of v_{calc}. Either highlight the corresponding cell or cell block (which should be a contiguous vertical column), or enter the corresponding address (or address range) in the window of the input box, then press the Enter key.

14 That is all. SolverAid will place the values for the standard deviations of the parameters to the immediate right of those parameters if there is free or over-writable space there; otherwise it will present the results in the form of a series of message boxes. SolverAid will also provide the standard deviation of the dependent variable, in the cell to the right of SRR.

3.7 Summary

In this chapter we have seen that unweighted least-squares analysis on a spreadsheet is a cinch. When no standard deviations are required, the trend-line (callable only from a graph) is very convenient. When standard deviations are needed, the Regression analysis routine in the Analysis ToolPak can be used. Both can be used for a line as well as for multi-parameter fitting, and for fitting to a polynomial, with or without a requirement that the curve goes through the origin. There is no need to struggle with equations such as (3.1-1) through (3.1-4), or (3.2-1) through (3.2-4): the general software takes care of it all. The above methods require no initial guess values for any of the parameters to be determined.

For so-called equidistant data sets (where equidistance applies to the independent variable), least-squares fitting is even simpler, and takes a form tailor-made for an efficient moving polynomial fit on a spreadsheet, requiring only access to a table of so-called convoluting integers, or software (such as described in section 10.9) where these integers are automatically computed.

When the data are transformed before they are fitted, a weighted least squares is usually called for. Again, this is no big deal, since an add-in macro is provided here for that purpose. It can even be used as a general-purpose least-squares routine: when no weights are specified, unit weights are assigned automatically, i.e., weighting is omitted. Moreover, because its source code is provided in chapter 10, this program can readily be modified to suit specific needs. For example, when more statistical parameters are required, appropriate additional output statements can easily be incorporated in the macro. Again, a weighted least squares need no initial guess values.

Finally we have seen how non-linear least-squares work. Excel provides a collection of competent, multi-parameter approaches in its Solver. When linear least-squares fitting is possible, that should always be the first choice, because it yields unambiguous answers, without the possibility of false minima. On the other hand, many problems do not allow such a straight-forward solution, in which case one may have to turn to non-linear least-squares fitting. Non-linear least-squares algorithms use (very sophisticated) trial-and-error methods, and therefore need initial estimates of the parameter values they are asked to determine. The answer provided can depend on the initial guess values given. Although it can therefore yield non-optimal (and sometimes even completely incorrect) answers, it can be an extremely powerful general-purpose tool when used with appropriate precautions. Solver does not provide estimates of the precision of its results, but SolverAid does. Table 2.6-1 summarized many of the salient aspects of the various methods available in Excel.

In this and the previous chapter we have emphasized least-squares methods. Because computers can facilitate their implementation, such methods have become part and parcel of quantitative science. The emphasis must now shift to the appropriate choice of functions to be fitted, and to a careful consideration of the nature of the experimental uncertainties. Unfortunately, the latter topic is beyond the scope of this short book.

CHAPTER **4**

ACIDS, BASES, AND SALTS

4.1
The mass action law and its graphical representations

The fundamental law of chemical equilibria was formulated in 1864 by two Norwegian brothers-in-law, the theoretician Cato Guldberg and the experimentalist Peter Waage, and was refined by them and others in the following decades; it is now known as the **Guldberg–Waage** or **mass action law**. It has been amply confirmed in the time since it was discovered, and it can be understood in term of thermodynamics; here we will simply state and use it. When chemical species are involved in an equilibrium of the form

$$a\text{A} + b\text{B} + c\text{C} + \cdots \rightleftharpoons p\text{P} + q\text{Q} + r\text{R} + \cdots \tag{4.1-1}$$

then the mass action law specifies that there exists a fixed relationship between the corresponding **concentrations** [A], [B], [C], ..., [P], [Q], [R], ..., and their **stoichiometric coefficients** $a, b, c, ..., p, q, r, ...$, namely that the ratio

$$\frac{[\text{P}]^p[\text{Q}]^q[\text{R}]^r\cdots}{[\text{A}]^a[\text{B}]^b[\text{C}]^c\cdots} \tag{4.1-2}$$

(or its inverse) is constant. Such a ratio is called an equilibrium constant, because it specifies the relation between the concentrations of the various participating chemicals *at equilibrium*. As far as the chemicals involved are concerned, it is arbitrary whether we formulate such an equilibrium as in (4.1-1) or, alternatively, as $p[\text{P}] + q[\text{Q}] + r[\text{R}] + \cdots \rightleftharpoons a[\text{A}] + b[\text{B}] + c[\text{C}] + \cdots$, because chemicals neither know nor care how we write their equilibrium. On the other hand, a specific way of writing may have mnemonic value for us, humans. It is here that convention comes in. For acid–base equilibria in aqueous solution, the area with which we will be mostly concerned in this chapter, it is customary to define the equilibrium in terms of acid *dissociation*, whereby the acid loses one or more protons. The corresponding constants are called **dissociation constants**. For instance, for the dissociation of a weak monoprotic acid we write

$$HA \rightleftharpoons H^+ + A^- \qquad (4.1\text{-}3)$$

$$K_a = \frac{[H^+][A^-]}{[HA]} \qquad (4.1\text{-}4)$$

where K_a has the dimension of a concentration, i.e., with the unit M. Where a logarithm is taken, as in pH or pK_a, the dimension is always deleted, but standard dimensions are implied; in other words, the property is thought to be divided by the same property in its standard state, with a numerical value of 1.

In contrast, for equilibria involving the formation of complexes, we typically define equilibria in terms of the *formation* of such complexes, in which case the constants involved are called **formation constants**. Here a typical example would be

$$Ag^+ + Cl^- \rightleftharpoons AgCl \qquad (4.1\text{-}5)$$

$$K_{s0} = \frac{[AgCl]}{[Ag^+][Cl^-]} \qquad (4.1\text{-}6)$$

By convention, solvation is usually left out of these expressions, on the assumption (useful for dilute solutions, but not necessarily valid for concentrated ones) that the solvent concentration is so large that it is essentially constant, and can therefore be included in the numerical value of the equilibrium constant. Here we will follow this convention, and in aqueous solutions we will therefore not distinguish formally between H_2CO_3 and CO_2, or between NH_3 and NH_4OH. Likewise, the hydrated proton will be represented simply by H^+. Perhaps the most obvious example of this convention is the autoprotolysis of water,

$$H_2O \rightleftharpoons H^+ + OH^- \qquad (4.1\text{-}7)$$

where, assuming $[H_2O]$ to be essentially constant, the equilibrium expression $K_a = [H^+][OH^-]/[H_2O]$ is conventionally simplified to

$$K_w = [H^+][OH^-] \qquad (4.1\text{-}8)$$

In many experimental situations, one controls the **total analytical concentration**, C, i.e., the sum of the concentrations of the various possible forms present. For example, we may weigh a certain amount of acetic acid and/or sodium acetate, and dissolve it in a given amount of water, in which case we know the total analytical concentration $C = [HAc] + [Ac^-]$. In situations like that it is often useful to define the **concentration fraction** of a particular species in the mixture. For the simple monoprotic acid defined in (4.1-3) we have

$$\alpha_{HA} = \alpha_1 = \frac{[HA]}{[HA] + [A^-]} = \frac{[H^+]}{[H^+] + K_a} \qquad (4.1\text{-}9)$$

$$\alpha_{A^-} = \alpha_0 = \frac{[A^-]}{[HA] + [A^-]} = \frac{K_a}{[H^+] + K_a} \tag{4.1-10}$$

where the right-hand form of the equations is obtained after substituting the definition (4.1-4) of the dissociation constant, and where we have introduced the formalism of labeling the concentration fractions with the number of attached, exchangeable protons: 1 for HA, 0 for A^-. The concentrations of HA and A^- then follow as

$$[HA] = C_a \alpha_1 = \frac{C_a [H^+]}{[H^+] + K_a} \tag{4.1-11}$$

and

$$[A^-] = C_a \alpha_0 = \frac{C_a K_a}{[H^+] + K_a} \tag{4.1-12}$$

It is convenient to treat bases as if they were acids. Consequently, instead of describing NH_3, or its hydrated form NH_4OH, as dissociating into $NH_4^+ + OH^-$, we consider NH_4^+ as the acid, dissociating into $H^+ + NH_3$. For a monoprotic base B such as ammonia we therefore write

$$HB^+ \rightleftharpoons H^+ + B \tag{4.1-13}$$

$$K_a = \frac{[H^+][B]}{[HB^+]} \tag{4.1-14}$$

$$\alpha_{HB^+} = \alpha_1 = \frac{[HB^+]}{[HB^+] + [B]} = \frac{[H^+]}{[H^+] + K_a} \tag{4.1-15}$$

$$\alpha_B = \alpha_0 = \frac{[B]}{[HB^+] + [B]} = \frac{K_a}{[H^+] + K_a} \tag{4.1-16}$$

Below we will show how such equilibria can be visualized in logarithmic concentration diagrams. The corresponding spreadsheets are often very useful for pH calculations.

Instructions for exercise 4.1

1 Open a new Excel spreadsheet.

2 In cell A1 enter the label C =, and in cell A2 a corresponding numerical value, such as 0.1. Also deposit, in cell A4, the label Ka =, and a corresponding value in A5, such as $= 10^\wedge - 4.76$, the K_a of acetic acid. Name the contents of cells A2 and A5 (but use Ca, conc, or some other name instead of the invalid name C).

3 In row 11, starting with cell A11 (and thereby leaving space near the top of the spreadsheet for small diagrams), enter the labels pH, $[H^+]$, $\log[H^+]$, $[OH^-]$, $\log[OH^-]$, αHA, $\log[HA]$, αA^-, and $\log[A^-]$. In order to type the symbol α, type an a, highlight it, then select and enter the symbol font in the font box on the left of the formatting toolbar.

4 In cell A13 deposit the value 0, in cell A14 the value 0.1, then highlight both, and drag the common handle down to cell A153. Alternatively, in cell A14 deposit the instruction = A13 + 0.1, then copy this down to row 153.

5 In column B calculate the corresponding values for $[H^+]$ (i.e., in cell B13 calculate = 10^{-A13}), and in column C compute the values of $\log[H^+]$, e.g., $-A13$ in cell C13.

6 Likewise, in column D, calculate the corresponding values of $[OH^-]$, e.g., in D13 as either = $10^{(A13-14)}$ or = $(10^{-14})/B13$, and in column E calculate the corresponding logarithm, $\log[OH^-] = pH - 14$.

7 Use (4.1-9) to compute α_1 from K_a and the corresponding value of $[H^+]$.

8 In column G calculate $\log[HA]$, keeping in mind that $[HA] = C\,\alpha_{HA} = C\,\alpha_1$.

9 In columns H and I likewise compute the corresponding values for αA^- and $\log[A^-]$ on the basis of (4.1-10).

10 We will now make two graphs commonly used in considering ionic equilibria. The first of these is the **distribution diagram**, in which we plot the concentration fractions α as a function of pH. (In acid–base problems, pH will often be our dependent variable, i.e., will serve as our x-axis.) The distribution diagram is readily made by plotting columns A, F and H. (Remember: in order to highlight non-contiguous columns in Excel, first highlight the first column, then release the shift key and depress the Ctrl key, move the cursor to the next column to be highlighted, switch over to the Shift key, highlight the second column, etc.)

11 Anchor the graph on cell B1, and specify an xy plot. Label the graph as a distribution diagram, and label its axes as pH (with a range from 0 to 14) and alpha (range: 0 to 1). Place this graph on the spreadsheet using the area otherwise occupied by cells B1:E10. (If you so prefer, make it on a separate chart, or do both.) The graph shows the regions in which each of the species is dominant. At $pH < pK_a$, the dominant species is HA, whereas A^- has the higher concentration at $pH > pK_a$.

12 By changing the numerical values in A2 and A5 you can explore how the distribution diagram depends on these two parameters. As can also be seen from (4.1-9) and (4.1-10), the concentration *fractions* are independent of the total analytical concentration, but they do depend strongly on the value of K_a. In fact, the point where the two curves intersect is defined by $pH = pK_a$; the two α's there have the value 0.5. Table 4.1 lists several acids and bases, and their pK_a's.

13 Another graph is the **logarithmic concentration diagram**. Its meaning is initially perhaps somewhat less intuitive, because it uses a double-logarithmic scale, but it is usually more informative because it shows minor as well as major concentrations; moreover, it has very simple limiting slopes. The logarithmic concentration diagram is obtained by plotting columns A, C, E, G and I. (In order to specify its range, start by highlighting cell A31, then use Shift + End + ↓ to cell 153, release the Shift key and depress Ctrl, use the mouse to move the cursor over to cell C153, release Ctrl and depress Shift, then go up with Shift + End + ↑, release Shift in favor of Ctrl, move to cell

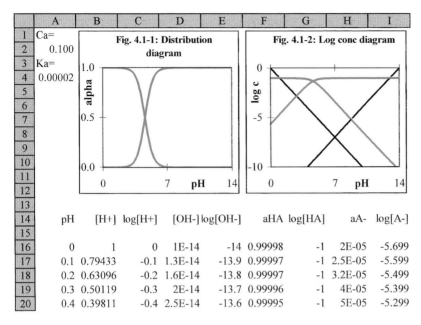

Fig. 4.1-1: Distribution diagram

Fig. 4.1-2: Log conc diagram

pH	[H+]	log[H+]	[OH-]	log[OH-]	aHA	log[HA]	aA-	log[A-]
0	1	0	1E-14	-14	0.99998	-1	2E-05	-5.699
0.1	0.79433	-0.1	1.3E-14	-13.9	0.99997	-1	2.5E-05	-5.599
0.2	0.63096	-0.2	1.6E-14	-13.8	0.99997	-1	3.2E-05	-5.499
0.3	0.50119	-0.3	2E-14	-13.7	0.99996	-1	4E-05	-5.399
0.4	0.39811	-0.4	2.5E-14	-13.6	0.99995	-1	5E-05	-5.299

Fig. 4.1 The top of the spreadsheet of exercise 4.1.

E13, switch back to Shift, then Shift + End + ↓ to cell E153, and so on, until you have highlighted all five columns.

14 Go to Insert ⇨ Chart etc. to make an *xy* graph. Using the mouse and the vertical slide-bar on the right of the spreadsheet, move the cursor to the top of the sheet, and deposit the chart symbol in cell F1. Plot the (vertical) log c scale from -10 (at the bottom) to 0 (at the top), the (horizontal) pH scale from 0 to 14. Alternatively you can use the pc $= -\log c$ scale by clicking on the vertical axis, then on Format Axis, and selecting Scale Minimum: 0, Maximum: 10, and check-marking Values in reverse order.

15 If you want the numbers on the pH scale to show below the graph, double-click on the figure, then on one of these numbers, then right-click on Format Axis. In the corresponding dialog box, on the page Patterns, select to have the Tick-Mark Labels Low for a log c scale, High for pc $= -\log c$.

16 Print the first page of the spreadsheet containing the graphs (it should look similar to Fig. 4.1) or, when you made the graphs on separate sheets, print these. When you use on-sheet graphs, you may want to reduce the widths of all columns in order to print it all on one sheet of paper. You can do that as follows. First select the entire sheet, by clicking on the gray rectangle where the row with the column labels ABCDEFG… meets the column with the row numbers 1234567… Then use Format ⇨ Columns ⇨ Width and enter an appropriate numerical value for Column Width.

17 Again vary the numerical values of *C* and K_a to see their effects. Note that *C* moves some (though not all) curves up or down, while K_a displaces the same curves sideways.

Table 4.1: *The* pK_a*-values for some monoprotic acids and bases.*

acid	pK_a	base	pK_a
acetic acid	4.76	ammonia	9.24
benzoic acid	4.20	aniline	4.60
butyric acid	4.82	benzylamine	9.35
dinitrophenol	4.11	butylamine	10.64
formic acid	3.74	ethanolamine	9.50
hydrofluoric acid	3.17	histamine	5.96
hydrochloric acid	<-1	hydroxylamine	5.96
lactic acid	3.86	lithium hydroxide	13.6
nitric acid	<-1	methylamine	10.64
nitrous acid	3.15	morpholine	8.49
perchloric acid	<-1	1,10-phenanthroline	4.68
phenol	9.98	potassium hydroxide	14.5
trichloroacetic acid	0.66	pyridine	5.23
trifluoromethylsulfonic acid	<-1	sodium hydroxide	14.1

Use order-of-magnitude changes, since small changes hardly show on logarithmic scales. Identify (and annotate as such) the four curves shown as depicting the logarithms of the concentrations of H^+, OH^-, HA and A^- respectively.

18 The lines representing $[H^+]$ and $[OH^-]$ are straight, while those for HA and A^- are bent yet contain two straight asymptotes. Verify mathematically (using (4.1-9) and (4.1-10)) and numerically (from the data in the spreadsheet, or by plotting the corresponding asymptotes) that, for log[HA], these asymptotes are $\log C$ and $\log(C[H^+]/K_a) = \log C - \log K_a - pH$, whereas those for $[A^-]$ are $\log(CK_a/[H^+]) = \log C + \log K_a + pH$, and $\log C$ respectively. Incidentally, those asymptotes form the basis of the **stick diagram**, which is easy to sketch, and very convenient for quick-and-dirty pH estimates when you don't have a computer handy, or when an approximate answer is all you need.

19 Save the spreadsheet as Monoprotic Acid.

For the single, monoprotic acids and bases in the above example, the distribution and logarithmic concentration diagrams are rather simple, yet they clearly show the relative and absolute concentrations respectively of the various species present. Such diagrams become all the more useful when we consider more complicated systems, such as polyprotic acids and bases, where it otherwise becomes increasingly difficult to envision what happens as a function of pH. We will do so in exercises 4.5 and 4.6.

As can be seen from Table 4.1, several acids have pK_a's for which only a lower limit is listed, as <-1. These acids are fully dissociated in water, and are therefore called **strong acids**. Examples are HCl, $HClO_4$ and HNO_3. Likewise there are bases that are almost completely dissociated in water, such as NaOH and KOH, as indicated by their large pK_a's. In this workbook we will consider these as fully dissociated, and therefore treat them as **strong bases**.

4.2 Conservation laws, proton balance, and pH calculations

Imagine being asked to calculate the pH of a 0.1 M solution of acetic acid in water. What principles would you use to find the answer to such a question, and how would you go about it? Obviously, the pH will depend on how much acid is used (0.1 M), and on the strength of the particular acid, i.e., on the total analytical concentration C and the dissociation constant K_a. Here is how these numbers can be used to arrive at the answer.

We start with two types of general relationships, the law(s) of **conservation of mass**, and the law of **conservation of charge**. We then combine these into one, the **proton balance**, which always can be derived from the conservation of mass and charge but which, fortunately, often can be written down merely *by inspection*. Combining the proton balance with the relations already encountered in section 4.1 then yields the answer.

In terms of the specific example used in the preceding paragraph, the mass balance requires that

$$[HA] + [A^-] = 0.1 \text{ M} \tag{4.2-1}$$

where we will usually delete the dimension, which is fine as long as we always use standard units, such as liters (but *not* mL!) for volumes, moles for amounts, and moles per liter (M) for concentrations.

The charge balance ensures that the solution is macroscopically electroneutral, by counting all cations and anions per liter of solution, i.e.,

$$[H^+] = [A^-] + [OH^-] \tag{4.2-2}$$

In this particular example, we can also interpret (4.2-2) in a different way, namely as a **proton balance**, an expression in which we specify which solution species have gained protons by comparison with the starting materials (here H_2O and HA) and which have lost protons. The proton balance focuses on H^+, the species of interest in pH calculations. Likewise, in complexation equilibria, we will encounter an analogous ligand balance, and in electrochemical equilibria an electron balance.

In order to answer the above question regarding the pH of 0.1 M acetic acid, we consult the logarithmic concentration diagram for a monoprotic weak acid with $C = 0.1$ M and $K_a = 10^{-4.76}$ M, and find the pH for which (4.2-2) is satisfied. Below we will use the spreadsheet to find that pH.

Alternatively we can use a stick diagram, which is a simplified version of the logarithmic concentration diagram, to obtain a quick-and-dirty pH estimate which, nonetheless, is usually correct to within 0.3 pH units. In the present example, either method yields a pH of 2.88.

We note that $pH = 2.88$ is where, in the logarithmic concentration diagram, the lines for $[H^+]$ and $[A^-]$ intersect. This is no accident: at this pH, the proton balance $[H^+] = [A^-] + [OH^-]$ can be approximated to $[H^+] \approx [A^-]$, because at the intersection of the lines for $[H^+]$ and $[A^-]$, $[OH^-]$ is more than eight orders of magnitude (i.e., a factor of more than $100\,000\,000$ times) smaller than $[A^-]$. That is good enough reason to neglect $[OH^-]$ with respect to $[A^-]$ in the expression $[A^-] + [OH^-]$.

What would be the pH of 0.1 M sodium acetate? Again, we use the mass balance (4.2-1), as well as a second mass balance, which specifies that

$$[Na^+] = 0.1 \text{ M} \tag{4.2-3}$$

whereas the charge balance now must include sodium ions, and therefore reads

$$[H^+] + [Na^+] = [A^-] + [OH^-] \tag{4.2-4}$$

Here, then, we have two mass balance equations plus the electroneutrality condition; there is always only *one* electroneutrality condition, while the number of mass balance equations depends on the number of non-interconvertible ionic species present. Since all three conditions must be met simultaneously, it is convenient to combine them in one expression,

$$[H^+] + [HA] = [OH^-] \tag{4.2-5}$$

which you may recognize as a proton balance, starting from undissociated H_2O and the fully dissociated salt NaA. To wit: H_2O can gain a proton to become (hydrated) H^+, A^- can gain a proton to become HA, and H_2O can lose a proton to yield OH^-. In this case, then, the required solution is found as that pH where (4.2-5) is satisfied: $pH = 8.88$. And, again, it corresponds to an intersection in the logarithmic concentration diagram, since $[H^+] \ll [HA]$ so that $[H^+] + [HA] = [OH^-]$ for all practical purposes reduces to $[HA] \approx [OH^-]$. These matters are discussed in much greater detail in my *Aqueous acid–base equilibria and titrations*, Oxford Univ. Press, 1999.

Instructions for exercise 4.2

1 Recall spreadsheet Monoprotic Acid.

2 Copy the instructions in row 153 to row 155, then modify the instructions in cells G155 and I155 to yield [HA] and $[A^-]$ instead of their logarithms.

3 Consider the standard form of (4.2-1), i.e., with all non-zero terms moved to the left-hand side of the expression: $[H^+] - [A^-] - [OH^-] = 0$. Use cell J155 to calculate that left-hand side of the standard form, which in this case would read = B155–I155–D155. The proton condition corresponds to the pH that will make J155 zero.

4 The easiest way to find that pH is by using the Goal Seek routine of Excel. Select $\underline{T}$ools $\Rightarrow$ $\underline{G}$oal Seek; in the resulting dialog box you $\underline{S}$et cell J155 To $\underline{v}$alue 0 By $\underline{c}$hanging cell

A155, then click on OK. Bingo, the spreadsheet will have found the pH, and will show it in cell A155.

5 Goal Seek is based on the principle of the Newton–Raphson algorithm, discussed in more detail in section 8.1, which finds a first estimate by determining the slope of the function, then extrapolates this to find the zero, uses this new estimate to obtain a second estimate, and so on.

6 You can also use Goal Seek without first having made an elaborate spreadsheet table; all that is needed is a blank spreadsheet and a single, analytical expression in terms of a single adjustable parameter.

7 To see that for yourself, somewhere else on the spreadsheet use three cells, which we will here call cells a, b and c. In cell a place a reasonable guess value for the $[H^+]$; this is the value that will be adjusted to make the proton condition zero. The proton condition in standard form should go in cell b, i.e., it should code for $[H^+] - [A^-] - [OH^-] = [H^+] - CK_a / ([H^+] + K_a) - 10^{-14} / [H^+]$. The third cell, c, is merely for your convenience, to calculate the pH from $[H^+]$. Note that we here do not need to use any approximations.

8 Engage Goal Seek and <u>S</u>et cell b To <u>v</u>alue 0 By <u>c</u>hanging cell a, at which point the answer appears in cell c.

9 The Newton–Raphson method often works, especially when the first estimate is close to the final value. However, one can generate situations where it will either fail to yield an answer, or produce an incorrect one. This is so because the Newton–Raphson method relies on the derivative of the function at the estimate, which may differ significantly from the derivative at the desired root. It may then be necessary first to use an approximate method to obtain an initial estimate for the variable parameter. In the present case, the Newton–Raphson method will typically work as long as the initial pH estimate is physically realizable for the given values of C and pK_a, in which case the proton balance written in standard form is a monotonic function of $[H^+]$.

10 On your own, now find the pH of 0.1 M sodium acetate, based on the proton condition (4.2-5) in its standard form, i.e., $[H^+] + [HA] - [OH^-] = 0$.

11 On the logarithmic concentration diagram for 0.1 M acetic acid made in exercise 4.1, note the above answers for the pH of 0.1 M acetic acid, and for 0.1 M sodium acetate, and see whether they indeed correspond with an intersection of two curves.

12 Find the pH of 0.1 M NH_3 in water (the pK_a of ammonia is 9.24, see Table 4.1), and that of 0.1 M NH_4Cl. For NH_3 and H_2O, the proton balance is $[H^+] + [NH_4^+] - [OH^-] = 0$; for $NH_4Cl + H_2O$ it is $[H^+] - [NH_3] - [OH^-] = 0$. Verify these proton balances, by deriving them from the mass and charge balance relations, and by considering proton gains and proton losses.

13 Also plot the corresponding logarithmic concentration diagram, and on it indicate the pH values just calculated. Rationalize why they again correspond with specific intersections, and note these on the plot.

14 Save the spreadsheet as Monoprotic Acid.

4.3

Titrations of monoprotic acids and bases

In a typical titration, known volumes of a reagent of known concentration are added to a known volume of a sample of *unknown* concentration, and addition is continued at least until an equivalent amount of reagent is added, at which point some measurable physical or chemical property indicates that a so-called **equivalence point** has been reached. The unknown concentration can then be calculated. Numerous properties can be used as indicators; historically, the first equivalence point indicator was the observation that bubble formation (effervescence) upon addition of potassium carbonate to vinegar would stop once the equivalence point had been reached or passed. Nowadays, the progress of most acid–base titrations is monitored either with a color indicator or, preferably, with a pH meter.

In a titration, then, the assignment is to determine the volume of the titrant necessary to reach the equivalence point, while one measures the pH (or a related quantity, such as the color of the added indicator) to monitor how close one is to that equivalence point. It is therefore logical to express the titration curve in terms of the titrant volume as a function of pH or $[H^+]$. This approach, which will be followed here, leads to simple theoretical expressions for the course of the titration, expressions that can all be described in terms of a simple master equation, and that allow easy and direct comparison with experimental data.

Alternatively one can try to express pH or $[H^+]$ as a function of titrant volume, reflecting how the typical volumetric experiment is performed (by measuring volumes) rather than what problem the titration tries to answer. Unfortunately, this more traditional approach leads to much more complicated mathematics, thereby hampering comparison between theory and experiment. We will not use it here, since it is poorly suited to *quantitative* analysis.

Below we will first consider monoprotic acids and bases. One of the simplest titrations is that of a strong monoprotic acid, such as HCl, with a strong monoprotic base, such as NaOH. Say that we titrate a sample of volume V_a of hydrochloric acid of unknown concentration C_a by gradually adding strong base of concentration C_b, the total volume of base added to the sample being V_b. Conservation of mass and charge then leads to the following expression for the **progress** of this titration:

$$\frac{V_b}{V_a} = \frac{C_a - \Delta}{C_b + \Delta} \tag{4.3-1}$$

where we have used the abbreviation

$$\Delta = [H^+] - [OH^-] = [H^+] - K_w / [H^+] \tag{4.3-2}$$

with K_w, the ion product of water (of the order of $10^{-14} M^2$), as defined in (4.1-8). The derivation of (4.3-1) is so simple that we will briefly indicate it here for the above specific example of titrating HCl with NaOH. We start from the charge balance equation

$$[H^+] + [Na^+] = [Cl^-] + [OH^-] \tag{4.3-3}$$

and then express $[Na^+]$ and $[Cl^-]$ in terms of the original sample concentration C_a and sample volume V_a, and the concentration C_b of the titrant used:

$$[Cl^-] = C_a V_a / (V_a + V_b) \tag{4.3-4}$$

$$[Na^+] = C_b V_b / (V_a + V_b) \tag{4.3-5}$$

where the dilution factors $V_a / (V_a + V_b)$ and $V_b / (V_a + V_b)$ take into account the mutual dilution of sample and titrant. Substitution of (4.3-4) and (4.3-5) into (4.3-3), and collection of terms in V_a and V_b, then yields (4.3-1).

For the titration of single weak acid with strong base, such as that of acetic acid with NaOH, we have instead of (4.3-1)

$$\frac{V_b}{V_a} = \frac{C_a \alpha_0 - \Delta}{C_b + \Delta} \tag{4.3-6}$$

where the concentration fraction of A^- is defined in (4.1–10). Note that (4.3-1) is the special case of (4.3-6) for $\alpha_0 = \alpha_{A^-} = 1$, i.e., for $K_a \gg [H^+]$. Indeed, a strong acid can be defined as one for which $K_a \gg [H^+]$ for all physically realizable values of $[H^+]$, a definition equivalent to stating that a strong acid is virtually completely dissociated at all realizable pH values.

The derivation of (4.3-6) is equally simple: instead of (4.3-3) and (4.3-4) we now use

$$[H^+] + [Na^+] = [A^-] + [OH^-] \tag{4.3-7}$$

$$[A^-] = C_a \alpha_0 V_a / (V_a + V_b) \tag{4.3-8}$$

and combine this with (4.3-5). Note that replacing $[Cl^-]$ by $[A^-]$ only results in the multiplication of C_a by α_0, so that multiplication of C_a in (4.3-1) by α_0 directly yields (4.3-6).

For the titration of a mixture of acids with a single strong base, (4.3-6) can be generalized to

$$\frac{V_b}{V_a} = \frac{\sum C_a \alpha_0 - \Delta}{C_b + \Delta} \tag{4.3-9}$$

where the summation extends over all (weak or strong) acids in the sample.

For the titration of a strong base with a strong acid we have

$$\frac{V_a}{V_b} = \frac{C_b + \Delta}{C_a - \Delta} \tag{4.3-10}$$

and for the titration of mixture of bases with a single strong acid

$$\frac{V_a}{V_b} = \frac{\sum C_b \alpha_1 + \Delta}{C_a - \Delta} \qquad (4.3\text{-}11)$$

Finally, the titration of a mixture of acids with a mixture of bases leads to the relation

$$\frac{V_b}{V_a} = \frac{\sum C_a \alpha_0 - \Delta}{\sum C_b \alpha_1 + \Delta} \qquad (4.3\text{-}12)$$

while that of a mixture of bases with a mixture of acids is described by

$$\frac{V_a}{V_b} = \frac{\sum C_b \alpha_1 + \Delta}{\sum C_a \alpha_0 - \Delta} \qquad (4.3\text{-}13)$$

Equations (4.3-12) and (4.3-13) are general relations for acid–base titrations of *monoprotic* acids and bases, and specifically include the effects of dilution of sample by the added titrant, and vice versa, through the dilution factors $V_a / (V_a + V_b)$ and $V_b / (V_a + V_b)$ introduced in (4.3-4), (4.3-5) and (4.3-8). Note, however, that any additional dilution (such as might result from rinsing the inside of the titration vessel) is not taken into account. In using the above equations, it is therefore advisable to *flow in* the titrant, by making the tip of the buret touch the inside wall of the titration vessel, rather than to add the titrant dropwise. Falling drops often lead to splashing on the inside walls of the titration vessel, which then need to be rinsed down. Moreover, the use of droplets can lead to large reading errors, since half a droplet left hanging on a typical buret tip introduces a reading error of about 1/40 mL. Flowing in the titrant is done most conveniently while a magnetic stirrer mixes the incoming titrant with the sample, but is more difficult when titrant and sample must be mixed manually by swirling the titration vessel.

Instructions for exercise 4.3

1 Recall the spreadsheet Monoprotic Acid.

2 In cell A7 deposit the label Cb = , and in cell A8 a corresponding numerical value, e.g., 0.1.

3 In cell J11 deposit the column heading Vb/Va, and in cell J13 the formula = ($A$2*H13 − B13 + D13)/($A$8 + B13 − D13), which you will recognize as (4.3-6).

4 Copy this formula so that it will be used in J13:J153.

5 Make a plot of V_b/V_a vs. pH, which will show the **progress curve**. Restrict the range of V_b/V_a from 0 to 2, in order to avoid the very large positive and negative values calculated by the program at pH-values that are physically not realizable for the chosen values of C_a, C_b and pK_a.

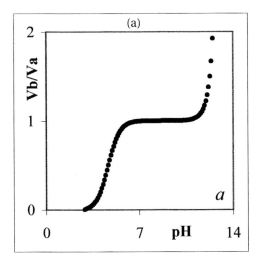
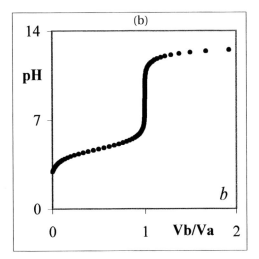

Fig. 4.3-1: (a) The progress curve, and (b) the titration curve, for the titration of 0.1 M acetic acid ($pK_a = 4.7$) with 0.1 M NaOH.

6 In a **titration curve** we instead plot the pH versus V_b / V_a. In order to plot such a curve, repeat the pH in column K (because Excel will automatically use the left-most selected column for the horizontal axis), then plot J13:J153 versus K13:K153. Alternatively, make a copy of the progress curve, double-click on it, then click on the data points. When (some of) these are highlighted, go to the formula bar, and there exchange all A's for J's, and vice versa, and deposit this change with the enter key. Finally, change the axis labels and titles around. This second method is more work, but does not require that column A be repeated in the spreadsheet.

7 Some representative curves are shown in Figs. 4.3-1a and 4.3-1b, as they might show when placed on the spreadsheet, in which case the actual calculations would of course have to be displaced downwards in order to remain visible.

8 Experiment with different K_a-values, such as 1, 10^{-2}, 10^{-4}, 10^{-6}, 10^{-8}, and 10^{-10} M. Also, experiment for a given K_a such as 10^{-4}, with various values of C_a and C_b, again changing them by, say, two orders of magnitude to see the effects.

9 Save as Monoprotic Acid.

Schwartz and Gran plots

Progress and titration curves are, typically, S-shaped. For the precise determination of the equivalence point it is often useful to rectify those curves. While this cannot always be done, rectification is possible for the titration of a single (weak or strong) monoprotic acid with a single strong monoprotic base, or vice versa, and leads to the Schwartz plot. A somewhat simpler but approximate rectification method yields the Gran plots. Here we will briefly explore both.

We again start from the electroneutrality condition

$$[H^+] + [Na^+] = [A^-] + [OH^-] \tag{4.3-7}$$

in which we now consider dilution:

$$[H^+] + \frac{C_b V_b}{V_a + V_b} = \frac{C_a \alpha_0 V_a}{V_a + V_b} + [OH^-] = \frac{C_a K_a V_a}{(V_a + V_b)([\text{II}^+] + K_a)} + [OH^-] \tag{4.4-1}$$

so that

$$\Delta(V_a + V_b)([H^+] + K_a) + C_b V_b([H^+] + K_a) = C_a K_a V_a \tag{4.4-2}$$

$$([H^+] + K_a)\{V_a + \Delta(V_a + V_b)/C_b\} = C_a K_a V_a / C_b = K_a V_{eq} \tag{4.4-3}$$

which yields the **Schwartz equation**

$$[H^+] V_b' = K_a (V_{eq} - V_b') \tag{4.4-4}$$

in which the **equivalence volume** V_{eq} is defined as

$$V_{eq} = C_a V_a / C_b \tag{4.4-5}$$

and where

$$V_b' = V_b + (V_a + V_b) \Delta / C_b \tag{4.4-6}$$

The equivalence volume is the primary goal of an analytical titration, because it allows us to determine the unknown concentration C_a from (4.4-5) as

$$C_a = C_b V_{eq} / V_a \tag{4.4-7}$$

At any time during the titration, the volume V_a of the original sample is known, as is the volume V_b and concentration C_b of the titrant used at any time during the titration. Moreover, because the pH is monitored, Δ is also known. Consequently, all terms on the right-hand side of (4.4-6) are known, and so is V_b'. The **Schwartz equation** (4.4-4) therefore allows us to rectify the entire titration curve simply by plotting $[H^+] V_b'$ versus V_b'. Such a **Schwartz plot** yields a single straight line, with slope $- K_a$, and ending at the horizontal axis precisely where $V_b' = V_{eq}$, making it very easy to determine the equivalence volume V_{eq} while, as an added benefit, the slope of the plot yields the value of K_a. Note that (4.4-4) applies to weak and strong acids alike; when the titrated acid is a strong one, the slope of the plot is (in theory) infinitely steep, i.e., the graph shows a vertical line.

Often, the second part on the right-hand side of (4.4-6) can be neglected, especially in the region just before the equivalence point, where

$\Delta = [H^+] - [OH^-]$ is small because, there, both $[H^+]$ and $[OH^-]$ are small. Under those conditions, $V_b' \approx V_b$, so that (4.4-4) reduces to

$$[H^+] V_b \approx K_a (V_{eq} - V_b) \qquad (4.4\text{-}8)$$

This is the basis of the **first Gran plot**, in which we make a graph of $[H^+] V_b$ versus V_b. The approximate sign $\approx$ is used in (4.4-8) because it is obvious by inspection that (4.4-8) cannot be reliable over the entire range of the titration. For example, at the beginning of the titration, $V_b = 0$, which would lead to $K_a V_{eq} = 0$, clearly an incorrect result for any finite value of K_a. Likewise, (4.4-7) yields a physically imposible result past the equivalence point, where $(V_{eq} - V_b)$ is negative, so that $[H^+] K_a$ would have to be negative, which it cannot be.

In a typical titration of an acid, the titrant is a strong base. In that case, the titration curve *past* the equivalence point is quite similar to that of the titration of a strong acid with a strong base. For that situation, another Gran plot is available, which actually is often the more useful one. We now start from (4.3-1) which we simplify by using the approximation $\Delta = [H^+] - [OH^-]$ $\approx - [OH^-]$. The result, $V_b / V_a \approx (C_a + [OH^-])/(C_b - [OH^-])$, can then be rearranged to

$$[OH^-] (V_a + V_b) = C_b V_b - C_a V_a = C_b (V_b - V_{eq}) \qquad (4.4\text{-}9)$$

where we have again used the definition (4.4-5) of the equivalence volume. Equation (4.4-9) is the basis for the **second Gran plot**, in which we plot $[OH^-] (V_a + V_b)$ versus V_b. Because of the assumption $\Delta \approx - [OH^-]$ the second Gran plot is only useful past the equivalence point. In that range it is often also convenient in the titration of polyprotic acids, where the Schwartz plot and the first Gran plot are usually no longer applicable.

Instructions for exercise 4.4

1 Recall the spreadsheet Monoprotic Acid.

2 In rows 13 through 153 of the next column, which will be either K or L depending on what option you used in exercise 4.3 under (6), calculate $V_b' / V_a = V_b / V_a + (1 + V_b / V_a)(\Delta / C_b)$, and next to it compute the quantity $[H^+] V_b' / V_a$, then make the Schwartz plot. It will save time and effort to scale both through division by the constant V_a, because you already have a column for V_b / V_a.

3 Next make a column for the product $[H^+] V_b / V_a$, and use it to make a Gran1 plot.

4 Finally, in the next column, calculate the quantity $[OH^-] (1 + V_b / V_a)$, then use it to make the Gran2 plot of $[OH^-] (1 + V_b / V_a)$ versus V_b / V_a.

5 Thumbnail sketches of such plots are shown as parts of Figs. 4.5-1 through 4.5-3.

6 Save your work as Monoprotic Acid.

Playing with theoretical expressions, as we do here, can sometimes be misleading. No matter how good the theory, experimental data never quite follow the theory, because of experimental uncertainties. It is therefore useful to make the theoretical data somewhat more realistic, by the addition of synthetic noise. For this we will use the Gaussian noise generator available in Excel. Keep in mind that such noise is again the ideal case, and that real noise will usually not be as well-behaved. Nonetheless, adding Gaussian noise is better than not adding any, and it can help us anticipate which methods are 'robust' and which are so sensitive to noise that their use will be 'of theoretical significance only'.

7 Recall the spreadsheet Monoprotic Acid.

8 Insert two new columns between those for V_b / V_a and V_b' / V_a, and label them 'noise' and '(Vb/ Va)n' respectively.

9 In the noise column enter Gaussian noise (using Tools ⇨ Data Analysis ⇨ Random Number Generation ⇨ OK ⇨ Distribution Normal, Mean = 0, Standard Deviation = 1), and in the next column add a fraction defined by a spreadsheet constant *na* (for noise amplitude) times this noise to the $(V_b / V_a)_n$ column.

10 Modify the spreadsheet code so that all columns to the right of the noise column refer to the new $(V_b / V_a)_n$ column rather than to V_b / V_a.

11 Likewise modify the charts to reflect the added noise. You can do this by first activating the charts by double clicking. Then click on a data point to get the = SERIES instruction showing in the formula bar. Modify the corresponding instructions in the formula bar to refer to the proper column, and finally use the enter key to deposit them.

12 You will notice that the Schwartz plot is somewhat more affected by the noise than the Gran plots, but that all three are still quite serviceable as long as the noise amplitude is not too large.

13 Save again as Monoprotic Acid.

4.5 The first derivative

Analytically useful acid–base titration curves are characterized by a rather fast pH change near the equivalence point. This suggests that the location of the equivalence point might be determined experimentally from that of the maximum in its first derivative, $d(pH)/dV_b$, or the zero-crossing of its second derivative, $d^2(pH)/dV_b^2$. The advantage of such an approach is that it does not rely on any particular theoretical model, but instead exploits the characteristic feature of the titration curve, i.e., its fast pH change in the region around the equivalence point. The method does not even require that the pH meter is carefully calibrated.

Unfortunately, there are two reasons why taking the derivative of the titration curve is usually not a recommended procedure for establishing the equivalence point. (a) The theory of titration curves shows that the equivalence point of a practical titration (i.e., one performed with a titrant of real rather than infinite concentration) does not quite coincide with the value of V_b where $d(pH)/dV_b$ is maximal. Fortunately, the resulting titration error is usually small enough to be negligible. (b) Much more importantly, taking the derivative of the titration curve greatly enhances the effect of experimental noise, often making it necessary to filter the experimental data first, which may introduce far greater titration errors. This sensitivity to noise is aggravated by the fact that experimental fluctuations are typically largest in the region of the equivalence point, where inadequate mixing speed, slow electrode response, and small buffer strength, can all conspire to generate experimental errors. As we will see in section 4.6, it is no longer necessary to consider the derivative to find an equivalence point, but for the sake of completeness we will briefly describe this procedure nonetheless. To this end we will use the Savitzky–Golay method you may already have encountered in section 3.3 (where it was used for smoothing rather than differentiation). This method requires that the data are equidistant in the independent variable. Since our spreadsheet data are based on constant pH increments, we will calculate the values of $d(pH)/d(V_b/V_a)$ as $1/\{d(V_b/V_a)/d(pH)\}$, first with a moving five-point parabola.

Instructions for exercise 4.5

1 Recall the spreadsheet Monoprotic Acid.

2 Add yet another column, this one labeled 'deriv', in which you calculate the quantity $d(pH)/d(V_b/V_a) = 1/\{d(V_b/V_a)/d(pH)\}$. Because the derivative is computed at the midpoint of a five-point polynomial, the first calculated point will yield a result for the third point in the column, so that the column with results will start two rows lower than the column with input data, and likewise will end two rows earlier.

3 The instruction to be entered on the third row should read $= 1/(-2*a - b + d + 2*e)$ where a, b, d, and e refer to the first, second, fourth and fifth cells in the column for $(V_b/V_a)_n$. (There are two other factors that, fortuitously, cancel each other in this case since their product is involved: a normalizing factor of 10, and the data spacing of 0.1 pH units.) Copy the instruction down to the third-from-the-bottom row of this deriv column.

4 Make a graph of $d(pH)/d(V_b/V_a)_n$ versus V_b/V_a. Display V_b/V_a only between 0.99 and 1.01, and leave the vertical scale undefined (so that the spreadsheet can scale it).

5 Vary the amplitude na of the noise, and observe the results. While the value of na will also affect the progress and titration curves, and the Schwartz and Gran plots, you will notice that it has a much more dramatic effect on the first derivative. In fact, the theoretical shape of the first derivative is visible only for $na < 0.001$, whereas the linear (Schwartz and Gran) plots barely show the effects of noise. At $na = 0.01$ the linear plots are noisy but can still be used, especially when combined with a least-squares line, but the derivative fails miserably to indicate the position of the equivalence point. At $na = 0.03$, only the Gran2 plot is still serviceable.

6 The derivative curve can be made less sensitive to noise by using a larger number of points. Try this by using a thirteen-point (instead of a five-point) parabola, with the coefficients $-6a, -5b, -4c, -3d, -2e, -f, h, 2i, 3j, 4k, 5l$ and $6m$, where a through m are the first through 13th values in the row for $(V_b / V_a)_n$. The proper scale factor in this case is no longer 1, but $182 \times 0.1 = 18.2$. Note that a 13-point parabola will calculate the seventh point in the deriv column; you must therefore leave the first and last six points open in the calculation (but still include them in defining the graph). As can be seen in Figs. 4.5-1 through 4.5-3, a relatively small amount of random noise can dominate the plot of the first derivative $d(pH)/dV_b$ vs. V_b or, as shown here, of $d(pH)/d(V_b / V_a)$ vs. V_b / V_a, even when a 13-point moving parabola is used to calculate it.

Figures 4.5-1 through 4.5-3 show thumbnail sketches such as you might embed at the top of your spreadsheet, for three different amounts of added noise. In viewing these figures you should keep in mind that the added noise is somewhat artificial, because it only affects the volume axis, and not the pH axis. (If desired you can of course add noise to the pH data as well.) Nonetheless it indicates that the Gran plot past the equivalence point (i.e., for a strong base) is the most robust. Note that noise of magnitudes such as that shown for $na = 0.03$ is experimentally completely unnecessary and unacceptable, and is used here only to emphasize the point that the various data analysis methods have quite different sensitivities to noise. Not everything that looks good in theory works well in practice, and the spreadsheet is a good tool to find this out.

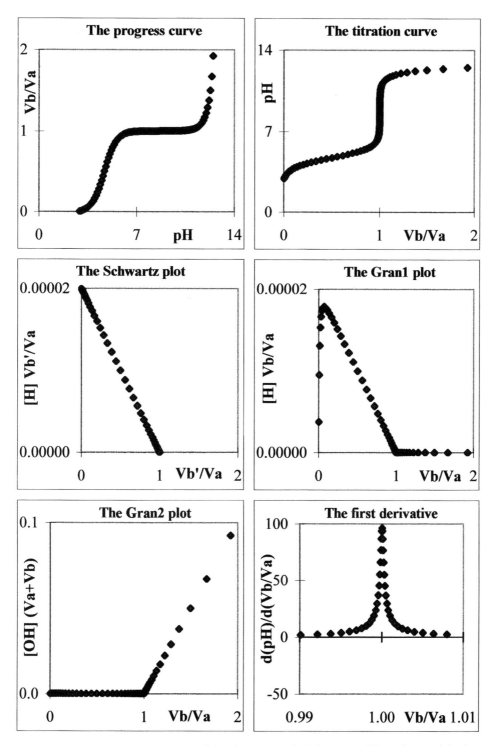

Fig. 4.5-1: The progress and titration curves, the Schwartz and Gran plots, and the first derivative of the titration curve, for the titration of 0.1 M acetic acid ($pK_a = 4.7$) with 0.1 M NaOH, without Gaussian noise added, $na = 0$. The first derivative is computed with a 13-point moving parabola.

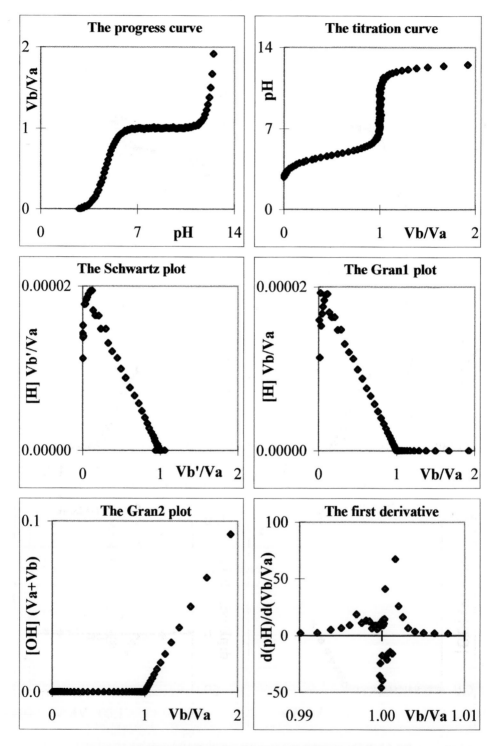

Fig. 4.5-2: The progress and titration curves, the Schwartz and Gran plots, and the first derivative of the titration curve, for the titration of 0.1 M acetic acid ($pK_a = 4.7$) with 0.1 M NaOH, with some Gaussian noise added, $na = 0.005$. The first derivative is computed with a 13-point moving parabola.

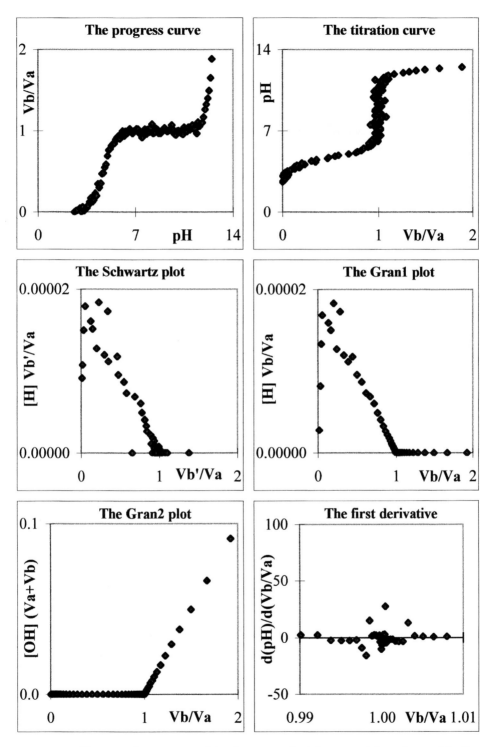

Fig. 4.5-3: The progress and titration curves, the Schwartz and Gran plots, and the first derivative of the titration curve, for the titration of 0.1 M acetic acid ($pK_a = 4.7$) with 0.1 M NaOH, with more Gaussian noise added, $na = 0.03$. The first derivative is computed with a 13-point moving parabola.

4.6 A more general approach to data fitting

The traditional way to finding the equivalence point, by observing when the color of an indicator dye changes, is equivalent to associating that equivalence point with a fixed pH, and monitoring when that pH has been passed. In that case we typically use one drop to distinguish between being 'before' and 'beyond' the equivalence point, and our resolution will be determined by the size of the drop used, which is usually of the order of about 0.05 mL.

The approaches we have discussed so far to determine the precise location of the equivalence point use more than just one point, and are therefore in principle less prone to experimental error. The Schwartz and Gran plots rely on a linearization of the titration curve; unfortunately, for samples that contain more than one monoprotic acid or base, linearization is no longer possible, nor is it (in general) for polyprotic acids and bases. And as for the alternative, we have seen that taking the derivative is easily overwhelmed by experimental noise. Is there no more robust yet general way to determine the equivalence volume with better precision?

Fortunately, there is such a method, which is both simple and generally applicable, even to mixtures of polyprotic acids and bases. It is based on the fact that we have available a closed-form mathematical expression for the progress of the titration. We can simply compare the experimental data with an appropriate theoretical curve *in which the unknown parameters* (the sample concentration, and perhaps also the dissociation constant) *are treated as variables*. By trial and error we can then find values for those variables that will minimize the sum of the squares of the differences between the theoretical and the experimental curve. In other words, we use a least-squares criterion to fit a theoretical curve to the experimental data, using the entire data set. Here we will demonstrate this method for the same system that we have used so far: the titration of a single monoprotic acid with a single, strong monoprotic base.

Despite its 'trial and error' nature, such a method is easily implemented on a spreadsheet. We make two columns, one containing the experimental data, the other the theoretical curve as calculated with assumed parameter values. In a third column we calculate the squares of the residuals (i.e., the differences between the two), and we add all these squares to form the sum of squares, SRR. This sum of the residuals squared, SRR, will be our data-fitting criterion. We now adjust the various assumed parameters that define the theoretical curve, in such a direction that SRR decreases. We keep doing this for the various parameters until SRR has reached a minimum. Presumably, this minimum yields the best-fitting parameter values. Incidentally, the third column is not needed when we use the command =SUMXMY2(*experimental data, theoretical data*).

Instructions for exercise 4.6

1 Recall the spreadsheet Monoprotic Acid, and add a few more columns to it. Alternatively, when the computer gets bogged down (as may happen when it has a relatively small memory and/or low cpu frequency), start with a new sheet. In that case, follow instructions (1a) and (1b); otherwise go directly to (2).

1a Open a new spreadsheet, label and assign spaces for the constants C_a, C_b, K_a, and K_w, and for the noise amplitude N.

1b Enter columns for pH, $[H^+]$, V_b/V_a, Gaussian noise, and $(V_b/V_a)_n$. Calculate V_b/V_a directly from (4.3-6) together with (4.1-10) and (4.3-2), then compute $(V_b/V_a)_n$ from V_b/V_a by adding na times the noise.

2 Now that we have a noisy data set $(V_b/V_a)_n$ that we will consider as our 'experimental' data, make separate labels and spaces for the *variables* C_{aa} and K_{aa} to be used in the theoretical curve. Also find a space for the label and value of the sum SRR.

3 In the next column, calculate a noise-free, theoretical curve, based on equations (4.3-6), (4.1-10), and (4.3-2), and using C_{aa}, C_b, K_{aa}, and K_w. Label the resulting column $(V_b/V_a)_t$, where the subscript t stands for *theoretical*.

4 In a final column, labeled RR for residuals squared, calculate the squares of the differences between the data in corresponding rows in columns $(V_b/V_a)_n$ and $(V_b/V_a)_t$.

5 In the space reserved for the value of SRR, enter the sum of the squares of the differences between the 'experimental' curve $(V_b/V_a)_n$ and the 'theoretical' curve $(V_b/V_a)_t$.

6 Before we are ready to do the 'trial-and-error' adjustment, we must make one modification. This is necessary because our experimental data set is not very realistic, in that it contains non-realizable, negative numbers, reflecting the fact that $(V_b/V_a)_n$ is not really a true experimental data set but has been generated artificially. We exclude such non-realizable numbers by modifying the RR column using an IF statement. Say that the V_b/V_a data are in column C (or wherever V_b/V_a is listed), the $(V_b/V_a)_n$ data in column E, and the $(V_b/V_a)_t$ data in column F. Then a line in, e.g., row 87 of column RR should read = IF(C87 < 0, 0, IF(B87 > Kw/Cb, 0, (E87−F87)^2)). This will add mere zeros to the sum Σ, and therefore contribute nothing to it, whenever the data are outside the physically realizable range. The upper limit specifies that $[OH^-]$ cannot exceed C_b.

7 Finally, in order to see what you are doing, plot *on the spreadsheet* both $(V_b/V_a)_n$ (as data points) and $(V_b/V_a)_t$ (as a line) versus pH.

8 With everything in place, enter some guess values for the variables C_{aa} and K_{aa} (make sure that they differ from C_a and K_a), and observe the plot as well as the value of SRR. Now change one of the variables, see whether SRR increases or decreases, then change it further in the direction of the decreasing SRR value until the change becomes minor. (At this point you can take rather large steps, first perhaps by 0.1 in C_{aa} or by one order of magnitude in K_{aa}.) Then adjust the other variable. You will notice in the plot that the

two variables affect the graph in mutually independent, orthogonal ways: C_{aa} moves the progress curve sideways, whereas K_{aa} changes the height of the step. This makes it easy to see in what direction to adjust the variables.

9 After this initial, crude adjustment, repeat the process with smaller increments, say by 0.01 in C_{aa} or by a factor of 0.1 in K_{aa}. After that, a third round will yield a result to within 0.001 in C_{aa} and a factor of 0.01 in K_{aa}, most likely good enough to stop.

10 Examine the values of C_{aa} and K_{aa} you have found; they should be rather close to their theoretical counterparts, C_a and K_a. When you used a small value for the noise amplitude na, the agreement should be quite good; there will of course be less agreement in the presence of a significant amount of noise.

In practice, doing this least-squares minimization by hand may be interesting once (and may then serve to illustrate the principle of the method), but it pretty soon will become tedious. Fortunately, computer algorithms have been designed to perform this search efficiently, and Excel contains several of them.

11 Click on Tools ⇨ Solver to get the Solver Parameters dialog box. There, Set Target Cell: to where you display the value of SRR, Equal to: Min Value, By Changing Cells: to where you display the values of C_{aa}, K_{aa}, then select Options. This opens the Solver Options dialog box, in which you select Show Iteration Results, then push OK to bring you back to the earlier dialog box, where you now press Solve.

12 By having instructed Solver to show its iteration results, you can see the process as it progresses. Whenever the Show Trial Solution box appears, press Continue to keep going. When the Show Trial Solution box gets in the way of your graph, just pick it up by its blue top edge and place it somewhere where it does not block your sight.

13 Chances are that you will *not* find a satisfactory solution. How come? For the answer, look at what we ask the program to do: to adjust C_{aa} by 0.1, and K_{aa} by less than 0.0001. Since the program adjusts its step size to the largest variable that it must adjust, it will grossly overshoot its K_{aa} target any time, and therefore cannot find a solution!

14 Understanding the cause of the problem is, as usual, most of its solution. The problem should disappear when we adjust pK_{aa} instead of K_{aa}. Indeed, it will, as you can verify by inserting a location for pK_{aa}, referring to that location instead of to K_{aa} in the solver dialog box, and entering the value of $10^\wedge-K_{aa}$ in the space for K_{aa}. When you now try the solver, it will work like a charm! Figure 4.6 illustrates the method. Note that RR reaches a minimum, but does not become zero.

In the above exercise we have on purpose used a rather excessive amount of artificial noise in order to illustrate the method. The example shows that this method deals quite competently with noise, even at levels where the

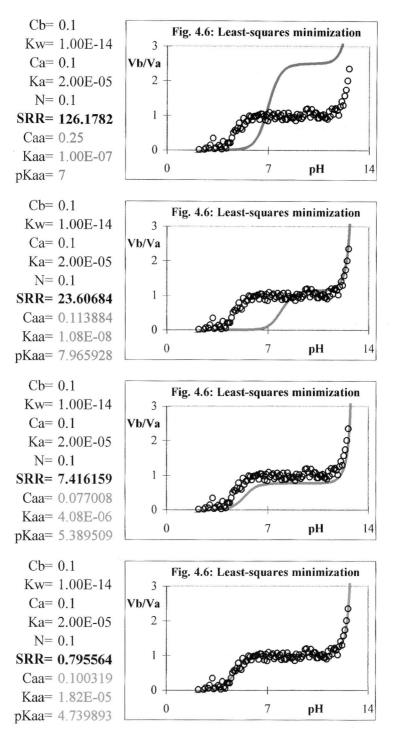

Cb= 0.1
Kw= 1.00E-14
Ca= 0.1
Ka= 2.00E-05
N= 0.1
SRR= 126.1782
Caa= 0.25
Kaa= 1.00E-07
pKaa= 7

Cb= 0.1
Kw= 1.00E-14
Ca= 0.1
Ka= 2.00E-05
N= 0.1
SRR= 23.60684
Caa= 0.113884
Kaa= 1.08E-08
pKaa= 7.965928

Cb= 0.1
Kw= 1.00E-14
Ca= 0.1
Ka= 2.00E-05
N= 0.1
SRR= 7.416159
Caa= 0.077008
Kaa= 4.08E-06
pKaa= 5.389509

Cb= 0.1
Kw= 1.00E-14
Ca= 0.1
Ka= 2.00E-05
N= 0.1
SRR= 0.795564
Caa= 0.100319
Kaa= 1.82E-05
pKaa= 4.739893

Fig. 4.6: Four successive stages in the automatic curve fitting of a very noisy progress curve ($na = 0.1$) for the titration of 0.1 M weak acid ($C_a = 0.1$ M, $K_a = 2 \times 10^{-5}$ M) with 0.1 M NaOH. The initial guess values used in the fitting were $C_a = 0.25$ M, $K_a = 1 \times 10^{-7}$ M. The spreadsheet parameters are shown to the left of the graphs.

derivatives have collapsed completely, and the Schwartz and Gran plots are no longer useful. In practice, a progress curve as noisy as the one shown indicates faulty equipment or some other serious problem, and should never be accepted! Even so, Solver recovered C_a to within 0.3%. When na is smaller, the errors in C_a and K_a are reduced proportionally. For a typical titration of a single monoprotic acid or base, the concentration error in using Solver is usually less than 0.1%, at which point the data-analysis method itself is no longer contributing to the uncertainty in the answer.

The above example used an artificial curve, but the same approach can be applied directly to experimental data. Take a set of titration data, enter them in the spreadsheet. (For convenience of displaying the data graphically on the spreadsheet, enter the pH as the first column, the titrant volume V_b in the next column.) Then calculate columns for $[H^+]$, $(V_b/V_a)_t$, and RR, and let Solver do the rest. There are no special requirements on data spacing or range; the higher the quality of the input data, the better the resulting values for C_{aa} and K_{aa}. The only requirement is that the appropriate theoretical model is used.

An unweighted least-squares analysis such as illustrated here usually suffices for chemical analysis, but for the most precise K_a-values one might want to use weighted least squares as developed by G. Nowogrocki *et al.*, *Anal. Chim. Acta* 122 (1979) 185, G. Kateman *et al.*, *Anal. Chim. Acta* 152 (1983) 61, and H. C. Smit *et al.*, *Anal. Chim. Acta* 153 (1983) 121.

4.7 Buffer action

As we have seen in section 4.3, the pH of a sample being titrated usually changes rapidly near the equivalence point, and much less so before and after that point. In section 4.5 we paid attention to the steep part of the curve, but we can also exploit its shallow parts, where the pH appears to resist change. A quantitative expression of such pH-stabilizing buffer action is the **buffer strength**, which is denoted by the symbol B, and is given by

$$B = [H^+] + C\alpha_0\alpha_1 + [OH^-] \tag{4.7-1}$$

where α_1 and α_0 are the concentration fractions of the monoprotic acid and its conjugate monoprotic base respectively, and C is its total analytical concentration.

A closely related measure is the **buffer index** or **buffer capacity** β, which is the parameter originally introduced by Van Dyke. The two differ only by the factor $\ln(10)$, i.e., the relation between B and β is

$$\beta = B\ln(10) \approx 2.30\,B \tag{4.7-2}$$

Strong acids and strong bases are the most powerful buffers, but their buffer action is restricted to the extreme ends of the pH scale. For a concentrated strong acid, the buffer strength is simply $[H^+]$; for a concentrated strong base, it is $[OH^-]$. At intermediate pH values, weak acids and their conjugate bases must be used to provide buffer action.

The buffer strength for the aqueous solution of a single monoprotic acid and its conjugate base follows from (4.7-1) as

$$B = [H^+] + C\alpha_0\alpha_1 + [OH^-] = [H^+] + \frac{C[H^+]K_a}{([H^+]+K_a)^2} + [OH^-] \qquad (4.7\text{-}3)$$

The product $\alpha_0\alpha_1$ has a maximum at $pH = pK_a$, where $\alpha_0 = \alpha_1 = \frac{1}{2}$, so that the product has the value ¼. For a single monoprotic buffer mixture (i.e., the mixture of a monoprotic acid and its conjugated base), the pH in the region of maximal buffer action, $pH \approx pK_a$, can often be estimated from the Henderson approximation

$$pH \approx pK_a + pC_a' - pC_b' \qquad (4.7\text{-}4)$$

where C_a' and C_b' are the concentrations of acid and conjugate base (e.g., acetic acid and sodium acetate) used to make the buffer mixture, their values being computed as if they were independent, non-interconverting species. The corresponding buffer strength is

$$1/B \approx 1/C_a' + 1/C_b' \qquad (4.7\text{-}5)$$

For a mixture of monoprotic acids or bases, the buffer strength is

$$B = [H^+] + \sum C\alpha_a\alpha_b + [OH^-] \qquad (4.7\text{-}6)$$

where each pair of conjugated acid and base comprises one term in the summation. So-called 'universal' buffer mixtures, such as those associated with the names of Prideau & Ward and Britton & Robinson, are carefully selected to allow the pH to be varied continuously with strong acid or base, while still providing buffer action at each pH.

Instructions for exercise 4.7

1 Recall the spreadsheet Monoprotic Acid, or start a new one with columns for pH and $[H^+]$, and fixed locations for C and K_a.

2 Make a column for B using (4.7-3).

3 Plot the buffer strength as a function of pH. Vary C and K_a and observe their effects. Figure 4.7 illustrates such a curve, and emphasizes the contribution to B of the buffer strength by separately and in color displaying the quantity $B - [H^+] - [OH^-] = C\alpha_0\alpha_1$.

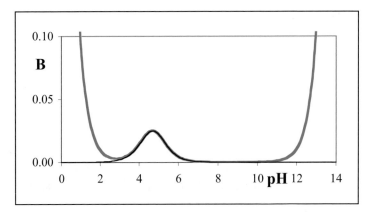

Fig. 4.7: The buffer strength of a single buffer mixture, $C = 0.1$ M, $pK_a = 4.7$. The black curve shows the contribution to B of $C\alpha_0\alpha_1$.

4.8 Diprotic acids and bases, and their salts

So far we have only considered monoprotic acids and bases. Fortunately, the corresponding relations for diprotic and polyprotic acids and bases are quite similar. The relations for diprotic acids and bases will be given here, while those for their polyprotic counterparts (with three or more dissociable protons) can be found in section 4.9.

For a diprotic acid H_2A such as oxalic acid we have

$$H_2A \rightleftharpoons H^+ + HA^- \tag{4.8-1}$$

$$K_{a1} = \frac{[H^+][HA^-]}{[H_2A]} \tag{4.8-2}$$

$$HA^- \rightleftharpoons H^+ + A^{2-} \tag{4.8-3}$$

$$K_{a2} = \frac{[H^+][A^{2-}]}{[HA^-]} \tag{4.8-4}$$

$$\alpha_{H_2A} = \alpha_2 = \frac{[H_2A]}{[H_2A] + [HA^-] + [A^{2-}]} = \frac{[H^+]^2}{[H^+]^2 + [H^+]K_{a1} + K_{a1}K_{a2}} \tag{4.8-5}$$

$$\alpha_{HA^-} = \alpha_1 = \frac{[HA^-]}{[H_2A] + [HA^-] + [A^{2-}]} = \frac{[H^+]K_{a1}}{[H^+]^2 + [H^+]K_{a1} + K_{a1}K_{a2}} \tag{4.8-6}$$

$$\alpha_{A^{2-}} = \alpha_0 = \frac{[A^{2-}]}{[H_2A] + [HA^-] + [A^{2-}]} = \frac{K_{a1}K_{a2}}{[H^+]^2 + [H^+]K_{a1} + K_{a1}K_{a2}} \tag{4.8-7}$$

while we write for a diprotic base such as carbonate (where $B^{2-} = CO_3^{2-}$)

$$H_2B \rightleftharpoons H^+ + HB^- \tag{4.8-8}$$

$$K_{a1} = \frac{[H^+][HB^-]}{[H_2B]} \tag{4.8-9}$$

$$HB^- \rightleftharpoons H^+ + B^{2-} \tag{4.8-10}$$

$$K_{a2} = \frac{[H^+][B^{2-}]}{[HB^-]} \tag{4.8-11}$$

$$\alpha_{H_2B} = \alpha_2 = \frac{[H_2B]}{[H_2B] + [HB^-] + [B^{2-}]} = \frac{[H^+]^2}{[H^+]^2 + [H^+]K_{a1} + K_{a1}K_{a2}} \tag{4.8-12}$$

$$\alpha_{HB^-} = \alpha_1 = \frac{[HB^-]}{[H_2B] + [HB^-] + [B^{2-}]} = \frac{[H^+]K_{a1}}{[H^+]^2 + [H^+]K_{a1} + K_{a1}K_{a2}} \tag{4.8-13}$$

$$\alpha_{B^{2-}} = \alpha_0 = \frac{[B^{2-}]}{[H_2B] + [HB^-] + [B^{2-}]} = \frac{K_{a1}K_{a2}}{[H^+]^2 + [H^+]K_{a1} + K_{a1}K_{a2}} \tag{4.8-14}$$

Likewise, for $Ca(OH)_2$ we can write

$$Ca^{2+} + H_2O \rightleftharpoons H^+ + CaOH^+ \tag{4.8-15}$$

$$K_{a1} = \frac{[H^+][CaOH^+]}{[Ca^{2+}]} \tag{4.8-16}$$

$$CaOH^+ + H_2O \rightleftharpoons H^+ + Ca(OH)_2 \tag{4.8-17}$$

$$K_{a2} = \frac{[H^+][Ca(OH)_2]}{[CaOH^+]} \tag{4.8-18}$$

$$\alpha_{Ca^{2+}} = \alpha_2 = \frac{[Ca^{2+}]}{[Ca^{2+}] + [CaOH^+] + [Ca(OH)_2]} = \frac{[H^+]^2}{[H^+]^2 + [H^+]K_{a1} + K_{a1}K_{a2}}$$
$$\tag{4.8-19}$$

$$\alpha_{CaOH^+} = \alpha_1 = \frac{[CaOH^+]}{[Ca^{2+}] + [CaOH^+] + [Ca(OH)_2]} = \frac{[H^+]K_{a1}}{[H^+]^2 + [H^+]K_{a1} + K_{a1}K_{a2}}$$
$$\tag{4.8-20}$$

$$\alpha_{Ca(OH)_2} = \alpha_0 = \frac{[Ca(OH)_2]}{[Ca^{2+}] + [CaOH^+] + [Ca(OH)_2]} = \frac{K_{a1}K_{a2}}{[H^+]^2 + [H^+]K_{a1} + K_{a1}K_{a2}}$$
$$\tag{4.8-21}$$

One of the advantages of using a formalism that treats acids and bases the same (by considering bases in terms of their conjugate acids) is that it makes it very easy to represent otherwise ambiguous species that can act as either acid or base, such as acid salts (e.g., bicarbonate) and amino acids. For example, for an amino acid such as alanine, which we will represent here in its protonated form as H_2A^+, we have

$$H_2A^+ \rightleftharpoons H^+ + HA \tag{4.8-22}$$

$$K_{a1} = \frac{[H^+][HA]}{[H_2A^+]} \tag{4.8-23}$$

$$HA \rightleftharpoons H^+ + A^- \tag{4.8-24}$$

$$K_{a2} = \frac{[H^+][A^-]}{[HA]} \tag{4.8-25}$$

$$\alpha_{H_2A^+} = \alpha_2 = \frac{[H_2A^+]}{[H_2A^+] + [HA] + [A^-]} = \frac{[H^+]^2}{[H^+]^2 + [H^+]K_{a1} + K_{a1}K_{a2}} \tag{4.8-26}$$

$$\alpha_{HA} = \alpha_1 = \frac{[HA]}{[H_2A^+] + [HA] + [A^-]} = \frac{[H^+]K_{a1}}{[H^+]^2 + [H^+]K_{a1} + K_{a1}K_{a2}} \tag{4.8-27}$$

$$\alpha_{A^-} = \alpha_0 = \frac{[A^-]}{[H_2A^+] + [HA] + [A^-]} = \frac{K_{a1}K_{a2}}{[H^+]^2 + [H^+]K_{a1} + K_{a1}K_{a2}} \tag{4.8-28}$$

Please note that equations (4.8-5) through (4.8-7), (4.8-12) through (4.8-14), (4.8-19) through (4.8-21), and (4.8-26) through (4.8-28) are identical when the concentration fractions are written merely as α_2, α_1, and α_0, where α_2 is the concentration fraction of the fully protonated form (H_2A for a diprotic acid, HA^+ for a diprotic amino acid, H_2B^{2+} for a diprotic base), α_0 that of the fully deprotonated form, while α_1 is the concentration fraction of the intermediate form. For the buffer strength of the solution of a diprotic acid and/or its conjugate bases we then have

$$B = [H^+] + C(\alpha_2\alpha_1 + \alpha_1\alpha_0 + 4\alpha_2\alpha_0) + [OH^-] \tag{4.8-29}$$

which can be compared with (4.7-1) for a monoprotic acid and/or base. When the pK_a's differ by 3 or more, the overlap between the two dissociation processes is small, in which case the term $4\alpha_2\alpha_0$ in (4.8-29) can be neglected, so that the system can be approximated as behaving as two separate monoprotic acids, each of concentration C but with pK_{a1} and pK_{a2} respectively.

For the titration of the fully protonated form of a diprotic acid with a strong monoprotic base we have

$$\frac{V_b}{V_a} = \frac{C_a(\alpha_1 + 2\alpha_0) - \Delta}{C_b + \Delta} \tag{4.8-30}$$

where C_a $(\alpha_1 + 2\alpha_0)$ denotes the number of moles of protons that are removed from the fully diprotonated acid. For the intermediary form (the acid salt, the basic salt, or the neutral or zwitterionic form of an amino acid) we likewise have

Table 4.8-1: *The pK_a-values for some diprotic acids, bases, and aminoacids.*

Acid or base	pK_{a1}	pK_{a2}	Amino acid	pK_{a1}	pK_{a2}
carbonic acid	6.35	10.33	alanine	2.35	9.87
catechol	9.40	12.8	asparagine	2.14	8.72
fumaric acid	3.05	4.49	glutamine	2.17	9.01
hydrogen sulfide	7.02	13.9	glycine	2.35	9.78
8-hydroxyquinoline	4.9	9.81	isoleucine	2.32	9.75
maleic acid	1.91	6.33	leucine	2.33	9.75
malic acid	3.46	5.10	methionine	2.20	9.05
malonic acid	2.85	5.70	phenylalanine	2.20	9.31
oxalic acid	1.25	4.27	proline	1.95	10.64
phthalic acid	2.95	5.41	serine	2.19	9.21
piperazine	5.33	9.73	threonine	2.09	9.10
salicylic acid	2.97	13.74	tryptophan	2.35	9.33
sulfuric acid	<-1	1.99	valine	2.29	9.72

$$\frac{V_b}{V_a} = \frac{C_a(\alpha_0 - \alpha_2) - \Delta}{C_b + \Delta} \qquad (4.8\text{-}31)$$

Similarly, the titration of the fully deprotonated base with a strong monoprotic acid is described by

$$\frac{V_a}{V_b} = \frac{C_b(\alpha_1 + 2\alpha_2) + \Delta}{C_a - \Delta} \qquad (4.8\text{-}32)$$

and the titration of the intermediary, half-deprotonated base (or, depending on your point of view, the half-protonated acid) with a strong acid by

$$\frac{V_a}{V_b} = \frac{C_b(\alpha_2 - \alpha_0) + \Delta}{C_a - \Delta} \qquad (4.8\text{-}33)$$

Instructions for exercise 4.8

1 Open a new spreadsheet, Diprotic acid.

2 A diprotic acid has two dissociation constants. To reflect this, make labels and assign spaces for C_a, C_b, K_w, K_{a1}, and K_{a2}. Either use numerical values for K_{a1} and K_{a2} that correspond to a given diprotic acid (of which several are shown in Table 4.8-1), or use your imagination. In the latter case, just make sure that $K_{a2} < K_{a1}$ (i.e., $pK_{a1} < pK_{a2}$).

3 Make columns for pH, $[H^+]$, $[OH^-]$, denom, α_2, α_1, and α_0, where denom represents the common denominator $[H^+]^2 + [H^+]\,K_{a1} + K_{a1}\,K_{a2}$.

4 Plot the distribution diagram.

5 Change the chart to single-logarithmic, and plot the logarithmic concentration diagram for $C = 1$ M, and/or

6 Change the chart to double-logarithmic, and plot α_2, α_1, and α_0 versus $[H^+]$, which will also produce the logarithmic concentration diagram for $C = 1$ M.

7 Make a column for the buffer strength B, then plot B vs. pH.

8 Make a column for the progress curve, V_b/V_a, for the titration of the diprotonated form with strong acid.

9 Plot the resulting progress curve, V_b/V_a versus pH.

10 Also plot the corresponding titration curve, pH vs. V_b/V_a.

11 Add a column of noise, as well as space for a noise amplitude parameter na. Fill the column with Gaussian noise of zero mean and unit standard deviation.

12 In the next column, $(V_b/V_a)_n$, add na times that noise to V_b/V_a. This will simulate 'experimental' data. Then add the adjustable variables C_{aa}, K_{aa1}, and K_{aa2}, and with these (as well as K_w and C_b) calculate the corresponding theoretical data $(V_b/V_a)_n$.

13 Use Solver to recover close approximations for the values of C_a, K_{a1}, and K_{a2}, by starting from different initial values for C_{aa}, K_{aa1}, and K_{aa2} with, say, $na = 0.01$. Keep in mind that you should vary the pK_a's rather than the K_a-values themselves. Then try different numerical values for C_a, K_{a1}, K_{a2}, and na, and observe what happens.

4.9 Polyprotic acids and bases, and their salts

In order to extend the discussion to polyprotic acids and bases, we first consider a triprotic acid such as orthophosphoric acid,

$$H_3A \rightleftharpoons H^+ + H_2A^- \tag{4.9-1}$$

$$K_{a1} = \frac{[H^+][H_2A^-]}{[H_3A]} \tag{4.9-2}$$

$$H_2A^- \rightleftharpoons H^+ + HA^{2-} \tag{4.9-3}$$

$$K_{a2} = \frac{[H^+][HA^{2-}]}{[H_2A^-]} \tag{4.9-4}$$

$$HA^{2-} \rightleftharpoons H^+ + A^{3-} \tag{4.9-5}$$

$$K_{a3} = \frac{[H^+][A^{3-}]}{[HA^{2-}]} \tag{4.9-6}$$

$$\alpha_{H_3A} = \alpha_3 = \frac{[H_3A]}{[H_3A] + [H_2A^-] + [HA^{2-}] + [A^{3-}]}$$

$$= \frac{[H^+]^3}{[H^+]^3 + [H^+]^2 K_{a1} + [H^+] K_{a1} K_{a2} + K_{a1} K_{a2} K_{a3}} \tag{4.9-7}$$

$$\alpha_{H_2A^-} = \alpha_2 = \frac{[H_2A^-]}{[H_3A] + [H_2A^-] + [HA^{2-}] + [A^{3-}]}$$

$$= \frac{[H^+]^2 K_{a1}}{[H^+]^3 + [H^+]^2 K_{a1} + [H^+] K_{a1} K_{a2} + K_{a1} K_{a2} K_{a3}} \tag{4.9-8}$$

$$\alpha_{HA^{2-}} = \alpha_1 = \frac{[HA^{2-}]}{[H_3A] + [H_2A^-] + [HA^{2-}] + [A^{3-}]}$$

$$= \frac{[H^+] K_{a1} K_{a2}}{[H^+]^3 + [H^+]^2 K_{a1} + [H^+] K_{a1} K_{a2} + K_{a1} K_{a2} K_{a3}} \tag{4.9-9}$$

$$\alpha_{A^{3-}} = \alpha_0 = \frac{[A^{3-}]}{[H_3A] + [H_2A^-] + [HA^{2-}] + [A^{3-}]}$$

$$= \frac{K_{a1} K_{a2} K_{a3}}{[H^+]^3 + [H^+]^2 K_{a1} + [H^+] K_{a1} K_{a2} + K_{a1} K_{a2} K_{a3}} \tag{4.9-10}$$

For a general n-protic acid we have, analogously,

$$\alpha_m = \frac{[H^+]^m K_1 K_2 \cdots K_{n-m}}{[H^+]^n + [H^+]^{n-1} K_1 + [H^+]^{n-2} K_1 K_2 + \cdots + K_1 K_2 \cdots K_n} \tag{4.9-11}$$

where $m = 1, 2, \ldots, n$, and where we have deleted the a in K_a in order to make the notation somewhat more readable.

The buffer strength of a triprotic acid, as a function of its total analytical concentration and of pH, is given by

$$B = [H^+] + C(\alpha_3\alpha_2 + \alpha_2\alpha_1 + \alpha_1\alpha_0 + 4\alpha_3\alpha_1 + 4\alpha_2\alpha_0 + 9\alpha_3\alpha_0) + [OH^-] \tag{4.9-12}$$

where the coefficients 1, 4, and 9 of the alpha products are the squares of the differences between the indices of these alpha's. Indeed, (4.7-5), (4.8-25), and (4.9-12) can be generalized for an arbitrary n-protic buffer to

$$B = [H^+] + C \sum_{i=0}^{n} \sum_{j=i+1}^{n} (j - i)^2 \alpha_i \alpha_j + [OH^-] \tag{4.9-13}$$

Again, when the pK_a's are well-separated, the terms with coefficients higher than 1 can usually be neglected. For a mixture of k different (non-interconvertible) buffers, their individual contributions should be added, i.e.,

$$B = [H^+] + \sum_{k} C_k \sum_{i=0}^{n_k} \sum_{j=i+1}^{n_k} (j - i)^2 \alpha_{ik} \alpha_{jk} + [OH^-] \tag{4.9-14}$$

For the titration of the triprotonated acid with a monoprotic strong base we have

$$\frac{V_b}{V_a} = \frac{C_a(\alpha_2 + 2\alpha_1 + 3\alpha_0) - \Delta}{C_b + \Delta} \tag{4.9-15}$$

or, in general,

$$\frac{V_b}{V_a} = \frac{C_a F_a - \Delta}{C_b + \Delta} \tag{4.9-16}$$

where $F_a = \alpha_2 + 2\alpha_1 + 3\alpha_0$ for the titration of the acid, $F_a = \alpha_1 + 2\alpha_0 - \alpha_3$ for that of the diprotic salt (as in monosodium phosphate), and $F_a = \alpha_0 - 2\alpha_3 - \alpha_2$ for the titration of the monoprotic salt (e.g., disodium phosphate), all the time using a strong monoprotic base as the titrant. Similarly, we have, for the titration of a fully deprotonated triprotic base with a strong monoprotic acid,

$$\frac{V_a}{V_b} = \frac{C_b F_b + \Delta}{C_a - \Delta} \tag{4.9-17}$$

with $F_b = \alpha_1 + 2\alpha_2 + 3\alpha_3$. For the equivalent titration of the monoprotic salt we have, likewise, $F_b = \alpha_2 + 2\alpha_3 - \alpha_0$, and for the titration of the diprotic salt with, say, HCl, $F_b = \alpha_3 - \alpha_1 - 2\alpha_0$. For tetraprotic, pentaprotic, hexaprotic etc., acids, (4.9-16) remains valid but the expression for F_a must be extended; similarly, (4.9-17) applies in general to all bases titrated with a single, strong monoprotic acid although the definition of F_b must reflect the particular base.

The above expressions can be generalized further to encompass an arbitrary mixture of acids, titrated with any mixture of bases, or vice versa, in which case they take the form

$$\frac{V_b}{V_a} = \frac{\sum C_a F_a - \Delta}{\sum C_b F_b + \Delta} \tag{4.9-18}$$

and

$$\frac{V_a}{V_b} = \frac{\sum C_b F_b + \Delta}{\sum C_a F_a - \Delta} \tag{4.9-19}$$

respectively. These, then, are general master equations for acid–base titrations. Their availability makes it possible to use the general data fitting method described in section 4.6 to analyze any acid–base titration.

Instructions for exercise 4.9

1 Either open a new spreadsheet, Triprotic acid, or extend the spreadsheet you made in Section 4.8.

2 The new spreadsheet should store C_a, C_b, K_w, and three K_a-values. Again, select either literature values for a triprotic acid, or use made-up numbers, as long as $K_{a1} > K_{a2} > K_{a3}$.

3 It should have columns for pH, $[H^+]$, $[OH^-]$, denom, and four alphas. Here, denom should be the common denominator of (4.9-7) through (4.9-10).

4 Plot the resulting distribution diagram.

5 Plot the corresponding logarithmic concentration diagram for $C = 1$ M by using a single-logarithmic scale. Note how the slopes of the various acid species vary between -3 and $+3$, and change every time the lines pass a pH value equal to a pK_a.

6 Make a column for the buffer strength B, and plot this quantity as a function of pH.

7 Make a column for the progress V_b/V_a of the titration, and plot the resulting progress and titration curves.

8 Add Gaussian noise, and make columns for $(V_b/V_a)_n$ and $(V_b/V_a)_t$ respectively, where the latter column requires a set of guess values for the acid concentration and its pK's.

9 See under what conditions use of Solver can recover a reasonably close value of the acid concentration, and of its pK's.

4.10 Activity corrections

Our treatment of acid–base equilibria so far has been based on the mass action law, i.e., on the constancy of the equilibrium constants. Comparison with experiment shows that this relatively simple model is by and large correct, just as it would be essentially correct to say that the earth rotates around the sun according to Kepler's laws. If one looks much closer, one will find that it is not quite so, but that the influence of the moon must be taken into account as a small correction if a more precise description is required. In fact, there is a hierarchy of corrections here, starting with the influence of the moon, then that of the planets, and eventually that of all other heavenly bodies. Although it might appear to be a hopeless task to include an almost endless number of stars and galaxies, in practice the list of effects we need to include is restricted by the limitations on the experimental precision of our measurements, and a simple hierarchy of corrections suffices for all practical purposes. A similar situation applies to acid–base equilibria.

The mass action law formalism, through its equilibrium constants, takes into account the interactions of the solvent with the various acids, bases, and salts; these certainly are the dominant effects, comparable to Kepler's law in the above analogy. However, the formalism of the mass action law does not explicitly consider the *mutual* interaction of the solute particles, nor the effect of these solutes on the concentration of the solvent. **Activity coefficients** f have therefore been introduced in order to incorporate such secondary effects; they are individual correction factors that multiply concentrations, somewhat analogous to the individual weights w introduced in least squares methods (see section 3.4).

The hierarchy of corrections starts with long-range effects, which are noticeable even in rather dilute solutions, where the average distances between the interacting particles are relatively large. The interactions are then predominantly coulombic, resulting from the mutual attraction of ions with charges of opposite sign, and the corresponding repulsion of ions with charges of the same sign. Interestingly, although an ionic solution is electroneutral, the attractive and repulsive forces do not quite compensate each other, but result in a *net attraction*, just as they do in an ionic crystal. Fortunately, this effect depends only on the charges involved rather than on the chemical nature of the solutes, and therefore can be described in rather general terms, as is done in the Debye–Hückel theory.

The dominant parameter in the Debye–Hückel theory is the **ionic strength** I, defined as

$$I = \frac{1}{2}\sum_i z_i^2 c_i \tag{4.10-1}$$

where z_i is the ionic valency, and c_i the ionic concentration. This definition is such that, for the simple case of the solution of a single, strong 1,1-electrolyte, the ionic strength is equal to the salt concentration (just as, for a single, strong, concentrated monoprotic acid, the buffer strength is equal to its concentration). To a first approximation, the Debye–Hückel result is

$$\log f_i = \frac{-0.5\, z_i^2 \sqrt{I}}{1 + \sqrt{I}} \tag{4.10-2}$$

which describes the deviations from 'ideal' behavior (as described by the mass-action law) in dilute ($I \leq 1$ mM) solutions quite well.

In more concentrated solutions, additional mutual interactions must be considered, which can only be described in terms of ion-specific parameters. We will not do that here, but instead use an expression that, again, does not require any species-specific parameters, yet tends to yield a reasonably good description for the *average* behavior of more concentrated solutions (even though it may not represent any *particular* solution very well). This is the so-called Davies expression,

$$\log f_i = -0.5\, z_i^2 \left(\frac{\sqrt{I}}{1 + \sqrt{I}} - 0.3\, I \right) \tag{4.10-3}$$

which, again, is restricted to ions, because $\log f_i = 0$ hence $f_i = 1$ when $z_i = 0$, i.e., for neutral species.

It is convenient to separate the ionic valency z_i from the remainder of (4.10-3), which we do by rewriting (4.10-3) as

$$\log f = -0.5 \left(\frac{\sqrt{I}}{1 + \sqrt{I}} - 0.3\, I \right) \tag{4.10-4}$$

or

$$f = 10^{-0.5\{[\sqrt{I}/(1+\sqrt{I})]-0.3\,I\}}$$ (4.10-5)

with

$$\log f_i = z_i^2 \log f = \log f^{z_i^2}$$ (4.10-6)

After these formal preliminaries we will now describe how activity corrections should be applied. In order to do so, we must distinguish between (a) the role played by the activity coefficient in equilibrium calculations, and (b) the particular measurement technique used. Below we will consider these various aspects in turn.

(a) Activity coefficients must be included in the definitions of the equilibrium constants, such as K_a and K_w. However, they should *not* be incorporated in the mass and charge balance equations, or in the related proton balance, since these are purely bookkeeping devices in terms of particle densities, i.e., concentrations. Consequently, the influence of activity corrections on distribution and logarithmic concentration diagrams is relatively simple, as they only affect the pK-values used.

(b) Then there is the effect of the experimental method. Electrochemical methods, such as using a pH electrode, yield values that are approximately equal to the ionic activity (i.e., the product of concentration and activity coefficient) of the sensed species rather than its concentration, whereas most other analytical methods (such as spectrometry) directly respond to concentrations. Therefore, when dealing with electrochemical measurements, we must make an additional activity correction for the measured quantity. Since the pH can be measured either electrometrically or spectrometrically, we will leave the pH scales in these diagrams in terms of concentrations, and only apply activity corrections when we specifically deal with *electrometric* pH measurements.

Finally we note that, in the Davies approximation, the value of $\log f$ depends only on the ionic strength of the solution, and that f_i is always 1 for neutral species. The mathematical analysis of acid–base measurements can therefore be simplified by performing them **at constant ionic strength**, which can be achieved for all practical purposes by adding a sufficiently large excess of non-participating, so-called **inert electrolyte**, such as $LiClO_4$. In that case, all activity coefficients can be considered to be constant during the experiment, which simplifies the data analysis.

We are now ready to put the various pieces together. First the general correction of the equilibrium constants. Here we use (4.10-5) and (4.10-6) to find the corrected, *t*hermodynamic equilibrium constants K^t as illustrated below for a few examples:

$$HA \rightleftharpoons H^+ + A^- \qquad K_a^{\,t} = \frac{[H^+]f_+[A^-]f_-}{[HA]f_0} = \frac{f_+f_-}{f_0}\,K_a = f^2\,K_a$$ (4.10-7)

where $f_+ = f_- = f$, since $(+1)^2 = (-1)^2 = 1$, and $f_0 = 1$;

$$HB \rightleftharpoons H^+ + B \qquad K_a^t = \frac{[H^+]f_+[B]f_0}{[HB^+]f_+} = \frac{f_+f_0}{f_+}K_a = K_a \qquad (4.10\text{-}8)$$

$$H_2O \rightleftharpoons H^+ + OH^- \qquad K_w^t = [H^+]f_+[OH^-]f_- = f_+f_-K_w = f^2K_w \qquad (4.10\text{-}9)$$

$$H_3A \rightleftharpoons H^+ + H_2A^- \qquad K_{a1}^t = \frac{[H^+]f_+[H_2A^-]f_-}{[H_3A]f_0} = \frac{f_+f_-}{f_0}K_{a1} = f^2K_{a1} \qquad (4.10\text{-}10)$$

$$H_2A^- \rightleftharpoons H^+ + HA^{2-} \qquad K_{a2}^t = \frac{[H^+]f_+[HA^{2-}]f_{2-}}{[H_2A^-]f_-} = \frac{f_+f_{2-}}{f_-}K_{a2} = f^4K_{a2} \qquad (4.10\text{-}11)$$

where $f_{2-} = f^4$;

$$HA^{2-} \rightleftharpoons H^+ + A^{3-} \qquad K_{a3}^t = \frac{[H^+]f_+[A^{3-}]f_{3-}}{[HA^{2-}]f_{2-}} = \frac{f_+f_{3-}}{f_{2-}}K_{a3} = f^6K_{a3} \qquad (4.10\text{-}12)$$

where $f_{2-} = f^4$ and $f_{3-} = f^9$ so that $ff_{3-} / f_{2-} = ff^9 / f^4 = f^6$.

Most tables of equilibrium constants list values for K_a^t, as obtained by extrapolation towards $I \rightarrow 0$, but the spreadsheet calculation requires the uncorrected K_a's. Below we will indicate how to make the activity correction.

Instructions for exercise 4.10

1 Start a new spreadsheet, Activity correction.

2 Make or copy the spreadsheet for the titration of a triprotic acid, with storage spaces for C_a, C_b, K_w, and three K_a-values, and with columns for pH, $[H^+]$, denom, and V_b/V_a. (This spreadsheet can also be used for a monoprotic or diprotic acid, by setting the unused K_a's to values smaller than 10^{-24}, i.e., the pK_a's to values larger than 24.)

3 Select literature values for orthophosphoric acid for the various K_a's.

4 First let us assume that the ionic strength is constant at, say, 0.1 M. Note that this can only be realized when C_a is much smaller than 0.1 M, so that the ionic strength can be determined by an inert electrolyte in excess.

5 Make columns labeled I and f. In the first enter a constant, say, 1. In the second column calculate the value of f according to (4.10-5).

6 Make new columns for $(denom)_t$ and $(V_b/V_a)_t$ (where t now stands for thermodynamic) which are similar to the earlier ones except that you should use K_{a1}/f^2 instead of K_{a1}, K_{a2}/f^4 instead of K_{a2}, and K_{a3}/f^6 instead of K_{a3}.

7 Plot both the uncorrected and the corrected progress curves, V_b/V_a and $(V_b/V_a)_t$, as a function of pH.

8 In order to plot the corresponding titration curves, copy (using Paste Special ⇨ Values) the column for $(V_b/V_a)_t$ below that for V_b/V_a, and make a second, double-length column for pH to the right of the existing data in the spreadsheet to facilitate plotting the pH versus (V_b/V_a) and $(V_b/V_a)_t$. Plot the titration curves.

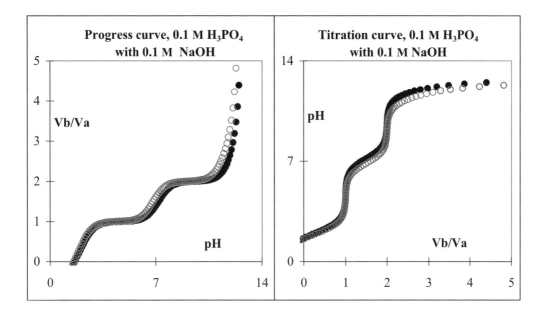

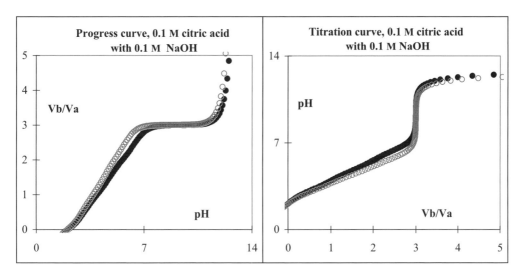

Fig. 4.10: Progress and titration curves for the titration with 0.1 M NaOH of 0.1 M orthophosphoric acid (top) and citric acid (bottom) respectively, calculated with (colored open circles) or without (black solid circles) taking into account activity corrections according to the Davies equation. Electrochemical pH detection is assumed (i.e., with a glass electrode), as well as the absence of other salts, so that the ionic strength varies during the experiment. Equilibrium constants used: for orthophosphoric acid $pK_{a1}{}^t = 2.15$, $pK_{a2}{}^t = 7.20$, $pK_{a3}{}^t = 12.15$; for citric acid $pK_{a1}{}^t = 3.13$, $pK_{a2}{}^t = 4.76$, $pK_{a3}{}^t = 6.40$.

9 We must apply an additional correction when the pH is determined electrochemically, in which case the measured pH is approximately $-\log f[H^+]$. In that case the second column for pH should be converted accordingly.

> When the ionic strength is not kept constant, its value changes during the titration, and must be calculated. We first compute I from the uncorrected data. In principle, the results must thereafter be recalculated based on the corrected parameters. Fortunately, since we deal here with a rather minor correction, such an iterative approach is seldom needed, and a single pass almost always suffices.

10 Insert a column computing the ionic strength I to the left of the column for f. For the calculation of the ionic strength of, say, the titration of C_a M H_3PO_4 with C_b M NaOH, use

$$I = \tfrac{1}{2}\{[H^+] + [Na^+] + [OH^-] + [H_2PO_4^-] + 4[HPO_4^{2-}] + 9[PO_4^{3-}]\}$$
$$= \tfrac{1}{2}\{[H^+] + C_b V_b / (V_a + V_b) + K_w / [H^+]$$
$$+ ([H^+]^2 K_{a1} + 4[H^+] K_{a1} K_{a2} + 9 K_{a1} K_{a2} K_{a3}) C_a V_a / (\text{denom} \times V_a + V_b))\}$$

11 Use these values for I to calculate f in the next column.

12 Plot the corresponding progress and titration curves, with and without activity correction, for orthophosphoric acid.

13 Do the same for citric acid. All you have to do here is to change the three values for K_{a1} through K_{a3} to those for citric acid, and plot new graphs.

> Figure 4.10 (on the previous page) illustrates the resulting progress and titration curves for orthophosphoric acid and citric acid respectively, showing both the uncorrected (black) and corrected curves (color) for the case in which the ionic strength is not controlled, and the pH monitor is an electrode. These are examples to showcase the effect: with triprotic acids the activity corrections are rather large since trivalent ions are involved. Even so, the effect is still relatively minor. This is why activity corrections are seldom made. Note that the values of the *equivalence volume* are not affected at all. This is true in general, since the definition of V_{eq} in (4.4-5) contains no factors subject to activity corrections. The same applies to polyprotic acids and bases, and to mixtures.
>
> In the cases illustrated, the ionic strength varied during the titration. When the ionic strength is kept constant, the only changes in the curves are a constant shift in the pH scale (assuming electrochemical pH detection) plus separate shifts for the various pK_a's. This is the reason why it is often wise to leave the pK_a-values to be determined by the data-fitting algorithm, even though they could have been preset with literature values.

4.11

A practical example

At the end of this chapter devoted entirely to acid–base equilibria it is appropriate to show a practical example. Titrations are usually restricted to relatively concentrated solutions, and are therefore unsuitable for, say, trace analysis. However, the precision, ease of use, and low cost of titrations can make them the method of choice when the concentrations to be analyzed are sufficiently high, and the measured acid–base properties can be related to valuable information. The example to be used below comes from Finland, where silage is routinely analyzed this way, with some 35 000 titrations performed annually at the Valio Finnish Cooperative Dairies Association in Helsinki. Careful analysis of an acid–base titration curve, inexpensively obtained with automated equipment in a matter of minutes, can provide quantitative information regarding the amounts of total amino acids, of lactic and acetic acid, of total soluble nitrogen, and of reducible sugars, all from a single titration! In northern countries, the dairy industry heavily relies on such analyses for quality control, since poor silage preservation can lead to flavor defects in both milk and cheese during the period in which the cattle are fed indoors.

Since this example uses a rather large data set, we will import it rather than copy it into the spreadsheet by hand, which would be an error-prone waste of time. The data can be downloaded from the website, http://uk.cambridge.org/chemistry/resources/delevie After downloading, store them as a data file on your hard disk or on a diskette. These data were kindly provided by Dr. M. Heikonen of Valio Ltd, and pertain to the analysis of a 5.0 mL sample of silage press-juice diluted with about 10 mL of water to a total sample volume of 15.0 mL, titrated with 1 M NaOH at a rate of about 3 mL/min; the total titration took about two minutes. The first column represents the pH, the second the titrant volume, in microliters. Adjacent data in the two columns of each line (representing pH and V_b respectively) are separated by commas, i.e., in computer jargon, they are **comma delimited**, a method of data storage that avoids problems due to deleted zeros and other variations in number of digits per data point.

The following introduction to the spreadsheet exercise is a simplification of the discussion in T. Moisio & M. Heikonen, *Animal Feed Sci. Technol.* 22 (1989) 341. The titration curve of silage press-juice is shown in Fig. 4.11-1. It shows three distinct regions: between pH 2 and pH 5 the carboxylic acid groups of amino acids are deprotonated, as well as weak acids such as lactic and acetic acid, so that the pH increases only gradually. There is much less buffer strength between pH 5 and 8, where the pH rises more rapidly. The pH range from 8.5 to 10.5 involves the neutralization of amines and ammonium, and can be taken as a measure of the available nitrogen. At pH values above 11, reducing sugars are titrated, and the buffering action of the titrant becomes important. All in all, this titration curve lacks prominent features,

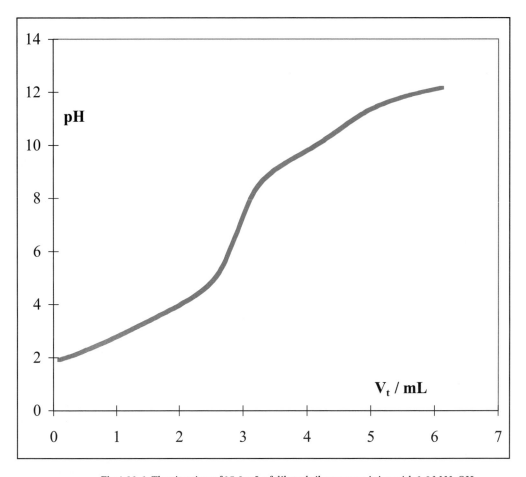

Fig 4.11-1: The titration of 15.0 mL of diluted silage press-juice with 1.0 M NaOH.

and therefore does not appear very promising for quantitative analysis. On the other hand, it contains information on three crucial quality criteria, viz. the concentrations of acids, protein degradation products, and sugars, and can be obtained quickly, inexpensively, and (equally important) with high precision.

It is possible to refine the above analysis, because the (literature values of the) pK_a's of amino acids mostly lie in the range from 1.9 to 2.4, while those of lactic and acetic acid are 3.86 and 4.76 respectively. These numbers are sufficiently far apart to allow their separate numerical analysis. Because of activity effects on both the equilibrium constants and the pH measurements, the precise values of the pK_a's depend on the ionic strength of the sample, and it is therefore best to treat the pK_a's as adjustable parameters, that are allowed to vary within rather narrow pH ranges. Here, then, is the analysis protocol we will follow initially. We will consider five separate ranges, one each for the amino acids (pH 1.9 to 2.7), lactic acid (3.4 to 4.0) and acetic acid (4.4 to 5.0), plus one for total nitrogen (8.8 to 10.8) and one for reducing sugars (11.3 to 11.9). We will confine the pK_a-values to their respective ranges. The equation to be fitted follows directly from (4.3-9) as

$$V_b = V_a \times \frac{C_{a1}K_{a1}/([H^+]+K_{a1}) + C_{a2}K_{a2}/([H^+]+K_{a2}) + \cdots - [H^+] + K_w/[H^+]}{C_b + [H^+] - K_w/[H^+]}$$

(4.11-1)

where V_a, C_b, and K_w are constants.

Instructions for exercise 4.11

1 Open a new spreadsheet.

2 Enter labels for the adjustable parameters con1, con2, con3, con4, con5, Ka1, Ka2, Ka3, Ka4, and Ka5, and the corresponding values of pKa1, pKa2, pKa3, pKa4, and pKa5. Anticipating our repeated use of Solver it is convenient to group the concentrations, K_a's, and pK_a's together, e.g., by putting the labels and values of each category in separate columns. For example, put all concentration labels in column A, all corresponding data in column B, all labels Ka in column C, their numerical values in column D, all labels pKa in column E, and in column F their values. Compute the K_a's from their pK_a-values: $K_{a1} = 10^{-pKa1}$ etc. Also make a space for the sum of the squares of the residuals, SRR.

3 Name five sample concentrations, con1 through con5, and assign them identical initial guess values, somewhere between 0.05 and 0.1 M. Similarly, name the K_a's, and assign the corresponding pK_a's values 2.3, 3.7, 4.7, 9.8, and 11.6 respectively.

4 In row 10 enter the following column headings: pH, Vb(exp), [H$^+$], Vb(calc), R, and RR.

5 Place the mouse pointer in cell A12 to identify where the data to be imported should be placed.

6 If you have stored the data file on a diskette, insert it now into its drive.

7 Select File ⇨ Open. In the resulting Open dialog box, push the triangular button to the right of the top window (labeled Look in:) and select the (floppy or 'hard') drive containing the data file.

8 Select the data file SilagepH and follow the instructions of the Text Import Wizard. The data are comma-delimited, therefore use Delimited, and as delimiter specify a Comma. The data should appear in your spreadsheet after you use the Finish key.

9 Now that you have imported the data, back to their analysis.

10 First, in column C calculate [H$^+$] from the pH in column A.

11 In column D calculate V_b using (4.11-1), i.e. on row 14 as = Va*(con1*Ka1/ (B14 + Ka1) + ⋯ + con5*Ka5/(B14 + Ka5)−B14 + Kw/B14)/(1 + B14−Kw/B14), since C_b = 1.0 M.

12 In column E compute the residuals, i.e., the differences between the values of $V_{b,exp}$ and $V_{b,calc}$ and, in column F, the squares of these residuals.

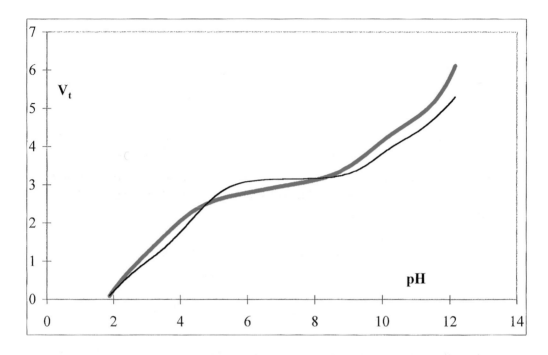

con1= 0.07   Ka1= 0.005012   pKa1= 2.3
con2= 0.07   Ka2= 0.000126   pKa2= 3.9
con3= 0.07   Ka3= 1.58E-05   pKa3= 4.8
con4= 0.07   Ka4= 1.58E-10   pKa4= 9.8
con5= 0.07   Ka5= 2.51E-12   pKa5= 11.6
    Va= 15   Kw = 1.00E-14   pKw = 14.0
                    SRR= 15.2365

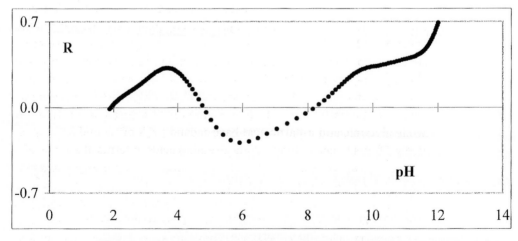

Fig. 4.11-2: The progress curve, and the residuals, before data fitting. Parameters to be varied are shown in color. The initial values of all concentrations have been set to 0.07 M.

13 Calculate the sum of those squares of the residuals, i.e., SRR.

14 On the spreadsheet, plot $V_{b,exp}$ and $V_{b,calc}$ versus pH. You may want to use a heavy black line for $V_{b,exp}$, and a thinner, colored line for $V_{b,calc}$, in which case you will see how the latter will nestle in the experimental curve once you achieve a good fit.

15 In a separate graph, also plot the residuals R versus the pH. At this point your charts may look like Fig. 4.11-2, except that we have incorporated copies of some of the top lines of the spreadsheet columns to show the various parameters involved.

16 Invoke Solver with Tools ⇨ Solver, and in the Solver Parameters dialog box instruct the computer to Set Target Cell to the address of SRR, Equal to Min, By Changing Cells B1:B5 (if that is where you put the five C_a's).

17 Push the Add… button, which will invoke the Add Constraint dialog box. This has three windows; in the first window type A2 (or whatever is the address of Ca1), in the middle window select the > = sign, and in the right-most window type 0.

18 Depress the Add button, and repeat for C_{a2}, etc. until all five concentrations have been constrained to positive values. Then depress the OK button, which will get you back to the Solver Parameters box.

19 Now you are ready: depress Solve to start the non-linear least-squares fitting process. When Solver is done, accept its results. Figure 4.11-3 illustrates them.

20 The fit may not please you, even though SRR will be about an order of magnitude smaller. The fit is especially poor in the middle range of pH values, where the program has little flexibility, since there are no pK_a-values between 4.7 and 9.8.

21 It is possible to get a somewhat better fit by letting the pK_a-values roam within narrow ranges, but this will not cure the poor fit at neutral pH, and the improvements in SRR are marginal.

At this point, then, we must make a choice: either we stick with a simplified model of silage, in terms of just five components, or we jettison the specific model and see whether we can obtain a much improved fit by incorporating more pK_a's. From the chemical analysis we already know that the silage press-juice contains (in order of decreasing concentration) leucine, alanine, glutamic acid, γ-aminobutyric acid, valine, glycine, lysine, proline, serine, and aspartic acid. Of these, γ-aminobutyric acid has a pK_a of 4.1, while glutamic and aspartic acid have second pK_a's of 4.4 and 3.9 respectively. Efforts to characterize all these amino acids in terms of a single pK_a will only be approximate at best. Similarly, the dominant protein degradation products, again in order of decreasing importance, are tyramine, cadaverine, putrescine, histamine, tryptamine, and phenylethylamine, with pK_a values between 9.8 and 10.7, while the pK_a of ammonia is 9.2. It is also useful to look at the residuals in Fig. 4.11-3. These show trends rather than random noise, and therefore suggest that it may be worthwhile to add a few more terms in the expression for V_b.

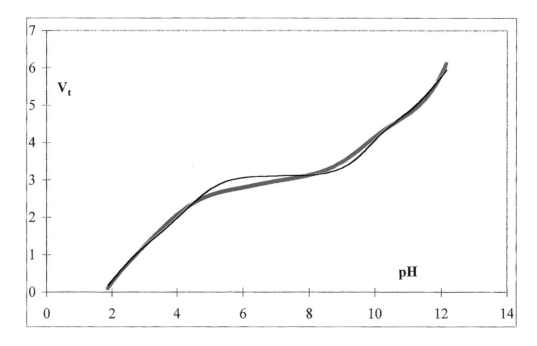

con1= 0.08926	Ka1= 0.005012	pKa1= 2.3
con2= 0.06681	Ka2= 0.000126	pKa2= 3.9
con3= 0.05126	Ka3= 1.58E-05	pKa3= 4.8
con4= 0.10172	Ka4= 1.58E-10	pKa4= 9.8
con5= 0.08797	Ka5= 2.51E-12	pKa5= 11.6
Va= 15	Kw = 1.00E-14	pKw = 14.0
	SRR= 1.249162	

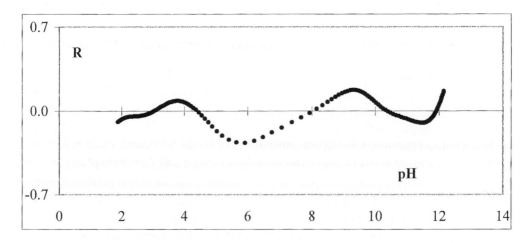

Fig. 4.11-3: The progress curve, and the residuals, after the data fit. The fitted concentration values C_a are shown in color.

Table 4.11-1: *The concentrations C_a (in mM), the values of pK_a, and the logarithm of the value of SRR, obtained for the various iterations on the data.*

Iteration	#1	#2	#3	#4		#1	#2	#3	#4
$C_{a1} =$	89	97	89	75	$pK_{a1} =$	2.3*	2.45	2.40	2.05
$C_{a2} =$	67	83	76	72	$pK_{a2} =$	3.9*	4.01	3.78	3.34
$C_{a3} =$	51		20	35	$pK_{a3} =$	4.8*		4.92	4.34
$C_{a4} =$		30	18	19	$pK_{a4} =$		6.87	6.64	6.26
$C_{a5} =$			17	14	$pK_{a5} =$			8.42	7.91
$C_{a6} =$	102	92	72	66	$pK_{a6} =$	9.8*	9.59	9.58	9.27
$C_{a7} =$			30	33	$pK_{a7} =$			10.77	10.20
$C_{a8} =$	88	129	153	79	$pK_{a8} =$	11.6*	11.90	12.26	11.68
					$pK_w =$	14.0*	14.0*	14.0*	13.49
					$f =$	1*	1*	1*	0.68
log SRR =	0.1	− 1.2	− 2.3	− 3.5			* : values fixed		

In view of the complex nature of the sample, and the apparent data quality, we will now explore what can be gained by adding several adjustable C_a-K_a combinations to our analysis, letting their pK_a's float within non-overlapping ranges. We will also account for activity effects by introducing an adjustable activity coefficient f, by modifying the instructions in column C to represent $[H^+]$ as $10^{-pH/f}$, and by letting pK_w float as well. Below we will see how far this approach will carry us.

22 In the spreadsheet, add the concentrations Ca6 through Ca8, and the labels and values for the corresponding Ka's and pKa's.

23 Introduce another adjustable constant, f, and modify the calculation of $[H^+]$ in column C, and that of the various pK's, as described in section 4.10.

24 Extend the instruction for Vb in column D to incorporate the three additional concentrations and equilibrium constants.

25 Return to Solver, extend the range of concentrations by changing the right-most number in the By Changing Cells from 5 to 8, then add D5:D8 for the pKa's, and finally, again separated by a comma, add the addresses of pKw and f. Quite a laundry list!

26 In the Solver Parameters dialog box, press Add... to get the Add Constraint box, and use it to set both lower and upper limits on the values for the pKa's. Constrain pKa1 to $>=$ 1.9 and, with a separate instruction, $<= 2.7$. Similarly constrain the other pKa's to their proper ranges: $3.2 \leq pKa2 \leq 4.0$, $4.2 \leq pKa3 \leq 5.0$, $6.0 \leq pKa4 \leq 7.0$, $7.8 \leq pKa5 \leq 8.6$, $9.0 \leq pKa6 \leq 9.8$, $10.0 \leq pKa7 \leq 11.0$, $11.2 \leq pKa8 \leq 13.0$, and $13.0 \leq pKw \leq 14.6$. Then go back to the Solver Parameters, and restart the curve fitting process using the Solve button.

The entire optimization uses a single criterion, minimizing the sum of squares of the residuals, SRR. As is readily seen from Table 4.11-1, SRR decreases by about an order of magnitude every time we add more adjustability to the curve-fitting procedure. This is so because we deal here with adjusting to the precise shape of the curve, not with noise reduction. Note that the values obtained for the various concentrations change from one try to another, indicating that these are fitting parameters rather than true chemical concentrations.

As illustrated in Figs. 4.11-4 through 4.11-6, one can indeed obtain a very close fit to the experimental data. While we have now abandoned any pretense of a simple one-to-one correlation with specific chemicals in the sample, we stand to gain substantially in usefulness of the analysis. In order to do so we must keep in mind the real purpose of the analysis, which is to be able to screen out spoiled silage, and perhaps even to suggest adjustments in order to prevent future spoilage. For that we do not need a precise correlation with specific chemicals, rather illusory anyway in such a complex sample. Instead, we need a correlation with silage condition as judged by other methods.

Comparison of such results with those of chemical analyses of a selected number of silage samples makes it possible to express the concentrations of amino acids, lactic acid, acetic acid, and reducing sugars, as well as the total nitrogen content, in terms of linear combinations of the eight computed C_a's. This, in turn, can be used to find a correlation with the quality of the silage. Such a correlation was indeed established: good silage should have a moderate concentration of lactic acid, and a much smaller concentration of acetic acid. Moreover, it should have little protein degradation, i.e., most soluble nitrogen should be in the form of amino acids. While the numerical values obtained for these concentrations from the titration curve are imprecise and somewhat interdependent, precision is not what is ultimately important here: good quality control typically deals with *ranges* of acceptable or unacceptable values rather than with single numbers. On the basis of such ranges, simple criteria of silage quality were formulated. The titration is readily automated, and can provide virtually instantaneous analyses of the samples in terms of these general criteria, which can then be used to advise farmers.

In the Moisio–Heikonen paper another, more elaborate but also more time-consuming data-analysis method was used – as detailed in *Fresenius J. Anal. Chem.* 354 (1996) 271 – but that is immaterial here. It is with fast and inexpensive methods such as these that enough samples can be analyzed to maintain a consistently high quality of the dairy products during the winter months.

You, reader, may well wonder why so much space is devoted to this particular example. The reason is simple: the availability of computers has changed the practice (though perhaps not yet the teaching) of analytical

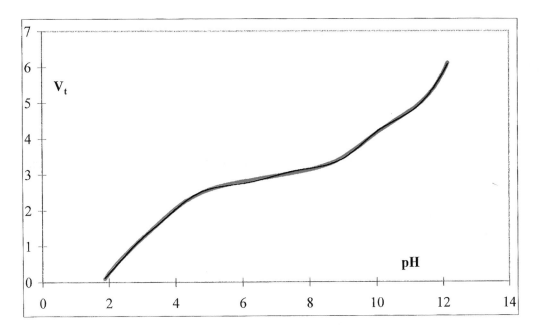

conl= 0.09743	Ka1= 0.003544	pKa1= 2.450
con2= 0.08313	Ka2= 9.7E-05	pKa2= 4.013
con3= 0.03014	Ka3= 1.35E-07	pKa3= 6.869
con4= 0.09241	Ka4= 2.55E-10	pKa4= 9.593
con5= 0.12946	Ka5= 1.26E-12	pKa5= 11.898
Va= 15	Kw = 1E-14	pKw = 14.0
	SRR= 0.052729	

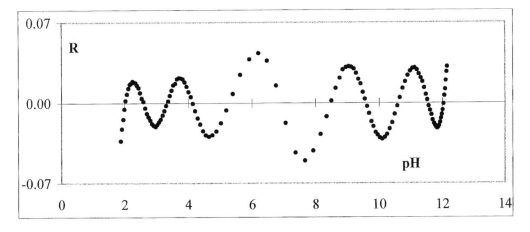

Fig. 4.11-4: The progress curve, and the residuals, during intermediate stages of the curve fitting. The fitted parameter values are shown in color.

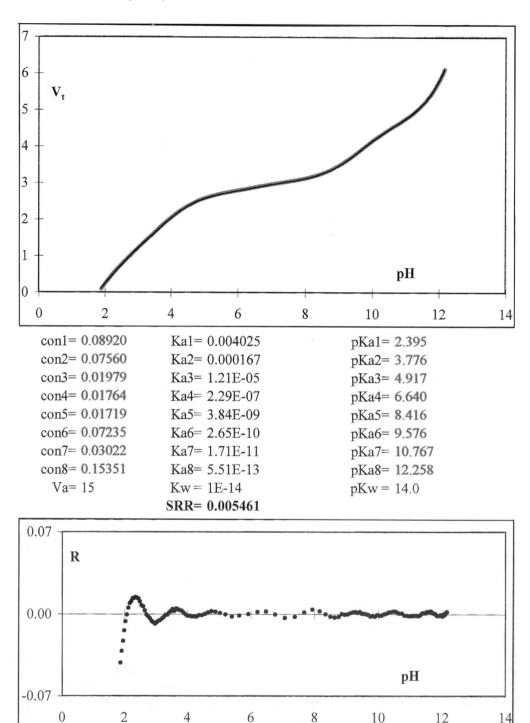

con1= 0.08920   Ka1= 0.004025   pKa1= 2.395
con2= 0.07560   Ka2= 0.000167   pKa2= 3.776
con3= 0.01979   Ka3= 1.21E-05   pKa3= 4.917
con4= 0.01764   Ka4= 2.29E-07   pKa4= 6.640
con5= 0.01719   Ka5= 3.84E-09   pKa5= 8.416
con6= 0.07235   Ka6= 2.65E-10   pKa6= 9.576
con7= 0.03022   Ka7= 1.71E-11   pKa7= 10.767
con8= 0.15351   Ka8= 5.51E-13   pKa8= 12.258
    Va= 15   Kw = 1E-14   pKw = 14.0
              SRR= 0.005461

Fig. 4.11-5: The progress curve, and the residuals, during intermediate stages of the curve fitting using more adjustable parameters. The fitted parameter values are shown in color.

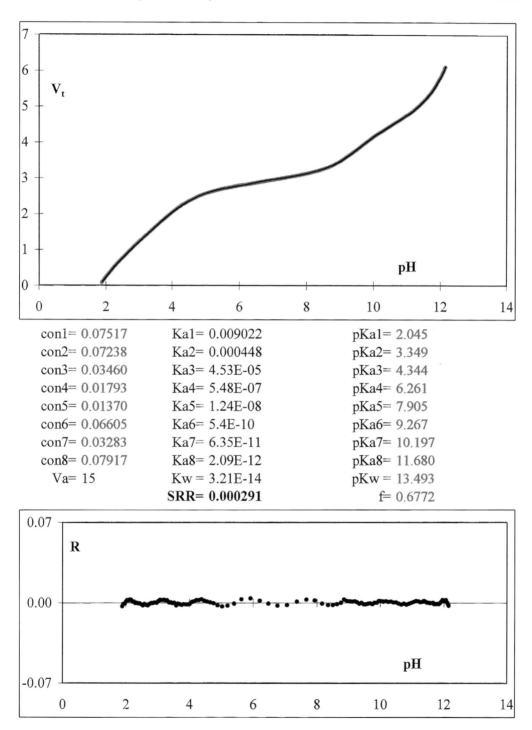

con1= 0.07517	Ka1= 0.009022	pKa1= 2.045
con2= 0.07238	Ka2= 0.000448	pKa2= 3.349
con3= 0.03460	Ka3= 4.53E-05	pKa3= 4.344
con4= 0.01793	Ka4= 5.48E-07	pKa4= 6.261
con5= 0.01370	Ka5= 1.24E-08	pKa5= 7.905
con6= 0.06605	Ka6= 5.4E-10	pKa6= 9.267
con7= 0.03283	Ka7= 6.35E-11	pKa7= 10.197
con8= 0.07917	Ka8= 2.09E-12	pKa8= 11.680
Va= 15	Kw = 3.21E-14	pKw = 13.493
	SRR= 0.000291	f= 0.6772

Fig. 4.11-6: The final, fitted progress curve, and the residuals. The fitted parameter values are shown in color.

chemistry. For example, in a method called near-infrared analysis (or NIR for aficionados of alphabet soup) a very similar approach is used: the spectra can be obtained quickly and with high precision, but are not easily interpreted in terms of specific chemical features. But they can be correlated with practically useful properties, such as the protein and water content of grain. Consequently, a consistent quality of flour can now be maintained even though the wheat derives from different soils, and growing conditions fluctuate from year to year.

Similar analyses are now common in the quality control procedures of many consumer products, such as beer and wine, coffee, fruit juice, and infant formula, where consistency is demanded even though the quality of the raw starting materials may vary with source and season. This is chemical analysis at its best, and the above example is emphasized here because it clearly illustrates this type of approach, which in general combines precise measurements with sophisticated numerical analysis to produce practical results.

4.12 Summary

In this chapter we have encountered the most important analytical aspects of acids and bases: (a) their individual speciation, as described by the mass action law, and as reflected in the distribution and logarithmic concentration diagrams, (b) their buffer action, and (c) their neutralization, as exploited in acid–base titrations.

In the past, much effort has been expended on finding the precise location of the equivalence point, using either differentiation of the titration curve, or its linearization. In our discussion of acid–base titration, we have used as our point of departure expressions for the progress of the titration rather than those for the titration curve itself. This was done primarily for mathematical convenience, since it is much more straightforward to express V_b/V_a explicitly in terms of $[H^+]$ than to achieve its converse. By insisting on solving for $[H^+]$ as a function of titrant volume, the traditional approach caters to the experimental *procedure* of volumetric analysis rather than to the real *analytical problem* addressed by it: how large a volume of titrant must be added in order to reach a given equivalence point, as defined by its pH or proton concentration, and as indicated by a color change or a pH reading. Parenthetically, since the traditional approach leads to rather intractable mathematics, it is also a poor choice of teaching tools if clarity and simplicity rather than obfuscation are our goals.

The availability of a master equation for acid–base titrations, and of convenient non-linear least-squares curve-fitting methods such as incorporated in Excel's Solver, have made the determination of the unknown sample concentration(s) relatively easy: a spreadsheet is all that is required for such an analysis. Of course, there is no guarantee that all component chemicals and their individual concentrations and pK_a's in a complicated mixture can always be resolved; that can only be expected for those concentrations and pK_a's that significantly affect the titration curve. This is illustrated in Fig. 4.10: even for a single acid such as orthophosphoric acid, the third pK_a can easily be missed, since its effect on the titration curve is often quite small.

The exercises with added simulated noise demonstrate the differences in robustness of the various data-analysis methods. Clearly, differentiation is easily affected by noise. Of the linear plots, the Gran2 plot is the more robust, because it uses that part of the titration curve that is determined largely by the strong base (or acid) typically used as titrant. In principle, the Schwartz plot is more linear than the Gran1 plot; in practice, it is also more sensitive to noise, at least to the type of volume noise we have simulated here, so that the linearity advantage disappears. For simple titration curves, non-linear least-squares data fitting appears to be the most robust of the methods discussed; it is also the most generally applicable of these methods. While non-linear least-squares methods are not always trouble-free, they seem to work well here, apparently because the progress curve is always monotonic.

We then encountered activity effects, the skunk at the party, because when activity effects must be taken into account, the mathematical description of acid–base equilibria becomes more complicated. Fortunately, *activity effects do not change the equivalence volumes*, which is the main reason why activity corrections are seldom made in analytical applications. On the other hand, when the primary goal of the titration is the precise determination of the pK_a-values, activity corrections can usually be restricted to mere changes in the pK_a-values by making the measurements at constant ionic strength.

Finally we have seen in section 4-11 how acid–base titrations can be used in practice, even without any preliminary separations or sample clean-up, and what trade-offs are made in such analyses. This example illustrates a rather radical departure from the traditional emphasis on titrations as methods of high precision. As illustrated in Table 4.11-1, even when precise concentrations of well-defined chemical species cannot be derived from such complex mixtures, they nonetheless can be made to yield very useful quantitative information.

As our last exercise of this chapter we will fit data to a small segment of a titration curve. It is, of course, poor analytical practice to use only part of a curve, but we will use it here merely to demonstrate the general power of computer fitting of experimental data. We will use data, given by Papanastasiou *et al.* in *Anal. Chim. Acta* 277 (1993) 119, for the titration with 0.1 M NaOH of a 50 mL sample containing both formic and propionic acid. The authors show the entire titration curve in their Fig. 6, but in their Table 6 only list numerical values for the buffer region of the curve. Nonetheless, 19 low-noise data pairs are given, while the solution involves only four parameters: the concentrations of the two monoprotic acids, and their acid dissociation constants. Therefore, the problem is well-defined, and is readily solved.

pH	V_b	pH	V_b	pH	V_b	pH	V_b
3.639	1.00	4.050	1.80	4.432	2.59	4.860	3.52
3.727	1.17	4.133	1.97	4.545	2.84	4.932	3.67
3.800	1.31	4.207	2.12	4.638	3.04	5.036	3.87
3.888	1.48	4.278	2.27	4.710	3.20	5.107	4.01
3.975	1.65	4.34	2.42	4.778	3.35		

Instructions for exercise 4.12

1 Open a spreadsheet.

2 Deposit labels for four columns, one each for pH, $[H^+]$, V_b, and $V_{b,\text{calc}}$.

3 Copy the above data in the appropriate columns.

4 In the column so labeled, calculate $[H^+]$ from the pH as $[H^+] = 10^{-pH}$.

5 Deposit labels and initial guess values for the following constants: C_1, C_2, K_1, and K_2, where 1 and 2 denote the two monoprotic acids.

6 With these values, calculate in the column for $V_{b,\text{calc}}$ the appropriate expression for the titration of a mixture of two monoprotic acids.

7 Also deposit a label for SRR, the sum of the squares of the residuals, and compute it from the data in columns V_b and $V_{b,\text{calc}}$ with the function = SUMXMY2.

8 Manually adjust the guess values to obtain an approximate fit between the experimental and calculated values.

9 The stage is now set for using Solver. Use it to minimize the value of SRR by letting it adjust the values of C_1, C_2, K_1, and K_2.

10 Now extend the columns for pH, $[H^+]$, and $V_{b,\text{calc}}$ to compute the remainder of the curve, up to pH 11, and compare your results with those in the above-quoted paper.

11 Even though the agreement between experimental and computed data should already be very good, it is of course possible to refine these data by taking activity effects into account, as was done by Papanastasiou et $al.$ Try it. In this case, the ionic strength I is simply given by $I = [Na^+] + [H^+]$. For a fair comparison with the results of Papanastasiou et $al.$ delete the Davies term $-0.3I$ from the expression for the activity coefficients.

CHAPTER 5

OTHER IONIC
EQUILIBRIA

Quantitative chemical analysis involves many types of ionic equilibria other than those between acids and bases, and the present chapter samples some of them. The *formation of metal complexes* takes place in homogeneous solution, and strongly resembles acid–base chemistry. In *extraction*, two different solvents are used, but both solutions are still homogeneous. Problems of *solubility* and *precipitation* involve two different physical forms of the compound of interest: one dissolved, the other a solid phase. *Electrochemical equilibria* also involve at least two phases, of which one is an electronic conductor, typically a metal, and the other an ionic conductor such as an aqueous solution. Despite these differences in their physics, we will encounter much analogy in the mathematical description of these equilibria, which is why the present chapter is best read after chapter 4.

5.1 Complex formation

The formal description of the equilibrium between a 1 : 1 complex of a metal ion M and a ligand L and its constituent parts,

$$M + L \rightleftharpoons ML \tag{5.1-1}$$

traditionally uses the **formation constant**

$$K_f = \frac{[ML]}{[M][L]} \tag{5.1-2}$$

where, for the sake of simplicity of notation, we have deleted all valencies. The dimension of such a formation constant is M^{-1}, conveniently indicating that we deal here with a formation constant rather than a dissociation constant (with the dimension M). This is one of several good reasons (another one is the ease of checking equations for their proper dimensionality, a most efficient way to catch many errors) to keep the dimensions of equilibrium constants rather than remove them through the introduction of standard states.

We will define the concentration fraction of metal ions without attached ligand as

$$\alpha_M = \alpha_0 = \frac{[M]}{[M] + [ML]} = \frac{1}{1 + K_f[L]} = \frac{1/K_f}{[L] + 1/K_f} \tag{5.1-3}$$

and the corresponding concentration fraction of metal ions complexed with one ligand as

$$\alpha_{ML} = \alpha_1 = \frac{[ML]}{[M] + [ML]} = \frac{K_f[L]}{1 + K_f[L]} = \frac{[L]}{[L] + 1/K_f} \tag{5.1-4}$$

where we present the right-most forms of (5.1-3) and (5.1-4) to emphasize the analogy with the acid–base formalism, with [L] taking the place of [H$^+$], and $1/K_f$ that of K_a.

The progress of the titration of a solution of volume V_M and concentration C_M of metal ions with a volume V_L of titrant solution containing a concentration C_L of free ligand L then follows as

$$\frac{V_L}{V_M} = \frac{C_M\alpha_{ML} + [L]}{C_L - [L]} \tag{5.1-5}$$

where [L] is the concentration of free ligand. This is fully analogous to the progress of the titration of a weak base with a strong acid,

$$\frac{V_a}{V_b} = \frac{C_b\alpha_{HB} + \Delta}{C_a - \Delta} \tag{4.3-11}$$

Instructions for exercise 5.1

1 Open a new spreadsheet. Place the labels Cm, Cl, and six labels Kf in cells A1 through H1, so that we can explore the effect of changing K_f.

2 In row 2 enter some corresponding numerical values, such as 0.1, 0.1, 0, 1, 100, 10^4 (as 1E4 or = 10^4; 10^4 is treated as if it were text), 10^6, and 10^8.

3 In row 4 enter the labels pL, [L], and six labels Vl/Vm.

4 In column A, starting in row 6, place the pL values 0 (0.1) 10, i.e., ranging from 0 to 10 with increment 0.1, and in column B compute the corresponding values of [L] = 10^{-pL}.

5 In cell C6 calculate V_L/V_M according to (5.1-4) and (5.1-5). In order to make it easy to copy this instruction to the adjacent columns, use the formula = ($A$2*C$2*$B6/(1 + C$2*$B6) + $B6)/($B$2-$B6), containing the partially absolute addresses C$2 and $B6.

6 Copy this instruction from C6 to the entire block C6:H106. Bingo!

7 Plot all these curves V_L/V_M versus pL. Figure 5.1-1 illustrates the result you should obtain with these parameter values. (Points are used to avoid the pesky vertical line due to physically unrealizable values calculated for pL ≤ 1.)

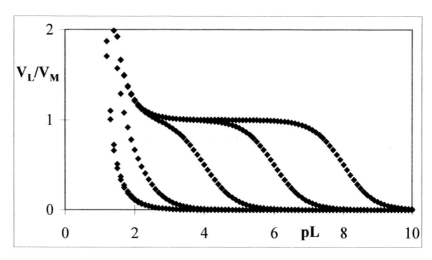

Fig. 5.1-1: Some progress curves for the titration of a metal ion with a ligand forming 1 : 1 complexes with that metal ion. Numerical values used: $C_M = C_L = 0.1$ M, and (from left to right) $K_f = 0$ and 1 (almost overlapping), 10^2, 10^4, 10^6, and 10^8 M^{-1}.

We see that only complexes with formation constants of the order of 10^6 M^{-1} or more will lead to titration curves with a sufficiently steep change in pL near the equivalence point (at $C_M V_M / C_L V_L = 1$) to be useful for volumetric analysis. None of the common monodentate ligands, such as the halide anions (Cl$^-$, Br$^-$, I$^-$) or the pseudohalides (CN$^-$, SCN$^-$, N$_3^-$), form such strong complexes, nor do the carboxylic acid anions (such as acetate) or ammonia (NH$_3$). However, in section 5.2 we will encounter special ligands, the chelates, that do form sufficiently strong 1 : 1 complexes.

Many metal ions don't stop at binding one ligand, but can surround themselves with several ligands, often up to four or even six. The maximum number of ligands usually depends on the ionic valencies of metal ion and ligand, on the coordination number of the metal ion, and on steric considerations. The latter are most pronounced for complexes of relatively small metal ions with rather bulky ligands.

Apart from the use of formation constants rather than dissociation constants, the formation of such poly-ligand complexes is quite analogous to, say, a phosphate anion binding up to three protons, and can likewise be described in terms of stepwise equilibrium constants. However, one should keep in mind that stepwise *formation* constants K_f start counting from the 'bare' metal ion, whereas stepwise *dissociation* constants such as K_a's start counting from the most highly 'complexed' species: K_{f1} for the chloride complexes of ferrous ions applies to the equilibrium Fe^{2+} + Cl$^-$ $\rightleftharpoons$ FeCl$^+$, whereas K_{a1} for phosphoric acid pertains to H$_3$PO$_4$ $\rightleftharpoons$ H$^+$ + H$_2$PO$_4^-$.

An ion such as Fe^{2+} can bind up to four chloride anions, or up to six thiocyanates. The formal description for the formation of such complexes is

$$K_{f1} = \frac{[ML]}{[M][L]} \tag{5.1-6}$$

$$K_{f2} = \frac{[ML_2]}{[ML][L]} \tag{5.1-7}$$

$$K_{f3} = \frac{[ML_3]}{[ML_2][L]} \tag{5.1-8}$$

etc., so that

$$[ML] = K_{f1}[M][L] \tag{5.1-9}$$

$$[ML_2] = K_{f1} K_{f2}[M][L]^2 \tag{5.1-10}$$

$$[ML_3] = K_{f1} K_{f2} K_{f3}[M][L]^3 \tag{5.1-11}$$

and so on, while the corresponding concentration fractions α are given by

$$\alpha_0 = [M] \ / \ \{[M] + [ML] + [ML_2] + [ML_3] + \cdots\} = 1 \ / \ \text{denom} \tag{5.1-12}$$

$$\alpha_1 = [ML] \ / \ \{[M] + [ML] + [ML_2] + [ML_3] + \cdots\} = K_{f1}[L] \ / \ \text{denom} \tag{5.1-13}$$

$$\alpha_2 = [ML_2] \ / \ \{[M] + [ML] + [ML_2] + [ML_3] + \cdots\} = K_{f1} K_{f2}[L]^2 \ / \ \text{denom} \tag{5.1-14}$$

$$\alpha_3 = [ML_3] \ / \ \{[M] + [ML] + [ML_2] + [ML_3] + \cdots\} = K_{f1} K_{f2} K_{f3}[L]^3 / \ \text{denom} \tag{5.1-15}$$

and so forth, where the common denominator is given by

$$\text{denom} = 1 + K_{f1}[L] + K_{f1} K_{f2}[L]^2 + K_{f1} K_{f2} K_{f3}[L]^3 + \cdots \tag{5.1-16}$$

8 In row 1 replace the labels Kf by Kf1, Kf2, Kf3, Kf4, Kf5, and Kf6.

9 In row 2 enter some corresponding numerical values. For example, for the six thio-cyanato complexes of Fe(III) we have $K_{f1} = 10^{1.96}$, $K_{f2} = 10^{2.02}$, $K_{f3} = 10^{-0.41}$, $K_{f4} = 10^{-0.14}$, $K_{f5} = 10^{-1.57}$, and $K_{f6} = 10^{-1.51}$ respectively.

10 In row 4 replace the labels Vl/Vm by denom, [M], [ML], [ML_2], [ML_3], and [ML_4], and add two more: [ML_5], and [ML_6].

11 In column C calculate denom as given by (5.1-16) or, more compactly, as $= 1 + K_{f1}[L]$ $(1 + K_{f2}[L] (1 + K_{f3}[L] (1 + K_{f4}[L] (1 + K_{f5}[L] (1 + K_{f6}[L]))))).$

12 In column D calculate $[M] = \alpha_0 C_M$ where α_0 is given by (5.1-12).

13 In column E calculate $[ML] = \alpha_1 C_M$, and likewise compute [ML_2], [ML_3], etc. in the next columns. Again, this can be simplified by coding, e.g., cell F6 as = C$2*$B6*E6, which can then be copied to G6 through K6.

14 Make the distribution diagram by plotting [M], [ML], [ML_2], etc. vs. pL $(= - \log[\text{SCN}])$ while setting C_M equal to 1, and compare your graph with Fig. 5.1-2. Note that, in this

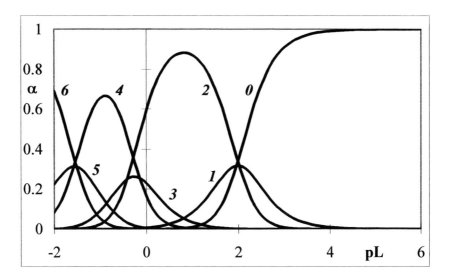

Fig. 5.1-2: The distribution diagram for the thiocyanato complexes of Fe(III) as computed with $K_{f1} = 10^{1.96}$, $K_{f2} = 10^{2.02}$, $K_{f3} = 10^{-0.41}$, $K_{f4} = 10^{-0.14}$, $K_{f5} = 10^{-1.57}$, and $K_{f6} = 10^{-1.51}$. The number of attached thiocyanate ligands L is indicated with each curve.

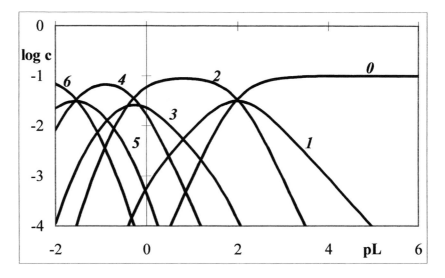

Fig. 5.1-3: The logarithmic concentration diagram for the thiocyanato complexes of Fe(III) for $C_M = 0.1$ M. The equilibrium constants are the same as in Fig. 5.1-2.

case, the species Fe^{3+}, $Fe(SCN)_2^+$, $Fe(SCN)_4^-$ and $Fe(SCN)_6^{3-}$ are the more important ones.

15 Make the corresponding logarithmic concentration diagram by using a semi-logarithmic frame and any value for C_M. Again, Fig. 5.1-3 shows the type of result you can expect.

16 Save the spreadsheet as Complexation, and close.

In the example shown in Fig. 5.1-2 there is relatively little differentiation between the various K_f-values. Moreover, none of the complex formation constants are particularly large. This is typical for metal complexes, and it is the reason that these equilibria are not very useful for volumetric analysis. However, we will see in section 5.3 how such equilibria can be exploited in metal-selective extractions.

5.2 Chelation

Although there are many examples of the formation of 1 : 1 complexes, their application in volumetric analysis is mostly restricted to those between metal cations and **chelating ligands**, which are complexing agents that, upon complexation, surround the metal ions and thereby satisfy all (or almost all) of its coordination positions. Typically, those coordination positions are then occupied by metal–oxygen or metal–nitrogen linkages, the coordinating groups being primarily carboxylic acids and amines.

The prototype of such a chelating ligand is EDTA (for *e*thylene *d*iamine *t*etraacetic *a*cid) which forms strong 1 : 1 complexes with many divalent and trivalent metal cations. EDTA has four carboxylic acid groups and two amino groups, and is therefore a hexaprotic acid, but its complex formation is due almost exclusively to its fully deprotonated anion, Y^{4-}. This brings in the pH, since the latter regulates the fraction α_{Y0} of EDTA that is fully deprotonated. In this case, then, the ligand concentration in solution, $[L] = [Y^{4-}]$, is given by $C_Y \alpha_{Y0}$, where C_Y is the total analytical concentration of EDTA, and α_{Y0} is a strong function of pH, see Fig. 5.2. We can rewrite (5.1-2) as

$$K_f = \frac{[ML]}{[M] C_Y \alpha_{Y0}} \tag{5.2-1}$$

which can be recast as

$$K_f' = \frac{[ML]}{[M] C_Y} \tag{5.2-2}$$

provided that the pH is kept constant. The parameter $K_f' = K_f \alpha_{Y0}$ is called a **conditional** formation constant, because it is indeed a constant on condition that the pH is kept constant, in which case α_{Y0} is also constant. Of course, the value of K_f' depends (through α_{Y0}) on the pH.

Equation (5.2-2) is a useful form for the EDTA complexes of metal ions such as magnesium and calcium that are rather strong bases, and therefore have little tendency to form hydroxides. However, many other metal ions that can be titrated with EDTA will often form hydroxy complexes, and these are usually titrated in the presence of complexing agents that keep hydroxide formation at bay. In that case, the expression for the conditional formation constant must take such complex formation of the metal ion into account as well, and then reads $K_f' = K_f \alpha_{Y0} \alpha_{M0}$, where α_{M0} is the fraction of

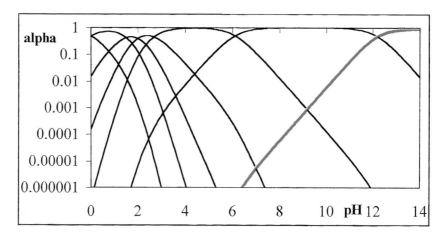

Fig. 5.2: The fractional concentrations α of the various forms of EDTA as a function of pH. The fractional concentration α_{Y0} of the chelating species Y^{4-} has been highlighted in color.

the metal not bound to either EDTA or the added complexing agent. The quantity α_{M0} will depend on the concentration of the added ligand, and so will the conditional constant K_f'.

Below we will determine the pH-dependence of α_{Y0}, which determines the feasibility of a successful EDTA titration.

Instructions for exercise 5.2

1 Open a new spreadsheet.

2 Place the labels Cm, Cy, and Ka1 through Ka6 in cells A1 through H1.

3 In row 2 enter some values for the concentrations C_M and C_Y, and use the following acid dissociation constants for EDTA: $K_{a1} = 1$ M, $K_{a2} = 10^{-1.5}$ M, $K_{a3} = 10^{-2.0}$ M, $K_{a4} = 10^{-2.68}$ M, $K_{a5} = 10^{-6.11}$ M, and $K_{a6} = 10^{-12.17}$ M.

4 In row 4 enter the labels pH, [H], denom, a6, a5, a4, a3, a2, a1, and a0.

5 In column A, starting in row 6, place the pH values 0 (0.1) 14, and in column B compute the corresponding values of $[H^+] = 10^{-pH}$.

6 In column C calculate the denominator of (4.9-11), i.e., $[H^+]^6 + [H^+]^5 K_{a1} + [H^+]^4 K_{a1} K_{a2} + \cdots + K_{a1} K_{a2} K_{a3} K_{a4} K_{a5} K_{a6} = [H^+]^6 (1 + (K_{a1}/[H^+]) (1 + (K_{a2}/[H^+]) (1 + (K_{a3}/[H^+]) (1 + (K_{a4}/[H^+]) (1 + (K_{a5}/[H^+]) (1 + (K_{a6}/[H^+]))))))).$

7 In columns D through J compute the concentration fractions α of the various protonation forms of EDTA, see (4.9-11). Again, you can simplify the coding by using $\alpha_6 = [H^+]^6/\text{denom} (= \$B6^6/\$C6)$, $\alpha_5 = K_{a1} \alpha_6 /[H^+] (= C\$2*D6/\$B6)$, which can then be copied to F6:J6.

8 Plot all these concentration fractions in a semi-logarithmic plot versus pH, so that the resulting graph will be double-logarithmic, as in Fig. 5.2.

The progress of the titration of a metal ion M with EDTA is analogous to (5.1-4) and (5.1-5) except that K_f must be replaced by $K_f' = K_f \, \alpha_{Y0}$, i.e.,

$$\frac{V_Y}{V_M} = \frac{C_M \alpha_{MY} + [Y]}{C_Y - [Y]} \qquad (5.2\text{-}3)$$

$$\alpha_{MY} = \frac{K_f'[Y]}{1 + K_f'[Y]} = \frac{K_f \alpha_{Y0}[Y]}{1 + K_f \alpha_{Y0}[Y]} \qquad (5.2\text{-}4)$$

which shows how we can manipulate chelation equilibria by changing the pH. For a satisfactory EDTA titration we have, just as in section 5.1, log $K_f' \geq 6$. By using so-called metallochromic indicators, or a mercury electrode, one can monitor $[Y^{4-}] = \alpha_{Y0}[Y]$ during the titration, analogous to how one uses acid–base indicators or a glass electrode to follow $[H^+]$ during an acid–base titration.

5.3 Extraction

In extractions one typically uses two solvents that have only limited mutual solubility, so that they form two separate phases. Typically, one solvent is polar (such as an aqueous solution), the other non-polar. Few ions will be extracted into the non-polar solvent, but neutral complexes will be, and this is often the basis for ion extraction. For example, only the neutral species $Fe(SCN)_3$ will be extracted into a non-polar phase from among the ferric thiocyanates illustrated in Fig. 5.1-2, even though $Fe(SCN)_3$ is only a minor species in aqueous solution.

Two parameters are used to characterize extractions. The more fundamental one is the **partition coefficient**, K_p, which describes the concentration ratio of the species common to both solvents, such as ($Fe(SCN)_3$ in the above example:

$$K_p = \frac{[Fe(SCN)_3]_{organic}}{[Fe(SCN)_3]_{aqueous}} \qquad (5.3\text{-}1)$$

When we define the volume ratio as

$$v = V_{organic} / V_{aqueous} \qquad (5.3\text{-}2)$$

the mass fraction of $Fe(SCN)_3$ extracted by equilibrating a sample of volume $V_{aqueous}$ with a volume $V_{organic}$ of extractant will be

$$\mu_{organic} = \frac{[Fe(SCN)_3]_{organic} V_{organic}}{[Fe(SCN)_3]_{aqueous} V_{aqueous} + [Fe(SCN)_3]_{organic} V_{organic}} = \frac{v K_p}{1 + v K_p} \qquad (5.3\text{-}3)$$

while the fraction remaining in the original sample is its complement, $\mu_{aqueous} = 1/(1 + v K_p)$.

A more practical parameter is the **distribution coefficient** D, which gives the corresponding ratio of the total analytical concentrations,

$$D = \frac{C_{Fe(III),organic}}{C_{Fe(III),aqueous}} \qquad\qquad (5.3\text{-}4)$$

The ligand concentration affects the fraction of ferric ions that is in the form of the neutral complex $Fe(SCN)_3$; on that basis one might want to use a thiocyanate concentration of about 2.5 M ($pL = -\log[SCN] \approx -0.4$), where $[(Fe(SCN)_3]$ goes through a maximum in Fig. 5.1-2. Indeed, Fe(III) can be extracted into oxygen-containing non-polar solvents such as diethyl ether or isobutyl alcohol, and can then be determined spectrometrically in that solvent.

In the spreadsheet exercise we will calculate the distribution of Fe(III) based on the data of Fig. 5.1-3. The parameter involved in the extraction is vK_p, where the volume ratio v depends on the experimental protocol, and K_p on the nature of the extractant used.

Instructions for exercise 5.3

1 Recall spreadsheet Complexation.

2 In cell I1 place the label vKp, and in cell I2 a numerical value for it.

3 Modify the instruction for denom in column C by multiplying the term involving $(Fe(SCN)_3$ by $(1 + vK_p)$. For example, the instruction in cell C6 might now read (with the modification printed in bold) = 1 + $C$2 + B6*(1 + $D$2*B6* (1 + $E$2*B6*(1 + **$I$2** + $F$2*B6*(1 + $G$2*B6*(1 + $H$2*B6)))))).

4 Change the label in G4 to read [ML3]aq, and in K4 and L4 add the labels [ML3]org and D respectively.

5 In column K calculate $[(Fe(SCN)_3]_{organic} = vK_p [(Fe(SCN)_3]_{aqueous}$ where vKp is stored in $I$2 while $[(Fe(SCN)_3]_{aqueous}$ is found in column G.

6 In column L compute D; in L6 this can be done as = K6/SUM(D6:K6).

7 Add the curve for $[(Fe(SCN)_3]_{organic}$ to the graph of Fig. 5.1-2. You can do this as follows: highlight K6:K116, and copy it. Now go to the graph, activate it, and paste. If necessary, click on the new curve until it is highlighted, right-click to get the Format Data Series, and format it to your taste.

8 Plot the resulting distribution diagram. Figure 5.3-1 shows an example.

9 Also make a graph of D versus pSCN, see Fig. 5.3-2.

Note that $Fe(SCN)_3$ is only a minor constituent of the set of ferri-thiocyanate complexes, yet that is sufficient for an extraction. For the (rather

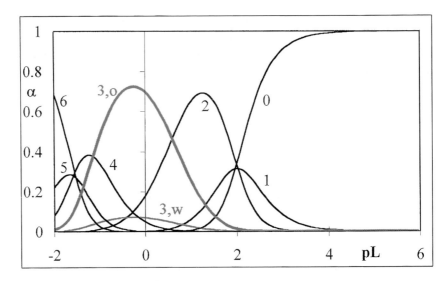

Fig. 5.3-1: The distribution diagram of an aqueous solution of Fe(III) in thiocyanate in equilibrium with an organic extractant, with the parameter values of Fig. 5.1-2, $vK_p = 10$, and $pL = -\log[SCN]$.

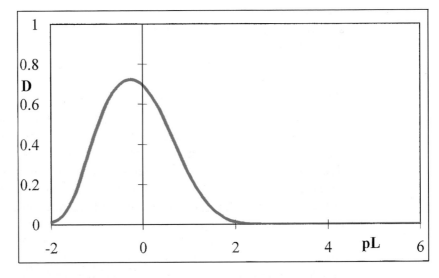

Fig. 5.3-2: The corresponding distribution D of Fe(III) between the organic and aqueous phases; $pL = -\log[SCN]$.

low) numerical value of vK_p used in Figs. 5.3-1 and 5.3-2, and with extraction from 1 M NH$_4$SCN, D is only about 0.7, and four successive extractions would be needed to remove at least 99% of Fe(III) from the aqueous phase. (One extraction would leave 30% or 0.3, hence four successive extractions will leave $(0.3)^4 = 0.008$ or 0.8%.)

Often, metals are extracted at low pH, in order to prevent the (often irreversible) formation of poorly soluble hydroxy complexes and polymers.

Thiocyanate extractions are therefore often performed in 0.5 M HCl, i.e., at a pH of 0.3. Furthermore, the acid used to enforce such a low pH may introduce competing ligands, such as Cl^-. In that case, we must also consider the acid–base equilibrium of thiocyanate (HSCN has a pK_a of 0.9) and the possibility of the formation of neutral ferric complexes of mixed ligand composition.

5.4 Solubility

So far we have used concentration as an open-ended variable, but it has an upper limit: each dissolved species has a solubility, above which the solvent can accommodate no more of it. While the solubilities for many salts are of the order of 0.1 to 10 M, they are sometimes much lower, and may then be of analytical use. In water, the solubility of ions is usually much higher than that of neutral compounds, in which case the solubility of salts can be described in terms of the equilibrium between the salt and its constituent ions. For example, for barium sulfate the equilibrium $BaSO_4 = Ba^{2+} + SO_4^{2-}$ can be described in terms of the equilibrium constant $K = [Ba^{2+}] [SO_4^{2-}] / [BaSO_4]$. When solid $BaSO_4$ is present, the solution is saturated with $BaSO_4$, so that $[BaSO_4]$ can usually be considered constant, in which case we can define the **solubility product** $K_{s0} = K[BaSO_4]$,

$$K_{s0} = [Ba^{2+}] [SO_4^{2-}] \tag{5.4-1}$$

The application of relations such as (5.4-1) is often so straightforward that it does not require a spreadsheet. However, complications may arise when one or more of the ionic species involved also participates in other equilibria. In this example that might occur when the pH is so low that the formation of HSO_4^- must be considered. In that case (5.4-1) must be combined with $K_{a2} = [H^+] [SO_4^{2-}] / [HSO_4^-]$ and with the applicable mass balance relations. When this leads to an equation of third or higher order, the Newton–Raphson method (Excel's Goal Seek) or a non-linear least-squares search (Excel's Solver) may be employed to find the solution. Although either can be used for such a one-parameter problem, Solver is often the more convenient one because it allows the user to set constraints (such as that the sought concentration cannot be negative), and thereby to avoid non-physical answers.

In our spreadsheet exercise we will consider a textbook example, the solubility of HgS as a function of pH, in a solution in equilibrium with solid HgS that contains no other sources of mercury and sulfur. HgS is quite insoluble, with a reported solubility product of $5 \times 10^{-54} M^2$. The case is complicated by the fact that the two participant ions, Hg^{2+} and S^{2-}, are both involved in acid–base equilibria. For Hg^{2+} these are the successive formation of three hydroxy complexes $HgOH^+$, $Hg(OH)_2$, and $Hg(OH)_3^-$, for S^{2-} the consecutive protonations to HS^- and H_2S. Consequently we must first formulate the problem to make it suitable for spreadsheet solution.

The two basic relations are the solubility product

$$K_{s0} = [Hg^{2+}] \, [S^{2-}] \tag{5.4-2}$$

and the mass balance

$$[Hg^{2+}] + [HgOH^+] + [Hg(OH)_2] + [Hg(OH)_3^-] = [H_2S] + [HS^-] + [S^{2-}] \tag{5.4-3}$$

In order to reduce the mathematical complexity we express the fractional concentration of Hg^{2+} in terms of $[OH^-]$ and the relevant formation constants, i.e.,

$$\alpha_{Hg^{2+}} = \frac{[Hg^{2+}]}{[Hg^{2+}] + [Hg(OH)^+] + [Hg(OH)_2] + [Hg(OH)_3^-]}$$

$$= \frac{1}{1 + K_{f1}[OH^-] + K_{f1}K_{f2}[OH^-]^2 + K_{f1}K_{f2}K_{f3}[OH^-]^3} \tag{5.4-4}$$

Likewise, we write for the fractional concentration of S^{2-}

$$\alpha_{S^{2-}} = \frac{[S^{2-}]}{[H_2S] + [HS^-] + [S^{2-}]} = \frac{K_{a1}K_{a2}}{[H^+]^2 + [H^+]K_{a1} + K_{a1}K_{a2}} \tag{5.4-5}$$

so that (5.4-3) can be rewritten as

$$\frac{[Hg^{2+}]}{\alpha_{Hg^{2+}}} = \frac{[S^{2-}]}{\alpha_{S^{2-}}} \tag{5.4-6}$$

By combining this with (5.4-2) we can now eliminate either $[Hg^{2+}]$ or $[S^{2-}]$, and obtain explicit solutions for either species, as in

$$[Hg^{2+}] = \frac{[S^{2-}]\alpha_{Hg^{2+}}}{\alpha_{S^{2-}}} = \frac{K_{s0}\alpha_{Hg^{2+}}}{[Hg^{2+}]\alpha_{S^{2-}}} \tag{5.4-7}$$

from which it follows that

$$[Hg^{2+}] = \sqrt{\frac{K_{s0}\alpha_{Hg^{2+}}}{\alpha_{S^{2-}}}} = \sqrt{\frac{K_{s0}\{[H^+]^2 + [H^+]K_{a1} + K_{a1}K_{a2}\}}{K_{a1}K_{a2}\{1 + K_{f1}[OH^-] + K_{a1}K_{f2}[OH^-]^2\}}} \tag{5.4-8}$$

This is an explicit expression for $[Hg^{2+}]$ in terms of $[H^+]$ and $[OH^-]$, ready for spreadsheet evaluation. Moreover, we can use it to calculate all other species involved, through

$$[HgOH^+] = K_{f1} \, [OH^-] \, [Hg^{2+}] \tag{5.4-9}$$

$$[Hg(OH)_2] = K_{f2} \, [OH^-] \, [HgOH^+] \tag{5.4-10}$$

$$[Hg(OH)_3^-] = K_{f3} \, [OH^-] \, [Hg(OH)_2] \tag{5.4-11}$$

$$[S^{2-}] = K_{s0} / [Hg^{2+}] \tag{5.4-12}$$

$$[HS^-] = [H^+] [S^{2-}] / K_{a2} \tag{5.4-13}$$

$$[H_2S] = [H^+] [HS^-] / K_{a1} \tag{5.4-14}$$

Finally, the solubility S_{Hg} of mercury is defined as the total analytical concentration of all mercury species in the saturated solution. Assuming that the concentration of dissolved HgS is negligible, we have

$$S_{Hg} = [Hg^{2+}] + [HgOH^+] + [Hg(OH)_2] + [Hg(OH)_3^-] \tag{5.4-15}$$

which, according to (5.4-3), is equal to $S_S = [H_2S] + [HS^-] + [S^{2-}]$.

Instructions for exercise 5.4

1 Open a new spreadsheet.

2 Place the labels Ks0, Ka1, Ka2, Kf1, Kf2, and Kf3 in cells A1 through F1.

3 In row 2 enter corresponding literature values, such as $K_{s0} = 10^{-53} M^2$, $K_{a1} = 10^{-7.02} M$, $K_{a2} = 10^{-13.9} M$, $K_{f1} = 10^{10.6} M^{-1}$, $K_{f2} = 10^{11.2} M^{-1}$, and $K_{f3} = 10^{-0.9} M^{-1}$.

4 In row 4 enter the labels pH, [H], pOH, [OH], pHg, pHgOH, pHg(OH)2, pHg(OH)3, pS, pHS, pH2S, and pS(Hg), where the latter terms denotes the solubility of mercury (which in this case happens to be equal to that of sulfur).

5 In column A, starting in row 6, place the pH values 0 (0.1) 14, and in columns B, C, and D compute the corresponding values of $[H^+] = 10^{-pH}$, $pOH = 14 - pH$, and $[OH^-] = 10^{-pOH}$.

6 In columns E through L calculate the negative logarithms of the concentrations of $[Hg^{2+}]$, $[HgOH^+]$, etc. using (5.4-8) through (5.4-15).

7 Plot the corresponding logarithmic concentration diagram. It should resemble Fig. 5.4-1.

A diagram as complicated as Fig. 5.4-1 requires some explanation. At low pH, the two dominant species (apart from H^+ and OH^-) are Hg^{2+} and H_2S; in the double-logarithmic representation used, the lines representing them coincide. In the range between pH 8 and 13, the dominant species are HS^- and $Hg(OH)_2$. In the middle range, at pH's between 4 and 7, the dominant species are H_2S and $Hg(OH)_2$, but at no pH are they Hg^{2+} and S^{2-}! The solubility (colored line) of both mercury and sulfur is minimal in this middle pH range, and even there does not get below 8×10^{-20} M, more than seven orders of magnitude (or, more specifically, a factor of 25 000 000) larger than the square root of K_{s0}. While both $Hg(OH)^+$ and $Hg(OH)_3^-$ are of little consequence for the solubility, all other species (including $Hg(OH)_2$) are crucial at one pH or another.

It is instructive to see what happens when, e.g., the equilibria involving the hydroxy species are deleted. This is readily simulated in the spreadsheet

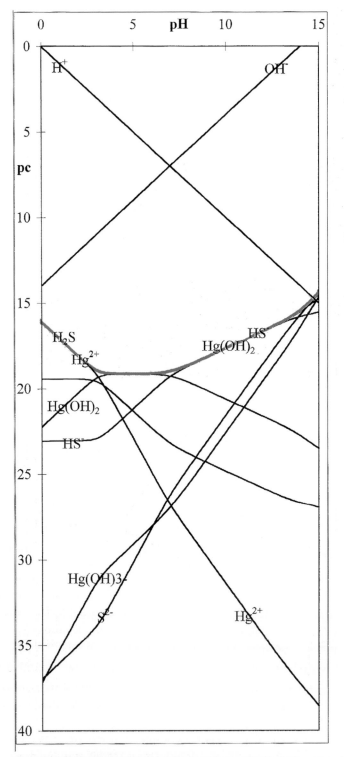

Fig. 5.4-1: The logarithmic concentration diagram for HgS. The colored line shows the solubility S of mercury and sulfur.

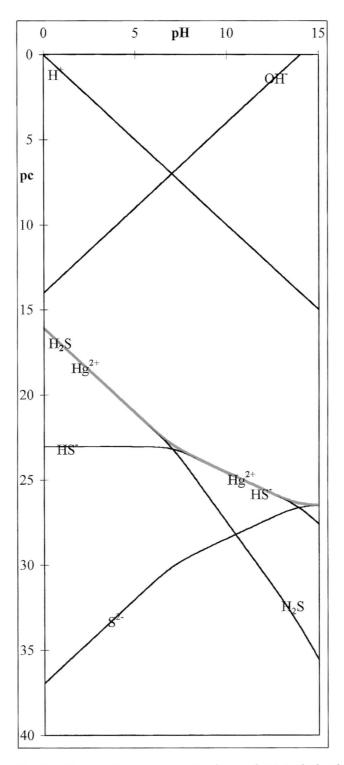

Fig. 5.4-2: The logarithmic concentration diagram for HgS calculated on the erroneous assumption that Hg^{2+} does not form hydroxy complexes.

by setting K_{f1}, K_{f2}, and K_{f3} to a sufficiently small value, say $\leq 10^{-20}$. (Just setting them equal to zero gives trouble, since the spreadsheet will then try to take the logarithms of zero concentrations.) The result is shown in Fig. 5.4-2. Such an (incorrect) diagram predicts that the solubility decreases at high pH, whereas the opposite occurs; in fact, the solubility calculated at pH 14 in Fig. 5.4-2, 4×10^{-27} M, is eleven orders of magnitude less than the value calculated in Fig. 5.4-1, 4×10^{-16} M! This clearly shows that one needs to have good information about the various chemical species involved before reliable predictions can be made. And that is precisely where we want to be: limited by real, chemical data, not by the mathematical means to manipulate them. The latter can be relegated to a spreadsheet.

5.5 Precipitation and dissolution

The equilibrium formation and dissolution of precipitates can be illustrated with silver chloride. Ag^+ forms a series of chloro complexes, at least up to $AgCl_4^{3-}$. In the presence of solid AgCl the formal description is most readily characterized by the following formalism:

$$AgCl_s = Ag^+ + Cl^- \qquad\qquad K_{s0} = [Ag^+][Cl^-]$$

$$AgCl_s = AgCl \qquad\qquad K_{s1} = [AgCl]$$

$$AgCl_s + Cl^- = AgCl_2^- \qquad\qquad K_{s2} = [AgCl_2^-]/[Cl^-]$$

$$AgCl_s + 2Cl^- = AgCl_3^{2-} \qquad\qquad K_{s3} = [AgCl_3^{2-}]/[Cl^-]^2$$

$$AgCl_s + 3Cl^- = AgCl_4^{3-} \qquad\qquad K_{s4} = [AgCl_4^{3-}]/[Cl^-]^3$$

where the constants K_s are *overall* formation constants rather than the *stepwise* formation constants K_f we have used so far. The two types of formation constants can be interconverted by

$$K_{s1} = K_{s0}\,K_{f1} \qquad\qquad K_{f1} = K_{s1}/K_{s0}$$

$$K_{s2} = K_{s0}\,K_{f1}\,K_{f2} \qquad\qquad K_{f2} = K_{s2}/K_{s1}$$

$$K_{s3} = K_{s0}\,K_{f1}\,K_{f2}\,K_{f3} \qquad\qquad K_{f3} = K_{s3}/K_{s2}$$

$$K_{s4} = K_{s0}\,K_{f1}\,K_{f2}\,K_{f3}\,K_{f4} \qquad\qquad K_{f4} = K_{s4}/K_{s3}$$

With the above soluble species, the solubility of silver is given by

$$S_{Ag} = [Ag^+] + [AgCl] + [AgCl_2^-] + [AgCl_3^{2-}] + [AgCl_4^{3-}]$$

$$= K_{s0}/[Cl^-] + K_{s1} + K_{s2}[Cl^-] + K_{s3}[Cl^-]^2 + K_{s4}[Cl^-]^3$$

$$= K_{s0}\{1/[Cl^-] + K_{f1} + K_{f1}K_{f2}[Cl^-] + K_{f1}K_{f2}K_{f3}[Cl^-]^2 + K_{f1}K_{f2}K_{f3}K_{f4}[Cl^-]^3$$

We are now ready to plot a logarithmic concentration diagram of all s er species as a function of pCl. We will first make such a diagram for a solution in equilibrium with solid AgCl.

Instructions for exercise 5.5

1 Open a new spreadsheet.

2 Place the labels pKs0, pKs1, pKs2, pKs3, and pKs4 in cells A1:E1.

3 In row 2 enter the literature values $pK_{s0} = 9.75$, $pK_{s1} = 6.05$, $pK_{s2} = 4.13$, $pK_{s3} = 3.3$, $pK_{s4} = 3.6$.

4 In row 4 enter the labels pCl, pAg, pAgCl, pAgCl2, pAgCl3, pAgCl4, and pS.

5 In column A, starting in row 6, place the pCl values -2 (0.1) 10.

6 In column B calculate pAg from $pAg = pK_{s0} - pCl$.

7 Likewise use $pAgCl = pK_{s1}$, $pAgCl_2^- = pK_{s1} + pCl$, $pAgCl_3^{2-} = pK_{s2} + 2\,pCl$, and $pAgCl_4^{3-} = pK_{s3} + 3\,pCl$ to compute the values in columns C through F.

8 Calculate $pS = -\log\{10^{-pAg} + 10^{-pAgCl} + 10^{-pAgCl2} + 10^{-pAgCl3} + 10^{-pAgCl4}\}$.

9 Plot the corresponding logarithmic concentration diagram. It should resemble Fig. 5.5-1.

The diagram of Fig. 5.5-1 illustrates the fact that the formation of complexes such as $AgCl_2^-$, $AgCl_3^{2-}$, and $AgCl_4^{3-}$ leads to a quite high silver solubility in concentrated chloride solutions. (The effect is exaggerated here by extending the calculation to pCl $= -2$ or $[Cl^-] = 100$, a rather unrealistically high value.)

Note that we made the assumption that the solution is at all times in equilibrium with solid AgCl. However, it quite often happens that the actual amount of precipitate is relatively small. For example, when we use a silver/silverchloride electrode, the AgCl is usually a thin, chocolate-brown coating on an otherwise shiny silver wire. In that case it does not take a large volume of a concentrated chloride solution to dissolve the AgCl, at which point the diagram is no longer applicable. (For that very reason, never refill a silverchloride reference electrode with simple NaCl or KCl solution, but instead use solutions presaturated with AgCl.)

How do we represent the solution when a precipitate can form or dissolve? We need one piece of additional information, namely the total analytical concentration of silver when all precipitate is dissolved. (This depends both on the amount of precipitate and on the volume of solution, but we only need the resulting analytical concentration in the absence of precipitate.)

Say that the amount of silver present in both solution and as a solid is such that the solution would have a concentration of 1 mM if all the precipitate

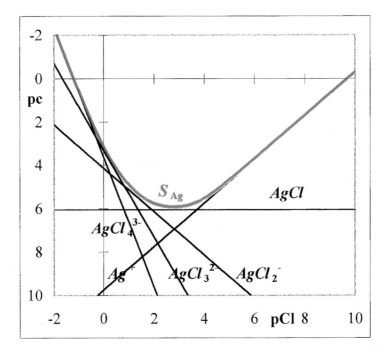

Fig. 5.5-1: The logarithmic concentration diagram for the chloro complexes of Ag(I) in an aqueous solution equilibrated with solid AgCl. The colored line shows the silver solubility S_{Ag}. Note that, in this case, the solubility of the "insoluble" neutral species, AgCl, is quite substantial; between a pCl of 2 and 3 it is the dominant component of the silver solubility S_{Ag}.

were dissolved. Then the precipitate will be present only in the range where the silver solubility (the colored line in Fig. 5.5-1) falls below 1 mM. Outside that range, all silver is dissolved, predominantly as Ag^+ and AgCl at low chloride concentrations, and as poly-chloro complexes at high chloride concentrations.

Now we are ready to make the corresponding logarithmic concentration diagram. We will use the IF statement of the spreadsheet to determine whether or not a precipitate will be present, and let the calculation self-adjust accordingly.

10 In row 1, starting with F1, add a number of new labels: pCmax, Cmax, Kf1, Kf2, Kf3, and Kf4. (In the K_f formalism it is easier to use K_f rather than pK_f.)

11 In F2 place a maximum value for pC, i.e., its value in the absence of any precipitate. In G2 calculate the corresponding value of pC_{max}, i.e., $= 10^{\wedge}-F2$.

12 In row 4 enter the following additional column labels: denom, pAg, pAgCl, pAgCl2, pAgCl3, pAgCl4. Yes indeed, the latter labels duplicate earlier ones, but don't worry about that.

13 In column H calculate [Cl⁻] from pCl in column A.

14 In column I compute the common denominator (as in section 5.1) as if it were simply a case of complex formation, without precipitation. For example, cell I6 might contain the instruction = 1 + $G$2*H6*(1 + $H$2*H6*(1 + $I$2*H6* (1 + $J$2*H6))).

15 In column J calculate pAg. Cell J6 might read = IF(G6 > $F$2,B6, − LOG($F$2/I6)). Or, in normal English: *if* pS exceeds pC_{max}, (in which case S is smaller than C_{max}), *then* keep the earlier answer (from column B), *otherwise* calculate pAg as − log(C_{max}/denom) where C_{max} is the total analytical concentration in the precipitate-free solution, see (5.1-12).

16 Likewise, in column K, compute pAgCl, using in K6 the command = IF(G6 > $F$2,C6, − log($F$2*$G$2*H6/I6)).

17 Similarly, calculate pAgCl2, pAgCl3, and pAgCl4 in columns L:N.

18 Plot the corresponding logarithmic concentration diagram, using columns A, H, and J through N. Compare with Fig. 5.5-2.

19 Vary the value of pS_{max} in cell F2 and observe what happens. When the amount of solid AgCl is too small to reach pK_{s1} in the solution volume used, no precipitate will form at any chloride concentration, as illustrated in Fig. 5.5-3.

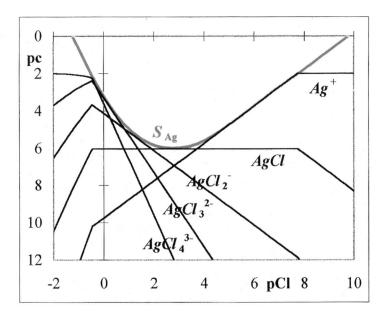

Fig. 5.5-2: The logarithmic concentration diagram for the chloro complexes of Ag(I) in aqueous solution, in the presence of a limited amount of silver. In this example, the total amount of silver would take a 10 mM solution (pS_{max} = 2) if it were all dissolved.

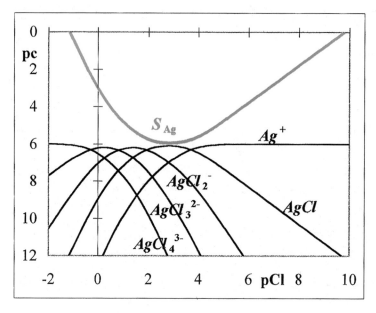

Fig. 5.5-3: The logarithmic concentration diagram for the chloro complexes of Ag(I) in aqueous solution, when the amount of silver is too low to exceed the silver solubility at any chloride concentration. In this example we have set $pS_{max} = 6$.

5.6 Precipitation titrations

The formation of precipitates can be used as the basis of a titration. Interestingly, the formal description of the progress of a precipitation titration resembles that of the titration of a strong base with a strong acid, and we will use the next few lines to show how this comes about. Imagine that a sample volume V_s of an iodide-containing solution of concentration C_s is titrated with a volume V_t of the titrant solution containing a concentration C_t of a soluble silver salt such as $AgNO_3$. For the sake of simplicity we will ignore the formation of poly-iodo complexes of silver, which are inconsequential to the titration, and whose inclusion would not much change our conclusions. The mass balance relations for iodide and silver will then be

$$[I^-] + [AgI] + P_{AgI} = C_s V_s / (V_s + V_t) \qquad (5.6\text{-}1)$$

$$[Ag^+] + [AgI] + P_{AgI} = C_t V_t / (V_s + V_t) \qquad (5.6\text{-}2)$$

where P_{AgI} is the concentration of iodide or silver removed by the precipitation of AgI. By subtracting (5.6-2) from (5.6-1) we *eliminate* P_{AgI} and obtain

$$[I^-] - [Ag^+] = (C_s V_s - C_t V_t) / (V_s + V_t) \qquad (5.6\text{-}3)$$

from which we derive the progress equation

$$\frac{V_t}{V_s} = \frac{C_s + [Ag^+] - [I^-]}{C_t - [Ag^+] + [I^-]} = \frac{C_s + [Ag^+] - K_{s0,AgI}/[Ag^+]}{C_t - [Ag^+] + K_{s0,AgI}/[Ag^+]} \tag{5.6-4}$$

which is indeed fully homologous with the progress of the titration of a strong base with a strong acid,

$$\frac{V_a}{V_b} = \frac{C_b + [H^+] - [OH^-]}{C_a - [H^+] + [OH^-]} = \frac{C_b + [H^+] - K_w/[H^+]}{C_a - [H^+] + K_w/[H^+]} \tag{4.3-10}$$

Instructions for exercise 5.6

1 Open a new spreadsheet and give it a name.

2 Place the labels Ct, Cs and Ks0,AgI in cells A1 through C1.

3 Deposit plausible values for C_t and C_s in A2 and B2 respectively, and in cell C2 place the literature value of $K_{s0,AgI}$, $8.3 \times 10^{-17}\,M^2$.

4 In row 4 enter the labels pAg, [Ag], and Vt/Vs.

5 In column A, starting in row 6, place the pAg values, starting with $-\log(Ct)$, using 0.1 increments, to say $-\log(Ct) + 20$, and compute the corresponding values of $[Ag^+]$ in column B. (Tying the pAg scale to pC_t avoids having an artificial near-horizontal line in the graph at physically unrealizable pAg values.)

6 In column C calculate Vt/Vs using the right-hand side of (5.6-4).

7 Plot the resulting progress curve of V_t/V_s vs. pAg.

8 Now we will indicate how you can convert a progress curve into a titration curve without reorganizing the spreadsheet. Highlight the progress curve. In the formula box you will then see the formula for that curve, such as = SERIES (,'Fig5.6'!$A$6:$A$206, 'Fig5.6'!$C$6:$C$206,1). The name 'Fig 5.6' identifies the name of your spreadsheet, and may of course be different in your case; if you have not yet given the spreadsheet a name, it will just be 'Sheet 1'.

9 In the formula box, now replace the two A's by C's, and vice versa, so that the revised formula will read = SERIES(,'Fig5.6'!$C$6:$C$206,'Fig5.6'!$A$6:$A$206,1). Depress the enter key.

10 This is all that is needed to exchange the x- and y-axes of the plot! You will now have the titration curve, apart from some necessary re-scaling and re-labeling to clean up the graph. Figure 5.6-1 shows what the result may look like.

Just as a pH electrode can monitor $[H^+]$, a silver electrode can report on $[Ag^+]$. This titration is therefore not only theoretically, but also experimentally fully analogous to that of a strong base with a strong acid.

Because the solubility products K_{s0} for AgI, AgBr, and AgCl are sufficiently different, it is possible to titrate a mixture of halides with silver,

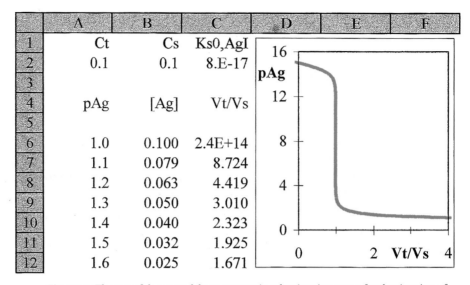

	A	B	C	D	E	F
1	Ct	Cs	Ks0,AgI			
2	0.1	0.1	8.E-17			
3						
4	pAg	[Ag]	Vt/Vs			
5						
6	1.0	0.100	2.4E+14			
7	1.1	0.079	8.724			
8	1.2	0.063	4.419			
9	1.3	0.050	3.010			
10	1.4	0.040	2.323			
11	1.5	0.032	1.925			
12	1.6	0.025	1.671			

Fig. 5.6-1: The top of the spreadsheet computing the titration curve for the titration of 0.1 M iodide with 0.1 M silver. The equivalence point has the coordinates $V_t/V_s = 1$, $pAg = -\frac{1}{2} \log K_{s0,\text{AgI}} = -0.5 \log(8.3E{-}17) = 8.04$.

simultaneously determining their individual concentrations. We therefore extend the spreadsheet calculations to simulate such a curve.

For the titration of a mixture containing C_I M iodide and C_{Br} M bromide the mass balance equations read

$$[I^-] + [AgI] + P_{AgI} = C_I V_s / (V_s + V_t) \tag{5.6-5}$$

$$[Br^-] + [AgBr] + P_{AgBr} = C_{Br} V_s / (V_s + V_t) \tag{5.6-6}$$

$$[Ag^+] + [AgI] + [AgBr] + P_{AgI} + P_{AgBr} = C_t V_t / (V_s + V_t) \tag{5.6-7}$$

The P's can again be eliminated by subtraction, whereupon we obtain

$$[I^-] + [Br^-] - [Ag^+] = ((C_I + C_{Br}) V_s - C_t V_t) / (V_s + V_t) \tag{5.6-8}$$

which leads directly to the expression for the progress of the titration

$$\frac{V_t}{V_s} = \frac{C_I + C_{Br} + [Ag^+] - [I^-] - [Br^-]}{C_t - [Ag^+] + [I^-] + [Br^-]}$$

$$= \frac{C_I + C_{Br} + [Ag^+] - (K_{s0,\text{AgI}} + K_{s0,\text{AgBr}})/[Ag^+]}{C_t - [Ag^+] + (K_{s0,\text{AgI}} + K_{s0,\text{AgBr}})/[Ag^+]} \tag{5.6-9}$$

When the titration starts, AgI is precipitated, because it is much less soluble than AgBr. The initial part of the titration is therefore described by (5.6-4). As more silver nitrate is added, more silver iodide precipitates, until almost all iodide has been precipitated, at which point the silver concentration increases rapidly. At a given moment (which in this case will occur before the titration curve reaches its first equivalence point) silver bromide starts to precipitate, and the formal description of the titration switches from (5.6-4) to (5.6-9), since the latter applies in the presence of both solids, AgI and AgBr.

The switch-over from one equation to the other will occur when the product $[Ag^+]$ times $[Br^-]$ exceeds $K_{s0,AgBr}$. The value of $[Ag^+]$ is directly available on the spreadsheet, in column B, and the value of $[Br^-]$ is given by C_{Br} times the dilution term $V_s/(V_s + V_t)$, i.e., $[Br^-] = C_{Br}/(1 + V_t/V_s)$. Again, V_t/V_s is already on the spreadsheet, in column C.

The spreadsheet representation of the titration will now involve switching from one equation to another, as soon as the product $[Ag^+][Br^-]$ exceeds the value of $K_{s0,AgBr}$. In this case, then, the titration curve really consists of separate pieces, not for reasons of mathematical convenience but as the direct consequence of the formation of a new precipitant phase. In the spreadsheet we can accomplish this change-over between the two formalisms by using IF statements. Note that it is still a completely straightforward calculation, without any circular reasoning.

11 In row 1 change the labels Ct and Cs into C(Ag) and C(I) respectively, and enter the additional labels C(Br) and Ks0,AgBr in cells D1 and E1 respectively.

12 In row 2 enter an appropriate value for C_{Br}, and the literature value for $K_{s0,AgBr}$, $5.2 \times 10^{-13}\,M^2$.

13 In cell D4 enter the column label Vt/Vs once more.

14 Now comes the working part. In cell D6 enter the following instruction: = IF(B6 < \$E\$2*(1 + C6)/\$D\$2,C6,(\$B\$2 + \$D\$2 + B6–(\$C\$2 + \$E\$2)/B6)/(\$A\$2–B6 + (\$C\$2 + \$E\$2)/B6)). This is longhand for 'if $[Ag^+] < K_{s0,AgBr}/[Br^-]$, use (5.6-4), otherwise use (5.6-9)'. Copy the instruction all the way down to row 206.

15 Go to the graph, and click on the curve until you get it highlighted, at which time its formula will reappear in the formula box. In it, change the C's into D's to make the formula read = SERIES(,'Fig5.6'!\$D\$6:\$D\$206,'Fig5.6'! \$A\$6:\$A\$206,1). Depress the enter key. Done.

The above treatment can be extended to mixtures including chloride. In the presence of all three precipitates, AgI, AgBr, and AgCl, the expression for the progress of the titration becomes

$$\frac{V_t}{V_s} = \frac{C_I + C_{Br} + C_{Cl} + [Ag^+] - [I^-] - [Br^-] - [Cl^-]}{C_t - [Ag^+] + [I^-] + [Br^-] + [Cl^-]}$$

$$= \frac{C_I + C_{Br} + C_{Cl} + [Ag^+] - (K_{s0,AgI} + K_{s0,AgBr} + K_{s0,AgCl})/[Ag^+]}{C_t - [Ag^+] + (K_{s0,AgI} + K_{s0,AgBr} + K_{s0,AgCl})/[Ag^+]} \qquad (5.6\text{-}10)$$

The description of the entire titration curve requires switching from (5.6-4) to (5.6-9) to (5.6-10) depending on the value of $[Ag^+]$. The spreadsheet

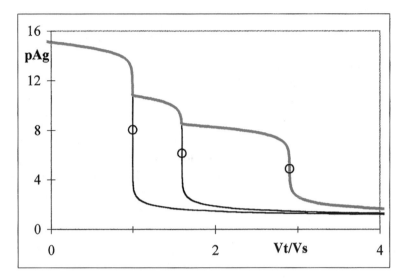

Fig. 5.6-2: The composite titration curve (colored line) for the titration of 0.1 M iodide + 0.06 M bromide + 0.13 M chloride with 0.1 M silver nitrate in the presence of an anti-coagulant salt. (In practice, these titrations are often performed on much more dilute solutions, typically about 1 mM.) The black lines show the titration curves for iodide, see eq. (5.6-4) and Fig. 5.6-1, and for iodide + bromide, eq. (5.6-9). Open circles show the equivalence points.

will again do this, automatically and ungrudgingly, when we use nested IF statements.

16 Add the labels C(Cl) and Ks0,AgCl to row 1, and corresponding values to row 2. The literature value for $K_{s0,AgCl}$ is 1.8×10^{-10} M^2.

17 In row 4 enter another label Vt/Vs, this time for column E.

18 Now comes the crowning glory: a *nested* IF statement. It will read, in words: if $[Ag^+] <$ $K_{s0,AgBr}/[Br^-]$, use (5.6-4), otherwise if $[Ag^+] < K_{s0,AgCl}/[Cl^-]$, use (5.6-9), otherwise use (5.6-10). This is not unlike a BASIC statement If... Then...ElseIf...Else... So here goes. In cell E6 deposit the following instruction: = IF(B6 < $E$2*(1 + C6)/$D$2,C6,IF(B6 < $G$2*(1 + D6)/$F$2,D6,($B$2 + $D$2 + $F$2 + B6 − ($C$2 + $E$2 + $G$2)/B6)/($A$2 − B6 + ($C$2 + $E$2 + $G$2)/B6))). Copy this instruction down to row 206.

19 Go back to the graph, activate the curve, and in the formula window change the column from D to E. Enter this change. You should now see a curve such as the colored one in Fig. 5.6-2.

When the halide titrations are performed without additives, the transitions between the various curve segments are often much less sharp than described here. This is a consequence of co-precipitation of AgBr in AgI, and of AgCl in AgBr, because these salts can form solid solutions and/or mixed crystals. Such co-precipitation can be suppressed by the presence of a coagulating agent such as aluminum nitrate (J. Motonaka, S. Ikeda, and N. Tanaka, *Anal. Chim. Acta* 105 (1979) 417), in which case the experimental titration curve indeed follows the theoretical one, and shows sharp breaks at the same places where the formalism switches from one equation to another.

In principle one could use the break points in the titration curve as approximations of the equivalence points, although these points do not quite coincide with the true equivalence points; moreover, co-precipitation (if not suppressed) often leads to a blurring of those points. At any rate, it is usually a better practice to avoid reliance on single points in a titration curves for the precise determination of the equivalence volumes, because such single readings are inherently rather vulnerable to experimental uncertainty.

One can of course fit experimental data to the entire, theoretical curve with a non-linear least-squares routine such as Solver. In this particular case, however, the direct, non-iterative method of using Gran plots provides a valid, simpler alternative. As illustrated below, such plots are quite linear, analogous to the Gran plots for the titration of *strong* acids and bases.

For the first part of the titration, when the only solid present is AgI, we start from the exact expression (5.6-4) for the progress. As long as there is still excess iodide present, $[Ag^+] \ll [I^-]$, in which case (5.6-4) reduces to $V_t / V_s = (C_I - [I^-]) / (C_t + [I^-])$, which we combine with $C_t V_{eq1} = V_s C_I$ to

$$[I^-] (V_t + V_s) \approx C_t (V_{eq1} - V_t) \tag{5.6-11}$$

In the region between the first and second equivalence points, we can often neglect the terms $[Ag^+]$ and $[I^-]$ in (5.6-9), which then reduces to $V_t / V_s \approx (C_I + C_{Br} - [Br^-]) / (C_t + [Br^-])$. Upon combining this with the expression for the second equivalence point, $C_t V_{eq2} = V_s (C_I + C_{Br})$, we obtain

$$[Br^-] (V_t + V_s) \approx C_t (V_{eq2} - V_t) \tag{5.6-12}$$

Likewise, between the second and third equivalence points, we can often neglect the terms $[Ag^+]$, $[I^-]$, and $[Br^-]$ in (5.6-10), in which case we can write $V_t / V_s \approx (C_I + C_{Br} + C_{Cl} - [Cl^-]) / (C_t + [Cl^-])$. Since the third equivalence point is given by $C_t V_{eq3} = V_s (C_I + C_{Br} + C_{Cl})$, we obtain

$$[Cl^-] (V_t + V_s) \approx C_t (V_{eq3} - V_t) \tag{5.6-13}$$

Finally, beyond the third equivalence point, neglect of $[I^-]$, $[Br^-]$ and $[Cl^-]$ leads to

$$[Ag^+] (V_t + V_s) \approx C_t (V_t - V_{eq3}) \tag{5.6-14}$$

So here we have a set of four Gran plots, three of the Gran1 type, and one Gran2, that are easy to apply and (as already indicated above) unusually linear, see Fig. 5.6-3. Note that these Gran plots are all of a similar type: a plot of $[Z] (V_t + V_s)$ versus V_t, or as shown here as $[Z] (V_t / V_s)$ versus V_t / V_s, where

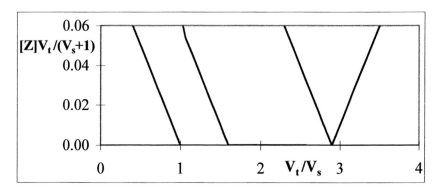

Fig. 5.6-3: The four Gran plots for the titration of a mixture of iodide, bromide, and chloride with a soluble silver salt in the presence of a coagulating agent, as in Fig. 5.6-2.

$[Z] = [I^-] = K_{s0,AgI}/[Ag^+]$, $\quad [Z] = [Br^-] = K_{s0,AgBr}/[Ag^+]$, $\quad [Z] = [Cl^-] = K_{s0,AgCl}/[Ag^+]$, and $[Z] = [Ag^+]$ respectively.

20 Add a fourth column to the spreadsheet, and in it calculate $[Z] (V_t/V_s + 1)$, where $[Z]$ is either $K_{s0}/[Ag^+]$ for one of the silver halide salts, or $[Ag^+]$ itself.

21 Plot $[Z] (V_t/V_s + 1)$ of column D versus V_t/V_s of column C for the various definitions of $[Z]$: $K_{s0,AgI}/[Ag^+]$, $K_{s0,AgBr}/[Ag^+]$, $K_{s0,AgCl}/[Ag^+]$, or $[Ag^+]$. Alternatively, if you want to graph all four Gran plots simultaneously, as in Fig. 5.6-3, you will need to make a column for each of them.

22 Save the spreadsheet.

5.7 The von Liebig titration

A venerable and very interesting titration, first described by von Liebig in 1851, is that of cyanide with silver. For obvious reasons you will not find it in any undergraduate laboratory manuals, but it is a good titration on which to practice your theoretical skills. It differs from the halide titration discussed in the preceding section in that the soluble complex $Ag(CN)_2^-$ is formed first, upon addition of half the equivalent amount of silver. Only when more silver is added will the sparingly soluble AgCN precipitate. The entire sequence is therefore a complexometric titration followed by a precipitation titration. Just this qualitative description might be enough to scare the fainthearted, but you, my reader, are by now a well-seasoned spreadsheeter, and you will encounter no unsurmountable problems in this challenge.

For the first part of the titration, i.e., as long as no solid AgCN has formed, we consider the cyanide complexes. At the beginning of the titration, cyanide is in great excess, and poly-cyano complexes may contribute significantly to the titration curve. We will therefore include them in the formal description. On the spreadsheet we do not need to make guesses as to what approximations to make, assumptions that can only be justified by a complete analysis anyway.

Silver forms cyano complexes up to $Ag(CN)_4^{3-}$; moreover, CN^- can be protonated to form HCN. Therefore we will consider the mass balance equations

$$[CN^-] + [HCN] + [AgCN] + 2[Ag(CN)_2^-] + 3[Ag(CN)_3^{2-}] + 4[Ag(CN)_4^{3-}]$$
$$= C_s V_s / (V_s + V_t) \tag{5.7-1}$$

$$[Ag^+] + [AgCN] + [Ag(CN)_2^-] + [Ag(CN)_3^{2-}] + [Ag(CN)_4^{3-}] = C_t V_t / (V_s + V_t) \tag{5.7-2}$$

and subtraction of (5.7-2) from (5.7-1) leads to

$$[CN^-] + [HCN] - [Ag^+] + [Ag(CN)_2^-] + 2[Ag(CN)_3^{2-}] + 3[Ag(CN)_4^{3-}]$$
$$= (C_s V_s - C_t V_t) / (V_s + V_t) \tag{5.7-3}$$

Substitutions of the type

$$[Ag^+] = C_t V_t \alpha_{Ag+} / (V_s + V_t) \tag{5.7-4}$$

and

$$[HCN] = [H^+] [CN^-] / K_a \tag{5.7-5}$$

then yield

$$\frac{V_t}{V_s} = \frac{C_s - [CN^-](1 + [H^+]/K_a)}{C_t F_f + [CN^-](1 + [H^+]/K_a)} \tag{5.7-6}$$

where

$$F_f = \alpha_{AgCN} + 2\alpha_{Ag(CN)_2^-} + 3\alpha_{Ag(CN)_3^{2-}} + 4\alpha_{Ag(CN)_4^{3-}}$$

$$= \frac{K_{f1}[CN^-] + 2K_{f1}K_{f2}[CN^-]^2 + 3K_{f1}K_{f2}K_{f3}[CN^-]^3 + 4K_{f1}K_{f2}K_{f3}K_{f4}[CN^-]^4}{1 + K_{f1}[CN^-] + K_{f1}K_{f2}[CN^-]^2 + K_{f1}K_{f2}K_{f3}[CN^-]^3 + K_{f1}K_{f2}K_{f3}K_{f4}[CN^-]^4} \tag{5.7-7}$$

Equations (5.7-6) and (5.7-7) describe the progress of the titration as long no precipitate is formed. As soon as solid AgCN appears in the titration vessel, we must instead use the analogue of (5.6-4) which, in this case reads

$$\frac{V_t}{V_s} = \frac{C_s - [CN^-](1 + [H^+]/K_a) + [Ag^+] - [Ag(CN)_2^-] - 2[Ag(CN)_3^{2-}] - 3[Ag(CN)_4^{3-}]}{C_s + [CN^-](1 + [H^+]/K_a) - [Ag^+] + [Ag(CN)_2^-] + 2[Ag(CN)_3^{2-}] + 3[Ag(CN)_4^{3-}]}$$

$$= \frac{C_s + [Ag^+] - (1 + [H^+]/K_a + K_{s2})K_{s0}/[Ag^+] - 2K_3/[Ag^+]^2 - 3K_4/[Ag^+]^3}{C_t - [Ag^+] + (1 + [H^+]/K_a + K_{s2})K_{s0}/[Ag^+] + 2K_3/[Ag^+]^2 + 3K_4/[Ag^+]^3} \tag{5.7-8}$$

where we have used the abbreviations $K_{s2} = K_{s0} K_{f1} K_{f2}$, $K_3 = K_{s0}{}^3 K_{f1} K_{f2} K_{f3}$, and $K_4 = K_{s0}{}^4 K_{f1} K_{f2} K_{f3} K_{f4} = K_3 K_{s0} K_{f4}$. The switch from (5.7-6) to (5.7-8) occurs when [Ag$^+$] as given by (5.7-4) equals $K_{s0} / [CN^-]$.

Instructions for exercise 5.7

1 Open a new spreadsheet and give it a name, e.g., von Liebig.

2 In the top row place labels for C_s, C_t, [H$^+$], K_a, K_{f1} through K_{f4}, as well as K_{s0}, K_3, and K_4.

3 Deposit plausible values for C_t, C_s, and [H$^+$], and enter the literature values $K_a = 6.2 \times 10^{-10}$ M, $K_{f1} = 8.3 \times 10^8$ M^{-1}, $K_{f2} = 4.2 \times 10^{11}$ M^{-1}, $K_{f3} = 5.0$ M^{-1}, $K_{f4} = 0.074$ M^{-1}, and $K_{s0} = 1.2 \times 10^{-16}$ M^2. Furthermore, calculate the numerical values of K_3 and K_4. Computing these here, once, rather than on every row of the spreadsheet, will speed up the calculation. Whether or not this has a noticeable effect will, of course, depend on the hardware used.

4 Enter labels for columns p[CN], [CN], denom, Ff, Vt/Vs, [Ag], denom*, Vt/Vs*, and pAg.

5 Let p[CN] range from 0 to 22 with increments of 0.1. In the next column, calculate the corresponding values for [CN$^-$].

6 In the column labeled denom compute the denominator of (5.7-7), e.g., in cell C6 as = 1 + \$E\$2*B6*(1 + \$F\$2*B6*(1 + \$G\$2*B6*(1 + \$H\$2*B6))). Alternatively you might want to name the individual constants, and refer to them by those names.

7 In the next column calculate F_f as given by (5.7-7), so that cell D6 might contain = (\$E\$2*B6*(1 + 2*\$F\$2*B6*(1 + 1.5*\$G\$2*B6*(1 + (4/3)*\$H\$2*B6)))) / C6.

8 Use (5.7-6) to calculate V_t / V_s in the column with that label.

9 In the next column compute [Ag$^+$], while at the same time testing whether solid AgCN can form. Since $\alpha_{Ag+} = 1/$denom and $V_t/(V_s + V_t) = (V_t/V_s)/(1 + V_t/V_s)$, the instruction might use the following logic: if $C_t \times (V_t/V_s) /$ (denom $\times (1 + V_t/V_s)$ is smaller than $K_{s0}/[CN^-]$, use $C_t \times (V_t/V_s) /$ (denom $\times (1 + V_t/V_s)$, otherwise use $K_{s0}/[CN^-]$. Or, in spreadsheet code for row 6: = IF (\$B\$2*E6/(C6* (1 + E6)) < \$I\$2/B6, \$B\$2*E6/(C6*(1 + E6)), \$I\$2/B6).

10 Now calculate the denominator of the right-most form of (5.7-8).

11 Calculate V_t/V_s*, for which you use either the earlier-computed values (as long as no solid AgCN is formed), or equation (5.7-8). An easy code for the latter uses $V_t/V_s* = (C_s + C_t -$ denom*) / denom*.

12 In the next column, calculate pAg from [Ag$^+$].

13 Plot pAg versus V_t/V_s. Figure 5.7-1 shows what type of result you can expect.

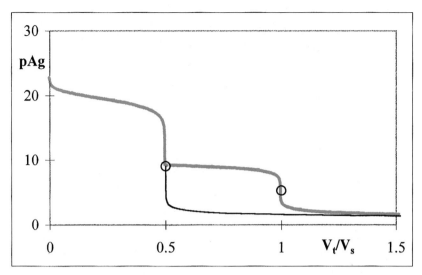

Fig. 5.7-1: The (colored) titration curve for the von Liebig titration of 0.1 M cyanide with 0.1 M silver, using the equilibrium constants given in the text. The black curve shows how the complexometric titration curve would have continued if somehow the precipitation of solid AgCN could have been prevented. Open circles show the equivalence points.

Analysis of experimental data can again be based on the break point in the potentiometric curve, or on the first appearance of a visible precipitate, both of which yield close approximations to the first equivalence point. Moreover we can use the first derivative of the potentiometric curve, Gran plots, or a non-linear least-squares fit of the experimental data to the theoretical model for the entire curve. It is clear how to do the overall fit, since we have just generated the theoretical curve. Below we will briefly examine the Gran plots.

In order to reduce the first, complexometric part of the curve, defined by (5.7-6) and (5.7-7), to a simple Gran plot, we must neglect the contributions of AgCN, $Ag(CN)_3^{2-}$, and $Ag(CN)_4^{3-}$ to the curve, in which case we obtain

$$[CN^-] (1 + [H^+]/K_a) (V_s + V_t) = 2C_t(V_{eq1} - V_t) \tag{5.7-9}$$

For the second part of the curve, we can either neglect all species other than $Ag(CN)_2^-$ (before the second equivalence point) or all cyanide-containing species (beyond that equivalence point) and obtain the Gran-plot approximations

$$K_{s0} K_{s2} (V_s + V_t) / [Ag^+] = C_t(V_{eq2} - V_t) \tag{5.7-10}$$

$$[Ag^+] (V_s + V_t) = C_t(V_t - V_{eq2}) \tag{5.7-11}$$

14 Calculate and plot the Gran plots according to (5.7-9) through (5.7-11). Your curves should look similar to those of Fig. 5.7-2.

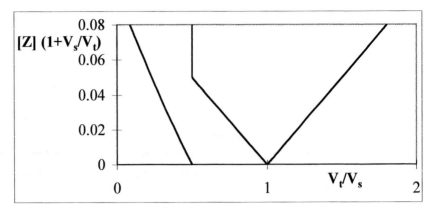

Fig. 5.7-2: Gran plots for a von Liebig titration of 0.1 M cyanide with 0.1 M silver, using the equilibrium constants given in the text.

The initial (colored) part of the titration curve in Fig. 5.7-1, and its (black) continuation, show that the complexometric titration curve is quite asymmetrical, the result of the initial involvement of the tri-cyano and tetra-cyano silver complexes. For that first part of the von Liebig titration, the Gran plot (which neglects those complexes) is noticeably curved, see Fig. 5.7-2. However, the next Gran plots are quite linear, reflecting the fact that, once AgCN starts to precipitate, $Ag(CN)_2^-$ is really the dominant species until the second equivalence point, and so is Ag^+ after that second equivalence point has been passed. Consequently, for a Gran-plot analysis, the second equivalence point would be much better suited, even though that will require the use of about twice as much silver in each titration.

5.8 The graphical representation of electrochemical equilibria

Traditionally, electrochemical equilibria are explained in terms of thermodynamic cell potentials. However, in electro*analytical* applications, such a description is of little use, because one almost always uses a *non-thermodynamic* measurement, with a reference electrode that includes a liquid junction. It is then more useful to go back to the basic physics of electrochemistry, i.e., to the individual interfacial potential differences that make up the total cell potential. This is the approach we will use here.

There are two types of interfacial potential differences: equilibrium and non-equilibrium potentials. (From now on we will use 'potential' as shorthand for 'potential difference'. Potentials of individual phases cannot be measured, but some potential differences can be.) The equilibrium potentials can again be subdivided into two categories: electron transfer and ion transfer potentials. The metal/metal ion potentials can be considered as either of these: we can look at a silver electrode in contact with an aqueous solution of silver nitrate as a system that allows the reduction of Ag^+ to Ag^0 and its reverse, the oxidation of Ag^0 to Ag^+, or as the transfer of a silver ion to and from the Fermi sea of metal ions and conduction electrons that make up a metal. Here, and in section 5.9, we will be concerned with electron transfer potentials. The potentials of a glass electrode, and of a fluoride electrode, are examples of ion transfer potentials, while the silver electrode illustrates a metal/metal ion electrode.

In all these cases of equilibrium potentials, the formal description is similar, and is based on the **Nernst equation**. For the transfer of n electrons between the oxidized species O and its reduced counterpart R, the Nernst equation reads

$$nf E = nf E_{OR}{}^\circ + \log \frac{[O]}{[R]} \tag{5.8-1}$$

where $f = F / (RT \ln(10))$, or about $16.9 \, V^{-1}$ at room temperature; its reciprocal, $1/f$, then has a value of about $0.059 \, V$. We write the Nernst equation in the above form to emphasize its homology with the mass action law for a weak monoprotic acid,

$$pH = pK_a + \log \frac{[A^-]}{[HA]} \tag{5.8-2}$$

Indeed, by introducing the definitions

$$h = 10^{-fE} \qquad\qquad k = 10^{-fE^\circ_{OR}} \tag{5.8-3}$$

we can rewrite (5.8-1) in a form that is fully isomorphous with (5.8-2), viz.

$$ph = pk + \log \frac{[O]}{[R]} \tag{5.8-4}$$

Consider, for example, the redox equilibrium $Fe^{3+} + e^- \rightleftharpoons Fe^{2+}$. When we define the total analytical concentration of iron in solution (regardless of its oxidation state) as C, it follows from the Nernst equation that the fractional concentrations of Fe^{3+} and Fe^{2+} are

$$\alpha_{Fe^{2+}} = \frac{h}{h+k} \qquad\qquad \alpha_{Fe^{3+}} = \frac{k}{h+k} \tag{5.8-5}$$

which are isomorphous with the acid–base relations

$$\alpha_{HA} = \frac{[H^+]}{[H^+] + K_a} \qquad\qquad \alpha_{A^-} = \frac{K_a}{[H^+] + K_a} \tag{5.8-6}$$

So far we have restricted the discussion to that of a single redox step, from Fe(III) to Fe(II), and a monoprotic acid–base system, but the analogy can be carried much further. For example, the aqueous redox behavior of vanadium involves the oxidation states V^{2+}, V^{3+}, VO^{2+}, and VO_2^+, all separated by the stepwise addition or extraction of one electron. This is fully analogous to a triprotic acid H_3A in which each form can be converted into its conjugate forms by the stepwise addition or extraction of single protons.

The vanadium redox chemistry involves protons, since we have the equilibria

$$V^{3+} + e^- \rightleftharpoons V^{2+} \tag{5.8-7}$$

$$VO^{2+} + 2H^+ + e^- \rightleftharpoons V^{3+} + H_2O \tag{5.8-8}$$

$$VO_2^+ + 2H^+ + e^- \rightleftharpoons VO^{2+} + H_2O \tag{5.8-9}$$

and this is reflected in the corresponding Nernst equations, which read

$$E = E^o_{V32} + \frac{1}{f}\log \frac{[V^{3+}]}{[V^{2+}]} \tag{5.8-10}$$

$$E = E^o_{V43} + \frac{1}{f}\log \frac{[VO^{2+}][H^+]^2}{[V^{3+}]} = E^*_{V43} + \frac{1}{f}\log \frac{[VO^{2+}]}{[V^{3+}]} \tag{5.8-11}$$

$$E = E^o_{V54} + \frac{1}{f}\log \frac{[VO_2^+][H^+]^2}{[VO^{2+}]} = E^*_{V54} + \frac{1}{f}\log \frac{[VO_2^+]}{[VO^{2+}]} \tag{5.8-12}$$

where

$$E^*_{V43} = E^o_{V43} + \frac{1}{f}\log[H^+]^2 = E^o_{V43} - \frac{2}{f}pH \quad \text{and}$$

$$E^*_{V54} = E^o_{V54} + \frac{1}{f}\log[H^+]^2 = E^o_{V54} - \frac{2}{f}pH \tag{5.8-13}$$

We now define

$$h = 10^{-fE} \qquad k_1 = 10^{-fE^o_{V32}} \qquad k_2 = 10^{-fE^*_{V43}} \qquad k_3 = 10^{-fE^*_{V54}} \tag{5.8-14}$$

so that

$$\alpha_{V^{2+}} = \frac{h^3}{h^3 + h^2 k_1 + h k_1 k_2 + k_1 k_2 k_3} \tag{5.8-15}$$

$$\alpha_{V^{3+}} = \frac{h^2 k_1}{h^3 + h^2 k_1 + h k_1 k_2 + k_1 k_2 k_3} \tag{5.8-16}$$

$$\alpha_{VO^{2+}} = \frac{h k_1 k_2}{h^3 + h^2 k_1 + h k_1 k_2 + k_1 k_2 k_3} \tag{5.8-17}$$

$$\alpha_{VO_2^+} = \frac{k_1 k_2 k_3}{h^3 + h^2 k_1 + h k_1 k_2 + k_1 k_2 k_3} \tag{5.8-18}$$

which can be compared with the analogous relations for a triprotic acid, (4.9-7) through (4.9-10). Below we will exploit this close analogy between redox and acid–base behavior to make logarithmic concentration diagrams.

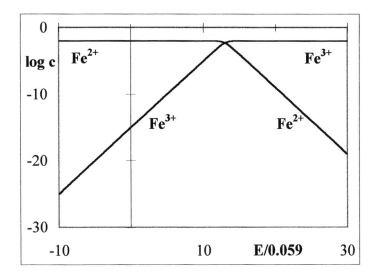

Fig. 5.8-1: The logarithmic concentration diagram for Fe^{2+}/Fe^{3+} at C = 0.01 M, calculated with $E^o_{Fe32} = 0.770$ V.

Instructions for exercise 5.8

1 Open a new spreadsheet.

2 Enter in the top row the labels C, f, and k, and below them the values for C, f = 16.904, and $k = 10^{-fE^o} = 10^ - (B2*0.77)$.

3 In row 4 place the labels fE, h, log[Fe2+], and log[Fe3+].

4 In cells A6:A406 deposit values for fE as −10 (0.1) 30.

5 In column B compute the corresponding values of $h = 10^{-fE}$.

6 In column C calculate log $[Fe^{2+}] = \log (C\alpha_{Fe2+}) = \log (hC/(h + k))$. Similarly, in column D, calculate log $[Fe^{3+}] = \log (kC/(h + k))$.

7 Plot log[Fe²⁺] and log[Fe³⁺] versus fE. Figure 5.8-1 shows such a graph.

8 For the vanadium redox system use a similar lay-out but including the pH and with three k-values, representing (in V) $E^o_{V32} = - 0.255$, $E^*_{V43} = 0.337 - 2$ pH/f, and $E^*_{V54} = 1.000 - 2$ pH/f.

9 In order to simplify the coding, you may also want to use a separate column to compute the common denominator in (5.8-15) through (5.8-18), in which case the labels in row 4 might read fE, h, denom, log[V(II)], log[V(III)], log[V(IV)], and log[V(V)].

10 Calculate the logarithms of the various concentrations (again assuming the absence of complexing ligands) and plot them versus fE for several pH-values. Figure 5.8-2 shows an example of such a plot.

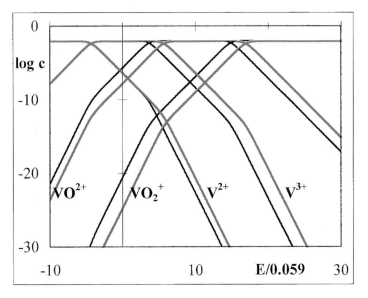

Fig. 5.8-2: The logarithmic concentration diagram for the vanadium redox system at $C = 0.01$ M and pH $= 0$ (colored lines) and pH $= 1$ (black lines), calculated with $E^o_{V32} = -0.255$ V, $E^o_{V43} = 0.337$ V, and $E^o_{V54} = 1.000$ V.

Note that we can consider *the potential of a solution* in the same way as we do the pH of a solution. The potential is defined by the presence of an oxidized and a reduced form of a redox couple, just as the pH is defined by the presence of an acid and its conjugated base. The potential is there regardless of whether we measure it or not, and therefore does not rely on contact with any electrodes. The potential of a solution should be no more mysterious than its pH, except that we happen to have a sense for acidity in our taste buds, but not one for potential. Still, we can imagine the pH of soil or blood without tasting it, and it should be no different for its potential. Of course, any *direct measurement* of the potential of a solution requires electrodes, but the potential can also be *calculated* from the concentrations of the redox partners when these are otherwise known or measurable, just as the pH can be calculated from, say, the spectrum of a solution containing pH-sensitive 'indicator' dyes.

In a number of cases, redox intermediates are rather unstable, and the equilibria are best described in terms of multi-electron transfers although, mechanistically, they most probably involve sequential one-electron transfers. A prime example is permanganate, MnO_4^-, which in neutral solution can be reduced to MnO_2 with an uptake of three electrons, and in acidic solution to Mn^{2+} according to

$$MnO_4^- + 8 H^+ + 5 e^- \rightleftharpoons Mn^{2+} + 4 H_2O \qquad (5.8\text{-}19)$$

The Nernst equation then takes the form

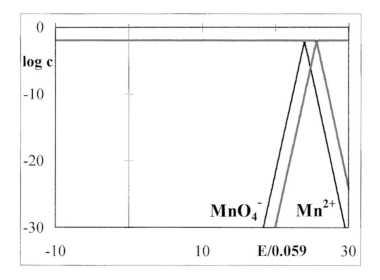

Fig. 5.8-3: The logarithmic concentration diagram for the MnO_4^-/Mn^{2+} couple at $C = 0.01$ M and pH = 0 (colored line) and pH = 1 (black line), calculated with $E^o_{Mn72} = 1.51$ V.

$$E = E^o_{Mn72} + \frac{1}{5f} \log \frac{[MnO_4^-][H^+]^8}{[Mn^{2+}]} = E^*_{Mn72} + \frac{1}{5f} \log \frac{[MnO_4^-]}{[Mn^{2+}]} \qquad (5.8\text{-}20)$$

where

$$E^*_{Mn72} = E^o_{Mn72} + \frac{1}{5f} \log[H^+]^8 = E^o_{Mn72} - \frac{8}{5f} pH \qquad (5.8\text{-}21)$$

In this case we then have

$$[Mn^{2+}] = \frac{h^5 C}{h^5 + k^5} \qquad (5.8\text{-}22)$$

$$[MnO_4^-] = \frac{k^5 C}{h^5 + k^5} \qquad (5.8\text{-}23)$$

where

$$k = 10^{-fE^*_{Mn72}} = 10^{-fE^o_{Mn72} + 8pH/5} \qquad (5.8\text{-}24)$$

The factors 5 multiplying f in the Nernst equation, and in the exponents of (5.8-23) and (5.8-24), reflect the 'simultaneous' exchange of five electrons.

11 Take a new spreadsheet, or modify the existing one, in order to calculate and plot the logarithmic concentration diagram for the permanganate/manganous ion couple, with $C = 10$ mM, and $E^o_{Mn72} = 1.51$ V at pH = 0, 1, and 2. Figure 5.8-3 shows two of these.

Matters are slightly more complicated for dichromate, $Cr_2O_7^{2-}$, a common oxidant which is reduced according to

$$Cr_2O_7^{2-} + 14\,H^+ + 6\,e^- \rightleftharpoons 2\,Cr^{3+} + 7\,H_2O \tag{5.8-25}$$

because of the dimeric nature of $Cr_2O_7^{2-}$. In this case we have

$$E = E_{Cr63}^o + \frac{1}{6f}\log\frac{[Cr_2O_7^{2-}][H^+]^{14}}{[Cr^{3+}]^2} = E_{Cr63}^* + \frac{1}{5f}\log\frac{[Cr_2O_7^{2-}]}{[Cr^{3+}]^2} \tag{5.8-26}$$

$$E_{Cr63}^* = E_{Cr63}^o + \frac{1}{6f}\log[H^+]^{14} = E_{Cr63}^o - \frac{14}{6f}pH \tag{5.8-27}$$

and

$$C = 2\,[Cr_2O_7^{2-}] + [Cr^{3+}] \tag{5.8-28}$$

$$[Cr^{3+}] = \frac{-h^6 + h^3\sqrt{h^6 + 8k^6 C}}{8k^6} \tag{5.8-29}$$

$$[Cr_2O_7^{2-}] = \frac{h^6 + 4k^6 C - h^3\sqrt{h^6 + 8k^6 C}}{8k^6} \tag{5.8-30}$$

$$k = 10^{-fE_{Cr63}^*} = 10^{-fE_{Cr63}^o + 14pH/6} \tag{5.8-31}$$

12 It will save many spreadsheet exponentiations to store in the top row of the spreadsheet the value of $k^5 = 10^{-5fE^o + 8pH}$ instead of $k = 10^{-fE^o + 1.6pH}$.

13 Take a new spreadsheet, or modify the existing one, to calculate and plot the logarithmic concentration diagram for the dichromate / chromous ion couple, with $C = 1$ mM and $E_{Cr63}^o = 1.36\,V$ at pH = 0, 1, and 2.

14 In this calculation you will encounter problems caused by the finite word length of the computer. Usually you are shielded from such problems by the 'double precision' of modern spreadsheets, but that is not enough for equations (5.8-29) and (5.8-30).

15 For example, for $h^6 \leqslant 8k^6 C$, we may reformulate (5.8-29) as $[Cr^{3+}] = \frac{1}{8}\,a\{\sqrt{(1 + 8C/a)} - 1\}$, where $a = (h/k)^6$. While this is *mathematically* identical to (5.8-29), it is *computationally* different, because h/k remains much larger than either h or k alone. Once you have calculated $[Cr^{3+}]$, you can find $[Cr_2O_7^{2-}]$ from (5.8-28).

16 At potentials E that are much more negative than E^* or, more precisely, for $8\,k^6 C \ll h^6$, you can use the series expansion $\sqrt{(1 + \delta)} \approx 1 + \delta/2 - \delta^2/8 + \delta^3/16 - \cdots$ to reduce (5.8-29) to $[Cr^{3+}] \approx C - 2\,k^6 C^2/\,h^6 + 8\,k^{12}C^3/\,h^{12} - \cdots$ and, subsequently, again compute $[Cr_2O_7^{2-}]$ from (5.8-28).

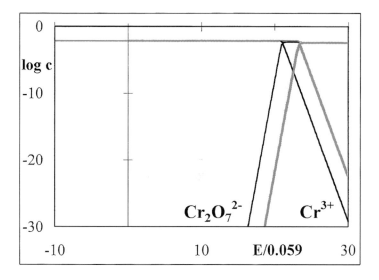

Fig. 5.8-4: The logarithmic concentration diagram for the $Cr_2O_7^{2-}/Cr^{3+}$ couple at $C = 0.01$ M and pH $= 0$ (colored line) and pH $= 1$ (black line), calculated with $E^o_{Cr63} = 1.36$ V.

Note that the limiting slopes in Fig. 5.8-4 for $[Cr^{3+}]$ are 0 and -3, and $+6$ and 0 for $[Cr_2O_7^{2-}]$, reflecting the fact that the reduction of $Cr_2O_7^{2-}$ takes up six electrons, while the oxidation of Cr^{3+} involves only three electrons.

We end this section with a caveat: what we have discussed so far are *equilibrium* properties, but many electrochemical systems are slow, in which case they are kinetically controlled. Equilibrium considerations specify what is possible, if the kinetics are fast enough; kinetics determine what is actually observable. When a redox process involves the breaking or making of chemical bonds, it is usually a slow process. Permanganate is one of the few exceptions, and even there, pure permanganate will react very slowly, but the reaction is speeded up considerably by catalysis from trace amounts of Mn^{2+}. But dichromate illustrates the limitation: in practice, it does not quite follow the above prediction.

5.9 Redox titrations

Because of the close analogy between acid–base and redox behavior, it will come as no surprise that one can use redox titrations, and also simulate them on a spreadsheet. In fact, the expressions for redox progress curves are often even simpler than those for acid–base titrations, because they do not take the solvent into account. (Oxidation and reduction of the solvent are almost always kinetically controlled, and therefore do not fit the equilibrium description given here. In the examples given below, they need not be taken into account in the range of potentials covered.) Although the number of useful redox titrations is limited by the requirement that they proceed with reasonable speed (at least comparable to that of adding titrant to the sample), those systems that do go fast enough often yield very sharp equivalence points, and are easily analyzed either by a total fit of the progress curve, or with Gran plots.

As our first example we will consider the titration of Fe(II) with Ce(IV). For the sake of simplicity we will neglect all complexes other than with the solvent, i.e., we will assume that the species involved are the aquo-ions Fe^{2+}, Fe^{3+}, Ce^{3+}, and Ce^{4+}. The oxidation of Fe^{2+} to Fe^{3+} consumes one electron, which can be provided by the reduction of Ce^{4+} to Ce^{3+}.

As with acid–base titrations, the mathematical description starts with the conservation of mass. For Fe^{2+} and Fe^{3+} we use (5.8-5) plus a dilution correction, since sample and titrant dilute each other during the titration. Consequently, when we titrate Fe^{2+} with Ce^{4+} we have

$$[Fe^{2+}] = \frac{hC_sV_s}{(h+k_s)(V_s+V_t)} \qquad [Fe^{3+}] = \frac{k_sC_sV_s}{(h+k_s)(V_s+V_t)} \qquad (5.9\text{-}1)$$

$$[Ce^{3+}] = \frac{hC_tV_t}{(h+k_t)(V_s+V_t)} \qquad [Ce^{4+}] = \frac{k_tC_tV_t}{(h+k_t)(V_s+V_t)} \qquad (5.9\text{-}2)$$

$$h = 10^{-fE} \qquad k_s = 10^{-fE^o_{Fe32}} \qquad k_t = 10^{-fE^o_{Ce43}} \qquad (5.9\text{-}3)$$

To this we must add a condition representing the conservation of charge. With acid–base titrations we saw that this was most readily done by invoking a proton balance; here we will likewise use an electron balance, i.e., an accounting of electrons consumed and electrons generated. In the present example, each Fe^{2+} oxidized to Fe^{3+} has released one electron, and each Ce^{4+} reduced to Ce^{3+} has accepted one. Therefore, assuming that we start with only Fe^{2+} and Ce^{4+}, we can write the electron balance as

$$[Fe^{3+}] = [Ce^{3+}] \qquad (5.9\text{-}4)$$

Substitution of (5.9-1) and (5.9-2) into (5.9-4) then leads directly to the expression for the progress curve

$$\frac{V_t}{V_s} = \frac{C_s\alpha_{Fe^{3+}}}{C_t\alpha_{Ce^{3+}}} = \frac{C_sk_s(h+k_t)}{C_th(h+k_s)} \qquad (5.9\text{-}5)$$

Finally, the equivalence point is given by $[Fe^{2+}] = [Ce^{4+}]$ or $h = \sqrt{k_sk_t}$.

The above formalism (R. de Levie, *J. Electroanal. Chem.* 323 (1992) 347) is similar to that of the titration of a weak acid with a weak base, or vice versa, see section 4.3, except that the terms in Δ are absent. However, this analogy only concerns the formalism, not the actual steepness of the titration. Redox titrations typically have a much wider span than acid–base titrations, and can therefore be quite steep.

Instructions for exercise 5.9-1

1 Either open a new spreadsheet, or modify the one used in exercise 5.8.

2 Enter in the top row the labels Cs, Ct, f, ks, and kt, and below them numerical values for C_s and C_t, $f = 16.904$, $k_s = 10^{-0.771f}$, and $k_t = 10^{-1.7f}$.

3 In row 4 place the labels fE, h, and Vt/Vs.

4 Starting in row 6 of column A, deposit values for fE as 5 (0.1) 35.

5 In column B compute the corresponding values of $h = 10^{-fE}$.

6 In column C calculate V_t/V_s according to (5.9-5).

7 Plot the progress curve, V_t/V_s vs. fE, and the titration curve, fE vs. V_t/V_s. Figure 5.9-1 shows such a graph.

The corresponding Gran plots use either h or $1/h$, which are both functions of the dimensionless potential fE during the titration, just as Gran plots in acid–base titrations use $[H^+]$ and $1/[H^+]$, both functions of the dimensionless acidity function pH. Indeed, h relates to fE in precisely the same way as $[H^+]$ relates to pH: $h = 10^{-fE}$ and $[H^+] = 10^{-pH}$. Specifically, in redox titrations, Gran1 plots are graphs of hV_t vs. V_t (or, in dimensionless quantities, hV_t/k_sV_s vs. V_t/V_s), while Gran2 plots use $1/h$ or k_t/h instead.

8 Calculate hV_t/k_sV_s, then plot it versus V_t/V_s.

9 Similarly, calculate and plot k_t/h versus V_t/V_s.

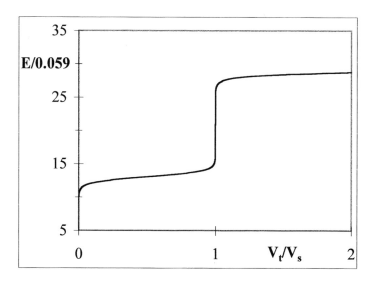

Fig. 5.9-1: The titration curve for the titration of 0.01 M Fe^{2+} with 0.01 M Ce^{4+}, calculated with $E^o_{Fe32} = 0.771$ V and $E^o_{Ce43} = 1.7$ V.

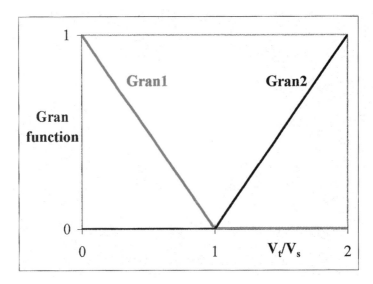

Fig. 5.9-2: The Gran plots for the titration curve of Fig. 5.9-1. The Gran functions are hV_t/k_sV_s for Gran1, and k_t/h for Gran2.

Figure 5.9-2 illustrates such Gran plots. They are quite linear, and are therefore well-suited for least-squares fitting of a line in order to find the equivalence point. The conversion from the measured quantity E to the derived quantity h is usually the major source of errors in such an analysis; a weighted least-squares fit is then called for, with weights proportional to h^2. For the Gran2 plot we likewise use $1/h$ with weights $1/h^2$.

For the titration of V^{2+} with Ce^{4+} we combine (5.9-2) with

$$[V^{2+}] = \frac{h^3 C_s V_s}{(h^3 + h^2 k_1 + h k_1 k_2 + k_1 k_2 k_3)(V_s + V_t)} \tag{5.9-6}$$

$$[V^{3+}] = \frac{h^2 k_1 C_s V_s}{(h^3 + h^2 k_1 + h k_1 k_2 + k_1 k_2 k_3)(V_s + V_t)} \tag{5.9-7}$$

$$[VO^{2+}] = \frac{h k_1 k_2 C_s V_s}{(h^3 + h^2 k_1 + h k_1 k_2 + k_1 k_2 k_3)(V_s + V_t)} \tag{5.9-8}$$

$$[VO_2^+] = \frac{k_1 k_2 k_3 C_s V_s}{(h^3 + h^2 k_1 + h k_1 k_2 + k_1 k_2 k_3)(V_s + V_t)} \tag{5.9-9}$$

where the vanadium species originate from the sample, and Ce^{4+} is the titrant. You will recognize (5.9-6) through (5.9-9) as derived from (5.8-15) through (5.8-18) by the incorporation of the dilution term $V_s/(V_s + V_t)$. The electron balance now reads

$$[V^{3+}] + 2\,[VO^{2+}] + 3\,[VO_2^+] = [Ce^{3+}] \tag{5.9-10}$$

because it takes one electron to oxidize V^{2+} to V^{3+}, two electrons to oxidize V^{2+} to VO^{2+}, etc. The progress equation now reads

$$\frac{V_t}{V_s} = \frac{C_s(\alpha_{V^{3+}} + 2\alpha_{VO^{2+}} + 3\alpha_{VO_2^+})}{C_t\alpha_{Ce^{3+}}} = \frac{C_s(h + k_t)(h^2k_1 + 2hk_1k_2 + 3k_1k_2k_3)}{C_th(h^3 + h^2k_1 + hk_1k_2 + k_1k_2k_3)} \quad (5.9\text{-}11)$$

where k_1 through k_3 are defined in (5.8-13) and (5.8-14).

Instructions for exercise 5.9-2

1 Open a new spreadsheet, or modify one used earlier.

2 Enter in the top row the labels Cs, Ct, pH, f, k1, k2, k3, and kt, and below them numerical values for C_s, C_t, and pH, $f = 16.904$, $k_1 = 10^{0.255f}$, $k_2 = 10^{-0.337f+2pH}$, $k_3 = 10^{-1.000f+2pH}$, and $k_t = 10^{-1.7f}$.

3 In row 4 place the labels fE, h, and Vt/Vs.

4 Starting in row 6 of the first column, place values for fE as -10 (0.1) 35.

5 In the second column compute the corresponding values of $h = 10^{-fE}$.

6 In the third column calculate V_t/V_s according to (5.9-11).

7 On the spreadsheet plot the progress curve, V_t/V_s vs. fE, and the titration curve, fE vs. V_t/V_s. To make the latter, either (1) enter a second column for fE to the right of that for V_t/V_s, and use these to make a new chart, or (2) click on the first graph, copy and paste it with Ctrl + c, Ctrl + v. In the copy, highlight the curve, then exchange the letters identifying the two columns in the formula box, and enter. Clean up by adjusting the axis scales and labels.

8 Figures 5.9-3 and 5.9-4 show such graphs.

You will recognize how easy it is to represent these titration curves: a formula such as (5.9-11) is readily derived, and it takes only three columns to plot it. Fitting experimental data to such a curve is equally simple: add your experimental data, calculate the sum of the squares of their residuals, and use Solver to minimize that sum. You can also use (5.9-11) to derive several Gran plots by retaining only one of the three alpha's in (5.9-11) and simplifying it accordingly. Instead of fE you may want to use E, in volts. We have used fE in order to emphasize the analogy with pH measurements: one unit of fE is precisely equivalent to one pH unit: about 59 mV at room temperature.

In the above examples we moved from one sample, Fe^{2+}, to another, V^{2+}, by replacing (5.9-1) by (5.9-6) through (5.9-9). In our final example of this section we will use another titrant, and illustrate how to generate the progress or titration curves for, say, the titration of Fe^{2+} or V^{2+} with permanganate in acidic solution. In that case, all you need to do is to replace (5.9-2). Adding dilution terms to (5.8-22) and (5.8-23) yields

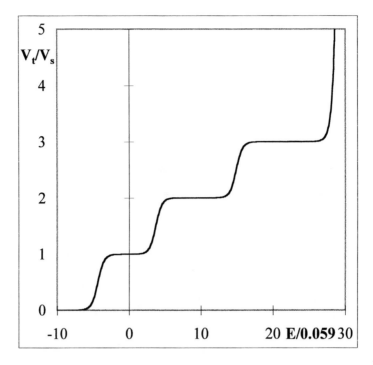

Fig. 5.9-3: The progress curve for the titration of 0.01 M V^{2+} with 0.01 M Ce^{4+} at pH 1.

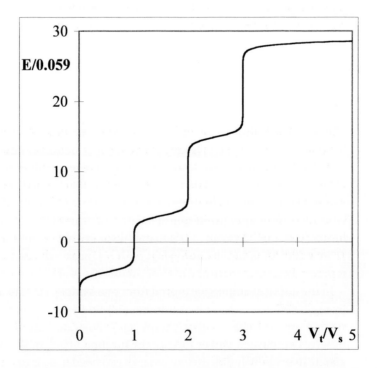

Fig. 5.9-4: The titration curve for the titration of 0.01 M V^{2+} with 0.01 M Ce^{4+} at pH 1.

$$[Mn^{2+}] = \frac{h^5 C_t V_t}{(h^5 + k_t^5)(V_s + V_t)} \qquad [MnO_4^-] = \frac{k_t^5 C_t V_t}{(h^5 + k_t^5)(V_s + V_t)} \tag{5.9-12}$$

$$k_t = 10^{-fE^*_{Mn72}} = 10^{-fE^o_{Mn72} + (8/5)pH} \tag{5.9-13}$$

while the electron balance for the titration of Fe^{2+} will read

$$[Fe^{3+}] = 5\,[Mn^{2+}] \tag{5.9-14}$$

or, for the titration of V^{2+},

$$[V^{3+}] + 2\,[VO^{2+}] + 3\,[VO_2^+] = 5\,[Mn^{2+}] \tag{5.9-15}$$

so that the progress of these titrations will be given by

$$\frac{V_t}{V_s} = \frac{C_s \alpha_{Fe^{3+}}}{5 C_t \alpha_{Mn^{2+}}} = \frac{C_s k_s (h^5 + k_t^5)}{5 C_t h^5 (h + k_s)} \tag{5.9-16}$$

or

$$\frac{V_t}{V_s} = \frac{C_s(\alpha_{V^{3+}} + 2\alpha_{VO^{2+}} + 3\alpha_{VO_2^+})}{5 C_t \alpha_{Mn^{2+}}} = \frac{C_s(h^5 + k_t^5)(h^2 k_1 + 2hk_1 k_2 + 3k_1 k_2 k_3)}{5 C_t h^5 (h^3 + h^2 k_1 + hk_1 k_2 + k_1 k_2 k_3)} \tag{5.9-17}$$

respectively. While it is unavoidable that the algebra gets a little messy when there are more species involved, the general form of the progress equation remains the same. It is therefore straightforward to extend the present formalism to other redox systems, and to fit their redox titrations as long as they exhibit equilibrium behavior.

5.10 Redox buffer action

The simultaneous presence of the oxidized and reduced form of a redox couple can stabilize the redox potential of a solution, just as the presence of an acid and its conjugate base can stabilize the pH. The formalism (R. de Levie, *J. Chem. Educ.* 76 (1999) 574) is quite similar to that of section 4.7, except that there are no terms for the oxidation or reduction of the solvent, because these are typically non-equilibrium processes which, moreover, are insignificant in the usual range of potentials considered. By analogy to (4.7-1) we write, for the **redox buffer strength** B of a one-electron redox couple $Ox + e^- \rightleftharpoons Red$, such as $Fe^{3+} + e^- \rightleftharpoons Fe^{2+}$ or $Ce^{4+} + e^- \rightleftharpoons Ce^{3+}$,

$$B = C\alpha_{Ox}\alpha_{Red} \tag{5.10-1}$$

where C is the total analytical concentration, $C = [Ox] + [Red]$. Because $C\alpha_{Ox} = [Ox]$ and $C\alpha_{Red} = [Red]$, we can rewrite (5.10-1) as $B = C^2 \alpha_{Ox}\alpha_{Red}/C = [Ox]$ $[Red] / ([Ox] + [Red])$ or

$$\frac{1}{B} = \frac{1}{[Red]} + \frac{1}{[Ox]} \tag{5.10-2}$$

For the $VO_2^+/VO^{2+}/V^{3+}/V^{2+}$ system the redox buffer strength is analogous to (4.9-12),

$$B = C(\alpha_{VO_2^+}\alpha_{VO^{2+}} + \alpha_{VO^{2+}}\alpha_{V^{3+}} + \alpha_{V^{3+}}\alpha_{V^{2+}} + 4\alpha_{VO_2^+}\alpha_{V^{3+}} + 4\alpha_{VO^{2+}}\alpha_{V^{2+}}$$
$$+ 9\alpha_{VO_2^+}\alpha_{V^{2+}}) \tag{5.10-3}$$

which reduces to

$$B \approx C(\alpha_{V^{2+}}\alpha_{V^{3+}} + \alpha_{V^{3+}}\alpha_{VO^{2+}} + \alpha_{VO^{2+}}\alpha_{VO_2^+}) \tag{5.10-4}$$

when the various standard potentials E^o and conditional potentials E^* are sufficiently far apart. In that case, a simple relationship such as (5.10-2) applies in each **buffer region**, i.e., the range of potentials around E^o or E^*, just as the acid–base buffer region occurs at $pH \approx pK_a$. Note that the numerical coefficients 1, 4 and 9 in the products of the alpha's are the *squares* of the numbers n of electrons involved in the transition from one form into the other, just as they are the squares of the number of protons involved in (4.9-12).

For permanganate the intermediate redox states are not stable, and we go almost directly from MnO_4^- to either MnO_2 (in neutral or basic media) or Mn^{2+} (in acidic solutions). In that case the general expression reduces to

$$B \approx 9C\alpha_{MnO_4^-}\alpha_{MnO_2} \tag{5.10-5}$$

$$\frac{9}{B} = \frac{1}{[MnO_4^-]} + \frac{1}{[MnO_2]} \tag{5.10-6}$$

for the three-electron step at neutral or basic pH, or, at low pH, for a five-electron transition,

$$B \approx 25C\alpha_{MnO_4^-}\alpha_{Mn^{2+}} \tag{5.10-7}$$

$$\frac{25}{B} = \frac{1}{[MnO_4^-]} + \frac{1}{[Mn^{2+}]} \tag{5.10-8}$$

All redox titration curves we have discussed here are independent of the total analytical concentration C of the redox couple. (This is not always the case: in the $Cr_2O_7^{2-}/Cr^{3+}$ couple the reduction of *one* $Cr_2O_7^{2-}$ generates *two* Cr^{3+} ions, which leads to a concentration-dependent redox titration curve.) Therefore, the above expressions precisely give C_s/ln (10) times the first derivative of the progress curve of the corresponding redox titration. You can convince yourself that this is so in exercise 5.10-1.

Instructions for exercise 5.10

1 Go back to the spreadsheet used for exercise 5.9-1.

2 Add two columns, one labeled Bs, the other 2.3Csderiv.

3 In column Bs calculate the buffer strength using (5.10-2), i.e., based the formula $C_s h k_s / (h + k_s)^2$.

4 In column 2.3Csderiv compute $C_s/\ln(10)$ times the first derivative of the progress curve. Calculate latter with the same simple formula used in section 4.5, i.e., the derivative dy/dx at row i is equal to $(-2y_{i-2} - y_{i-1} + y_{i+1} + 2y_{i+2})/(10\delta)$, where the data spacing δ is 0.1 so that $10\delta = 1$.

5 Plot the two resulting curves versus fE. Figure 5.10-1 shows what such a curve should look like.

6 Similarly extend exercise 5.9-2 to include two columns, labeled Bs and 2.3Csderiv respectively.

7 Calculate B_s from (5.10-4) as $C_s h \{ k_1/(h + k_1)^2 + k_2/(h + k_2)^2 + k_3/(h + k_3)^2 \}$.

8 In the next column compute $C_s/\ln(10)$ times the first derivative of the progress curve, as under instruction (4).

9 Again plot these two curves of redox buffer strength as a function of the dimensionless potential fE, and compare your results with Fig. 5.10-2.

The redox buffer strength serves the same role for the potential of a solution as the acid–base buffer strength serves for its pH. In both cases it is assumed that the corresponding equilibria are established quickly on the time scale of the experiment. With redox equilibria, which often involve bond breaking, this condition is less often met than with acid–base equilibria, where fast establishment of equilibrium is the norm.

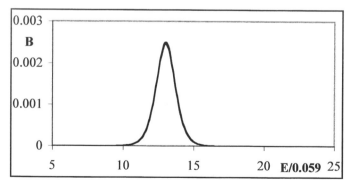

Fig. 5.10-1: The redox buffer strength of a strongly acidic aqueous iron solution of 0.01 M analytical concentration, calculated from (5.10-1) or by differentiation of the progress curve (5.9-5); the two agree to within 2%.

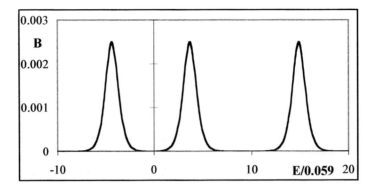

Fig. 5.10-2: The redox buffer strength of an aqueous vanadium solution of 0.01 M analytical concentration at pH = 0, calculated from (5.10-3) or by differentiation of the progress curve of Fig. 5.9-3. Again the two agree to within the computational accuracy of the differentiation algorithm used.

5.11 Summary

In this chapter we have applied the methods of chapter 4 to ionic equilibria other than those between acids and bases. Of course, complexation, extraction, solubility, precipitation, and redox equilibria may also involve acid–base equilibria, which is why we treated acid–base equilibria first. The examples given here illustrate that the combination of exact theory with the computational power of a spreadsheet allows us to solve many problems that occur in quantitative chemical analysis, and to analyze experimental data accordingly. Even quite complicated titrations, such as the multi-component precipitation titrations, the von Liebig titration, and redox titrations involving many species and complicated stoichiometries, can be handled with ease.

Again, we have not included activity corrections, because (both didactically and computationally) these are best added afterwards whenever such corrections are required. The principles involved are the same as those explained in section 4.10: activity corrections apply to the equilibrium constants (such as K_{fi}, K_{si}, and k) but not to the mass and charge balance relations and their derivatives, such as a ligand balance or an electron balance. Furthermore, electrometric measurements must be corrected for activity effects, but spectroscopic measurements should not be. At any rate, as the example of HgS in section 5.4 illustrates, the proper *chemistry* of including all important species is always far more important than the proper *physics* of making activity corrections.

The reader may wonder why the discussion in this chapter is restricted to ionic equilibria. Are there not many other equilibria, not involving ions, solubilities of non-ionic compounds, electrochemical processes involving neutral species, etc.? The reasons for focusing exclusively on ions are (1) that the formalism of equilibria is often more complicated when ions are involved, therefore making them more in need of spreadsheet computation, and (2) that many equilibria involving neutral species are established so slowly that their kinetics must be taken into account. Such slow processes are often of little analytical value, and they therefore fall outside the reach of this book. However, chapter 9 discusses one aspect of slow reaction rates where spreadsheets can be helpful, namely in the simulation of reaction kinetics.

Non-equilibrium behavior may also affect some ionic reactions. In our examples we have therefore emphasized processes involving substitution-labile ions rather than substitution-inert ones. Problems of slow kinetics are especially common with ionic redox reactions, in which case equilibrium considerations indicate what is theoretically feasible, but not necessarily what is truly factual. This is why so many quantitative electrometric methods are based on either silver or mercury, two metals on which the metal/metal ion equilibrium is usually established so rapidly that the underlying kinetics can be neglected in routine analytical measurements, and on platinum, where the same applies to many electron transfer processes between soluble redox couples.

CHAPTER **6**

SPECTROMETRY, CHROMATOGRAPHY, AND VOLTAMMETRY

This short chapter contains a somewhat disparate collection of topics, commonly treated in either quantitative or instrumental chemical analysis. They illustrate a variety of methods, but their order is of no particular importance. The reader should therefore feel free to pick and choose from them, in any order.

In the first examples, spectrometry is used as an excuse to revisit some of the problems and methods encountered in earlier chapters, such as the determination of the pK_a of an indicator dye, and multicomponent analysis. We also illustrate the still little-known absorbance–absorbance diagrams.

We use chromatography as our pretense to simulate the action of a differential equation. In a second chromatographic exercise, we show how the van Deemter plot can be linearized.

Our example in polarography illustrates how a spreadsheet can be used to simulate a rather complex curve, in this case reflecting the interplay between the Nernst equation, Fick's law of diffusion, and drop growth. The first two factors also play a role in cyclic voltammetry, where we introduce semi-integration as an example of deconvolution.

6.1 Spectrometric pK_a determination

Spectrometry can be used to determine the pK_a of a weak acid. Here we will show this with a data set that is already more than 60 years old, at the same time illustrating that the quality of the data usually depends much more on the experimental care taken in obtaining them than on the availability of the latest instrumentation.

The data were taken from H. von Halban & G. Kortüm, *Z. Elektrochem.* 40 (1934) 502. Weighed amounts of 2,4-dinitrophenol were dissolved in carbonate-free water, and were compared by differential spectrometry with similarly weighed solutions of the same dye in 5 mM NaOH, in which

Table 6.1-1: *The total analytical concentration C of 2,4-dinitrophenol in water, and the resulting concentration [A–] of 2,4-dinitrophenolate as measured by differential spectrometry.*

$C/\mu M$	$[A^-]/\mu M$	$C/\mu M$	$[A^-]/\mu M$	$C/\mu M$	$[A^-]/\mu M$
701.5	204.9	358.3	136.3	180.1	87.70
677.3	200.7	353.3	135.0	153.6	78.85
618.6	190.0	285.2	118.0	136.4	72.75
598.7	186.5	262.1	111.8	122.6	67.58
550.5	177.5	239.5	105.6	114.4	64.38
441.8	155.1	212.8	97.83	92.45	55.41
399.1	145.7				

2,4-dinitrophenol is fully dissociated. The measurements were made at 436 nm, where the acid form has negligible absorption. The temperature was controlled at 25 °C, the dye and the water were carefully purified, and the glassware was meticulously cleaned. The measurements were made in a stationary cuvet; the various solutions were entered and removed by flushing. The resulting experimental data are listed in Table 6.1-1.

Instructions for exercise 6.1

1 Open a new spreadsheet.

2 Enter column headings for C, $[A^-]$, α, K_a, $I^{1/2}$, pK_a, and $pK_{a,calc}$.

3 Enter the data for C and $[A^-]$ (in M, not μM!) from Table 6.1-1.

4 Calculate α as $[A^-]/C$, K_a as $\alpha^2 C/(1-\alpha)$, $\sqrt{I}$ as $\sqrt{[A^-]}$, and pK_a as $-\log K_a$.

5 With a linear regression (e.g., Trendline, or Tools ⇨ Data Analysis ⇨ Regression) obtain the value of pK_a extrapolated to infinite dilution.

6 Compute $pK_{a,calc}$ based on the least-squares parameters found.

7 Plot the original data as well as the least-squares line through them. The plot should look more or less like Fig. 6.1-1.

Note that we can use a linear extrapolation because, at the low ionic strengths of these solutions, the Debye–Hückel limiting law is quite sufficient. If all experimental data were of this quality, we could have pK_a-values listed to three significant decimal places!

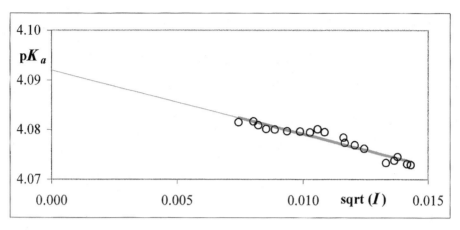

Fig. 6.1-1: The pK_a-values calculated from the data (open circles) and the straight line drawn through them by least-squares, extrapolated (thin colored line) to yield the 'infinite dilution' value.

6.2 Multi-component spectrometric analysis 1

According to Beer's law, the measured absorbance of a solution of a single light-absorbing species is directly proportional to its concentration. For a solution containing a mixture of absorbing species, the measured absorbance is then simply the linear combination of the absorbances of all species in that solution, each measured at the same wavelength. When the different species in the mixture have different spectra, we can do a multi-component analysis and extract the concentrations of the individual species.

For a mixture of two species, the minimum requirement would be to make measurements at two wavelengths; for a mixture of three, at least three wavelengths must be used, etc. Here we will show how Beer's law can be used to determine the concentrations of four species by making absorbance measurements at four different wavelengths. As our example we will consider the mixture of ethylbenzene and the three xylenes in cyclohexane; we will use infrared absorption data listed in R. P. Bauman, *Absorption Spectroscopy*, Wiley 1962 p. 408, and reproduced in Table 6.2-1.

We notice that the wavenumbers have been chosen judiciously, in that each compound has a significant absorbance at one of these, while contributing relatively little at the other wavenumbers. Indeed, the four components have absorption peaks at precisely one of these wavenumbers, and the mixture shows four almost baseline-separated absorption peaks, one for each component. This, then, is an 'ideal' example, of which the solution is, at least in principle, quite straightforward: we need to solve four simultaneous equations in four unknowns.

Table 6.2-1: *The absorbances (per gram of compound in 50 mL cyclohexane) of ethylbenzene and the three xylenes, at the wavenumbers listed in the first column, and (in the last column) the absorbance of a test mixture of these compounds, all as measured in cyclohexane.*

Wavenumber /cm^{-1}	ethyl-benzene	*o*-xylene	*m*-xylene	*p*-xylene	unknown mixture
696.3	1.6534	0.0	0.1289	0.0641	0.07386
741.2	0.5524	4.7690	0.0668	0.0645	0.22036
768.0	0.1544	0.0	2.8542	0.0492	0.08676
795.2	0.0768	0.0	0.0968	2.8288	0.07721

We now denote the various wavenumbers by the index i, and the concentrations of the four species in the mixture by the index j. We have four equations of the type

$$x_1 = a_{1,1} c_1 + a_{1,2} c_2 + a_{1,3} c_3 + a_{1,4} c_4 \tag{6.2-1}$$

$$x_2 = a_{2,1} c_1 + a_{2,2} c_2 + a_{2,3} c_3 + a_{2,4} c_4 \tag{6.2-2}$$

$$x_3 = a_{3,1} c_1 + a_{3,2} c_2 + a_{3,3} c_3 + a_{3,4} c_4 \tag{6.2-3}$$

$$x_4 = a_{4,1} c_1 + a_{4,2} c_2 + a_{4,3} c_3 + a_{4,4} c_4 \tag{6.2-4}$$

where the x_i represent the absorbances of the mixture at the four wavenumbers i, each divided by the optical pathlength b through the solution. The terms $a_{i,j}$ are the measured absorbances of the pure reference compounds at those same wavenumbers, and the c_j's are the concentrations of the four components in the mixture, i.e., the concentrations to be determined by the experiment. These four equations can be written much more compactly, in terms of matrix notation, as

$$\mathbf{X} = \mathbf{A} \mathbf{C} \tag{6.2-5}$$

where **X** and **C** are vectors, and **A** is a matrix,

$$\mathbf{X} = \begin{bmatrix} x_1 \\ x_2 \\ x_3 \\ x_4 \end{bmatrix}, \quad \mathbf{A} = \begin{bmatrix} a_{1,1} & a_{1,2} & a_{1,3} & a_{1,4} \\ a_{2,1} & a_{2,2} & a_{2,3} & a_{2,4} \\ a_{3,1} & a_{3,2} & a_{3,3} & a_{3,4} \\ a_{4,1} & a_{4,2} & a_{4,3} & a_{4,4} \end{bmatrix}, \quad \mathbf{C} = \begin{bmatrix} c_1 \\ c_2 \\ c_3 \\ c_4 \end{bmatrix} \tag{6.2-6}$$

Using the rules of matrix algebra (which are briefly reviewed in section 8.9) we left-multiply both sides of (6.2-5) by the inverse of **A**,

$$\mathbf{A}^{-1}\mathbf{X} = \mathbf{A}^{-1}\mathbf{A}\mathbf{C} = \mathbf{I}\mathbf{C} = \mathbf{C} \tag{6.2-7}$$

(where $\mathbf{A}^{-1} \mathbf{A} = \mathbf{I}$, the unit matrix) so that we can immediately obtain the desired result **C**, which contains the concentrations in the mixture, simply from $\mathbf{C} = \mathbf{A}^{-1}\mathbf{X}$. The spreadsheet will do the mathematics for us.

Instructions for exercise 6.2

1 Open a new spreadsheet.

2 In cell B4 deposit a label for the matrix **A**, in cell B11 a label for the matrix **A**$^{-1}$, in F4 place a label for the vector **X**, and in F11 a label for the vector **C**.

3 In A6:D9 enter the numbers from the middle four columns of Table 6.2-1, in the order in which they appear there, i.e., with 1.6534 in cell A6, and 2.8288 in cell D9.

4 In F6:F9 enter the numbers from the last column of Table 6.2-1, i.e., starting with 0.07386 in F6, and ending with 0.07721 in F9.

Now that the experimental data are in place, here comes the matrix algebra.

5 Highlight A13:D16, then type = MINVERSE(A6:D9), and press Ctrl + Shift + Enter (i.e., hold down the Control and Shift keys while depressing the Enter key), in order to inform the spreadsheet that you intend this formula for the *entire* block. You will see the inverse matrix appear in that block.

6 Now highlight F13:F16, type = MMULT(A13:D16,F6:F9), and again depress Ctrl + Shift + Enter to enter this instruction in the highlighted block. That's it: F13:F16 now contains the four sought concentrations.

7 In order to validate your answer, in H6 compute the absorbance at 696.3 cm^{-1} as = A6*$F$13 + B6*$F$14 + C6*$F$15 + D6*$F$16.

8 Copy this instruction to cells H7:H9. What do they show?

For your information: the 'unknown' mixture had been made up from the pure components, and contained 0.420, 0.398, 0.271, and 0.248 g/50 mL of ethylbenzene and of ortho-, meta-, and para-xylene respectively. The (relatively small) differences between these numbers and your results in F13:F16, of less than $\pm 2.5\%$, are *not* caused by computational errors, but instead reflect uncertainties in the measured absorbances. Although the infrared absorbances are listed in Table 6.2-1 to three or four digits, they were (conservatively) rated as most likely good to $\pm 5\%$ only.

Also for your information: in 1962, when Bauman wrote his book, neither personal computers nor spreadsheets were available, and he commented on page 411 of his book that calculating the inverse matrix **A**$^{-1}$ "represents roughly an hour and a quarter of work, including checking". Thank you, personal computer; thank you, spreadsheet.

6.3 Multi-component spectrometric analysis 2

In the preceding section we determined the concentrations of four chemical substances by making measurements on the unknown mixture and on four single-component standards at four different wavelengths. Four measurements are clearly the minimum requirement for determining four unknown concentrations. While use of the minimum number of wavelengths may save some time, it makes the results critically dependent on the quality of the measurements, because it leaves no margin for experimental error. It is therefore preferable to determine the entire spectrum of the mixture (or, more precisely, measure a large number of absorbance data over a fairly wide range of wavelengths), to be compared with the corresponding spectra of the single components, each of course obtained under otherwise identical conditions in the same solvent. Here we will illustrate this approach for a simulated mixture of three different species. As you will see, generating the synthetic spectra will take more time than analyzing them!

To keep things simple, we will use Gaussian curves to generate our mock spectra, but you are welcome to modify the instructions by substituting other shapes, or even by 'drawing them by hand' by entering numbers.

Figure 6.3-1 illustrates three such made-up spectra, and the spectrum of a mixture of arbitrary amounts of these three. It does not matter that the spectrum of the mixture does not show much structure.

Instructions for exercise 6.3

1 Open a new spreadsheet.

2 In cell A1 deposit label ampl =, in A2 the label center =, and in A3 the label width =. Copy A1:A3 and paste it into cell A4.

3 In some out-of-view place like N10:Q40, enter some Gaussian noise (using Tools ⇨ Data Analysis… ⇨ Random Number Generation, OK ⇨ Distribution: Normal, Mean = 0, Standard Deviation = 1, Output Range: N10:Q10, OK).

4 In cell G1 deposit the label na =, and in H1 its value, say 0.005.

5 In cells C1:C6 deposit some numbers, such as 0.7, 470, 2000, 0.4, 600, and 500. Right-align them.

6 In row 8 deposit the labels wavelength, unknown, spectrum 1, spectrum 2, and spectrum 3.

7 Fill A10:A40 with the numbers 400 (10) 700 representing the wavelengths of the visible region of the spectrum, in nanometers.

8 For the time being, skip column B, and in C10 deposit the instruction = C\$1*EXP(− ((\$A10 − C\$2)^2)/C\$3) + \$H\$1*O10.

9 Highlight cell C10, go to the formula box, there highlight the part (C\$1*EXP(− ((\$A10 − C\$2)^2)/C\$3)), copy it with Ctrl + c, then go to the end of that same instruction, add a plus sign, and paste it in.

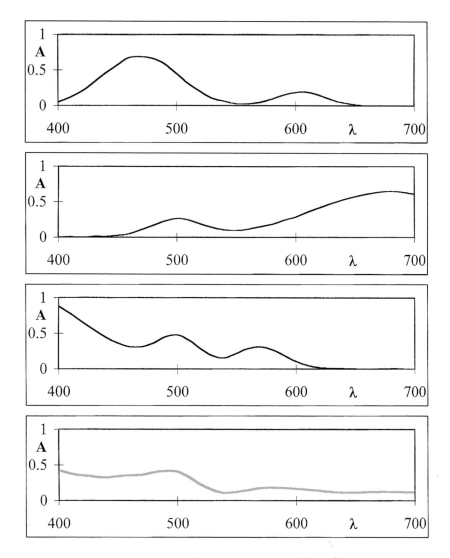

Fig. 6.3-1: Three made-up spectra of pure species (top), and that of their mixture.

10 In the part of the instruction just added, replace the numbers 1, 2, and 3 by 4, 5, and 6 respectively, so that it reads = C$1*EXP(–(($A10 – C$2)^2)/C$3) + $H$1*O10 + C$4*EXP(– (($A10 – C$5)^2)/C$6). Depress the Enter key.

11 Copy the modified instruction down to row 40.

12 Make a graph of C10:C40 vs. A10:A40. You will recognize the values of $C$1 and $C$4 as the amplitudes, $C$2 and $C$5 as the center wavelengths, and $C$3 and $C$6 as determining the peak widths.

13 Place a set of constants (different from those in C1:C6) in D1:D6.

14 Copy the instruction from cell C10 to cell D10, then copy it to row 40.

15 Plot the resulting spectrum #2 as D10:D40 versus A10:A40. Alternatively you can high-light D10:D40, copy it with Ctrl + c, activate the inner frame of the figure made under point (11), then paste the new curve in with Ctrl + v.

16 Repeat steps (13) through (15) to generate a third fantasy spectrum in column E. If you feel like it, you can of course make these spectra more varied, e.g., by letting one have only one absorption peak, while another gets three or more. Suit your fancy.

17 Now you are ready to simulate the spectrum of the 'unknown' mixture. In cell B10 deposit the instruction = 0.3*(C10 + $H$1*O10) + 0.2*(D10 + $H$1*P10) + 0.45*(E10 + $H$1*Q10) + 0.005*N10 to simulate the spectrum of the unknown.

18 Plot the simulated 'unknown' spectrum.

You are free, of course, to pick three characteristic wavelengths, and to solve the resulting three simultaneous equations. The added noise will then (rather strongly) affect your results, but you will have no way of knowing by how much. The method illustrated below is not only much less sensitive to noise, but also provides error estimates and, most importantly, is much easier to implement. Of course it uses matrix algebra, just as you did in section 6.2, but that will be completely invisible to you, the user. The entire analysis comes prepackaged with the spreadsheet.

We will use the standard linear least squares routine, with column B as the dependent variable y, and columns C, D, and E as the independent variables x_1, x_2, and x_3. After all, the absorbance of the mixture, A_{mixture}, is given by Beer's law as

$$A_{\text{mixture}} = b\,(\,a_1\,c_1 + a_2\,c_2 + a_3\,c_3) \tag{6.3-1}$$

where b is the optical path-length, the a_i's are the absorbances of the individual species i, and the c_i's their concentrations in the mixture. Since the spectra of the individual species represent b times the a_i's, the entire problem has only three unknowns, c_1, c_2, and c_3. The a_i's of course are functions of the wavelength, as is A_{mixture}, which is why we have entire columns for them. But the whole problem is simply one of a multiple-parameter fitting, which we already encountered in section 3.1. (Now you understand why we initially left column B free, so that we could fit the standard format of the regression routine.) So here we go.

19 Select Tools ⇨ Data Analysis... ⇨ Regression, OK ⇨ Input Y Range: B10:B40, Input X Range: C10:E40, Output Range: A44, OK). Now sit back, it is done, the problem is solved. You will find the results in B61:C63. The intercept should be insignificant, i.e., smaller than its standard deviation.

20 Print out the results, together with the concentrations assumed in the simulation, for various amounts of added noise (as selectable in H1).

In the above example, using the data illustrated in Fig. 6.3-1, we find $c_1 = 0.305 \pm 0.004$, $c_2 = 0.198 \pm 0.005$, $c_1 = 0.453 \pm 0.008$. You will recognize that these results are all right on target. Moreover, when we make the noise amplitudes zero, we recover the exact numbers we had put in, 0.3, 0.2, and 0.45 in our example. Clearly, the uncertainty is purely the result of the added noise, introduced on purpose to simulate more realistic data. Because this method uses 120 data points, much of the random noise will now cancel.

You will have noticed that you spent much more time setting up the problem (by generating phantom spectra) in exercise 6.3 than you spent solving it. In practice, you will of course use existing spectra, generated by a spectrophotometer. In that case, the analysis merely requires that you import the data arrays in the proper order (unknown mixture first, reference spectra of pure compounds next), and then invoke the Regression analysis. It really is that simple.

6.4 The absorbance–absorbance diagram

When a compound has a spectrum that is a complicated function of pH (or of some other variable, such as a ligand concentration), as the result of successive protonation (or complexation) steps, as in Fig. 6.4-1, we first need to establish *how many species* are involved in the optical behavior. To this end it is often useful to analyze the set of spectra in terms of an absorbance–absorbance diagram. In such a diagram we plot the absorbance at one wavelength against that at another wavelength, at constant optical pathlength b and total analytical concentration C of the absorbing species, using the pH (or pX) as the implicit variable. The resulting plot consists of a series of connected, near-linear line segments, one fewer than the number of different species formed. The points where the extrapolated linear sections intersect yield the absorbances of the intermediate species at the two wavelengths used, see Fig. 6.4-2. Midway between these special points we have $A = (A_i + A_{i+1})/2$ or pH $\approx$ pK_a. The book by J. Polster and H. Lachmann, *Spectrometric Titrations* (VCH, 1989) contains a number of fine examples of such diagrams. Below we simulate an absorbance–absorbance diagram in order to introduce the reader to this interesting method.

We will use as our example a weak diprotic acid, H_2A. We will label the three species by the number of attached, dissociable protons, i.e., H_2A will be denoted by the subscript 2, HA^- by 1, and A^{2-} by 0. The corresponding concentration fractions α_2, α_1, and α_0 are given in (4.8-5) through (4.8-7) respectively. Furthermore we denote two different wavelengths by $'$ and $''$ respectively, and associate different molar absorptivities with each species.

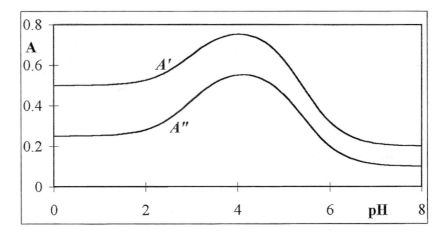

Fig. 6.4-1: The absorbances A' and A'' as a function of pH.

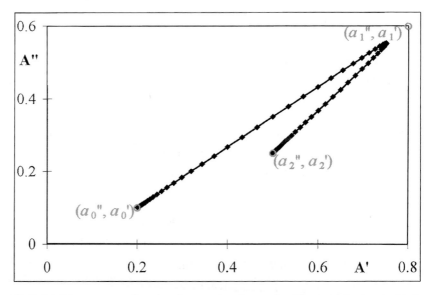

Fig. 6.4-2: The corresponding absorbance–absorbance plot. The special points (a_0'', a_0'), (a_1'', a_1'), and (a_2'', a_2') are shown and labeled in color.

We then have

$$A'_{\text{mixture}} = (a_2'\alpha_2' + a_1'\alpha_1' + a_0'\alpha_0')bC$$

$$= \left(\frac{[H^+]^2 a_2' + [H^+]K_{a1}a_1' + K_{a1}K_{a2}a_0'}{[H^+]^2 + [H^+]K_{a1} + K_{a1}K_{a2}}\right)bC \qquad (6.4\text{-}1)$$

$$A''_{\text{mixture}} = (a_2''\alpha_2'' + a_1''\alpha_1'' + a_0''\alpha_0'')bC$$

$$= \left(\frac{[H^+]^2 a_2'' + [H^+]K_{a1}a_1'' + K_{a1}K_{a2}a_0''}{[H^+]^2 + [H^+]K_{a1} + K_{a1}K_{a2}}\right)bC \qquad (6.4\text{-}2)$$

We now plot $A'_{mixture}$ versus $A''_{mixture}$ as a function of $[H^+]$. We could consider the resulting plot in terms of mathematics, but here we will take the less rigorous approach of merely illustrating it, using the spreadsheet. Since the total analytical concentration C and the optical path-length b must be kept constant, they can both be set to unity; they will only affect the scale of the resulting plots, not their shape.

Instructions for exercise 6.4

1 Open a new spreadsheet.

2 In cells A1 and A2 place the labels $a_2' =$ and $a_2'' =$ respectively, and right-align them. Similarly, in cells C1:C2 deposit the labels $a_1' =$ and $a_1'' =$, in cells E1:E2 the labels $a_0' =$ and $a_0'' =$, and in G1:G2 the labels $K_{a1} =$ and $K_{a2} =$.

3 In cells B1 and B2 deposit the values 0.5 and 0.25, in D1:D2 0.8 and 0.6, in F1:F2 0.2 and 0.1, and in H1:H2 $= 10^ - 3$ and $= 10^ - 5.4$. Left-align them.

You may wonder why we select these specific numbers. They are roughly those that can be read from the data on phthalic acid at 286 and 290 nm respectively, as published by R. Blume *et al.*, *Z. Naturforsch.* 30B (1975) 263 and in the earlier-mentioned book by Polster & Lachmann on *Spectrometric Titrations*, and reproduced, for those without easy access to the original sources, on page 503 of my *Principles of Quantitative Chemical Analysis*. If you prefer to use other numbers, by all means feel free to use your own; the method does not hinge on the specific numbers used.

4 In row 4 deposit the labels pH, $[H^+]$, A', and A''.

5 In A6:A86 calculate the numbers 0 (0.1) 8.

6 In B6:B86 compute the corresponding values of $[H^+]$.

7 In C6:C86 use (6.3-1) to calculate A', the absorbance at 286 nm.

8 In D6:D86 likewise compute A'', at 290 nm.

9 Plot A' and A'' versus pH, see Fig. 6.4-1.

10 Plot A'' versus A', compare Fig. 6.4-2.

This example is a close image of the experimental data at 290 and 286 nm for phthalic acid, as published by R. Blume *et al.* and Polster & Lachmann. Upon extrapolation, the two straight-line segments will intersect in the point (a_1'', a_1'). Such absorbance–absorbance plots are superb diagnostics of the *number* of species involved, especially when the analysis uses several different wavelengths. The plots require high-quality data at constant total analytical concentration C, which are most readily obtained by titrating the solution with a much more concentrated titrant (to keep dilution effects negligibly small) or by using (dilution-less) electrochemical titrant generation, and by circulating the test solution through a stationary, firmly seated cuvet. Removing and reinserting a cuvet is almost always a prescription for low-precision absorbance measurements. Extrapolation of the linear parts of the plot yields values of the absorbances a_1'' and a_1' of the monoprotonated species.

6.5 Chromatographic plate theory 1

Chromatography is the most versatile chemical separation method we have available at present. In it, a mixture is separated into various components on the basis of differences in the speed at which these components move through a chromatographic column. What differentiates their speeds is that the column slows them down by one mechanism or another. Many different retarding effects can be exploited, based on such diverse molecular properties as solubility, charge, size, adsorption, or biochemical affinity. Here we will consider **partition chromatography**, which is based on differences in (solvation) energy.

In a chromatographic column there are always two phases: one moving (the *mobile* phase), the other *stationary*. The mobile phase can be either a gas or a liquid, and chromatographies are often characterized on that basis as either gas or liquid chromatography. The stationary phase can be a wall coating or, in a so-called packed column, a coating on and inside porous particles. Here we will model the continuous column as composed of a large number N_p of very small sections or "plates". We will describe the continuous motion of the mobile phase as proceeding in discrete installments, in which the mobile phase from one plate displaces that from the next plate, and then stays in place long enough to establish local equilibrium.

The centerpiece of our model is the **partition coefficient** $K_p = c_s / c_m$, the equilibrium constant that describes the equilibrium distribution of a chemical species between the two contacting phases. Here c_s is the concentration in the stationary phase, and c_m that in the mobile phase. Just as in the treatment of extraction in section 5.3, we now define a **mass fraction** in the mobile phase, μ, as

$$\mu = \frac{c_m v_m}{c_m v_m + c_s v_s} = \frac{1}{1 + K_p v_s / v_m} \tag{6.5-1}$$

where v_s and v_m are the plate volumes of the stationary and mobile phase respectively. In each plate the ratio v_s and v_m is the same, as is K_p, so that we can consider μ as our primary model parameter.

Now consider the total mass $m_{p,t}$ of the chemical species considered in plate p at time t. It is composed of two parts: the mobile phase in that plate contains the fraction μ of the total mass $m_{p,t}$, i.e., $\mu\, m_{p,t}$, while the stationary phase holds a mass $(1 - \mu)\, m_{p,t}$. Likewise we have a mass $\mu\, m_{p-1,t}$ in the mobile phase just upstream from plate p.

We now move the mobile phase from plate $p-1$ to plate p while, simultaneously, moving the mobile phase in plate p to plate $p+1$. The total mass in plate p at time $t+1$ will then be

$$m_{p,t+1} = \mu \, m_{p-1,t} + (1-\mu) \, m_{p,t} \tag{6.5-2}$$

which you will recognize as a *recursion formula*. We will use (6.5-2) to describe the separation process in partition chromatography, in which the chemical species traveling down the column hops in and out of the stationary phase. Since the sample will move through the column only when it resides in the mobile phase, the more time it spends in the stationary phase, the slower it moves through the column, and the longer will be its **retention time** t_r, i.e., the time it takes that species to travel the entire length of the column. We will assume that the mobile phase moves with a constant linear velocity through the column (which is correct for a liquid, but must be corrected for the compressibility of gases and vapors when these constitute the mobile phase) and therefore spends a fixed time τ in each plate. We will measure the elapsed time t in units of τ.

Instructions for exercise 6.5

1 Open a new spreadsheet.

2 In cell A1 deposit the label $\mu=$, and in cell A2 a corresponding value between 0 and 1 (because μ is a fraction).

3 In cell B4 place the label p=, and enter the labels 1, 2, 3, etc. in cells C4, D4, E4, etc., all the way till Z4. Of course you will only enter the first two, and will let the spreadsheet put the other numbers in place, by dragging them sideways by the handle. Tricks that work vertically also work horizontally.

4 In cell A5 place the label t/τ= in the left-most corner of the cell.

5 In cell A6 deposit the value 0, in cell A7 place the number 1, in cell A8 the instruction =A7 + 1, and copy this down till row 206. This will be our time scale, i.e., the numbers in column A will be our simulated times t/τ.

6 In cell C6 deposit the instruction = \$A\$2*B5 + (1–\$A\$2)*C5 expressing the recursive relation (6.5-2).

7 Copy this instruction to the entire block C6:Z206. Since you have not yet introduced any sample into the model, the entire block will show zeros.

8 Finally, introduce your sample by overwriting the instruction in cell C6 with the number 1, equivalent to injecting your sample at the end of the column connected to the injection port. Bingo! The spreadsheet fills, showing the distribution of the species in the various plates (each plate being represented by a separate column) and at different times (in the rows, with time increasing as you move down). The top of the spreadsheet (except for cells E1:H2) will now look like Fig. 6.5-1, with numbers that depend on the value of μ used.

	A	B	C	D	E	F	G	H
1	$\mu =$			$A =$	$t_r =$	$\mu\, t_r =$	$\sigma_r =$	$N_p =$
2	0.8			1.00	48.00	24.00	6.95	23.88
3								
4		$p =$	1	2	3	4	5	6
5	$t =$							
6	0		1	0	0	0	0	0
7	1		0.2	0.8	0	0	0	0
8	2		0.04	0.32	0.64	0	0	0
9	3		0.008	0.096	0.384	0.512	0	0
10	4		0.0016	0.026	0.1536	0.410	0.4096	0
11	5		0.00032	0.0064	0.0512	0.2048	0.4096	0.32768
12	6		0.00006	0.00154	0.01536	0.08192	0.24576	0.39322
13	7		0.00001	0.00036	0.00430	0.02867	0.11469	0.27525

Fig. 6.5-1: The top of the spreadsheet, except for the contents of cells D1 through H2.

9 So far the spreadsheet is just as useful as a chromatograph without a detector: we have yet to find out when and how much of the sample emerges from the column. Do this by entering in cell AA6 the instruction $= \$A\$2*Z6$, which represents the amount of material $\mu\, m_{Np,t}$ moving out of the last plate, $p = N_p$, at time t. You will recognize the above instruction as corresponding to the absence of a stationary phase in the detector 'plate'.

10 Plot AA6:AA206 versus A6:A206. This will be the chromatogram (provided the detector response is proportional to the amount of material that passes it) of a species that travels with a mobile mass fraction μ through a column containing N_p plates.

11 Now make a composite chromatogram, showing how the elution differs for various mixture components that have different values of μ. In order to organize such a plot, label cells AB4 through AI4 with values for μ in the sequence 0.2 (0.1) 0.9.

12 Highlight the data in AA6:AA206, copy that column with Ctrl + c to the clipboard, highlight the appropriate target cell in row AB6:AI6, then use Edit ⇨ Paste Special ⇨ Values to copy the *values* in column AA to the column for that particular value of μ.

13 Change the value of μ in A2, copy the resulting values from column AA to their proper places, then plot them. Figure 6.5-2 shows an example of such a graph for just three values of μ.

14 For each elution curve, note the time t at which the maximum amount of sample passes the detector. You can find the maximum *value* in a given column by scanning them visually or, easier, by letting the spreadsheet do it for you with an instruction

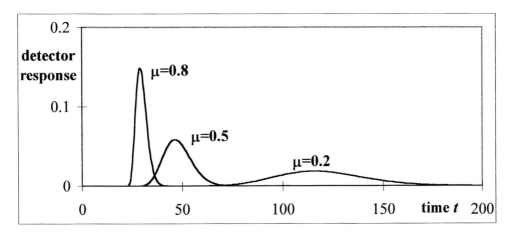

Fig. 6.5-2: Three simulated chromatograms, for $\mu = 0.2$, 0.5, and 0.8 respectively.

such as = MAX(AE6:AE206). In order to find the corresponding *time* use the instruction = MATCH(MAX(AE6:AE206), AE6:AE206,0) which has the syntax MATCH(*value to be found, array to be searched, in what order*), with commas to separate the three different pieces of information. This will find the (integer) value of t/τ where the curve has a maximum. When the maximum lies between two integer t/τ-values, it will pick the time associated with the higher value; when the two values near the maximum are identical, as in Fig. 6.5-2, it will pick one of them. We will use the resulting t/τ-value (which is only good to ± 0.5) as a rough, first estimate of t_r/τ. That is good enough for now; in section 6.7 we will see how t_r/τ should actually be determined.

15 Use these data to plot the estimated value of t_r/τ versus $1/\mu$. This should result in a linear graph through the origin, i.e., a proportionality.

16 Save the spreadsheet for further use in subsequent sections.

Why do we get this proportionality between t_r and $1/\mu$? Consider again the mechanism: the sample moves only when it is in the mobile phase, so that it will emerge from the column after a retention time $t_r = t_m/\mu$, where t_m is the time needed for the pure mobile phase itself to travel through all N_p plates in the column (so that $t_m/\tau = N_p$), and μ is the fraction of time that the sample spends in the mobile phase. The value of t_m can often be estimated by injection of a compound that is not or barely retarded; in gas chromatography, methane is often used as such.

We now consider the entire column rather than a small section thereof, as was done in the plate model. Let the column contain a total volume V_s of stationary phase, and a total volume V_m that can be occupied by the mobile phase. Since the column is uniformly coated or "packed", the ratio V_m/V_s will be equal to the ratio v_m/v_s we used earlier for a single plate. Therefore the fraction of time that the sample spends in the mobile phase, μ, can also be expressed in terms of these macroscopic, directly measurable volumes, because

$$\mu = \frac{1}{1 + K_p \, v_s / v_m} = \frac{1}{1 + K_p \, V_s / V_m} \tag{6.5-3}$$

Upon combining (6.5-4) with $t_r = t_m / \mu$ we obtain

$$t_r = t_m \, (1 + K_p \, V_s / V_m) \tag{6.5-4}$$

and after multiplication by the volume velocity v (e.g., in mL/s) of the mobile phase, so that $V_r = v \, t_r$ and $V_m = v \, t_m$,

$$V_r = V_m \, (1 + K_p \, V_s / V_m) = V_m + K_p \, V_s \tag{6.5-5}$$

which expresses the **retention volume** V_r in terms of the **void volume** V_m of the column, the volume V_s of stationary phase in the column, and the partition coefficient K_p. Equation (6.5-5) is the basic law of partition chromatography, and is often rewritten as

$$V_r' = V_r - V_m = K_p \, V_s \tag{6.5-6}$$

where V_r' is called the **adjusted retention volume**, which is here seen to be directly proportional to both K_p and V_s. The linear relation you observed 'experimentally' in the results of the simulation was nothing but the combination of (6.5-3) and (6.5-4).

The velocity v has the dimension of volume per time; $v = V_m / t_m$. We can also define a linear velocity v' defined as $v' = L / t_m = N_p H / t_m = H / \tau$, where L is the column length, and H is the corresponding length of a single plate; H stands for 'height'. For a vapor, these velocities vary with pressure along the column, and are usually referred to the column outlet; for liquids, which are essentially incompressible, v and v' are constant throughout the column.

The notion of a chromatographic 'plate' comes from the oil refinery, where distillation columns indeed have identifiable platforms or plates. However, a continuous column has no such discernible subdivisions. Still, the model defines a plate as a column segment of such length that it corresponds, on average, with one equilibration of the sample between the mobile and the stationary phase. We will come back to this matter in sections 6.8 and 6.9. In practice, the 'number of theoretical plates' is an empirical quantity, determined from the shape of the chromatographic peak. Typically, one plate corresponds with a column length of the order of 0.1 mm. Note that a relatively small number of plates N_p suffices to simulate the characteristic chromatographic behavior of individual peaks, even though any self-respecting chromatographic column will have thousands of theoretical plates. As we will see in section 6.6, a large number of theoretical plates is needed for the partition-chromatographic *separation* of compounds with very similar K_p-values.

There is something amazing about the above simulation: all it takes to represent the equilibrium distribution of the sample in the column, in any plate and at any time, is *one simple recursion formula*, (6.5-2), repeated over the entire array C6:Z206. Even the expression for the mass passing the detector comes from that same equation, merely by deleting the term representing the (absent) stationary phase in that detector. The entire process can of course be described mathematically, which in this case will yield a response similar to a binomial distribution, see below. However, the molecules know no mathematics, and merely follow the simple rules of probability that determine their partitioning behavior. Likewise, the simulation uses no higher math, but finds the same result through the repeated application of the simple recursion formula for partitioning between a stationary and a mobile phase.

6.6
Chromatographic plate theory 2

The same logic that leads to the simulation of the previous section can of course be used to obtain a closed-form expression for the chromatographic peak. As derived by, e.g., Fritz & Scott in *J. Chromat.* 271 (1983) 193, and again assuming (as in section 6.5) unit sample size and detector sensitivity, we have

$$\frac{(t/\tau-1)!}{(t/\tau-t_m/\tau)!(t_m/\tau-1)!}\,\mu^{t/\tau}\,(1-\mu)^{(t-t_m)/\tau}$$

$$= \frac{t_m}{t}\,\frac{(t/\tau)!}{(t/\tau-N_p)!N_p!}\,\mu^{t/\tau}\,(1-\mu)^{(t/\tau-N_p)} \tag{6.6-1}$$

where t is time, $t_m = \tau N_p$ is the time it takes a non-retained species to elute from the column, and μ (equal to $1/(1+k)$ in the notation of Fritz & Scott) is the mass fraction in the mobile phase.

A major advantage of (6.6-1) over a simulation is that the equation can generate results for a variety of conditions. This is especially useful for large numbers of plates and correspondingly long elution times, where the simulation becomes unwieldy; the equation has no such limitations. However, for large times t there may be a *computational* problem with the numerical evaluation of (6.6-1), because the factorials can quickly grow too large for the spreadsheet. At that point, the spreadsheet will show the error message #NUM! rather than the hoped-for numerical result.

The culprit is digital overflow, a problem discussed in section 8.12, where we also describe a way to bypass it. Our remedy is based on the observation that the quantity described in (6.6-1) is always well within the capabilities of the spreadsheet (after all, the total area under the curve is 1, and the peaks are not all that narrow), while individual terms in its numerator and denominator may lead to overflow. The trick, then, is to avoid calculating those individual terms, and instead compute the logarithms of the individual terms in (6.6-1), to combine these to form the logarithm of (6.6-1), and only then to exponentiate to find its direct value. Below we will first show that the results of the simulation indeed coincide with the predictions of (6.6-1).

Instructions for exercise 6.6-1

1 Reopen the spreadsheet of exercise 6.5.

2 In cell AB7 deposit the instruction $= LN(A7)$, and in cell AB8 the command $= AB7 + LN(A8)$. Copy the latter instruction down to the bottom row (306) of the computation. This will calculate the values of ln $(N!)$.

3 Now compute the binomial coefficients $(t/\tau - 1)! / [(t/\tau - N_p)!(N_p - 1)!]$ in cell AC30 as $= EXP(AB29 - AB6 - \$AB\$29)$. Copy this down. Note that we start in cell AC30, i.e., at $t = t_m$.

4 Calculate the detector output in cell AD30 with $= AC30*(\$A\$2^\wedge A30)*$ $(1/\$A\$2 - 1)^\wedge(A30 - 24)$. Again copy this down.

5 Compare the result in column AD with the simulated detector response in column AA. A simple way to do so is to calculate the sum of squares of the differences, using the instruction $= SUMXMY2(AA30:AA306,AD30:AD306)$. This will show the complete coincidence between simulation and theory, i.e., with differences of the order of the computational and truncation errors of the spreadsheet.

So far we have considered a single compound traveling through the column. Any **separation** by partition chromatography must involve at least two different compounds, with different values of K_p, otherwise the partitioning process cannot distinguish between them. For realistic values of V_s/V_m, and for K_p-values differing by less than a few percent, we usually need a far larger number of plates than we can simulate realistically on a spreadsheet. We will use (6.6-1) to calculate chromatograms of binary mixtures for 10 to 1000 plates, whereas the spreadsheet can only hold 256 columns.

For an analysis of the chromatographic separation of a binary mixture we will again assume the same, constant linear flow rate $v' = L/t_m = H/\tau$ throughout all plates, as one can expect with a non-compressible mobile phase.

Instructions for exercise 6.6-2

1 Open a new spreadsheet.

2 In cells A1 and B1 enter labels N and N! respectively.

3 In cells B3 and B4 enter labels for μ and N_p, and place some corresponding values (e.g., 0.9 and 10) in cells C3 and C4.

4 In cell A7 deposit the number 0, in cell A8 the instruction = A7 + 1, and copy this instruction down to row 2007.

5 In cell B7 deposit a zero, in cell B8 the value LN(A8), in cell B9 the instruction = A8 + LN(A9), then copy this instruction down to row 2007.

6 In cell C4 you have specified a value for N_p. Skip (starting with row 7) the first N_p cells in column C, i.e., for N_p = 10 start in cell 17, and there compute the function (6.6-1) as = (EXP($B16 – $B7 – $B$16))*(C$3^$A17)*(1/C$3 – 1) ^$A17 – C$4).

7 Copy this instruction down to row 157; going down much further makes little sense for this particular parameter choice.

8 Plot C7:C157 versus A7:A157 to show the resulting peak position and shape.

9 Copy the value for N_p from cell C4 to cell D4, and in cell D3 deposit a different value of μ, such as 0.8.

10 Copy cell C17 to cell D17, and then down column D to row 157. Note that the dollar signs in the instruction in cell C17 do not protect the C, so that the copied instruction will work without further modification in column D.

11 Highlight D7:D207, copy it (with Ctrl + c), then activate the inner frame of the graph you made under point (8), and paste (Ctrl + v) the new data into the graph.

12 In column E calculate the sum of the two signals, i.e., in cell E7 deposit = C7 + D7, and copy this down to row 157.

13 Following the procedure under point (11), copy and paste E7:E157 into the graph. Differentiate by line width and/or color between the total detector response E7:E157 and its individual components, C7:C157 and D7:D157.

14 You should now have a graph that resembles the left-hand part of the top panel in Fig. 6.6-1.

15 Now repeat this for the same values of μ but different values of t_m, keeping in mind that you should start the computations at row 7 + N_p, i.e., at row 32 for N_p = 25, etc. When you copy the instructions from column C into the new columns, make sure that the middle term in the exponential is set to $B7 in the first row of that calculation. And that, in plotting the data, you always start from row 7.

16 As you calculate for larger N_p-values, you must start further down in the spreadsheet, and you should also extend the computation further. For example, for N_p = 100, you

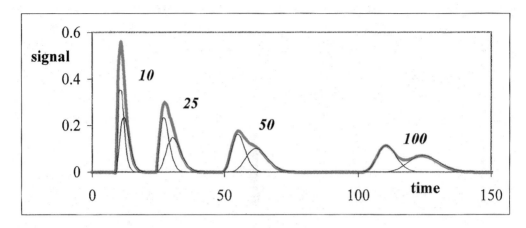

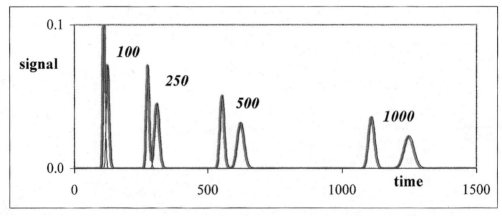

Fig. 6.6-1: Calculated chromatograms for two sample components, with $\mu = 0.9$ and $\mu = 0.8$ respectively, and their combined response (in color), for various values of the number N_p of theoretical plates, as indicated with the curves.

start at row 107, and you might want to calculate the response through row 207; for $N_p = 1000$, start in row 1007, and compute all the way down through row 1507. Of course, the horizontal scale of the graph must be adjusted to this longer range. Figure 6.6-1 illustrates how this can be done by splitting the display into two parts.

17 At some point the factor exp [ln $(t/\tau)!$ − ln $(t/\tau − N_p)!$ − ln $N_p!$] will become too large, and the spreadsheet will warn you with #NUM!. But now you know the trick: merely change the instruction to yield exp [ln $(t/\tau)!$ − ln $(t/\tau − N_p)!$ − ln $(N_p)!$ + (t/τ) ln μ + $(t/\tau − N_p)$ ln $(1 − \mu)$], and continue. You can of course use this form everywhere in the spreadsheet.

For the rather large differences in μ-values ($\mu_1 = 0.8$, $\mu_2 = 0.9$) shown in Fig. 6.6-1, baseline separation requires almost 500 plates; to resolve peaks with a smaller difference in μ, a larger number of plates will be needed.

Fortunately, most columns contain many thousands of plates, and can therefore separate compounds that differ only little in their μ-values.

The above example illustrate that a separation between two compounds requires *two* factors: (1) that their retention times are different, and (2) that the resulting peaks are sufficiently narrow so that, despite their different retention times, they do not overlap appreciably. Consequently, both peak position (as defined by t_r or V_r) and peak width (as characterized by σ_r) are important for a chromatographic separation. Retention times are proportional to N_p, whereas peak widths are (approximately) proportional to $\sqrt{N_p}$. This is why, everything else being the same, larger values of N_p yield better peak separations.

6.7 Peak area, position, and width

We now return to the simulated curves, in order to show how to extract the area, position, and width from a chromatographic peak. Several simple methods are available for symmetrical peaks (and even more for a special subset of these, Gaussian peaks), but since chromatographic peaks are often visibly asymmetric (and in that case obviously non-Gaussian), we will here use a method that is *independent* of the particular shape of the peak. It is a standard method that, in a chromatographic context, is described, e.g., by Kevra *et al.* in *J. Chem. Educ.* 71 (1994) 1023.

Let the detector response as a function of time t be described by the function $f(t)$. The area A under the response curve can then be computed as

$$A = \int_0^\infty f(t)\,\mathrm{d}t \tag{6.7-1}$$

while the time at which the peak goes through its maximum, and the peak width σ, follow from its first and second moments about the origin. Specifically, the first moment yields the retention time t_r as

$$t_r = \frac{\int_0^\infty t f(t)\,\mathrm{d}t}{\int_0^\infty f(t)\,\mathrm{d}t} = \frac{1}{A}\int_0^\infty t f(t)\,\mathrm{d}t \tag{6.7-2}$$

and the variance σ^2 of the peak is given by

$$\sigma^2 = \frac{\int_0^\infty t^2 f(t)\,\mathrm{d}t}{\int_0^\infty f(t)\,\mathrm{d}t} - \left(\frac{\int_0^\infty t f(t)\,\mathrm{d}t}{\int_0^\infty f(t)\,\mathrm{d}t}\right)^2 = \frac{1}{A}\int_0^\infty t^2 f(t)\,\mathrm{d}t - \left(\frac{1}{A}\int_0^\infty t f(t)\,\mathrm{d}t\right)^2 \tag{6.7-3}$$

The area under a chromatographic peak is routinely calculated by integration, through (6.7-1), and is used here to verify that the integration works; since we "injected" 1, we should obtain an area of 1.00. It is often advocated that the value of t_r be read off directly from the chromatographic peak, as the value of t at the peak maximum, but this is correct only for symmetrical peaks. In general, for symmetrical and asymmetrical peaks alike, it is better to use (6.7-2) instead.

Likewise, it is sometimes advocated that the standard deviation σ of the peak be determined as half its width at 60.7% of its maximum peak height, implying a Gaussian peak shape. A far worse suggestion is to draw tangents to the peak at that height, and extrapolate these tangents to determine the difference of their intercepts with the baseline, which difference is then taken to be 4σ. Integration based on (6.7-3) is more generally applicable, more accurate, and more precise, and should therefore be used whenever the chromatogram is available in digital form.

Instructions for exercise 6.7

1 Reopen the spreadsheet of exercise 6.5.

2 We will now use the three columns AB through AD that we had kept open for later use. (If you did not do that, highlight the column labels AB through AD, right-click, then in the resulting menu click on Insert to insert three new columns.) We will use the first column to compute the area A using (6.7-1), the second to find t_r with (6.7-2), and the third to obtain σ^2 from (6.7-3).

3 In cells AB7 through AD7 deposit the instructions $= (AA6 + AA7)/2$, $= (A6 + A7)*AB7/2$, and $= (A6 + A7)*AC7/2$ respectively, and copy these down to row 205, where $(A6 + A7)/2$ represents the average value of time t in this interval.

4 In cell AB3 compute the peak area A by trapezoidal integration, with the instruction $= SUM(AB6:AB205)$.

5 Likewise, in cell AC3, calculate the retention time t_r with $= SUM(AB6: AB205)/AB3$.

6 Use cell AD3 to compute σ_p^2 as $= SUM(AB6:AB205)/AB3 - (AC2)^2$.

7 Verify in cell AC4 that the product of μ and t_r/τ is indeed constant using the instruction $= A2*AC3$, and in cell AD4 compute the number of theoretical plates from (6.5-10) as $= AC3*(AC3 - 24)/AD3^2$.

8 Copy these results to the top left corner of the spreadsheet, so that you can readily change μ and read the results. For example, in cells D2 through H2 place the instructions $= AB3$, $= AC3$, $= AC4$, $= SQRT(AD3)$, and $= AD4$ respectively, and label these appropriately, as illustrated in Fig. 6.5-1.

9 Make a table of A, t_r/τ, $\mu t_r/\tau$, σ_p, and N_p as a function of μ for μ-values ranging from 0.2 through 0.9.

These numerical examples indicate that the plate model predicts both $t_r = \mu t_m$ and $\sigma_p = t_r(t_r - t_m)/N_p$. This latter result suggests that the number of theoretical plates can be obtained experimentally from

$$N_p = \frac{t_r(t_r - t_m)}{\sigma_p{}^2} \tag{6.7-4}$$

The numerical agreement between simulation and (6.7-4) is quite good, but not perfect. Below we illustrate how we can obtain more complete agreement by using a better integration algorithm, such as a Newton–Cotes integration described in section 8.7.

10 Change the instruction in cell AB8 to = (7*AA6 + 32*AA7 + 12*AA8 + 32*AA9 + 7*AA10)/90, and copy this instruction down the column.

11 Likewise change the instructions in columns AC and AD. For instance, the instruction in cell AC8 should read = (7*A6*AA6 + 32*A7*AA7 + 12*A8*AA8 + 32*A9*AA9 + 7*A10*AA10)/90, while the instruction in cell AD8 should instead have terms with 7*A6^2*AA6 etc.

12 Go to the top left corner of the worksheet, change the value of μ in cell A2, and tabulate your results as a function of μ as they will appear in cells D2:H2. The new data will reflect the improved integration accuracy.

Determining the number of theoretical plates

For a sufficiently large number of theoretical plates, the closed-form plate theory result (6.6-1) for the chromatographic curve can be approximated by the Gaussian distribution

$$\frac{1}{\sigma\sqrt{2\pi}} \exp\left[\frac{-(t - t_r)^2}{2\sigma^2}\right], \quad t_r = \frac{t_m}{\mu}, \quad \sigma^2 = \frac{t_r(t_r - t_m)}{N_p} \tag{6.8-1}$$

Instructions for exercise 6.8

1 Reopen the spreadsheet of exercise 6.6-2.

2 Highlight the column heading F, click on it, and in the resulting menu select Insert.

3 In the new cell F3 copy the value for μ from cell C3, and likewise copy the value of N_p from C4 to F4. In cell F5 deposit the value for σ^2 as = F4*(1 − F3)/ (F3*F3).

4 In cell F17 deposit = (1/SQRT(F$5*2*PI()))*EXP(− ($A17 − F$4/F$3)* ($A17 − F$4/F$3)/(2*F$5)), and copy this down to row 157.

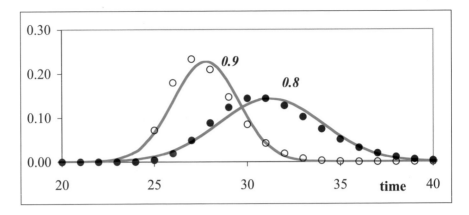

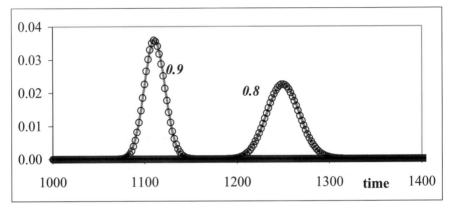

Fig. 6.8-1: Theoretical partition chromatograms (open circles) for $\mu = 0.9$ and $\mu = 0.8$ (as indicated with the curves), for $N_p = 25$ (top panel) and $N_p = 1000$ (bottom panel) respectively. The Gaussian curves (colored) are calculated from (6.8-1).

5 Plot C7:C157 and F7:F157 versus A7:A157, identifying one data set as discrete points, the other as a continuous curve.

6 Repeat the above for some other values of μ and N_p.

7 Figure 6.8-1 illustrates what you might find.

The agreement between the theoretical result and a Gaussian curve is fairly poor for $N_p = 25$, especially for μ-values close to 1 (where the plate model predicts quite asymmetrical peaks), but it is already quite good for $N_p = 1000$. For columns with many thousands of theoretical plates, it is clearly justified to treat the resulting peaks as Gaussians, with the values of t_r and σ_r as defined in (6.8-1).

There are several independent factors that can cause broadening of chromatographic peaks, and the resulting variances σ^2 are additive. For example we may have

$$\sigma^2 = \sigma_p^{\ 2} + \sigma_i^{\ 2} + \sigma_t^{\ 2} + \sigma_d^{\ 2} + \sigma_k^{\ 2} + \sigma_e^{\ 2} \qquad (6.8\text{-}2)$$

where $\sigma_i^{\ 2}$ describes broadening when the sample is not injected as a suffi-ciently narrow plug, $\sigma_t^{\ 2} + \sigma_d^{\ 2} + \sigma_k^{\ 2}$ represent the peak broadening effects of tortuosity, diffusional, and interfacial kinetics considered in the van Deemter equation (see section 6.9), and $\sigma_e^{\ 2}$ is any extra-columnar peak broadening, such as that due to a non-zero detector volume.

Of the variances listed in (6.8-2), the contribution from slow phase trans-fer kinetics ($\sigma_k^{\ 2}$) is directly proportional to the flow rate of the mobile phase through the column, whereas diffusional broadening ($\sigma_d^{\ 2}$) is inversely pro-portional to that flow rate. The contributions of these two terms can there-fore be distinguished, as described in the next section. The magnitude of $\sigma_e^{\ 2}$ can be kept small by proper instrument design, and that of $\sigma_i^{\ 2}$ by proper experimentation. We can therefore obtain N_p by measuring σ^2 with an assumption-free method based on (6.7-3), and by then subtracting esti-mates for all non-negligible terms beyond $\sigma_p^{\ 2}$ in (6.8-2). Finally, we can use (6.7-4) to obtain N_p from $\sigma_p^{\ 2}$ through $N_p = t_r (t_r - ct_m)/\sigma^2_p$. Unfortunately this is a rather elaborate procedure.

But here we must tread with some caution. Despite its venerable place in the development of chromatography, there is an obvious problem with applying plate theory to chromatography, because actual chromatographic columns do not contain discrete, identifiable plates. Alternative models have been used that do not involve the assumption of discrete plates, but instead start with an infinitesimally thin slice of the column to obtain a diffe-rential equation describing chromatographic behavior. Such models may differ from the present approach in assuming that the mobile phase does not move uniformly through the column, but that only a fraction of the mobile phase moves on. Such models predict a Gaussian peak with the same retention time t_r but with a different peak width σ, and lead to a different expression for N, viz. $N = (t_r / \sigma)^2$.

Which expression should one use, $\sigma_p^{\ 2} = t_r (t_r - t_m)/N_p$ or $\sigma^2 = t_r^2/N$? The expression for $\sigma_p^{\ 2}$ clearly represents band broadening by the partitioning process, in which only a fraction of the sample transfers to the stationary phase, so that the sample spreads out over many plates. This is most readily seen by considering what happens with a non-retained sample: when $t_r = t_m$, plate theory predicts $\sigma^2_p = t_r (t_r - t_m)/N_p = 0$, i.e., a sample that does not partition into the stationary phase should show no corresponding broadening. On the other hand, in continuum models that ascribe the broadening to a partially stagnant mobile phase, all samples are affected, independent of their partition coefficients. This leads to broadening even for non-retained compounds, i.e., for $t_r = t_m$.

Which model better describes the experiments? The answer to this ques-tion is not yet clear, and may well depend on the type of column used. A valid criterion would seem to be that a constant value is found for either N_p or N

for members of a homologous series of compounds chromatographed at the same time on the same column, after correction for the other peak-broadening effects. There are too few experiments of this type in the literature to support one model or the other, and it may not even be an either/or question: the flow of mobile phase in a packed column cannot be uniform, and even in a capillary column it will exhibit a Poiseuille profile. Ultimately, the usefulness of mathematical models to describe experimental data must be determined by experiment. Sections 6-5 through 6-8 are mainly meant to illustrate how much a spreadsheet can help one evaluate the consequences of a mathematical model.

6.9 Optimizing the mobile phase velocity

Chromatography is a *dynamic* process; in practice, equilibrium between the mobile and the stationary phase (as assumed in plate theory) is seldom established. In principle, equilibrium might be approached by moving the mobile phase slowly enough, but in that case all peaks may broaden because the sample stays so long in the column that thermal motion (diffusion) might spread them out. This is especially the case in gas chromatography, because of the considerable molecular motion in the vapor phase, where the mean free path between collisions is much larger than in solution. On the other hand, when the mobile phase moves too fast through the column, there may not be enough time for establishing partition equilibrium of the sample molecules between the stationary and mobile phase. For a practical separation, the optimum speed v of the mobile phase is therefore a compromise, going fast enough to keep *diffusional broadening* at bay, but not so fast that *interfacial transfer kinetics* (i.e., the lack of equilibrium) become dominant. That balancing act is described by the **van Deemter equation**

$$H = A + B/v + Cv \qquad (6.9\text{-}1)$$

where H is an abbreviation for the "height equivalent to a theoretical plate". The parameter A is often associated with the tortuosity of a packed column (in which case one would set $A = 0$ for a capillary column), B reflects diffusional broadening, and C the kinetic (i.e., non-equilibrium) behavior, that is, the finite rates of interfacial processes. While closed-form expressions for B and C can be derived, all three parameters in (6.9-1) are usually treated as adjustable parameters. In order to express H in terms of directly measurable quantities one often uses

$$H = L/N_p \qquad (6.9\text{-}2)$$

where L denotes the length of the column, and N_p the number of plates it contains. We have seen in section 6.8 that there is some question whether to use $N = t_r(t_r - t_m)/\sigma^2$ or $N_p = t_r^2/\sigma^2$, but in the present context this is immaterial. Either definition will do, or any other practical definition of the reciprocal of the square of the peak width, the parameter we want to minimize.

The van Deemter equation contains three adjustable parameters, A, B, and C, while H and v are the observable parameters: v is calculated from the measured linear carrier velocity at the column exit (in gas chromatography, after correction for the pressure-dependent compressibility of the gas in the column), and H is found from the peak shape and position. It will not take you long to figure out that (6.9-1) can be fitted to a second-order polynomial of the form $Y = A_0 + A_1 X + A_2 X^2$ when we make the following substitutions: $Y = H v$, $X = v$, $A_0 = B$, $A_1 = A$, and $A_2 = C$. Therefore, a linear least-squares routine can be used. The only remaining question concerns the nature of the experimental uncertainties.

Of the experimentally accessible parameters, the column length and the retention time t_r (or the retention volume V_t) can usually be determined with sufficient precision. The same applies to the mobile phase velocity v. The weakest link is usually the peak width, which yields σ, because the peaks are typically quite narrow. When the peak width $w_{1/2} = \sigma \sqrt{(8 \ln 2)}$ is taken from a strip-chart recorder (as was done in much of the older work, before computers or microprocessors were used), the line-width of the recorder trace may be a significant part of the peak width. Similarly, when σ was estimated by drawing tangents to the peak, the relative uncertainties in the resulting value are often considerable. We therefore assume that the major source of experimental uncertainty lies in σ. Since H is proportional to σ^2, so that $\partial H / \partial \sigma \propto \sqrt{H}$, a weighted second-order least-squares routine is indicated, with weights given by $w = 1/\{\partial Y / \partial \sigma_v\}^2 = 1/\{\partial (Hv)/\partial \sigma_v\}^2 \propto 1/Y$. Note that we only need the *relative* weights w, because multiplying all weights by an arbitrary constant is inconsequential to the result. We therefore use the proportionality $\propto$ to avoid much unnecessary algebra.

As our example we will use data taken from Fig. 1a of a paper by R. Kieselbach in *Anal. Chem.* **33** (1961) 23, which shows H as a function of v for air, butane, and cyclohexane, flowing through a gas-chromatographic column (1 m long, 6 mm inner diameter) packed with 35-80 mesh Chromosorb coated with 10% Dow-Corning 200 silicone oil. The carrier gas was helium, the column was kept at room temperature, and a thermistor was used as the detector. Table 6.9-1 lists the data as estimated from the graph.

Instructions for exercise 6.9

1 Open a new spreadsheet.

2 Enter labels for v and H in row 4, and enter the corresponding data from Table 6.9-1 for air in A6:B21. This completes the data entry stage.

3 For the data analysis, enter additional labels in row 4 for Y, w, X, and XX.

Table 6.9-1: *The height equivalent to a theoretical plate, H, as a function of the linear, pressure-corrected mobile phase velocity v for air, butane, and cyclohexane for a 1 m long, 6 mm ID glass column packed with 35-80 mesh Chromosorb coated with 10% DC-200, using He at room temperature as the mobile phase.*

v / cm s^{-1}	H / cm air	butane	v / cm s^{-1}	H / cm air	butane	v / cm s^{-1}	H / cm cyclohexane
2.14	0.417	0.201	16.7	0.071			
4.20	0.226	0.120	18.8	0.066	0.094	2.46	0.161
7.02	0.142	0.093	18.8	0.075	0.101	4.76	0.090
9.20	0.104	0.087	21.6	0.065	0.101	8.64	0.075
11.5	0.099	0.077	21.6	0.069	0.108	11.0	0.073
13.8	0.076	0.082	26.6	0.075	0.108	13.8	0.069
16.3	0.060	0.089	26.6	0.085	0.115	16.1	0.083
16.3	0.071	0.092	31.6	0.069	0.134	18.5	0.082

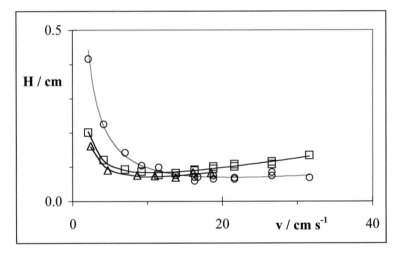

Fig. 6.9-1: The experimental data for air (circles), butane (squares) and cyclohexane (triangles) together with the curves fitted through them.

4 In columns C through F compute $Y = Hv$, $w = 1/Y$, v, and v^2 respectively.

5 Highlight C6:F21, and call the weighted least-squares routine. It will deposit the results for A_0 through A_2 in D22:F22, with the corresponding standard deviations below it, in row 23. This completes the analysis stage.

6 Label column G for H_{calc}, and in it compute H as $A_0/v + A_1 + A_2 v$.

7 Plot the experimental data points (from columns A and B) together with the line calculated in column G. This completes the graphing stage.

8 Repeat the same procedure for the data for butane and cyclohexane. Figure 6.9-1 shows you what the results should look like.

In this example, a numerical fit of the data is perhaps overkill, because the optimal flow rate can be estimated directly from the graph. Its precise value is not crucial, and one might even elect to use a velocity that is slightly higher than optimal, thereby trading a small amount of extra peak broadening for a considerably shorter analysis time.

6.10 Polarography

Polarography is an example of voltammetry that is quickly disappearing from the undergraduate laboratory, the victim of what Foster, Bernstein, & Huber have aptly called a *Phantom Risk* (MIT Press, 1993). At room temperature, mercury (melting point $-39\,°C$, boiling point $+357\,°C$) is a rather innocuous liquid metal with a high surface tension (which makes its spills difficult to clean up, since it causes mercury droplets to form non-wetting, highly mobile spheres) and a very low vapor pressure (0.0012 Torr or 0.16 Pa at $20\,°C$). However, when mercury is heated, its vapor pressure becomes considerable, at which point it can do much neurological harm, eventually leading to mad-hatters disease. Another nasty one is when mercury is freely discarded, in which case bacteria can convert it into the highly poisonous methylmercury, the cause of Minamata disease. Fortunately, use of elemental mercury at room temperature, in a well-ventilated room, with distillation and recycling of used mercury rather than disposal into the sink, is quite harmless both to the experimenter and to the environment.

A polarographic experiment involves recording a current–voltage curve on a mercury electrode immersed in a test solution. The electrical circuit is completed by a reference electrode (and, usually, a separate 'auxiliary' electrode), but these are of no consequence to the discussion to follow.

A typical mercury electrode is made from a mercury reservoir, some connecting tubing, and a glass capillary, often some 25 cm (about 10″) long and 5 mm (0.2″) wide, with an internal diameter of about 0.1 mm. Under a hydrostatic pressure of about 1 atmosphere (from some 75 cm of mercury head) the liquid metal flows through the vertical capillary at a rate of the order of a milligram per second. It forms a near-spherical droplet at the bottom end of the capillary, held together by interfacial tension. Eventually, as mercury continues to flow into the droplet, its weight will become too large for the interfacial tension to carry, the droplet will fall off, and the process will start anew; typical drop times range from 5 to 15 seconds. Since electrical contact with the drop is made through the capillary, the measurement is always made on the drop that is still attached to the capillary. The entire assembly is called a **dropping mercury electrode**, where 'dropping' pertains to the mercury, not to the glass capillary.

In chapter 5 we have used the Nernst equation in terms of the solution concentrations. That was permissible because, in potentiometry, the currents are negligible, so that there is no perceptible difference between the interfacial concentrations and those in the bulk of the solution. However, when there are substantial currents flowing through the electrode, as in voltammetry, the interfacial and bulk concentrations are no longer the same, in which case the Nernst equation shows its true nature as a *local* law, describing the relation between the potential and the *interfacial* concentrations, as long as the electron transfer is fast enough.

The relation between the interfacial and bulk concentrations depends on mass transport, most often by diffusion (i.e., thermal motion) and/or convection (mechanical stirring). Often a stationary state is reached, in which the concentrations near the electrode can be described approximately by a diffusion layer of thickness δ. For a constant diffusion layer thickness the Nernst equation takes the form

$$E = E^o + \frac{RT}{nF} \log \frac{i - \overleftarrow{i}}{\overrightarrow{i} - i} \tag{6.10-1}$$

where $\overrightarrow{i}$ and $\overleftarrow{i}$ are the **limiting currents** associated with reduction and oxidation respectively. Equation (6.10-1) describes the stationary current–voltage curve and, also, the *envelope* of the polarographic current–voltage curve. The limiting currents are the characteristic features of (6.10-1): when the potential is sufficiently far from E^o the currents are simply the limiting currents, i.e., they are completely determined by the speed at which the redox reagents can reach the electrode interface, or its products be removed from it. We can invert (6.10-1) so that it expresses the current i in terms of the potential E, as

$$i = \frac{\overrightarrow{i} \exp[nF(E - E^o)/RT] + \overleftarrow{i}}{\exp[nF(E - E^o)/RT] + 1} \tag{6.10-2}$$

In polarography, not enough time is available for the diffusion layer to reach its stationary thickness. Instead, the current per unit electrode area decreases with the square root of time, the signature time-dependence for diffusion. On the other hand, the area of the growing drop expands, proportional to the two-thirds power of drop age τ (i.e., time elapsed since the previous mercury drop fell off). These two counteracting effects, diffusion currents per area proportional to $\tau^{-1/2}$, and area growth as $\tau^{2/3}$, combine to yield polarographic current-time curves with a time dependence of $\tau^{-1/2} \times \tau^{2/3} = \tau^{1/6}$, as expressed in the **Ilkovič equation**.

Finally, since polarography involves mercury, which often solvates metals as amalgams, there can be a significant difference between the potential E^o of standard tables and the appropriate value of E^o for an amalgam-forming metal. Polarography therefore uses so-called **half-wave potentials** $E_{1/2}$ instead of standard potentials E^o. All the above effects are incorporated in spreadsheet exercise 6.10.

Instructions for exercise 6.10

1 Open a new spreadsheet.

2 In row 1 deposit labels for $E_{1/2}$, $\vec{i}\ \tau^{-1/6}$, $\overleftarrow{i}\ \tau^{-1/6}$, F/RT, and the starting potential E_{init}.

3 In row 2 place corresponding numerical values, such as $-0.6, 0, -1, 40$, and -0.3.

4 In row 4 place column headings for t (time), E, τ (drop age), and i_f. The subscript f on the current identifies this current as a **faradaic current**, i.e., one associated with electrochemical reduction or oxidation, as distinct from the charging current we will encounter soon. For τ either use 'tau' or type a 't', then highlight it, select Format ⇨ Cells, in the resulting Format Cells dialog box, select Fonts, then in the window for Font select Symbol, and press OK. If your spreadsheet shows the Formatting toolbar, the same can be accomplished simply by clicking on the arrow to the right of the Font window, and selecting the Symbol font in the left-most window of the Formatting toolbar instead.

5 In cell A6 place a very small number, such as 1E−10. The reason why we do not start with 0 will become clear with instruction (9).

6 In cell A7 calculate = A6 + 0.2, and copy this down to cell A806.

7 In cell B6 refer to E_{init} by depositing the instruction = $E$2.

8 In cell B7 calculate the gradually changing potential as = B6 − 0.002, and copy this down to row 806. This will generate a 1.6 volt scan of the applied potential. (In polarography the potential is typically scanned towards more *negative* potentials during the recording of a polarogram.)

9 In cell C6 simulate the drop age τ as = A6–INT(A6). This yields an output running from 0 (or, more precisely, 1E − 10) to 0.9, thereby simulating the drop age τ. If we had started in A6 with 0, the value of τ might occasionally run to 1.0, as the result of round-off errors in the values in column A. We prevent this by using in cell A6 a starting value such as 1E − 10, significantly larger than the accumulated round-off errors yet still insignificantly small for our calculations.

10 In cell D6 deposit the instruction for the faradaic current i_f = ($B$2*EXP($D$2*(B6 − $A$2)) + $C$2)*C6^(1/6)/(EXP($D$2*(B6 − $A$2)) + 1). You will recognize this as (6.10-2) times the time-dependency $\tau^{1/6}$.

11 Plot i versus E. For both axes, select Values in Reverse Order on the Scale page of the Format Axis dialog box. For the position of the scales use the options in the Format Axis dialog box, on the Pattern page under Tick-Mark Labels. For Reverse Order you should select to place them High.

Usually the current–voltage curves are recorded on a strip-chart recorder, which cannot follow the rapid fall of the current when the mercury droplet falls off. This effect can be mimicked by selecting a larger value for A6, say 0.04.

The sensitivity limit of polarography is typically around 1 and 10 μM, and is determined by the **charging current** i_c needed to charge the continuously changing electrode interface. This current is proportional to $(E - E_z)\, C\tau^{-1/3}$, where E_z is the potential of zero charge, and the integral capacitance C is to a first approximation independent of both potential and redox chemistry. Upon closer examination we find that the integral capacitance is often about a factor of 2 larger at $E > E_z$ than for $E < E_z$, reflecting the fact that solution anions can typically approach the electrode closer (because they tend to be less strongly hydrated) than cations. We can readily incorporate such a charging current, which should be added to the earlier-computed faradaic current i_f to yield the total current i.

12 In cell F1 place the label C, and in G1 the label Ez.

13 Enter numerical values for these, such as 0.1 and -0.4 respectively.

14 In cells F4 and G4 enter labels for i_c and i respectively.

15 In cell F6 store the instruction = IF(\$B6 > \$G\$2,2*(\$B6–\$G\$2)*\$F\$2 *C6^($-1/3$),(\$B6–\$G\$2)*\$F\$2*\$C6^($-1/3$)), in G6 calculate the algebraic sum $i_f + i_c$, and copy both down to row 806.

16 Plot F6:F806 vs. B6:B806 and G6:G806 vs. B6:B806. Your graphs should resemble those of Fig. 6.10-1.

At higher concentrations of the *electroactive* species, adsorption of components of the redox couple, coupled with the fluidity of the liquid–liquid interface, may lead to mechanical streaming of the solution at the interface, generating a **polarographic maximum**. Charging currents and polarographic maxima typically limit classical polarography to concentrations of the electroactive species between 10^{-3} and 10^{-5} M.

Most applications of polarography involve metal ion analysis, where only one valence state is present in solution, such as Tl^+, Cd^{2+}, Pb^{2+}, or Zn^{2+}. However, that is not always the case: one can analyze Fe^{2+} and Fe^{3+} simultaneously by polarography, in a complexing solution such as made with oxalate, in which case the polarogram shows (at $E < E_{1/2}$) a limiting reduction current $\overleftarrow{i}$ due to the reduction of Fe^{3+} to Fe^{2+}, and (at $E > E_{1/2}$) a limiting oxidation current $\overrightarrow{i}$ for the oxidation of Fe^{2+} to Fe^{3+}. We simulate this in Fig. 6.10-2.

It is possible to exploit the difference in the time-dependencies of i_f (proportional to $t^{1/6}$) and i_c (proportional to $t^{-1/3}$) to separate the two (J. N. Butler & M. L. Meehan, *J. Phys. Chem.* 69 (1965) 4051), and thereby to push back the lower limit somewhat. However, instrumental refinements such as **square wave polarography** (G. C. Barker & I. L. Jenkins, *Analyst* 77 (1952) 685) and, **pulse polarography** (G. C. Barker & A. W. Gardner, *Z. Anal. Chem.* 173 (1960) 79) are much more efficient in doing so, and are therefore the methods of choice for concentrations between 10^{-5} and 10^{-7} M.

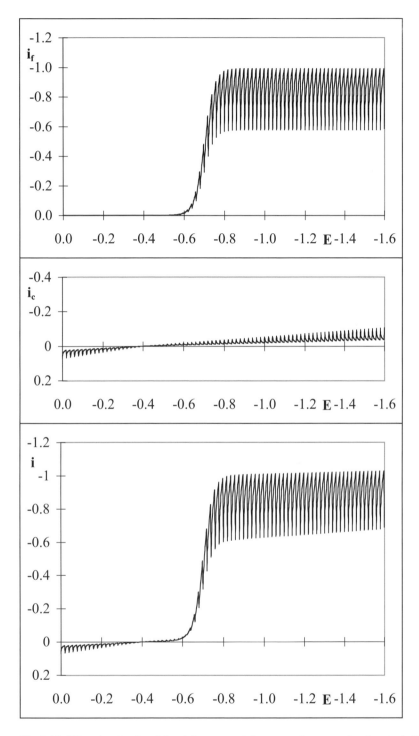

Fig. 6.10-1 Top: the simulated faradaic current i_f for a one-electron reduction at a half-wave potential of $-0.7\,V$. Middle panel: the simulated charging current i_c. Bottom panel: the total simulated polarographic current $i = i_f + i_c$. Parameters used: $E_{1/2} = -0.7$, $\overline{i}\ \tau^{-1/6} = 0$, $\widetilde{i}\ \tau^{-1/6} = -1$, $F/RT = 40$, $C = 0.03$, $E_z = -0.4$.

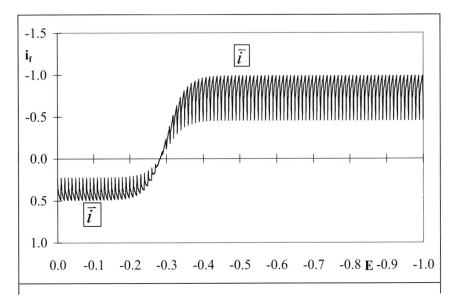

Fig. 6.10-2: The simulated polarogram for a one-electron redox couple with a half-wave potential of -0.3 V. The concentrations of the reduced and oxidized forms of the couple are directly proportional to their respective limiting currents, $\vec{i}$ and $\overleftarrow{i}$.

Finally we expand the simulated polarogram by the inclusion of a second reduction wave. This is easy because, in the absence of chemical interactions between the electroactive species, the polarographic currents are strictly additive.

17 Enter a positive value in B2 and observe the resulting curve for i_f vs. E. Figure 6.10-2 illustrates such a polarogram.

18 Highlight the row number 3, right-click on it, and click on Insert. This will insert a new row.

19 In the new row 3 enter numerical values for a second substance and its polarographic characteristics, e.g., $E_{\frac{1}{2}} = -0.7$, $\vec{i} \; \tau^{-1/6} = 0$, and $\overleftarrow{i} \; \tau^{-1/6} = -0.4$.

20 Copy the instruction from D7 to E7, then extend it to read $= (\$B\$2*EXP(\$D\$2*(B7 - \$A\$2)) + \$C\$2) *C7^(1/6)/(EXP(\$D\$2*(\$B7 - A\$2)) + 1) + (\$B\$3*EXP(2*\$D\$2*(\$B7 - \$A\$3)) + \$C\$3)*\$C7^(1/6)/(EXP(2*\$D\$2*(\$B7 - \$A\$3)) + 1)$ (for an added two-electron reduction wave, $n = 2$).

21 Plot the resulting polarogram, which should resemble Fig. 6.10-3.

When you compare graphs such as these with real polarograms of 'reversible' reductions (in electrochemistry, the term 'reversible' is meant to convey that, except for mass transport, equilibrium rather than kinetic

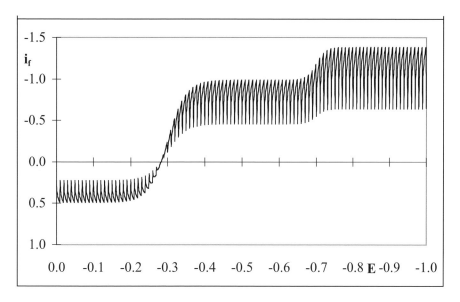

Fig. 6.10-3: The simulated polarogram for a one-electron redox couple with a half-wave potential of − 0.3 V plus a two-electron reduction of another species with a half-wave potential of − 0.8 V. Again, all concentrations are directly proportional to their respective limiting currents.

considerations determine the shape of the observed behavior) you will see that they are quite close representations. The message is that you can now simulate almost any curve for which you have an appropriate theoretical, closed-form description. And then you can use that simulation to design and/or test your analysis protocols.

6.11 Linear sweep and cyclic voltammetry 1

Most metals are solid at room temperature, in which case polarography (based on the reproducible formation and detachment of mercury droplets) cannot be used. When we apply to a *stationary* electrode the same experimental method (of recording the current resulting from scanning the applied voltage as a linear function of time) we obtain a current–voltage curve that is called a **linear sweep voltammogram**. Except when it is applied to microelectrodes, such current–voltage curves do not resemble polarograms (without the fluctuations in the current caused by the growth and fall of the mercury droplets) but exhibit a current maximum where a polarogram might have shown a S-shaped polarographic wave. When the direction of the voltage scan is subsequently traced in reverse, the resulting curve is a **cyclic voltammogram**.

The direct mathematical analysis of a linear sweep or cyclic voltammogram is much more complicated than that of a polarogram, because the rates of the electrochemical reactions (indirectly, through their dependence on potential), as well as the rate at which material can reach the electrode, are then time-dependent. (In polarography, the scan rate dE/dt, where E is the applied potential and t is time, is so small that its effect on the electrode reactions can be neglected; in linear sweep and cyclic voltammetry, much higher scan rates are typically used.) The resulting current–voltage curves can either be compared with theoretical models obtained by digital simulation, or they can be transformed into the shape of the corresponding polarogram (without those drop-caused fluctuations), which can then be analyzed by existing methods.

The currently available digital simulation packages make simplifying assumptions regarding the dependence of the rates of the electrode reactions on the applied potential, which they take to be exponential. The transform methods do not make such assumptions, and are therefore sometimes preferable. In this section we will illustrate how we can use the transform method to simulate a linear sweep voltammogram and a cyclic voltammogram. And in section 6.12 we will illustrate how to apply the transform to experimental data.

The complicated time dependence of the faradaic current is caused by mass transport, specifically by diffusion. (The other forms of mass transport, convection and electromigration, can both be kept negligibly small: convection by not using the method close to, say, a construction site, machine shop, or other source of mechanical vibrations; electromigration by using an excess of an otherwise **inert electrolyte**.) Oldham (*Anal. Chem.* 41 (1969) 1121; 44 (1972) 196) showed that a linear sweep voltammogram obtained on a planar electrode can be transformed into the corresponding polarogram by a method called **semi-integration**. Conversely, a polarographic wave can be transformed into the corresponding linear sweep voltammogram by the inverse operation, **semi-differentiation**. Strictly speaking, the transforms given here work only for planar electrodes with negligible edge effects (so that planar diffusion applies everywhere), in the absence of (or after correction for) charging current i_c. (Note that the charging current on a stationary electrode is due to the changing potential dE/dt, whereas at a dropping mercury electrode it is primarily caused by the changing electrode area dA/dt.)

We start from the stationary current–voltage curve (6.10.2) for a solution that contains one oxidizable electroactive species (so that $\overleftarrow{i} = 0$), which we write in dimensionless form as

$$\frac{i}{\overrightarrow{i}} = \frac{\exp[nF(E - E^\circ)/RT]}{\exp[nF(E - E^\circ)/RT] + 1} \tag{6.11-1}$$

For a linear sweep voltammogram, the applied voltage is a linear function of time, $E = E_1 + vt$, where $v = dE/dt$ is the sweep rate, in V s^{-1}. We then use the macro SemiDifferentiate to find the corresponding shape of the linear sweep voltammogram.

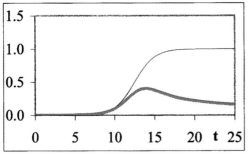 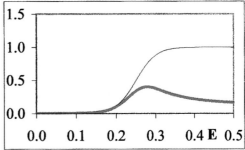

Fig. 6.11-1: The dependence of the current ratio (6.11-1) (black) and of its semi-differential (blue) on time t (left) and potential E (right).

Instructions for exercise 6.11

1 Start a new spreadsheet.

2 In the top two rows of the spreadsheet, enter labels and values respectively for the time increment Δt, the sweep rate v, the potentials E_1 and E_2, the standard potential E^o, and the values of n and F/RT, such as $\Delta t = 0.001$, $v = 0.02$, $E_1 = 0$, $E_2 = 0.5$, $E^o = 0.25$, $n = 1$, and $F/RT = 40$.

3 Below those (and separated by an empty row) enter labels for time t, potential E, and dimensionless current i/i_{lim}.

4 In the columns below these labels, enter values for t (from 0 to 25 in increments of 0.05), the corresponding values for $E = E_1 + vt$ (from 0 to 0.5 in increments of 0.001), and values of i/i_{lim} calculated from (6.11-1).

5 Call the SemiDifferentiate macro, and reply to the input boxes.

6 Plot the resulting curve, together with i/i_{lim}, as a function of t.

7 Also plot the resulting curve, together with i/i_{lim}, as a function of E. The two plots should be identical when the voltage axis from E_1 to E_2 has the same length as the time axis from 0 to $|E_2 - E_1|/v$.

8 Figure 6.11-1 illustrates the types of plot you should have made.

For the simulation of a cyclic voltammogram we retrace the potential after we have reached its final value, E_2, so that $E = E_1 + vt$ for $0 \le t \le |E_2 - E_1|/v$ while $E = E_2 - vt$ for $|E_2 - E_1|/v \le t \le 2|E_2 - E_1|/v$.

9 Now extend the spreadsheet by doubling the length of the time column to 0 (0.05) 50.

10 For $t \ge 25$ use the reversed direction of the voltage scan, so that E runs back from E_2 to E_1 as $E = E_2 - v(t - 25)$. A plot of E vs. t would now have the form of an isosceles triangle.

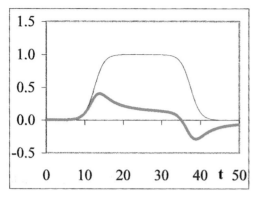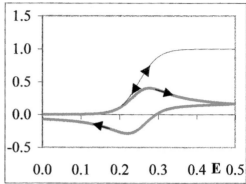

Fig. 6.11-2: The dependence of the current ratio (6.11-1) (black) and of its semi-differential (blue) on time t (left) and potential E (right). Note that the traces in the left-hand panel have, as it were, been folded back.

11 Again call the SemiDifferentiate macro, and apply it to the entire data set.

12 Plot the resulting curve, together with i/i_{lim}, as a function of t.

13 Also plot the resulting curve, together with i/i_{lim}, as a function of E.

14 Compare your results with Fig. 6.11-2.

The two plots in Fig. 6.11-2 are now quite different: the signal as a function of time shows oxidation first, followed (after reversal of the scan direction) by a reduction of the just-oxidized) material. The same signal in the cyclic voltammogram is folded back because it is plotted not versus time, but versus the applied potential.

15 Again double the length of the time axis, and make the potential retrace the 'up' and 'down' scans a second time.

16 Calculate the resulting semi-differential, and plot the resulting curve, together with i/i_{lim}, as a function of t and of E, respectively. Compare with Fig. 6.11-3.

Note that the second cycle does not quite retrace the first. However, the difference between subsequent cycles becomes rather small, and after a few cycles a steady-state cyclic voltammogram is approached. You can of course verify this by further extending the simulation. In that case you may want to reduce the data density by using, e.g., $\Delta t = 0.2$ instead, thereby changing the voltage increments to $\Delta t \times v = 0.2 \times 0.02 = 0.004$, as otherwise the macro might become quite slow.

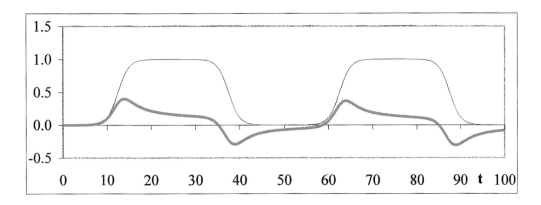

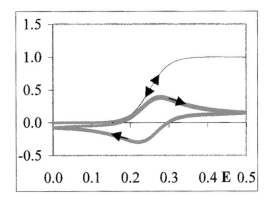

Fig. 6.11-3: The dependence of the current ratio (6.11-1) (black) and of its semi-differential (blue) on time t (top) and potential E (bottom).

6.12 Linear sweep and cyclic voltammetry 2

We will now illustrate the use of semi-integration to convert a linear sweep voltammogram into the shape of the equivalent stationary current–voltage curve. To this end we will use an experimental data set kindly provided by Hromadova & Fawcett, obtained for the reduction on Au(110) of 0.5 mM $[Co(NH_3)_6](ClO_4)_3$ in 0.093 M aqueous $HClO_4$ at 25 °C, using a scan rate of 20 mV s^{-1}, as described in *J. Phys. Chem.* 104A (2000) 4356.

Instructions for exercise 6.12

1 Start a new spreadsheet.

2 Import the data set labeled Hromadova & Fawcett.

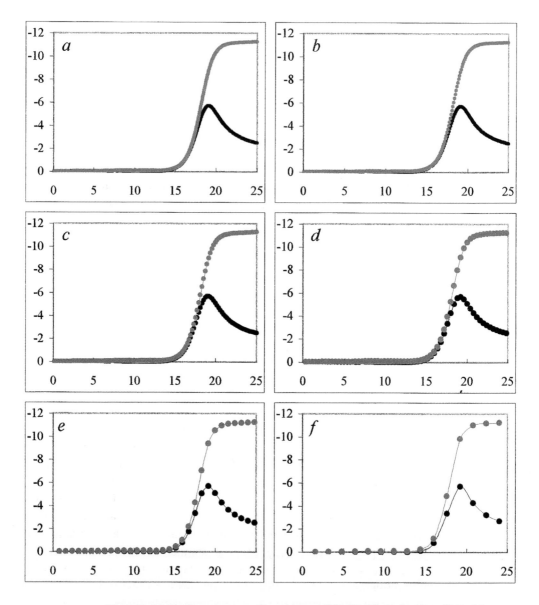

Fig. 6.12. (a): The linear sweep voltammogram (black) of the reduction of 0.5 mM $[Co(NH_3)_6](ClO_4)_3$ in 0.093 M aqueous $HClO_4$ at Au(110), at 25 °C, with a scan rate of 20 mV s^{-1}, and its semi-integral (blue). **(c)** through **(f)**: The same after the original data set has been thinned by successive factors of 2, finally (in panel f) ending up with only 15 equidistant data points. Data from Hromadova & Fawcett, *J. Phys. Chem.* 104A (2000) 4356.

3 Insert a few lines above the data table to make space for constants and column headings.

4 The first data column contains the time, in seconds, and the second column the current, in μA.

5 Use the macro SemiIntegrate to convert the linear sweep voltammogram to its steady-state counterpart. In this form it can readily be analyzed by the methods pioneered by Koutecký, and described in, e.g., *Principles of Polarography* by Heyrovský & Kůta (Academic Press 1966).

6 As in any numerical integration or differentiation, the data density must be sufficient. Since the data are given, their density cannot be increased, but we can see the effect of reducing the data density by culling data from the original set. An auxiliary macro, DataThinner, is provided for this purpose. Check it out. Figure 6.12 illustrates what happens with this data set when its density is reduced successively by factors of two. In this case of high-quality experimental data, the resulting data deterioration in the semi-integral is barely noticeable in the graphs, even after several repeated reduction steps.

6.13 Summary

This chapter showcases various ways in which spreadsheets can be applied to problems of instrumental chemical analysis, with examples taken from spectroscopy, chromatography, and electrochemistry.

We started with a simple set of spectrometric observations from which we could determine a pK_a by extrapolating its value to infinite dilution. Then we used matrix algebra to solve a set of simultaneous equations representing the spectrometric analysis of a mixture. In section 6.3 we showed the power and convenience of performing the same type of analysis using far more than the minimum number of input data, improving data precision (by diminishing the effects of experimental uncertainty) while at the same time simplifying the analysis (by using a standard least squares routine). Subsequently we used the spreadsheet to illustrate the general appearance of absorbance–absorbance diagrams, diagnostic plots that are most useful in establishing the number of separately identifiable species involved in equilibria of light-absorbing compounds.

In sections 6.5 through 6.7 we simulated chromatographic plate theory and used it to demonstrate some of the basic results of partition chromatography: the dependence of retention volume on the partition coefficient of the eluting species, and the volume ratio of the mobile and stationary phases in the column. In sections 6.8 and 6.9 we explored the predictions of plate theory for the width of the chromatographic peak – and found the theoretical plate to be a somewhat shaky concept. In this example, then, we used the spreadsheet to help us evaluate the numerical consequences of a theory, thereby allowing us to ask more fundamental questions regarding the underlying model.

The last section dealing with chromatography returned to the more mundane problem of using the spreadsheet to determine the parameters of the van Deemter equation. Alternatively we could have used Solver plus SolverAid.

In section 6.10, we used the spreadsheet to simulate polarograms, fairly complicated, time-dependent current–voltage curves observed on dropping mercury electrodes, with faradaic currents determined by diffusion coefficients, the Nernst equation, and drop growth, plus (background) charging currents. And in sections 6.11 and 6.12 we encountered transformations that can convert a linear sweep voltammogram onto the shape of a polarographic wave and vice versa, thereby making them more easily amenable to further mathematical analysis.

There are countless analytical uses of spreadsheets, and we could have picked many other examples. In fact, some of the cases incorporated in earlier chapters of this book would have fitted right in here, such as resolving different radiochemical half-lives, a problem addressed in section 3.6. Additional illustrations will be given in chapters 7 and 8. But even from the few examples given here you will get the idea: spreadsheets can be used profitably for many quantitative aspects of chemical analysis: for simulating mathematical relations, for extracting specific information from experimental data, and for general data fitting.

Many of the data-fitting procedures demonstrated so far are based on least-squares analysis. In the next chapter we will encounter Fourier transformation, which can often provide an alternative approach to solving such problems.

CHAPTER 7

FOURIER
TRANSFORMATION

7.1 Introduction to Fourier transformation

We can describe the sine wave $A \sin(2\pi ft)$ as a never-ending function of time t, forever oscillating between the limits $+A$ and $-A$. Alternately we can represent it in the frequency domain as a signal with a single frequency, f, and a given, fixed amplitude, A. The two descriptions are fully equivalent, and illustrate that, in this example, the same function can be represented either as a continuous function of time t, or as a single-valued function of frequency f. In general, *any time-dependent phenomenon can alternatively be expressed as a function of frequency, and vice versa*, and Fourier transformation allows us to transform data from the time domain to the frequency domain, and back. Since Fourier transformation is a mathematical operation, time and frequency are merely examples of two associated parameters x and $1/x$ that have a dimensionless product. In spectroscopy, wavelength λ and wavenumber $\bar{v}$ form another pair of such associated parameters.

Fourier transformation has found important applications in many branches of science; here we mention especially its use in various analytical instruments (such as nuclear magnetic resonance, infrared, and mass spectrometry), and in signal processing. Below we will illustrate some properties of Fourier transformation in the latter context.

The Fourier transform $F(f)$ of a time-dependent function $f(t)$ will be defined as

$$F(f) = \int_{-\infty}^{\infty} f(t) e^{-2\pi jft} dt \tag{7.1-1}$$

which can be combined with Euler's theorem

$$e^{-jx} = \cos(x) - j\sin(x) \tag{7.1-2}$$

to yield

265

$$F(f) = \int_{-\infty}^{\infty} f(t)\cos(2\pi ft)dt - j \int_{-\infty}^{\infty} f(t)\sin(2\pi ft)dt \qquad (7.1\text{-}3)$$

where $F(f)$ and $f(t)$ are continuous functions of time t and frequency f respectively, while $j = \sqrt{-1}$. Numerical computation has difficulty with infinity, and the **discrete Fourier transform** is therefore defined somewhat differently, namely as

$$F(f) = \sum_{k=0}^{N} f(t)\cos(2\pi k/N)dt - j \sum_{k=0}^{N} f(t)\sin(2\pi k/N)dt \qquad (7.1\text{-}4)$$

where the frequencies are restricted to the discrete set $k = 0, 1, 2, ..., N$. It is the discrete Fourier transform that is readily calculated on a computer, and that has revolutionized the way Fourier transformation is used in instrumentation and signal analysis. Consequently, it is the discrete Fourier transform which we will use here: from now on, when we mention Fourier transformation, we will mean *discrete* Fourier transformation.

In order to facilitate the calculations, a macro computing the Fourier transform of an array of complex numbers is provided; the program is described in some detail in section 9.5. (For details about the Fourier transform method itself you might want to consult books such as R. Bracewell, *The Fourier Transform and its Applications*, McGraw-Hill, 1965, or E. O. Brigham, *The Fast Fourier Transform*, Prentice-Hall, 1974, 2nd ed. 1988.) The macro has the following properties, restrictions, and requirements:

(a) The input data can be real, imaginary or complex. In fact, they will be treated as complex numbers; when they are real, their imaginary components just happen to be zero, and similarly the real components will be zero for imaginary input data. The same applies to the output data.

(b) The macro is restricted to k input data, where k is an integer power of 2, from $2^1 = 2$ through $2^{10} = 1024$. Allowable k-values are therefore 2, 4, 8, 16, 32, 64, 128, 256, 512, and 1024. (This limitation is linked to the version of Excel used; Excel 97, 98, and 2000 can handle larger arrays, and users of these versions may want to increase the maximum value of k in the macro accordingly.)

(c) The input data are preferably *centered* around $t = 0$ or $f = 0$.

(d) The macro requires a specific, *three-column format*. The first column contains the variable (such as time, frequency, etc.), the second the real parts of the input data, and the third their imaginary components. When the input data are real, the third column should contain zeros and/or blanks; likewise, when the input data are imaginary numbers, the second column should be filled with zeros and/or blanks.

(e) The macro writes its output again in the three columns immediately to the right of the input data. The output uses the same format as the input: first the variable, then the real components of the output data, and finally its imaginary components.

(f) After operation, the output will remain highlighted, thereby facilitating subsequent operations, such as inverse transformation.

(g) In order to use the macro, *first* highlight the three-columns-wide block containing the input data, *then* call the macro. The specific mechanism of calling the macro depends on how it has been installed. The general method is via <u>T</u>ools ⇨ <u>M</u>acro. In the resulting Macro dialog box you can then select either Forward() or Inverse(). It is convenient to assign these macros shortcut key combinations (e.g., Ctrl + F and Ctrl + f, or Ctrl + f and Ctrl + g in case you do not want to use the Shift key), and even more convenient to build them into the menu, or to insert an icon for them in a toolbar. The various methods of installing the macro are described in section 10.4, and the macro itself is detailed in section 10.5.

In what follows we will assume that the forward and inverse Fourier transform macros have been installed; if this is not the case, follow the installation instructions given in section 10.4 before proceeding with the exercises of the present chapter.

Instructions for exercise 7.1

1 Open a new spreadsheet.

2 Reserve the top 8 rows for thumbnail sketches.

3 In cell A9 place the label A =, in B9 enter the numerical value 3, in C9 place the label f =, and in D9 deposit the instruction = pi()/8. (In order to emphasize that these form pairs, use the align right and align left icons on the formula bar to shift the label to the right, and the associated value to the left.)

4 In row 11 deposit the labels time, Re, Im, frequency, Re, and Im.

5 In column A, starting with cell A13, place the numbers $-8, -7, -6, \cdots, -1, 0, 1, \cdots, 6, 7$, i.e., $-8\,(1)\,7$.

6 In B13 deposit the instruction = $B$9*COS(A13*$D$9), and copy this down to row 28.

7 Fill C13:C28 with zeros.

8 Plot A13:C28 on the sheet, in A1:C8, which should show one cycle of a cosine wave, centered around $t = 0$.

9 Highlight A13:C28, and call the macro Forward(). This is all it takes to perform a Fourier transformation.

10 Plot the output of the macro, D13:F28, in a thumbnail sketch at D1:F8. The spreadsheet should now resemble Fig. 7.1-1.

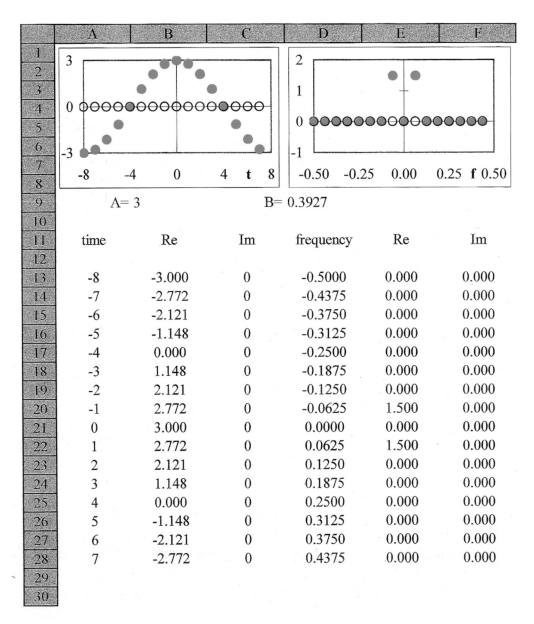

Fig. 7.1-1: The spreadsheet as it should look after Fourier transformation of a cosine wave $3\cos(\pi t/8)$. In the figures, the colored dots show the real components of the transform, the open circles its imaginary component.

The thumbnail sketches tell the story. The left graph shows the cosine wave, plotted as a function of time t. The Fourier transform is to its right, in a graph that has frequency f as its abscissa (horizontal axis). Just like the time scale, the frequency scale starts at a negative value. The Fourier transform generates results for both positive and negative frequencies, even though the latter have no apparent physical meaning.

Also note that, in the plot of the cosine wave, the point at $t=+8$ is missing. This is fully intentional: the data string shown is interpreted by the Fourier transform as a segment of an infinitely repeating sequence, and the first point of the next segment will start at $t=8$. The digital Fourier transformation requires precisely one repeat unit, then analyzes this unit as if the signal contained an infinite number of repeating units on either side.

Upon comparison of the thumbnail plots in Fig. 7.1-1 with those in many standard texts on Fourier transformation you may notice that the input data are customarily displayed starting at $t=0$, with the output data shown on a frequency axis with a discontinuity in its middle. The way we display the data here avoids this discontinuity in the frequency axis, but in all other respects is fully equivalent to the usual representation.

Now let's see how to read the transform. If we ignore the negative frequencies for the moment, we see that the Fourier transform of a cosine is a single point, since all other points (at positive frequencies) have the value zero. That single point has a frequency of 0.0625 Hz, where $0.0625 = 1/16$, where 16 is the period of the time segment used. Its amplitude is 1.5, which is half of the value 3 stored in $B$9. (The other half can be found at $f=-0.0625$ Hz.) In the next exercise we will change both the frequency and the amplitude to see what happens.

11 Change the value of the amplitude, and verify that the amplitude of the Fourier transform tracks that of the cosine wave. Note that the Fourier transform is not automatic: you must invoke the macro before you will see the consequent change in columns D through F, and in the right thumbnail sketch.

12 Change the time scale used, and look at what happens with the transform.

13 Also change the frequency in $D$9, say to $\pi/4$ or $\pi/2$. Observe the resulting change in the transform.

14 Now change the instruction in A13:A28 from cosine to sine, and transform the data. You will see that a sine wave has an imaginary Fourier transform, as illustrated in Fig. 7.1-2.

15 Place the label C = in cell E9, and D = in G9, together with some associated numbers, such as 2 in cell F9 and π in cell H9. Also, change the code in A13:A28 to the sum of a sine and a cosine, with different frequencies and amplitudes. Observe the transform: it should be the sum of the transforms of the individual cosines, each with its own frequency and amplitude, see Fig. 7.1-3.

16 Sines and cosines oscillate symmetrically around zero, and therefore have zero average over the time period considered in the transform. Add an offset to the function and see what happens in that case. Figure 7.1-4 gives it away: an offset only affects the point at zero-frequency in the transform. That zero-frequency contribution indeed shows the function average.

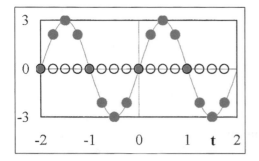 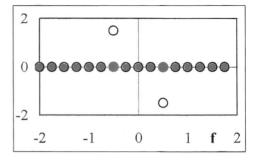

Fig. 7.1-2: The function $3\sin(\pi t)$ and its Fourier transform. The continuous line through the sine wave is shown in order to emphasize the underlying function, sampled here at discrete intervals. Again, the colored dots show the real components, the open circles the imaginary components.

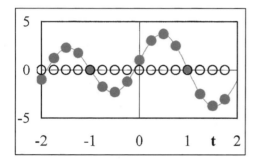 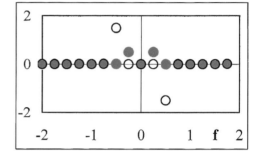

Fig. 7.1-3: The function $\cos(\pi t/2) + 3\sin(\pi t)$ and its Fourier transform.

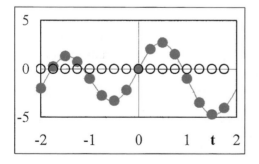 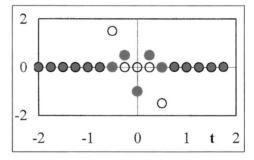

Fig. 7.1-4: The function $\cos(\pi t/2) + 3\sin(\pi t) - 1$ and its Fourier transform.

It is one of the main properties of Fourier transforms that they allow us to view the individual components of a complex mixture of sinusoidal signals. In Fig. 7.1-3 we see the cosine in the real part of the transform, at $f = 0.25$ Hz, and the sine in its imaginary component, with a frequency of 0.5 Hz and a three times larger amplitude.

Fourier transformation is a two-way street: there is a forward transform (which we have used so far) as well as an inverse transform. The latter brings us back from, say, the frequency domain to the time domain. This is possible because Fourier transformation leads to a **unique** relation between a function and its transform, and vice versa, so that we can recover the original function unambiguously with an inverse Fourier transform.

An example of a unique transform pair is the pH: to every value of $[H^+] > 0$ we can assign a corresponding pH, and likewise from every pH we can compute a specific $[H^+]$; no information is lost in going from one to the other. There is, of course, a difference: the pH is a single *number*, while the Fourier transformation involves an entire *function*. Nonetheless, the idea of uniqueness is applicable to both. But note that not all familiar transforms are unique: when we specify x, the quantity $\sin(x)$ is well-defined, but when we specify the value of $\sin(x)$ we cannot recover x without ambiguity: when $x = x_0$ is a solution, so is $x = x_0 \pm 2n\pi$, where n is an arbitrary integer.

17 Highlight D13:D28 and call the inverse Fourier transform. Plot it in G1:I8. You should get back a *replica* of the graph in A1:C8.

It is not necessary to use just sines or cosines. A square wave is also an infinitely repetitive signal, and we can use just one repeat unit of it. A new aspect of a square wave is that it has discontinuities. Where the function switches abruptly from, say, $+1$ to -1, you should use its average value at that discontinuity, i.e., $[(+1) + (-1)]/2 = 0$.

18 Extend A13:A28 to A44.

19 Enter 1 in B13:B20, 0 in B21, -1 in B22:B36, 0 in B37, and 1 in B38:B44. Now call the forward Fourier transform, and follow that up with an inverse Fourier transform to make sure that you recover the original signal.

Figure 7.1-5 illustrates a square wave symmetrical around $t = 0$, and its Fourier transform. Such a square wave can indeed be considered as the sum of a number of cosines,

$$sqw(\omega t) = \frac{4}{\pi}\left\{\cos(\omega t) - \frac{1}{3}\cos(3\omega t) + \frac{1}{5}\cos(5\omega t) - \frac{1}{7}\cos(7\omega t) + \cdots\right\}$$

$$= \frac{4}{\pi}\sum_{n=0}^{\infty}\frac{(-1)^n\cos[(2n+1)\omega t]}{2n+1} \tag{7.1-5}$$

Upon checking you will find that the coefficients produced by the 16-point Fourier transformation are not quite equal to $2/[(2n+1)\pi]$, because we do not use an infinite series. (Still, we get fairly close even with a 16-point transform, with only four cosines, the first coefficient being 0.628 instead of

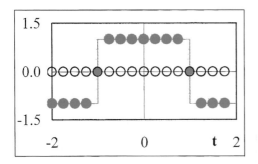 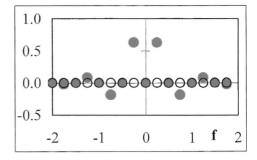

Fig. 7.1-5: The square wave and its Fourier transform.

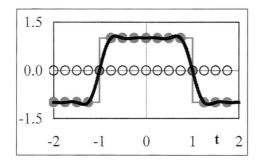

Fig. 7.1-6: The function $y = 2[a_1\cos(\omega t) + a_3\cos(3\omega t) + a_5\cos(5\omega t) + a_7\cos(7\omega t)]$, with $\omega = \pi/2$, and the coefficients $a_1 \approx 0.628$, $a_3 \approx -0.187$, $a_5 \approx 0.0835$, and $a_7 \approx -0.0249$, produced by the Fourier transform of a square wave, see Fig. 7.1-5. Even though just four terms provide a rather poor approximation of a square wave, the black line goes exactly through all 16 colored points of the square wave.

$2/\pi = 0.637$.) Interestingly, even though the Fourier series is truncated, the inverse transform *exactly* duplicates the original function. Figure 7.1-6 illustrates why this is so, with only four cosine terms: while the expression a_1 $\cos(\omega t) + a_3 \cos(3\omega t) + a_5 \cos(5\omega t) + a_7 \cos(7\omega t)$ does not provide a very good approximation of a square wave, it does go precisely through the discrete points of the original waveform.

Transients are usually non-repeating signals, but there is no harm in thinking about them as samples of an infinitely repeating set of them, as long as these repeat units do not overlap, i.e., as long as the repeat unit shows the transient reaching its final state, to within the required precision. Again, when the transient starts at $t = 0$ with a sudden transition from 0 to e^{-kt}, its value at $t = 0$ should be taken as $\frac{1}{2}(0 + e^0) = \frac{1}{2}(0 + 1) = 0.5$.

20 In B13 deposit the instruction = $B$9*exp(- $E$9*A13), and copy this down till row 28.

21 Set the value in B9 to 2, and that in E9 to 5.

22 Call the forward Fourier transform, then the inverse transform.

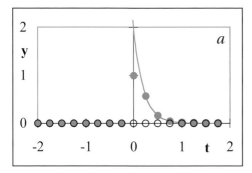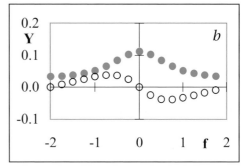

Fig. 7.1-7: The transient $2e^{-5t}$ and its Fourier transform.

23 Again see for yourself what happens when you change (one at the time) the values in cells B9 and E9 respectively.

Figure 7.1-7 illustrates that the Fourier transform method indeed also works for transients. This may not come as a surprise to an analytical chemist, since some of the major instrumental analytical methods that use Fourier transformation apply that method to transients, such as the free induction decay in FT-NMR, and the interferogram in FT-IR.

We will now summarize some of the most important properties of the digital Fourier transform. These properties are merely stated here. Readers interested in their mathematical proofs should consult textbooks dealing specifically with Fourier transformation.

1 Fourier transformation allows us to switch almost effortlessly between two *complementary aspects* of a function of time t, and its representation in terms of the corresponding frequencies f. Instead of time t and frequency f we can use other parameters that are inversely related to each other (so that their product is dimensionless), such as wavelength λ and wavenumber $\bar{\nu}$. Although we treat Fourier transformation here as a mathematical concept, such a transformation is not outside our normal, daily experience. When we hear a bird in the woods, we can easily follow the song by its pitch (frequency) even though there may be many other, simultaneous sounds at other frequencies, such as the noise of the wind in the trees. When we listen to an orchestra, we can focus on the flutes, or the cellos, by virtue of their specific frequency ranges, even though the sound of both is mixed with that of many other instruments as a function of time. In other words, for sound our brain performs the equivalence of a Fourier transform. The same applies to vision: when we experience a color, we will often know immediately its major components. Both of these skills can of course be honed by training, but apparently are already present in rudimentary form in our untrained brains. Digital Fourier transformation performs this service for us on numbers in data sets, explicitly and almost instantaneously.

2   The *continuous* Fourier transformation of a *periodic* (i.e., infinitely repeti-
    tive), continuous function f(t) of time *t* is a representation F(f) in the fre-
    quency domain that contains only *discrete* frequencies *f*. The Fourier
    transform of a *non-repetitive*, continuous function f(t), such as a single
    transient, is a *continuous* function F(f) of frequency *f*. Such distinctions
    disappear, of course, in *digital* Fourier transformation, where both the
    input and output arrays are discrete. Some consequences of the discrete
    nature of the input data in digital Fourier transformation will be discussed
    in section 7.4.

3   The Fourier transform provides a *unique* relation between a function and
    its transform, i.e., there is no loss of information when we replace a function
    by its Fourier transform, or vice versa. This property is, of course, crucial in
    Fourier transform spectrometry, since it allows us to measure a time-
    dependent function and obtain from it the *spectrum*, i.e., its representation
    in the frequency domain.

4   The Fourier transform assumes that its input constitutes one complete
    repeat unit of an infinitely repetitive signal. For non-periodic signals, many
    problems (of which some are detailed in section 7.4) can be avoided by
    making sure that f(t) starts and ends at the same value, and with the same
    derivatives. 'Same'ness here is not necessarily a mathematical identity, but
    is best defined in terms of the required precision.

5   Functions that are symmetrical around $t = 0$, i.e., functions such that
    $f(-t) = f(t)$, have a *real* Fourier transform. Such functions are called *even*;
    we started out with such an even function in our first sample function,
    $\cos(\omega t)$. For odd functions, i.e., where $f(-t) = -f(t)$, the resulting Fourier
    transform is imaginary, as was illustrated by $\sin(\omega t)$. In general, functions
    will be neither even nor odd, in which case their Fourier transform will be
    complex, i.e., with both real and imaginary components.

6   A *shift in time* in the time domain corresponds in the Fourier transform
    domain to a shift in *phase angle*, and vice versa. In mathematical terms,
    when f(t) is shifted in time by t_0 to $f(t + t_0)$, the corresponding Fourier trans-
    form will be shifted from F(f) to $F(f) \times \exp[-2\pi j f t_0]$, where $j = \sqrt{-1}$.
    Similarly, $F(f + f_0)$ has the inverse transform $F(t) \times \exp[2\pi j f_0 t]$.

7   Differentiation in the time domain corresponds to division by $2\pi j f$ in the
    frequency domain. Likewise, integration in the time domain corresponds to
    multiplication by $2\pi j f$ in the frequency domain. We will use these proper-
    ties in section 7.4.

8   One of the most useful properties of Fourier transformation is that it con-
    verts a **convolution** into a multiplication. (Convolution is one of many **cor-
    relations**, i.e., mathematical operations between functions, that can be
    greatly simplified by Fourier transformation.) Since convolution is a rather
    involved mathematical operation, whereas multiplication is simple, convo-
    lutions are often performed with the help of Fourier transformation. We will
    explore this property in more detail in sections 7.5 and 7.6.

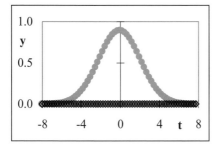
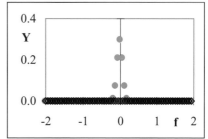
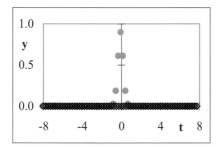
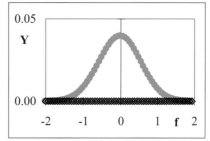

Fig. 7.1-8: The Gaussian $y = (0.9/c)\exp[-(t/c)^2]$ (left) and its Fourier transform (right) for $c = 3$ (top) and $c = 0.4$ (bottom).

To end this section we now illustrate some of the above properties. Many of the figures shown on the next two pages involve the Gaussian function $y = (a/c)\exp[-(t/c)^2]$, where the pre-exponential factor $(1/c)$ is used to compare peaks occupying the same area. A Gaussian is a convenient function because its Fourier transform is again a Gaussian, and the same applies to the inverse transform. The factor $a = 0.9$ is used here merely to obtain convenient scales. Gaussians curves are common in chromatograms. In Fig. 7.1-8 we see that the narrower is the original function, the wider is its transform.

Figure 7.1-9 shows the effect of moving the center of the Gaussian peak, at which point the function is no longer even (in the sense discussed above under point (5), i.e., symmetrical around $t = 0$), so that its transform has both real and imaginary components.

Figure 7.1-10 compares the Fourier transforms of a Gaussian and a Lorentzian peak. The Lorentzian peak has wider 'tails', and consequently a narrower Fourier transform. The comparison is made between the Gaussian $y_G = a\exp[-(t/c)^2]$ and a Lorentzian of equal area, $y_L = (a/\sqrt{\pi})/[1 + (t/c)^2]$, see section 8.6. Lorentzian curves are often encountered in spectroscopy.

As a convenience to the user, the Fourier transform macro will accept two types of *in*put: data that are properly centered (i.e., with sequence numbers ranging from $-2^{N/2}$ to $2^{N/2} - 1$), or data that start at zero (i.e., ranging from 0 to $2^N - 1$).

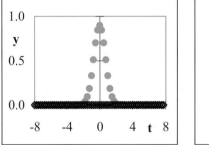
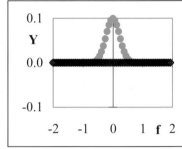

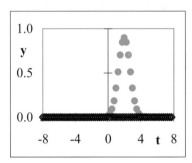
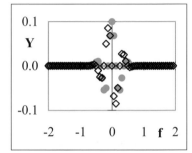

Fig. 7.1-9: The Gaussian $y = 0.9 \exp[-(t - t_0)^2]$ (left) and its Fourier transform (right) for $t_0 = 0$ (top) and $t_0 = 2$ (bottom).

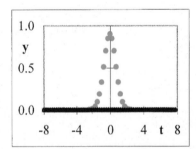
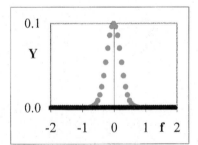

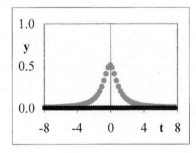
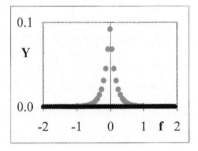

Fig. 7.1-10: The Gaussian $y_G = 0.9 \exp[-t^2]$ (top left) and a Lorentzian $y_L = (0.9/\sqrt{\pi})/(1 + t^2)$ of equal area (bottom left), together with their Fourier transforms (top right and bottom right respectively).

7.2 Interpolation and filtering

We will now illustrate a powerful pair of properties of Fourier transformation, namely its power to *interpolate* and to *filter* data. First we look at interpolation. The idea is as follows. Imagine that a periodic signal is completely described by a given set of frequencies, i.e., it does not contain any frequencies higher than a given value f_{max}. Its Fourier transform will reflect that. Now go to the Fourier transform, extend it to frequencies above f_{max}, and specify that the signal has zero contributions at those higher frequencies; this process is called **zero-filling**. When we apply an inverse transformation to this *extended* data set, we will recover the original signal with added, *interpolated* points, since the longer frequency record corresponds to a more detailed representation in time t.

Instructions for exercise 7.2-1

1 Start a new spreadsheet, and organize it like the earlier one, or just copy that earlier one.

2 Here we will use a rather minimal, four-point data set: for time in A13:A16 enter -2, -1, 0, and 1, for the real components of the signal in B13:B16 enter 0, -1, 0, and 1, and place zeros in C13:C16. There should be no data below row 16.

3 Run the Fourier transform of this data set. The graphs should look like those in Fig. 7.2-1. An inverse transform should yield a replica of the input data.

4 Copy the column headings in D11:F11 to G11:I11, and those in A11:C11 to J11:L11.

5 Copy D13:F16 to G27.

6 Highlight G29:G30, grab its common handle, and drag that handle down to G44. Likewise, activate G27:G28, grab the common handle, and extend the column *upwards* to G13. Yes, it will do that!

7 Now comes the trick: we add zeros to the higher frequencies by filling the blocks H13:I26 and H31:I44 with zeros.

8 Make a thumbnail graph of H13:H44 and I13:I44 vs. G13:G44 in G1:I8.

9 Highlight block G13:I44 and call in the inverse Fourier transform.

10 Make a thumbnail graph of K13:K44 and L13:L44 vs. J13:J44 in J1:L8. Your spreadsheet should now resemble that in Fig. 7.2-2.

What you have just done can be interpreted as follows. You had a sine wave with only four points per cycle. After Fourier transformation, you extended the transform to 32 frequencies, but without adding any content. By doing

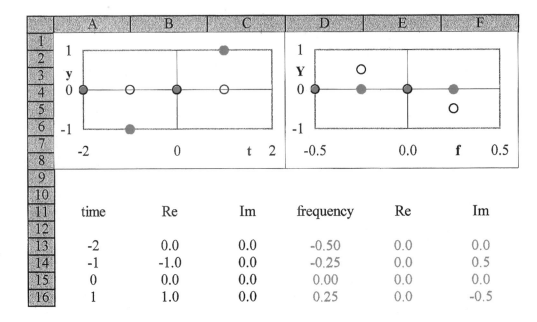

	time	Re	Im	frequency	Re	Im
	-2	0.0	0.0	-0.50	0.0	0.0
	-1	-1.0	0.0	-0.25	0.0	0.5
	0	0.0	0.0	0.00	0.0	0.0
	1	1.0	0.0	0.25	0.0	-0.5

Fig. 7.2-1: The spreadsheet for a four-point sine wave and (in color) its Fourier transform.

so, you have provided additional information, viz. that there is *no noise* at the higher frequencies added. This is equivalent to specifying that the curve is smooth between the original data points. Upon inverse transformation, you then reconstructed 32 data points of the original signal, eight times as many as it had originally. Those extra points show up as an interpolation.

As mentioned earlier, the real power of Fourier transformation lies in its application to general functions rather than to just sines and cosines. To illustrate this, we will now use a simple Gaussian curve as our input, then interpolate it.

11 Place an amplitude A in B9, a center value B in D9, and a peak width C for the Gaussian peak in F9.

12 Place the instruction $= \$B\$9*\exp(-(((A13-\$D\$9)/\$F\$9)^2))$ in B13, and copy this down to cell B44. Fill C13:C44 with zeros.

13 Adjust the numerical values of the constants in row 9 to give you a complete though sparse Gaussian peak. Just for the fun of it, select F9 such that the peak maximum falls in the range between two adjacent data points. The visible part of the Gaussian peak should occupy no more than about 20% of the data set.

14 Compute the Fourier transform of A13:C44.

15 Graph the input and output data, and compare these with Fig. 7.2-3.

frequency	Re	Im	time	Re	Im
-4.00	0.00	0.00	-2.000	0.000	0.000
-3.75	0.00	0.00	-1.875	-0.195	0.000
-3.50	0.00	0.00	-1.750	-0.383	0.000
-3.25	0.00	0.00	-1.625	-0.556	0.000
-3.00	0.00	0.00	-1.500	-0.707	0.000
-2.75	0.00	0.00	-1.375	-0.831	0.000
-2.50	0.00	0.00	-1.250	-0.924	0.000
-2.25	0.00	0.00	-1.125	-0.981	0.000
-2.00	0.00	0.00	-1.000	-1.000	0.000
-1.75	0.00	0.00	-0.875	-0.981	0.000
-1.50	0.00	0.00	-0.750	-0.924	0.000
-1.25	0.00	0.00	-0.625	-0.831	0.000
-1.00	0.00	0.00	-0.500	-0.707	0.000
-0.75	0.00	0.00	-0.375	-0.556	0.000
-0.50	0.00	0.00	-0.250	-0.383	0.000
-0.25	0.00	0.50	-0.125	-0.195	0.000
0.00	0.00	0.00	0.000	0.000	0.000
0.25	0.00	-0.50	0.125	0.195	0.000
0.50	0.00	0.00	0.250	0.383	0.000
0.75	0.00	0.00	0.375	0.556	0.000
1.00	0.00	0.00	0.500	0.707	0.000
1.25	0.00	0.00	0.625	0.831	0.000
1.50	0.00	0.00	0.750	0.924	0.000
1.75	0.00	0.00	0.875	0.981	0.000
2.00	0.00	0.00	1.000	1.000	0.000
2.25	0.00	0.00	1.125	0.981	0.000
2.50	0.00	0.00	1.250	0.924	0.000
2.75	0.00	0.00	1.375	0.831	0.000
3.00	0.00	0.00	1.500	0.707	0.000
3.25	0.00	0.00	1.625	0.556	0.000
3.50	0.00	0.00	1.750	0.383	0.000
3.75	0.00	0.00	1.875	0.195	0.000

Fig. 7.2-2: The part of the spreadsheet containing the zero-filled transform, and its inverse Fourier transformation. The seed from Fig. 7.2-1 is shown in block G27:I30.

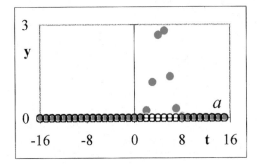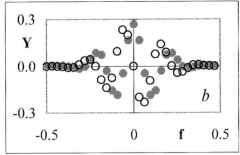

Fig. 7.2-3: A 32-point data set containing about six points of a Gaussian curve, $y = 3 \times \exp[-(x-4.56)^2 / (1.5)^2]$, and its inverse Fourier transform.

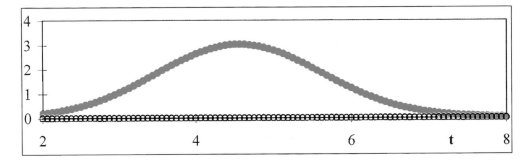

Fig. 7.2-4: The central part of the curve obtained after zero-filling the Fourier transform to 512 points, followed by inverse transformation.

16 Now copy block D13:F44 to G253.

17 Grab G253:G254 by the handle and draw back to G13. Likewise extend the frequency scale down to G524.

18 Fill H13:I252 and H286:I524 with zeros.

19 Call the inverse Fourier transform of the data in block G13:I524. You should now see the reconstructed input curve, but with 16 times as many points, 15 of them interpolated.

20 Graph the interpolated curve, and compare with Fig. 7.2-4.

While the original curve had only six points that rose significantly above the baseline, the interpolated curve clearly has quite a few more. Note that we have not added any real information to the original data set, but have merely interpolated it, without making any assumptions on the shape of the underlying data set, other than that it is noise-free. Consequently we can, e.g., determine the peak position by fitting a number of points near the peak maximum to a low-order polynomial such as $y = a_0 + a_1 x + a_2 x^2$, using the

Savitzky–Golay method, and then find the peak maximum from the coefficients as $x_{peak} = -a_1/2a_2$.

Now we will do the opposite. We will take a signal, add random noise, and Fourier transform it. If we know the main frequencies of the signal, we can remove a large part of the noise, simply by setting all signals at unneeded frequencies equal to zero. Upon inverse transformation we will then recover most of the original signal, while we discriminate against much of the noise. Filtering is usually such a trade-off: giving up some signal in return for losing much more noise.

As our signal we will take a set of Gaussian curves that might represent, e.g., a chromatogram or an NMR spectrum. We will first generate such a set, and examine its Fourier transform. We will see that most of the transform is localized in a rather small part of the frequency spectrum. This implies that we can delete much of the frequency spectrum with relatively little loss of signal, and we will test this.

Then we will be ready for a more realistic 'experiment', by adding random noise to the signal. Now that we know in what frequency range we can filter out the noise with relative impunity, we will do so, and observe the resulting, filtered data.

Instructions for exercise 7.2-2

1 Open a new spreadsheet.

2 In cells A1, B1, and C1 place labels for amplitude, center, and width.

3 In cells A3:A7 deposit some amplitudes, in B3:B7 some peak centers, and in C3:C7 some peak widths.

4 In row 11 place the labels time, Re, Im, freq., Re, Im, freq., Re, Im, time, Re, Im, noise. We will here use 'time' and 'frequency' as generic parameters, as might be appropriate for chromatography, but 'time' might also represent elution volume, NMR frequency, NMR magnetic field shift in ppm, potential in electrochemistry, distance in crystallography, etc., and 'frequency' its inverse.

5 In A13:A524 place the rank order numbers −256 (1) 255.

6 In B13 enter the instruction $= \$A\$3*EXP(-(((A13 - \$B\$3)/\$C\$3)^2)) + \cdots$, the dots representing similar expressions for the variables in rows 4 through 7.

7 Add a column of zeros for the imaginary component of the signal, in C13:C524, then Fourier transform the resulting data set.

8 Plot the input signal and its transform. Figure 7.2-5 illustrates what you might obtain for frequency spectrum.

9 Inspection of the low-frequency (middle) section of the spectrum suggests that frequencies above a given value contribute little to the transformed signal, and may

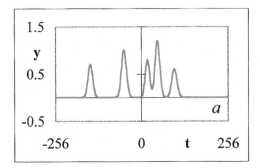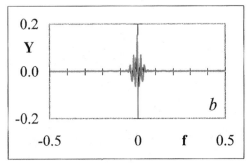

Fig. 7.2-5: (a) The input signal and (b) its Fourier transform.

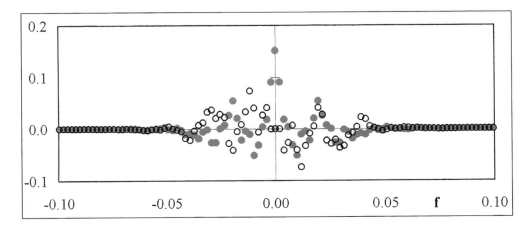

Fig. 7.2-6: The central section of Fig. 7.2-5b.

therefore be expendable. (The specific cut-off frequency depends, of course, on the functions used, and on your judgement as to what contribution is small enough to be neglected.) In the example of Fig. 7.2-6, an absolute cut-off frequency $|f|$ of about 0.07 appears reasonable.

10 In G13 place the instruction = D13, and copy this to G13:I524, thereby making a copy of D13:F524.

11 Then, *in that copy*, replace the data at higher frequencies by zeros. For example, for a cut-off at $|f| > 0.05$, zero the data in H13:I243 and H295:I524.

12 Highlight G13:I524 and inverse transform it. Plot the recovered signal; it should resemble the original closely. Upon closer inspection you will find that there are oscillatory differences, that can be made acceptably small by proper choice of the cut-off frequency, see Fig. 7.2-7. Looking at those residuals, as in Fig. 7.2-7b, is much more informative than a visual comparison of the curves.

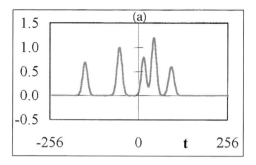
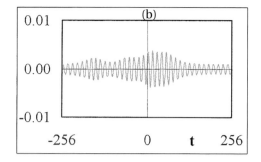

Fig. 7.2-7: (a) The filtered output signal, and (b) the differences between it and the original shown in Fig. 7.2a for a cut-off at $|f| \geq 0.07$.

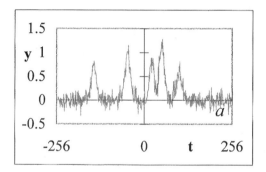
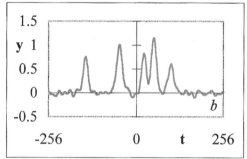

Fig. 7.2-8: (a) The noisy input signal, and (b) the same after Fourier transformation, removal of all frequency components with $|f| \geq 0.07$, and inverse transformation.

13 Now introduce a column of Gaussian noise (Tools ⇨ Data Analysis ⇨ Random Number Generation ⇨ Distribution Normal, Mean 0, Standard Deviation 1) and modify the signal in B13:B524 by adding to it *na* (for noise amplitude) times that noise. The value of *na* should of course be placed near the top of the spreadsheet, together with its label. By setting *na* equal to 0 you will recover the earlier signal.

14 Fourier transform the noise-containing signal, in A13:C524.

15 Filter the transformed signal.

16 Inverse transform this filtered output.

17 Figure 7.2-8 illustrates such a noisy input signal and its filtered output.

In order to remove noise by Fourier-transform filtering we can look at the transform, as in Fig. 7.2-6. However, it is often more convenient to inspect the **power spectrum**, which is a (usually semi-logarithmic) plot of the magnitude (i.e., of the square root of the sum of squares of the real and imaginary components) of the Fourier transform. Such a power spectrum is shown in Fig. 7.2-9, both for a noise-free signal, and for the same signal with noise. The power spectrum is symmetrical, i.e., the information at negative and

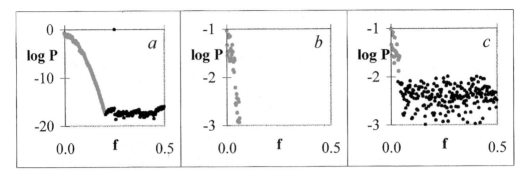

Fig. 7.2-9: (a) The power spectrum of Fig. 7.2-5 (b) the same enlarged to show its top in greater detail, and (c) the power spectrum of Fig. 7.2-8.

positive frequencies is identical, and it is therefore commonly displayed only for positive frequencies. Moreover, by using a logarithmically compressed vertical scale, orders-of-magnitude differences are emphasized, thereby more clearly delineating the frequency region where the signal dominates, and that where the noise is the more important factor. A power spectrum often allows us to make an informed choice of the cut-off frequency.

18 Calculate and plot the power spectrum, i.e., $\log P = \frac{1}{2} \log (\text{Re}^2 + \text{Im}^2)$ vs. frequency. Here Re and Im refer to the real and imaginary components of the Fourier transform, in columns E and F. Plot $\log P$ only for positive frequencies, as in Fig. 7.2-9.

The power of the noise-free signal falls off rapidly with frequency, until it hits the truncation noise at about 10^{-16} due to the finite number of digits used to represent numbers in Excel. After we add Gaussian noise (so-called 'white' noise because its average power is independent of frequency, i.e., it is represented in the power spectrum by a roughly horizontal line), that added noise becomes dominant at all but the lowest frequencies. This is the rationale for replacing the higher-frequency contributions by zeros. The power spectrum suggests where this is done most appropriately. In the above example, the cut-off should occur where the power falls below about −2, at f = 0.042. As can be seen in Fig. 7.2-6, by cutting off the frequency components at f > 0.042, we introduce some signal distortion. Clearly, the noise forces our hand here, since we would have preferred to keep the frequencies in the range between 0.042 and 0.07. There is, of course, also noise at the lower-frequency components, but zero-filling at those frequencies would remove more signal than noise, and would therefore be counterproductive.

The above example illustrates the trade-offs involved in filtering out noise. It is clear from Fig. 7.2-9 that zero-filling at higher frequencies works

only as long as the power of the low-frequency signal components significantly exceeds that of the corresponding noise. Otherwise, any information
about the signal will be 'buried' in the noise. In sections 8.3 and 8.4 we will
encounter some correlation methods that can sometimes still pull a buried
signal out of the noise.

The above examples have illustrated the use of a *sharp* cut-off filter, equivalent to multiplying the Fourier transform by 1 up to a cut-off frequency, and
by 0 at higher frequencies. A more *gradual* cut-off can of course be made by
multiplying the frequency spectrum by some frequency-dependent attenu-
ating function, instead of the abrupt pass-all-or-none approach of the
example. There are many such more gradual filtering functions, usually
named after their originators, such as Hamming, Hanning, Parzen, and
Welch. Unfortunately, with a gradual cut-off filter, it is not so clear what
compromise between noise-reduction and signal distortion is being struck.

In principle, Fourier transformation allows you to be much more specific,
and to pick out or reject, e.g., a single frequency or a specific set of frequencies. For example, one can tune in to a particular frequency (the digital
equivalent of a sharply tuned filter, or of a lock-in amplifier, see section 8.4),
or selectively remove 'noise' at, say, 60, 120, and 180 Hz, while leaving signals
at other frequencies unaffected. However, this requires that the selected frequency or frequencies precisely coincide with those used in the Fourier
transformation, in order to avoid the so-called leakage to be described in
section 7.4.

7.3 Differentiation

Differentiating a function in the Fourier transform domain is, in principle,
both straightforward and easy. Differentiation with respect to time t in the
time domain is equivalent to multiplication by $j\omega = 2\pi j f$ in the frequency
domain. (The product 2π times f is called the **angular frequency** ω.) This
is perhaps most readily illustrated using Euler's formula, $e^{\pm j\omega t} = \cos(j\omega)$
$\pm j\sin(j\omega)$, which upon differentiation yields $de^{\pm j\omega t}/dt = \pm j\omega e^{\pm j\omega t}$.
Therefore, differentiation of a function can be accomplished by taking its
Fourier transform, multiplying the resulting data by $j\omega$, and inverse transforming this. The Fourier transform of the original function is, in general,
a complex number, $a + jb$, and multiplication by $j\omega$ therefore results in $j\omega$
$(a + jb) = \omega(-b + ja)$. Consequently, in order to differentiate a function in
the time domain, we Fourier transform it, then in a copy exchange the
real and imaginary columns in the transform, multiply the data now in
the real column by $-\omega$ and those in the imaginary column by ω, then
apply an inverse transformation. Below we will illustrate this for a
Gaussian curve.

Instructions for exercise 7.3

1   Open a new spreadsheet.

2   Organize it the same way as the preceding spreadsheet, with the input signal in columns A-C, its transform in D-F, a copy of that transform in G-I, and the final result in J-L. This way you can play with the input data without having to change the instructions every time.

3   In A13:A524 place values for time t.

4   In B13:B524 place the instruction for one (or more) Gaussian(s).

5   In C13:C524 place zeros.

6   In G13 place the instruction = D13.

7   In H13 deposit the instruction = 2*PI()*D13*F13, and in I13 the instruction =−2*PI()*D13*E13.

8   Copy the instructions in G13:I13 down to row 524.

9   You are now ready to go. Highlight A13:C524 and call the forward Fourier transform. Then highlight G13:I524 and use the inverse Fourier transform instead. Voilà.

10 Plot the input data (B13:B524 versus A13:A524; there is nothing to see in C13:C524) and the final result where, likewise, you only need to show the real component. Figure 7.3-1 illustrates what you what you might see.

11 Note that we need not symmetrize the time scale of the input signal around zero, but that the output will be computed that way. If it bothers you, add 256 to the t-scale, as we have done in Fig. 7.3-2b.

12 Add some noise to the signal in B13:B524, and see how it tends to swamp the derivative. Figure 7.3-2 illustrates this. Of course you will see this only *after* you have again used the forward and inverse transforms, as in (9). The Fourier transform is performed by a macro. Macros do *not* update automatically, as the standard spreadsheet functions do.

13 Wow, that is not a pretty picture! The noise is obviously much more important in the derivative than it is in the original curve, for a reason that will be explained in section 8.8. Note the change in vertical scale in the plot of the derivative, in order to keep most of the data inside the picture frame.

14 But by now you know what to do: set the higher-frequency data in the transform equal to zero, as in section 7.2, and try again. Now, in one single operation, you both differentiate and filter out most of the noise, as illustrated in Fig. 7.3-3. *You* are now in control.

15 Apply the same to a set of peaks with noise, as in Fig. 7.3-4.

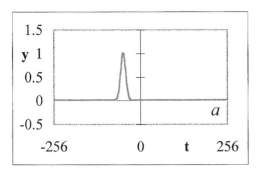 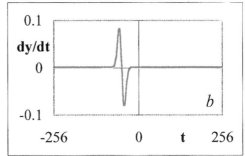

Fig. 7.3-1: (a) A Gaussian peak, and (b) its first derivative.

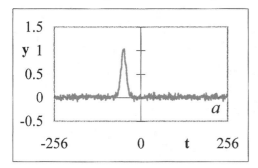 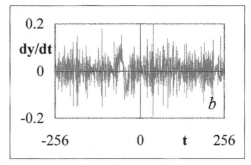

Fig. 7.3-2: (a) A Gaussian peak with noise, and (b) its first derivative.

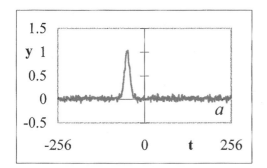 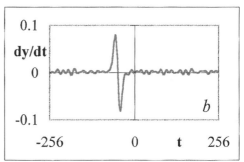

Fig. 7.3-3: (a) The same Gaussian peak with noise, and (b) its filtered first derivative.

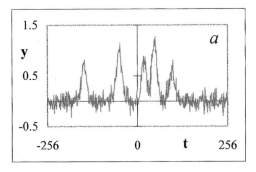
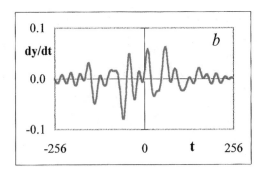

Fig. 7.3-4: (a) The signal of Fig. 7.2-8, and (b) its first derivative (after filtering by zero-filling at |f| > 0.042).

7.4 Aliasing and leakage

A discussion of the digital Fourier transform would be incomplete without a consideration of its inherent problems. Digital Fourier transformation has two weak spots that are both consequences of its digital nature. Because we only use a finite data set, some frequencies will fall *outside* the available frequency range, and will then be misrepresented; this is called aliasing. Another consequence of the finite set of frequencies is that some frequencies *inside* the range covered will not fit either, because they fall *in between* the available frequencies. This leads to leakage. Below we will illustrate aliasing on a small data set, and leakage on a large set. Once you understand what causes these problems you can often avoid them: homme averti en vaut deux, forewarned is forearmed.

Instructions for exercise 7.4-1

1 Open a new spreadsheet.

2 Organize it in the fashion of exercise 7.1 (see Fig. 7.1-1), with space at the top, in rows 1 through 8, for small graphs.

3 In A9 and C9 deposit the labels A = and B =, and in B9 and D9 place the corresponding numerical values 1.

4 In row 11 place the labels time, Re, Im, freq., Re, Im, time, Re, Im in A11:I11.

5 In A13:A20 deposit the numbers −4, −3, −2, −1, 0, 1, 2, and 3.

6 In B13 deposit = $B$9*COS($D$9*PI()*$A13/4). Note the dollar sign in front of the letter A for a mixed absolute/relative address.

7 In A22:A102 deposit the numbers −4.0, −3.9, −3.8, …, 3.9, 4.0.

8 Copy the instructions from A20 to J22. (This is where the dollar sign preceding the letter A comes in handy.) Copy the instruction down to J102.

9 In A1:C8 make a thumbnail graph of the signal. In it, display the data in A13:A102 as circles, those in D13:D102 as a line. (Note that A21:A102 and D13:D21 are left blank.) At this point, the top of your spreadsheet should look more or less like Fig. 7.4-1.

10 In cell J9 place the label B′ = , and in cell K9 the instruction = D9.

11 Copy D22:D102 to K22:K102.

12 Modify the instruction in K22 to refer to $K$9 rather than to $D$9, so that it will now read = $B$9*COS($K$9*PI()*$A13/4). Copy this instruction down to K102.

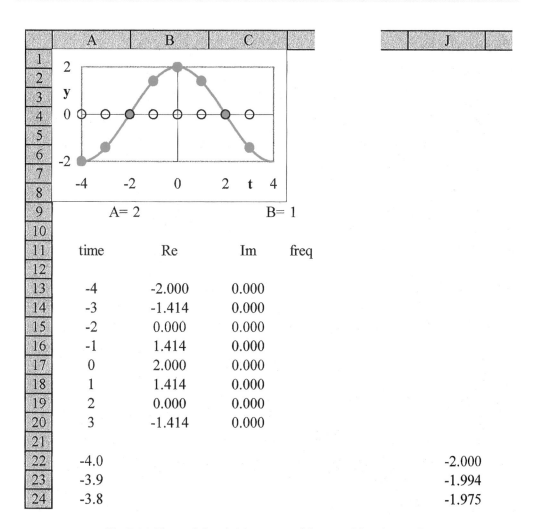

Fig. 7.4-1: The top left and right corners of the spreadsheet in exercise 7.4.

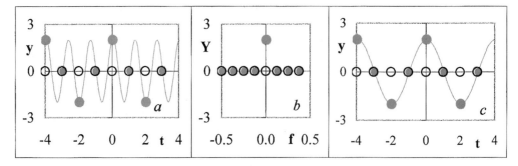

Fig. 7.4-2: From left to right: (a) the signal for B = 5, (b) its transform, and (c) the inverse transform of the latter. The line in (a) shows the cosine for B = 5, and that in (c) the cosine wave for B' = 3, drawn through the *same* data points.

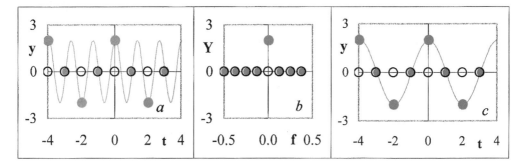

Fig. 7.4-3: From left to right: (a) the signal for B = 6, (b) its transform, and (c) the inverse transform of the latter. The line in (a) shows the cosine for B = 6, and that in (c) the cosine wave for B' = 2, drawn through the same data points.

13 Highlight block A13:C20, and Fourier transform it. The result will show in D13:F20.

14 Continue with the inverse Fourier transform of D13:F20. The result in G13:I20 should now contain a replica of A13:C20.

15 In D1:F8 make a graph of the data in block D13:F20, i.e., of E13:E20 and F13:F20 vs. D13:D20.

16 In G1:I8 make a graph of the data in block G13:K102. Then remove the curve for J13:J102 vs. G13:G102. The third panel should resemble the first.

17 Change the value of B in cell D9 to 2, then to 3, and see what happens.

18 We will for the moment skip B = 4, and jump directly to B = 5. Everything seems to be OK until you notice that the middle panel is the same as the one you found for B = 3. Indeed, when you substitute a 3 for B' in cell K9 (so that it no longer automatically traces cell D9) you will see that such a lower-frequency cosine fits the data just as well, see Fig. 7.4-2.

19 Try B = 6. Its transformation is identical to that for B = 2, as you can again verify by comparing the result in the third panel with B' = 2, see Fig. 7.4-3.

What you see here is that an eight-point Fourier transform apparently cannot count beyond 4, because it clearly confuses B = 5 with B = 3, and likewise 6 with 2, and 7 with 1. This is called aliasing: a cosine with B = 5 can masquerade under the alias B = 3.

For your consolation: the transform is still slightly smarter than those birds that count zero, one, many, so that a birdwatcher's blind entered by three observers is considered empty after the bird has seen two observers leave again: for such a bird brain, 'many' minus 'many' is zero!

As with most problems, aliasing is easily prevented once you understand what causes it. You will avoid aliasing by using a sufficiently large data set. Which immediately brings up the question: what is meant by 'sufficiently large'? The answer, called the Nyquist theorem, is that you need to *sample more than two points per cycle* of any periodic signal in order to avoid aliasing. Otherwise the sample is said to be undersampled.

Note that this has nothing to do with the Fourier transform per se, but everything with the more general problem of representing continuous functions by discrete samples of such functions. The Nyquist theorem specifies that the underlying, continuous, repetitive function cannot be defined properly unless one samples it more than twice per its repeat period.

Let's go back to the spreadsheet to experiment some more.

20 Now try B = 4. In the graph you might miss its transform, since the real point at f = 4 has the value 2, and therefore may require a change of vertical scale in the middle thumbnail sketch.

21 But the problem is really more serious than that: while the frequency is well-defined, its amplitude is not. This is illustrated in Fig. 7.4-4 with the signal $y = 3[\sin(2\pi ft) + \cos(2\pi ft)]$, which fits the data equally well. And there are many more such combinations.

Figure 7.4-4 illustrates the Nyquist criterion: the signal is sampled at exactly two points per cycle, which is just not good enough. In this borderline case the frequency is recovered, but the amplitudes of the possible sine and cosine components of the signal are not.

The above illustrates what happens when the signal frequency lies outside the range of frequencies used in the Fourier analysis, in which case the digital Fourier transform will misread that frequency as one within its range. As already indicated, another problem occurs when the frequency lies within the analysis range, and also satisfies the Nyquist criterion (i.e., is sampled more than twice during the repeat cycle of that signal), but has a frequency that does not quite fit those of the analysis, as illustrated below.

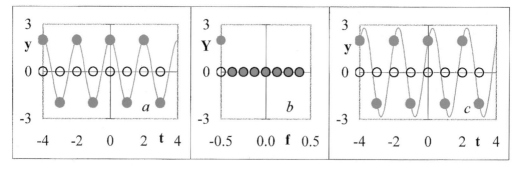

Fig. 7.4-4: From left to right: (a) the signal for B = 4, (b) its transform, and (c) the inverse transform of the signal in (b). The line in (c) shows the sum of a sine and cosine wave, with A = 3 and B = 4.

Instructions for exercise 7.4-2

1 Open a new spreadsheet, and organize it as that in exercise 7.4-1, with rows 1 through 8 reserved for small graphs, and with similar labels.

2 In A13:A140 deposit the numbers $-64, -63, \ldots, -1, 0, 1, \ldots, 62, 63$.

3 In B13 deposit the instruction = $B$9*COS(A13*$D$9) where $B$9 refers to the amplitude, and $D$9 to the frequency.

4 In B9 place the value 1, and in D9 the value = PI()/8, where $\pi/8 \approx 0.3927$.

5 In C13:C140 enter zeros.

6 Make a graph of the signal in A1:C8.

7 Compute the Fourier transform of the signal in D13:F140.

8 In D1:F8 make a graph of the transform or, better yet, of its middle half, for $-0.25 \le f \le 0.25$ (since there is nothing to be seen at higher frequencies).

9 The top of the spreadsheet should now look like Fig. 7.4-5.

10 Now change the value in D9 from $\pi/8$ to, say, 0.4, and repeat the transformation. Figure 7.4-6 shows what you should see.

In Fig. 7.4-5 the cosine wave fits exactly eight times, and this shows in its transform, which exhibits a single point at $f = \pm 8 \times (1/128) = \pm 0.0625$. On the other hand, the cosine wave in Fig. 7.4-6 does not quite form a repeating sequence, and its frequency, $\pi/(0.4 \times 128) \approx 0.06136$, likewise does not fit any of the frequencies used in the transform. Consequently the Fourier transform cannot represent this cosine as a single frequency (because it does not have the proper frequency to do so) but instead finds a combination of sine and cosine waves at adjacent frequencies to describe it. This is what is called **leakage**: the signal at an in-between (but unavailable) frequency as it were leaks into the adjacent (available) analysis frequencies.

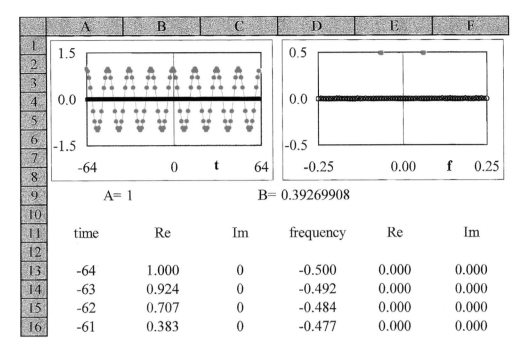

Fig. 7.4-5: The top of the spreadsheet showing the function $y = \cos(\pi t / 8)$ and its Fourier transform.

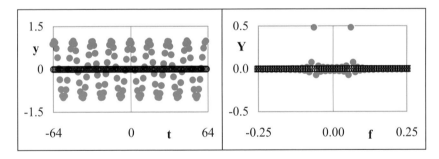

Fig. 7.4-6: The function $y = \cos(0.4\,t)$ and its Fourier transform.

11 Imagine that you take experimental data at a rate of 1 point per millisecond, a quite comfortable rate for an analog-to-digital converter, and quite sufficient for many analytical experiments, such as the output of a gas chromatograph. Unavoidably, the signal will also contain some 60 Hz emanating from the transformers in the power supplies of the instrument. (The main frequency is 60 Hz in the US; in most other countries, it would be 50 Hz.) In D9 enter the corresponding frequency, $= 1000/60$ (where the factor 1000 is used to express the frequency in the corresponding units of *per milli-second*) or $= 1000/50$ (outside the US). Figures 7.4-7 and 7.4-8 illustrate what happens with the Fourier transform of such a signal.

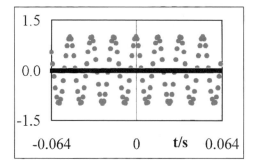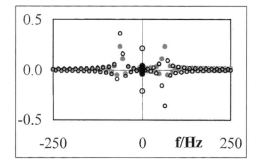

Fig. 7.4-7: The function $y = \cos(2\pi \times 60\ t)$ sampled at 1 ms intervals, and the central portion of its Fourier transform.

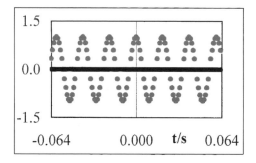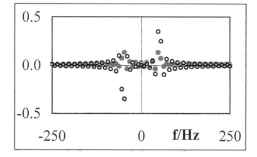

Fig. 7.4-8: The function $y = \cos(2\pi \times 50\ t)$ sampled at 1 ms intervals, and the central portion of its Fourier transform.

Keep in mind that leakage does *not* misrepresent the signal, but expresses it in terms of the available frequencies. It is successful in doing so. It is just that a *simpler* representation would be obtained if the appropriate frequency were available.

In both figures, leakage occurs because the Fourier transform of a data set for $-0.064 \le t < 0.064$ s has the frequencies $\pm nf$ where $n = 0, 1, 2, 3, \ldots, 64$ and $f = 1/0.128 = 7.8125$ Hz. Therefore, around 50 and 60 Hz, the closest the transform can come to represent this signal is with 46.875, 54.6875, or 62.5 Hz.

We could have avoided this leakage by selecting a slightly different time interval: at 60 Hz, an interval of 1.0416667 ms would have put the 60 Hz signal in a single frequency slot, and likewise an interval of, e.g., 1.25 ms would have put a 50 Hz signal in its unique place. A possible benefit of doing so would be that we can then simply filter out any unwanted mains signal by transforming, setting that *single* frequency to zero, and inverse transforming. Here you have the ultimate in frequency-selective filtering, where you just pick off the one frequency you want to remove. In practice there will

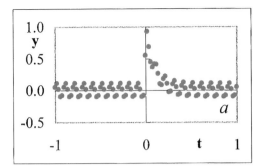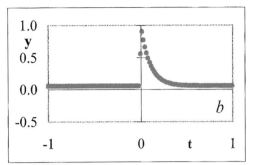

Fig. 7.4-9: The function $y = \exp(-10t)$ for $t > 0$ with superimposed 60 Hz pick-up of 0.1 $\sin(2\pi \times 60t) + 0.05 \cos(2\pi \times 60t)$, sampled at 20 ms intervals (left) and at $16\frac{2}{3}$ ms intervals (right).

usually be several related signals, e.g., at 120 and 180 Hz, but these harmonics can be removed simultaneously because they are exact multiples of the fundamental frequency to be removed. This is *not* to suggest that it is OK to be careless with so-called mains-frequency pick-up: prevention (by proper design and signal shielding) is always preferable to restoration.

Which brings us to a related point. Modern computer-based data acquisition methods sometimes show a low-frequency oscillation superimposed on the signal. Say that we have a transient signal decaying with a time constant of 0.1 s, and sampled with a 20 ms. Any 60 Hz pick-up will then show as a 10 Hz oscillation, i.e., as the beat frequency between the sampling rate of 50 Hz and the 60 Hz pick-up. Fig. 7.4-9a illustrates this.

Here, then, we have an example of aliasing that has nothing to do with Fourier transformation! The remedy is simple: first reduce the source of the pick-up as much as possible, by careful signal shielding. If that does not suffice, use a different sampling rate, so that the beat frequency disappears. In this case, sample at 60 rather than at 50 Hz, i.e., at $16\frac{2}{3}$ ms intervals. Figure 7.4-9b shows the result. In this case, the 60 Hz signal has been converted into a constant offset, because it is sampled always at exactly the same moment in its 60 Hz cycle. What happens when you cannot hit it quite at this sampling rate, but have to settle for, say, once every 16 ms? Try it out on the spreadsheet.

7.5 Convolution

When we record the spectrum of a compound, the data obtained reflect both the spectrum itself, *and* the properties of the spectrometer, such as its slit width. When the slits are wide open, we usually get plenty of light on the photodetector but the spectrum may become blurred, i.e., lose resolution.

On the other hand, when the slits are too narrow, we may have too little light on the photodetector to make a measurement. Clearly, we must often make an instrumental compromise between resolution and sensitivity. No matter what compromise is used, what we observe is, in mathematical terms, the **convolution** of the actual spectrum of the cell and the distortion of the spectrometer.

Similarly, when we listen to music through a speaker, what we hear is the original music, distorted by its passage through whatever recording device (microphone + amplifier), storing device (such as tape, casette, record, compact disk), and audio reproduction system (amplifier, speaker) is used. More precisely, what we hear is the convolution of the original music and the instrumental response.

Likewise, when we use a laser flash to excite fluorescence, we can observe a decay curve of that fluorescence. When the laser flash is not of negligible duration compared to the decay time, we actually observe the convolution of the flash intensity and the intrinsic fluorescent decay of the sample.

The list of examples can go on, since convolution is a quite general phenomenon every time an instrument is used to make a measurement. However, the list is by no means restricted to instrumental distortion. For example, in atomic absorption, the absorption process is typically observed at the relatively high temperatures of a flame or a plasma. Consequently, Doppler broadening (which leads to a Gaussian distribution) convolves with the absorption process (exhibiting a Lorentzian distribution) because the absorbing particles move with respect to the light source. Similarly, in the formation of precipitates, nucleation convolves with crystal growth, since no crystallite can grow before it has been nucleated. Yet another example is the effect of time-dependent diffusion in cyclic voltammetry, as described in section 6.12. In general, when more than one physical process must be taken into account, chances are that a convolution is involved.

What, precisely, is convolution? It is an integral that contains both the original signal (the true spectrum, the original music, the intrinsic fluorescent decay, the absorption spectrum of non-moving species, the growth process apart from nucleation, the stationary current–voltage curve without diffusion, etc.) as well as the 'distorting' effect. In the integral, the two are as it were forced to slide past each other. The mathematical definition of the convolution of two functions, $x(t)$ and $y(t)$, is

$$x(t) * y(t) = \int_{-\infty}^{\infty} x(\tau)\, y(t - \tau)\, \mathrm{d}\tau \tag{7.5-1}$$

where t is a variable (here we use the symbol t because it often represents time), and τ is a so-called dummy variable, which only has meaning inside the integral. A detailed discussion of the mathematical properties of

ABC	DEF	GHI	JKL			MNO	PQR
$x(t)$	$X(f)$	$y(t)$	$Y(f)$			$X \cdot Y$	$x * y$
	$\searrow$ FT $\nearrow$		$\searrow$ FT $\nearrow$	multiply X and $Y \rightarrow X \cdot Y$		$\searrow$ FT^{-1} $\nearrow$	

Fig. 7.5-1: The layout of the data in convolution exercise 7.5-1. The top row identifies the columns used, the middle row the functions encountered there, and the bottom row the mathematical operations between them. The data for $X \cdot Y$ are computed as the products of the complex numbers in $X(f)$ and $Y(f)$.

convolution lies outside the realm of this book, but can be found in many textbooks on mathematics or on mathematical physics. Here we merely mention one property, namely commutation: $x(t) * y(t) = y(t) * x(t)$.

The connection of convolution with Fourier transformation (and the reason to include it in this chapter) lies in the **convolution theorem**, which shows that a convolution in the time domain is equivalent to a simple multiplication in the frequency domain, and vice versa:

$$x(t) * y(t) \Leftrightarrow X(f) \cdot Y(f) \tag{7.5-2}$$

Consequently we can use Fourier transformation to perform a convolution. This is illustrated in spreadsheet exercise 7.5-1 by using a simulated spectrum of three almost baseline-separated peaks, and showing the broadening (i.e., loss of resolution) resulting from, e.g., using a spectrometer with wide-open slits. Alternatively one might think of that example in terms of, say, nearly baseline-separated chromatographic peaks, distorted by a detector with too large a dead volume. The mathematical symbols t and f merely denote complementary parameters, such that their product is dimensionless. We will here describe convolution generically in terms of time and frequency, but t could also stand for, say, wavelength, in which case f would denote wave number.

Before we embark on exercise 7.5-1 we will first sketch its spreadsheet layout, see Fig. 7.5-1. We will use two input functions, $x(t)$ and $y(t)$, to be placed in columns ABC and GHI respectively, say t in column A, the real part $\text{Re}(x)$ of x in B, and its imaginary part $\text{Im}(x)$ in C. Columns DEF and JKL are reserved for their Fourier transforms, $X(f)$ and $Y(f)$ respectively. In columns MNO we then multiply X and Y to form their product, $X \cdot Y$, and in columns PQR we finally calculate the convolution $x * y$ by inverse Fourier transformation of $X \cdot Y$. Note that the individual components of $X(f)$ and $Y(f)$ are complex numbers, which must be taken into account in computing their product.

Below we will convolve a synthetic signal (in A12:B75) with a window function (in G12:H75) and observe what happens. First we will use a Π-shaped window function, one that mimics the effect of, say, a narrow slit on a spectrum as long as edge diffraction can be neglected.

Instructions for exercise 7.5-1

1 Open a spreadsheet.

2 In cells A10, G10 and P10 deposit the label time; in cells D10, J10 and M10 the label freq., in cells B10, E10, H10, K10, N10, and Q10 the label real, and in cells C10, F10, I10, L10, O10, and R10 the label imag.

3 In cells A12:A75 and G12:G75 place the times − 4 (0.125) 3.875.

4 In cell B12 deposit the instruction = 0.4*EXP(− ((2.3 + A12)^2)/0.2) + 0.5* EXP(− ((0.7 + A12)^2)/0.2) + 0.3*EXP(− ((1.1 − A12)^2)/0.2) or a similar expression containing a number of almost baseline-separated peaks. (Here we use Gaussians, but you can take other functions.) Copy this instruction down to row 75.

5 In C12:C75 deposit zeros.

6 In A1:C8 place a thumbnail sketch of the function $x(t)$.

7 Fourier transform these data to generate $X(f)$ in columns DEF.

8 In D1:F8 place a small graph of $X(f)$.

9 Fill H12:I75 with zeros, then deposit the value 1 in cell H44.

10 Plot $y(t)$ in G1:I8.

11 Fourier transform G12:I75 to obtain $Y(f)$ in J12:L75.

12 Plot $Y(f)$ in J1:L8.

13 Copy the frequency scale (i.e., the contents of D12:D75 or J12:J75) to M12:M75.

14 Now comes the multiplication of the complex numbers. Consider $X(f)$ as a set of complex numbers $e + jf$ where $j = \sqrt{-1}$, and likewise $Y(f)$ as a set of complex numbers $k + j\ell$. We then have $(e + jf)(k + j\ell) = (ek - f\ell) + j(e\ell + fk)$. Therefore, in cell N12 deposit the instruction = E12*K12 − F12*L12 (for the real part of that product), and in cell O12 the instruction = E12*L12 + F12*K12 (for its imaginary part). This will generate the product $X \cdot Y$. Copy these two instructions down to row 75.

15 Make a graph of $X \cdot Y$ in M1:O8.

16 Finally, use the inverse Fourier transform of M12:O75 to compute the convolution $x * y$ in P12:R75, and show it graphically in P1:Q8. The top of your spreadsheet should now look similar to Fig. 7.5-2.

The operation does not modify the original signal beyond reducing its amplitude, as an ideal monochromator slit also would, by restricting the amount of light that reaches the photodetector. Below we will take care of that signal attenuation.

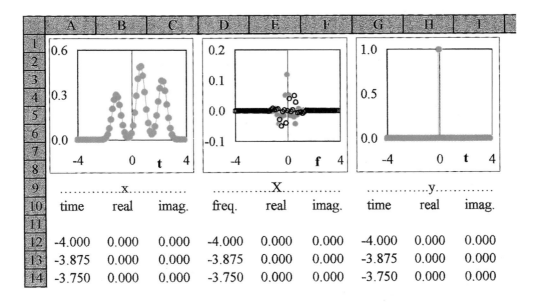

	......x......			......X......			......y......	
time	real	imag.	freq.	real	imag.	time	real	imag.
-4.000	0.000	0.000	-4.000	0.000	0.000	-4.000	0.000	0.000
-3.875	0.000	0.000	-3.875	0.000	0.000	-3.875	0.000	0.000
-3.750	0.000	0.000	-3.750	0.000	0.000	-3.750	0.000	0.000

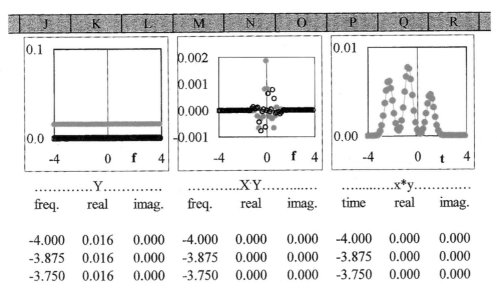

	......Y......			......X·Y......			......x*y......	
freq.	real	imag.	freq.	real	imag.	time	real	imag.
-4.000	0.016	0.000	-4.000	0.000	0.000	-4.000	0.000	0.000
-3.875	0.016	0.000	-3.875	0.000	0.000	-3.875	0.000	0.000
-3.750	0.016	0.000	-3.750	0.000	0.000	-3.750	0.000	0.000

Fig. 7.5-2: The top of the spreadsheet, shown here in two parts (since it is too wide to display in one piece), as it might look at this point.

17 Normalizing the window function is most readily done by using as normalizing factor the average value of the window function $y(t)$ used. Therefore, modify the instruction in N12 by dividing it by AVERAGE(\$H\$12:\$H\$75) or, equivalently, by \$K\$44, and similarly correct the instruction in O12. Copy both modifications down to row 75.

18 Repeat the inverse FFT of M12:O75. Now you should recover the original function of A12:C75 without attenuation, because we have effectively convolved it with the digital equivalent of a Dirac delta function, which is zero everywhere except at one place (here: $t = 0$), where it is so large that it has unit area.

We are now ready to explore the consequences of various window functions.

19 Replace the zeros by ones in H42, H43, H45, and H46.

20 Apply the Fourier transform to G12:I75, and the inverse Fourier transform to M12:O75.

21 You have now simulated opening the monochromator slits (without changing the amount of light falling on the photodetector, because of the normalization you built in at step 17), and you will see the resulting distortion of the output signal, as illustrated in Fig. 7.5-3.

22 Figure 7.5-3 also shows the effects of widening the window even further. Verify those results.

Other window functions, such as triangular windows of variable widths, can also be used. In fact, there is a whole bevy of window functions available, often named after their originators or proponents, such as Bartlett, Hamming, Hanning, Parzen (for the triangular window), and Welch. Some of these are discussed in Section 12.7 of the book *Numerical Recipes* (W. H. Press *et al.*, Cambridge University Press 1986).

23 Refill H12:H75 with zeros.

24 Replace the zeros in H41 through H47 by 0.25, 0.5, 0.75, 1, 0.75, 0.5, and 0.25 respectively. This constitutes a triangular window.

25 Fourier transform G12:I75, then inverse Fourier transform M12:O75.

We note that the convolution causes a broadening of the original peaks, reducing their heights and increasing their widths, thereby increasing their overlap. This smearing effect can be reduced by using a narrower (and therefore less distorting) window function $y(t)$, or worsened by making $y(t)$ broader. Incidentally, in this particular example, by using a

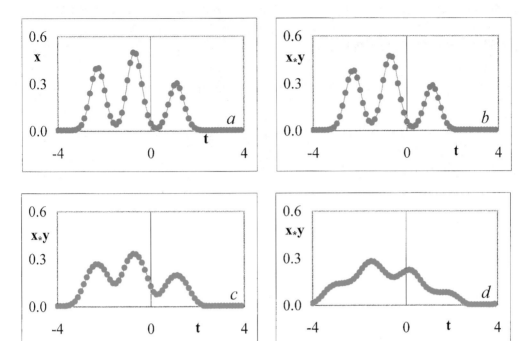

Fig. 7.5-3: Distortion resulting from a rectangular window of increasing width. (a): Original signal; (b): after convolution with a five-point rectangular window function; (c): after convolution with a nine-point rectangular window function; (d): after convolution with a 13-point rectangular window function.

rectangular window of 15 or more points, we may even generate phantom peaks!

While there will be other convolution exercises, we will postpone them until we have simplified the computation. To that end a simple macro is provided, called Convolution. For the sake of simplicity it assumes that the input functions have no imaginary components, and that the same applies to the resulting output function. Moreover, it does not display any of the intermediate transforms. These constraints greatly simplify the spreadsheet, since now only *four* columns are needed: one for the common time scale, two for the two real input functions, and one for the result. The macro handles all mathematical manipulations out-of-sight, but the method follows the same logic as exercise 7.5-1, and its VBA code is given in section 10.6. You will need this macro for the next few exercises; if it has not yet been installed on your computer, this is the time to copy it from the disk(ette) into your spreadsheet module.

For the convenience of the user, the Convolution macro has been written such as to accept a time scale that can start at any arbitrary value. Of course, the time increments must still be equidistant, and there must be 2^N input data, where N is a positive integer subject to the constraint $2 \leqslant N \leqslant 10$.

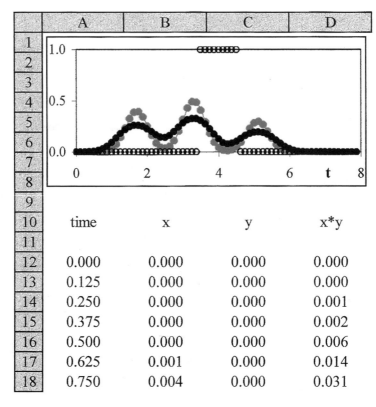

The table portion of the spreadsheet:

	A	B	C	D
10	time	x	y	x*y
11				
12	0.000	0.000	0.000	0.000
13	0.125	0.000	0.000	0.000
14	0.250	0.000	0.000	0.001
15	0.375	0.000	0.000	0.002
16	0.500	0.000	0.000	0.006
17	0.625	0.001	0.000	0.014
18	0.750	0.004	0.000	0.031

Fig. 7.5-4: The top of the spreadsheet showing a convolution. The original function x is shown in color, the window function y as open circles, and the convolution $x * y$ as solid black circles.

Instructions for exercise 7.5-2

1 Open a new spreadsheet.

2 In cells A10 through D10 deposit the labels time, x, y, and x*y respectively.

3 In cells A12:A75 place the times 0 (0.125) 7.875.

4 In cell B12 deposit the instruction $= 0.4*EXP(- ((1.7 - A12)\wedge 2)/0.2) + 0.5*$ EXP$(- ((3.3 - A12)\wedge 2)/0.2) + 0.3*EXP(- ((5.1 - A12)\wedge 2)/0.2)$ or a similar expression. Copy this instruction down to row 75.

5 In C12:C75 deposit the nine-point rectangular window function used in exercise 7.5-1. Make sure that it is centered around C44, i.e., that its ones are placed in C40:C48.

6 Highlight the area A12:C75, and call the macro Convolution.

7 The result $x * y$ will now appear in D12:D75.

8 In A1:D8 make a graph of x, y, and $x * y$ versus t. Figure 7.5-4 shows what the top of your spreadsheet might now look like.

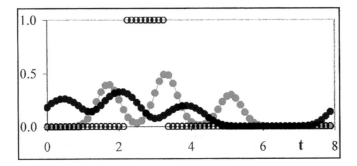

Fig. 7.5-5: Shifting the center of the window function (here a nine-point rectangular function) off-center causes a time shift of the resulting convolution (solid black circles).

Now, with the macro taking care of the mechanics of convolution, we can more conveniently explore its properties. First, move the window function y around on the time axis.

9 Move the set of nine contiguous 1's in C12:C75 a few spaces forward or back, and record the resulting convolution. Remember, the convolution macro does not update automatically, but must be invoked every time you want it updated.

10 Also, move it more boldly, i.e., more than a just a few spaces. Figure 7.5-5 shows such a result.

Clearly, shifting y with respect to x yields a *phase shift* of the resulting convolution $x * y$. It is as if the open circles drag the closed circles with them. In order to avoid such phase shifts we will below only use centered window functions.

11 Explore the effect of other window functions, such as the triangular function (see exercise 7.5-1 under instruction 24), a trapezoidal function, a Gaussian, a Lorentzian, or whatever.

12 Note that too wide a window function may make the convolution spill over an edge. You can avoid this by adding zeros to the beginning and end of columns B and C, with an accompanying extension of the time scale in column A. The total number of points must remain an integer power of 2, so that it is best in the present example to add 32 zeros at both the beginning and end.

7.6 Deconvolution

Why belabor convolution? Obviously, instruments or methods sometimes distort what we want to measure, and the answer would seem to be to build instruments or utilize methods that introduce less distortion. True enough, but we seldom can wait for the ideal instrument or procedure, and typically need to work with what is available to us now. Here is where convolution comes in or, rather, deconvolution.

We will first summarize what we have found so far. Let us assume that we have some **q**uantity q (such as a spectrum, a chromatogram, or a decay curve) that we want to measure, with a minimum of distortion. The observation convolutes this with a **r**esponse function r, which may be caused by, e.g., instrumental distortion. Consequently what we measure is the **s**ignal s, which is related to q by

$$s = q * r \tag{7.6-1}$$

In general, q, r, and s will be functions of wavelength (for a spectrum), of time (for an audio response, or a fluorescent decay), etc. We measure the signal s, but we are really interested in the underlying, undistorted quantity q. We have seen that Fourier transformation converts (7.6-1) into a simple multiplication,

$$S = Q \times R \tag{7.6-2}$$

where, as before, capital symbols denote the Fourier-transformed quantities, and the sign $\times$ stands for multiplication.

Let us assume for the moment that we can measure the instrument response function r by itself. We certainly can measure the signal s. We then take their Fourier transforms, which yields R and S. Equation (7.6-2) now allows us to calculate Q simply as $Q = S / R$. From there it is only an inverse Fourier transformation to calculate q, the quantity of interest, corrected for distortion! This process is called **deconvolution**. The same macro that can perform a convolution can also do the deconvolution. The relevance of deconvolution to spectrometry is illustrated in W. E. Blass and G. W. Halsey, *Deconvolution of Absorption Spectra*, Academic Press 1981, and P. A. Jansson, *Deconvolution with Applications in Spectroscopy*, Academic Press 1984.

You may think that this sounds too good to be true, and it often is: in practice it may not always be possible to recover the undistorted signal. But sometimes we can, and the following exercises will demonstrate under what conditions this may be the case.

Instructions for exercise 7.6-1

1 Reopen the same spreadsheet used for exercise 7.5-2.

2 Place the labels time, x * y, y, and x (or, if you prefer, s, r, and q) in cells E10 through H10.

3 Copy the column starting with A12 to cell E12.

4 Likewise, copy the data for x * y in column D to the new column F. And copy the column for y from C to G. Do not fill column H.

5 Highlight the data in columns E through G, and use the macro to deconvolve them.

6 Ignore for now the input box labeled Adjustable Hanning Window; when it shows, just enter 0.

7 In E1:H8 plot the data in columns F through H versus those in column E. You should obtain a copy of the graph in A1:D8, except that the input and output curves are exchanged!

8 Also compute, in column I, the difference between corresponding data (on the same row) in columns B and H. You may have to multiply the difference by quite a large factor to see non-zero numbers.

So far so good: the method works beautifully. Then why the warning on the previous page? Here is the problem: the presence of noise often makes deconvolution impractical or even impossible. Again, we will use the spreadsheet to demonstrate this.

9 Generate two columns of Gaussian noise of zero average and unit standard deviation, and in new columns add na times that noise to both the signal in column F and the response function in column G. (Of course add noise from a *different* set to each column.) The noise amplitude na is used to control the amount of noise added.

10 Now increase the value of na from, say, $1E-12$ to $1E-6$, $1E-3$ and $1E-2$, every time invoking the macro to perform the deconvolution.

As Fig. 7.6-1b shows, deconvolution is extremely sensitive to noise. While convolution smears out and smoothens the data, deconvolution does the opposite. As long as the noise is very much smaller than the signal, deconvolution works well; with relatively more noise, deconvolution may not be feasible.

Since our deconvolution method is based on Fourier transformation, a simple method for noise reduction suggests itself: take the Fourier transforms $X \cdot Y$ and Y of the noisy functions $x * y$ and y, and filter both (or just Y in case y is noise-free) by replacing their high-frequency parts by zeros before performing their division. Here we use a somewhat more gradual high-frequency

(a)

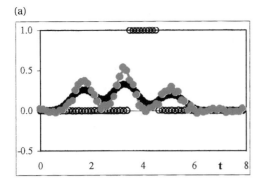

(b)

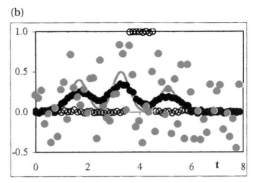

Fig. 7.6-1: Deconvolution of x ∗ y with y when both have added Gaussian noise of amplitude na = 0.001 (left) and 0.01 (right) respectively. Colored points: the deconvoluted data; colored line: the original curve.

cut-off function $[0.5 + 0.5 \; \cos(\pi f / f_{max})]^{w}$, based on the Hanning window $0.5 + 0.5 \; \cos(\pi f / f_{max})$, made adjustable with the parameter w so that we can select the minimum noise reduction needed.

Figure 7.6-3 illustrates some results of such filtering on the data of Fig. 7.6-1b. As can be seen, the presence of noise now generates oscillations, and filtering these out by increasing w (i.e., by narrowing the frequency range) introduces considerable distortion. Deconvolution clearly works best for signals with a very high signal-to-noise ratio.

In order to see what is happening here we go back to the power spectrum. Figure 7.6-2 shows the power spectra of the original function (the colored curves in Figs. 7.5-4 and 7.5-5) on two different scales, and the power spectrum of the same with added noise as was used in Fig. 7.6-1b. Clearly, when the noise gets larger, it will obscure all but a few low frequencies, and eventually these will be overwhelmed as well. At that point, the original signal obviously cannot be recovered any more by deconvolution.

Keep in mind that, because of the Nyquist criterion, the number of available frequencies is only half the number of data points. Or, to put it differently, all frequencies f in the Fourier transform have their negative counterparts at $-f$. The largest number you can therefore select for filtering is *half* the number of data points in the set, and at that point you would filter out everything!

In order to decide at what frequency to start filtering, we go back to the power spectrum. Figure 7.6-2 shows the power spectra of the original function (the colored curves in Figs. 7.5-4 and 7.5-5) on two different scales, and the power spectrum of the same with added noise as was used in Fig. 7.6-1b. Clearly, when the noise gets larger, it will obscure all but a few low frequencies, and eventually these will be overwhelmed as well. At that point, the original signal obviously cannot be recovered any more by deconvolution.

Unfortunately, the boundary between signal and noise is not sharp: there is signal and noise at *all* frequencies. The color in Fig. 7.6-2 suggests what

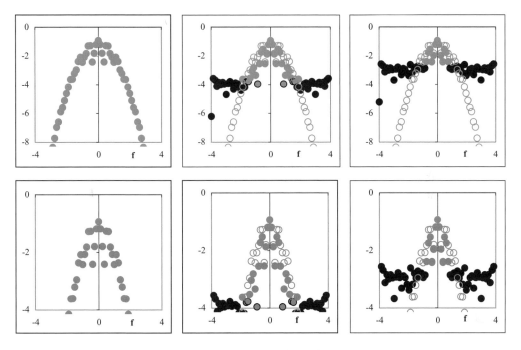

Fig. 7.6-2: The power spectrum ½ log (Re² + Im²) vs. f of the original spectrum (left) and that of the same with added Gaussian noise, *na* = 0.001 (middle panel) or *na* = 0.01 (right). The bottom panels show the same with an enlarged vertical scale. Large solid circles: the power spectrum; small open circles: the power spectrum of the noise-free signal. Color is used to indicate those data points that are mostly 'signal', while black signifies mostly 'noise'. As indicated by a few points in the middle panels, that distinction is somewhat ambiguous.

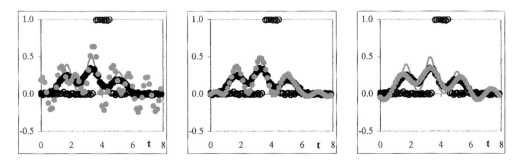

Fig. 7.6-3: Deconvolution of the noisy spectrum of Fig. 7.6-1b (with *na* = 0.01) while filtering out the contributions at the intermediate, Fourier-transformed signal for the top 25 (left panel), 26 (middle), or 28 (right) frequencies. The thin colored line shows the original, noise-free function.

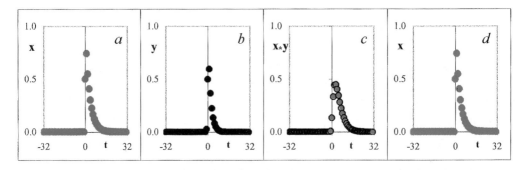

Fig. 7.6-4. From left to right, the various stages in exercise 7.6-2: the assumed fluorescence decay signal x, the assumed laser intensity curve y, the resulting measured signal $x * y$, and the recovered 'true' fluorescence decay signal after deconvolution of $x * y$ with y.

points are *predominantly* signal or noise, and even that distinction is not unambiguous. The deconvolution macro is set up to select those contiguous frequencies where the noise is predominant, and zero out the corresponding contributions before the final inverse Fourier transformation. Other arrangements, in which, e.g., the ambiguous data are given an intermediate weight, are of course possible. Here we merely illustrate the principle by using an easily implemented scheme. The bottom right panel in Fig. 7.6-2 suggests that, in that case, only six (of the total of 32) frequencies are worth keeping.

Here is another example. In laser-excited fluorescence, the fluorescence typically decays exponentially, and can then be characterized by a time constant τ. When that time constant is much larger than the length of the laser pulse, the resulting fluorescent signal is a simple exponential. However, when τ is not much larger than the time the laser light excites the sample, the resulting signal (assuming the fluorescence is a linear function of excitation light intensity) will be the convolution of the two. Figure 7.6-4 illustrates this with a simulated sequence in which the fluorescence signal x is convoluted with the laser intensity y to generate the measured output $x * y$. When the laser intensity is known, the measured signal can be deconvoluted to recreate the underlying fluorescent decay curve. As before, the method can readily be overwhelmed by the presence of noise. In experiments where subsequent use of deconvolution is anticipated, one should therefore strive for minimal noise levels in both the signal and the window fraction.

Instructions for exercise 7.6-2

1 Open a new spreadsheet.

2 In column A enter the label time, and values for a time scale of 2^N numbers, such as -32 (1) 31.

3 In column B, below the label x, enter a hypothetical fluorescence decay curve, such as $x = e^{-0.3t}$ for $t > 0$ (and $x = 0$ for $t < 0$, $x = 0.5$ for $t = 0$).

4 In column C, under the label y, enter a hypothetical laser intensity curve. In the example shown in Fig. 7.6-4 we have used the most asymmetric curve shown in Fig. 8.6-3, given by $y = 1/(e^{-3.5x} + e^{0.5x})$.

5 In column D we then use the macro to compute the convolution $x * y$.

6 In column E copy the corresponding contents of column A.

7 Likewise, in column F copy $x * y$, and in column G the function y.

8 Deconvolution of the contents of columns E:G will now generate x.

The take-home message of this section is that, as long as the signals involved are relatively noise-free, deconvolution is possible. In that case we can correct our observations for artifacts that are reproducibly measurable or theoretically predictable.

7.7 Summary

Fourier transformation has become a ubiquitous method in chemical methodology and instrumentation. Molecular structures are solved by Fourier transformation of their X-ray diffractograms, and when you see a scanning tunneling microgram in which the atoms or molecules look like smooth balls, you are almost surely looking at a picture that has been filtered by two-dimensional Fourier transformation. Virtually all modern NMR instruments are based on Fourier transformation, and the same applies to most infrared spectrometers. Outside chemistry, Fourier transformation plays a role in many other areas, e.g., in the solution of partial differential equations, the design of antennas, and the processing of satellite pictures. We have devoted this entire chapter to Fourier transformation because of its general importance to modern instrumental methods of chemical analysis.

The concept of Fourier transformation is the representation of a time-domain function f(t) in the frequency domain as F(f), and vice versa. Such transformations are firmly based on human experience: for instance, we hear sound as a sequential phenomenon (i.e., a function of time), yet the brain also analyzes it in terms of pitch, i.e., as a function of frequency. In our description of Fourier transformation we have kept the mathematics to a minimum, but instead have used graphics to demonstrate some of its main principles. Consider this, therefore, as a visual introduction to the topic, as a means to whet your appetite for it, to demonstrate its power, and to alert you to its limitations. With a fast and convenient Fourier transform macro, the method is now so easy to implement that we can use the spreadsheet to learn about Fourier transformation and its properties in an intuitive, non-mathematical way. This can then be followed by a more formal description if and when we are ready for that.

In using Fourier transformation one should be aware that there are several conflicting conventions. For example, many engineering texts (as well as the *Numerical Recipes*) use conventions for forward and inverse transforms that are the opposite from those used in mathematics and physics; the latter convention is used here. Likewise, there are different conventions for how to distribute the scale factors between the forward and inverse transform, and for whether the time and frequency axes should be centered around zero (as used here) or start at zero. (The Fourier transform macro will accept either input format).

When the use of Fourier transformation in data analysis is anticipated, it is best to design the experiment such that 2^N data points will be taken, in order to take advantage of the speed of 2^N-based Fourier transform algorithms. Also make sure that the data can be represented as a segment of an infinitely repeating chain, otherwise you may introduce artifacts in the transform. In practice, the latter requires that the underlying function and its first few derivatives at the beginning and end of the sample match each other to within experimental error. For transient phenomena this can most readily be achieved by taking a sufficiently long sample so that, at the end of the sampling interval, the signal has fully returned to its baseline value.

CHAPTER 8

STANDARD MATHEMATICAL OPERATIONS

In this chapter we will encounter a number of standard mathematical operations that are conveniently performed and/or illustrated on a spreadsheet. We start with a brief description of the logic underlying the Goal Seek and Solver methods of Excel. Then we consider two methods often encountered in spectroscopy, viz. signal averaging and lock-in amplification. Subsequently the focus shifts toward numerical methods, such as peak fitting, integration, differentiation, and interpolation, some of which we have already encountered in one form or another in the context of least squares analysis and/or Fourier transformation. Finally we describe some matrix operations that are easy to perform with Excel.

8.1 The Newton–Raphson method

The Newton–Raphson method is often used to solve problems involving a single variable, and is implemented in Excel as Tools ⇨ Goal Seek. The method requires that a function $F(x)$ can be formulated as an explicit mathematical expression in terms of a variable x. We now want to know for what value of x the function $F(x)$ has a particular value, A. The Newton–Raphson approach then searches for a value of x for which $F(x)$ is equal to A. Often one selects $A = 0$, in which case the corresponding value of x is called a **root** of the function $F(x)$.

The Newton–Raphson algorithm must start with a reasonably close first estimate, x_0, of the desired value x_A for which $F(x_A) = A$. If the function $F(x)$ were linear between $x = x_0$ and x_A, we could find x_A simply from

$$\frac{\mathrm{d}F(x)}{\mathrm{d}x} = \frac{F(x_0)}{x_0 - x_A} \qquad \text{or} \qquad x_A = x_0 - \frac{F(x_0)}{\mathrm{d}F(x)/\mathrm{d}x} \tag{8.1-1}$$

In general, of course, $F(x)$ will not be linear in the interval from x_0 to x_A, but as long as the non-linearity is not too severe, we can use (8.1-1) as a first step

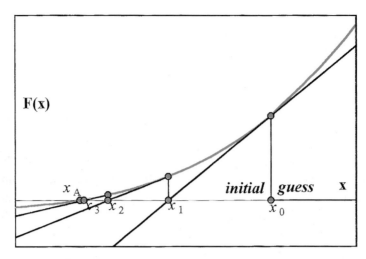

Fig. 8.1-1: The Newton–Raphson algorithm finds the root of an equation through iteration, of which the first three steps are shown here.

in an iterative procedure, obtaining an improved estimate, x_1, that we can
then use as the starting point for further refinement, etc. As we move closer
to the value x_A the linearity of the function will usually improve (since, over a
sufficiently small interval, most physically well-behaved functions
approach linearity) so that we will quickly home in on the correct answer.
The first three steps in such a sequence are depicted in Fig. 8.1-1.

Equation (8.1-1) contains the derivative $dF(x)/dx$, but the spreadsheet
does not need to determine that derivative in a formal, mathematical way:
instead, it uses $\Delta F(x)/\Delta x$, just as we did in exercise 2.3.

When the initial estimate is far off, the Newton–Raphson method may not
converge; in fact, when the initial value is located at an x-value where $F(x)$
goes through a minimum or maximum, the denominator in (8.1-1) will
become zero, so that (8.1-1) will place the next iteration at either $+\infty$ or $-\infty$.
Furthermore, the Newton–Raphson algorithm will find only one root at a
time, regardless of how many roots there are. On the other hand, when the
method works, it is usually very efficient and fast. Exercise 8.1 illustrates how
the Newton–Raphson algorithm works.

Instructions for exercise 8.1

1 Open a new spreadsheet.

2 In column A deposit x-values in the range 0 (0.2) 8.

3 In column B compute sin(x).

4 Plot the sine wave; it will serve you as reference.

5 In cell C1 deposit an x-value, say, 2.

6 In cell D1 deposit = sin(C1).

7 Select <u>T</u>ools ⇨ <u>G</u>oal Seek… and in the Goal Seek dialog box specify to <u>S</u>et cell: D1 To <u>v</u>alue: 0 By <u>c</u>hanging cell: C1 ⇨ OK.

8 Note that the cell to be changed should contain a numerical value, not an instruction, because the latter will prevent Goal Seek from adjusting the value.

9 Cell C1 should now show a value close to π (≈ 3.141593), and D1 a value close to 0.

10 The agreement may not be very good. This will happen when the algorithm does not use a sufficiently high precision. In that case, click on <u>T</u>ools ⇨ Options…, select the Calculation tab, then in <u>I</u>teration see what is listed for Maximum <u>C</u>hange. If it is something like 0.001, add a few zeros behind the decimal point, click on OK, and try Goal Seek again. You should now get a much closer approximation of the true zero crossing.

11 Instead of 2, place other values in C1 and try Goal Seek again. Make a crude map of the results obtained for a handful of initial values. What do they show?

8.2 Non-linear least squares

The operation of the Solver is much harder to illustrate, because it is a multi-parameter adjustment. Moreover, it is a much more sophisticated routine, capable of using several different optimizing algorithms. It can even include constraints on the variables. In the previous chapters we already used Solver extensively, and we will here only add a few comments about it.

Solver finds a minimum in the sum of the squares of the residuals very much like precipitation on mountains finds its way to the ocean: it does not know where to go, but just follows the local slope down. And just as some of it may end up in a lake without outlet to the ocean, Solver may end up in a local (rather than the global) minimum. Which way the solution will go depends, for both running water and non-linear least squares, on the point of departure. When the initial conditions in Solver are close to the final ones, chances of getting stuck in a local minimum are greatly reduced.

The sum of squares of the residuals makes a multi-dimensional surface, with mountain tops and valleys. The most popular algorithm to slide down that surface and find its lowest point is associated with the names of Levenberg and Marquardt, and is described in detail in, e.g., chapter 14 of the *Numerical Recipes*.

In Solver you can assign up to two constraints per cell, and up to 100 additional constraints. The constraints are $<=$ (for $\leq$), $>=$ (for $\geq$), $=$ (for equality), int (for integer), and bin (for binary). Constraints are useful to avoid physically unrealistic solutions, such as those with negative concentrations or equilibrium constants.

As a practical matter, with more than a few adjustable parameters it is often advisable to use Solver *incrementally*, starting with just a few variables, and successively adding more of them. In either case you need reasonable guess values for all parameters. We illustrated this guided, gradual approach in section 4.11.

Unlike water that flows smoothly from high to low, a computer must take discrete steps. The step size in Solver is determined by the largest adjustable variables. When the smaller variables are of a quite different magnitude, such as the K_a's in many polyprotic acids, use their pK_a's instead as the adjustable parameters, in order to make a more even playing field. The same can be achieved when you Use Automatic Scaling in the Solver Options.

Whether Solver performs a weighted or unweighted least-squares optimization is under full control of the user. In case you want Solver to do a weighted least squares, simply multiply each of the squares of the residuals by their individual weights before adding them as SRR, in which case you minimize the sum of the squares of the weighted residuals. Weighting can be applied for various reasons: (1) you may know the variances of the various points, (2) you may want to correct for some earlier transformation, or (3) you may want to downplay some parts of the data set, and emphasize others. While the latter is rather arbitrary, it may sometimes be necessary to reduce the effect of some extreme points which otherwise might overwhelm all other data.

You can interrupt Solver by pressing the *Esc*ape button on your keyboard. This will abort the calculation being done at that time, and show the previous result calculated by Solver. Help ⇨ Answer Wizard contains a large number of useful comments and hints concerning Solver, as does the Answer Wizard Index. Consult these if you want to know more about it.

Solver provides values for the parameters of a non-linear least-squares fit, but no estimates of their precision. The latter can be obtained with the macro SolverAid described in chapter 10.

8.3 Signal averaging

In pushing an experimental method to its maximum sensitivity, one often runs into random noise as the limiting factor. When such noise is indeed random, and is not correlated with the signal, one can sometimes use **signal averaging** (also called **co-addition**) to reduce the effect of the noise. Below we will illustrate the method, using as our example a set of Gaussian peaks with added Gaussian noise.

Instructions for exercise 8.3

1 Open a new spreadsheet.

2 In row 1 deposit labels for nine constants, A through I, plus a noise amplitude na, and (in cell K1) an offset for plotting the results.

3 Use row 2 for the corresponding numerical values.

4 In A6:A106 deposit the numbers 1 (1) 100.

5 In B110:Q210 deposit Gaussian noise ("normal" Distribution) of zero Mean and unit Standard deviation.

6 In B6 deposit an expression for the sum of three Gaussians, of the form $A \exp[-B(x-C)^2] + D \exp[-E(x-F)^2] + G \exp[-H(x-I)^2]$, and to this add some noise amplitude *na* times the Gaussian noise stored in cell B110.

7 Copy the instruction from B6 to the entire block B6:Q106.

8 In S6 deposit the instruction = (B6 + C6 + D6 + E6)/4 + \$K\$2, where \$K\$2 contains the offset constant. Copy this instruction to all cells in S6:S106.

9 Similarly, in columns T, U, and V compute the averages of columns F through I, J through M, and N through Q respectively, plus the same offset.

10 Likewise, in column X, calculate the average of columns S through V, again together with the offset \$K\$2.

11 In column Z calculate the noise-free function $A \exp[-B(x-C)^2] + D \exp[-E(x-F)^2] + G \exp[-H(x-I)^2]$, to which you add an offset, such as + 3*\$K\$2.

12 Plot B6:B106, S6:S106, X6:X106, and Z6:Z106 versus A6:A106 to display, in sequence, a noisy curve, the average of four or sixteen of such curves, and the original, noise-free function. Figure 8.3-1 shows such a graph.

At best, signal averaging yields an improvement of the signal-to-noise ratio proportional to $\sqrt{N}$, where N is the number of signals averaged. Consequently, the method is rather inefficient and time-consuming, except when the entire curve can be obtained in a very short time, as with fluorescence transients following a short laser pulse.

Successful signal averaging also requires that the experimental conditions are very reproducible. When some measurement parameters experience appreciable drift from one sample to the next, averaging them may not lead to any improvement in the signal-to-noise ratio.

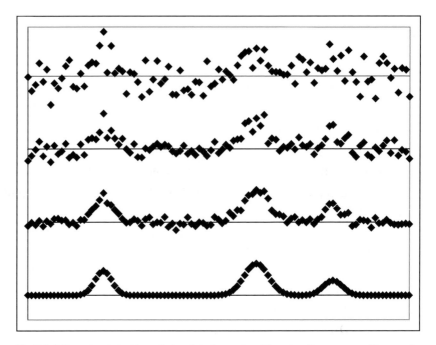

Fig. 8.3-1, from top to bottom: A simulated sample with noise, the average of four and sixteen such samples, and the noise-free signal.

8.4 Lock-in amplification

The terms **lock-in amplification** and **synchronous detection** describe a correlation method commonly used in spectrometry. It requires that the light source be modulated at a given, known frequency. The signal is then analyzed for the component of the same frequency and the same phase, or with a fixed phase shift. In this way, random fluctuations, *even those at the same measurement frequency*, will be attenuated, because random fluctuations will also be random with respect to the phase of the signal source. Lock-in amplification is also used to distinguish, say, atomic absorption from atomic emission, since only the absorption signal will be encoded by modulating the amplitude of the external light source.

Lock-in amplification can be understood mathematically in terms of the multiplication of a sine wave and a square wave. Below we will use a graphical approach, which illustrates rather than derives the result.

Instructions for exercise 8.4

1 Open a new spreadsheet.

2 In row 1 deposit labels for π, phase angle ϕ, offset ε, noise amplitude na, and average.

3 Use A2 through C2 for associated numerical values.

4 In row 4 deposit labels for x, ref. (for the reference sine wave), sqw(x) (where sqw is shorthand for square wave), signal, and rect.signal.

5 Fill A6:A406 with 0 (0.01) 4 times π, using π as stored in $A$2.

6 In B6:B406 compute the sine of the angle listed in column A. This will simulate the reference signal.

7 In C6 calculate the corresponding reference square wave. Conceptually the simplest way to do so would be with = B6/ABS(B6), but this may cause trouble at the zero-crossings of the sine wave. It is therefore preferable to use = IF(ABS(B6) < 1E − 10,0,B6/ABS(B6)) or = IF(B6 > 1E − 10,1,IF(B6 < − 1E − 10, − 1,0)) instead. Copy this down to row 406.

8 In column D deposit Gaussian noise.

9 In column E compute the simulated signal as $\sin(x + \phi) + \varepsilon$ (where ϕ is the phase angle in $B$2 and ε the offset in $C$2) together with some added noise (of amplitude controlled by na in $D$2).

10 In row F calculate the synchronously rectified signal as the product of the terms in columns C and E.

11 In cell D2 compute the average of the synchronously rectified signal as = SUM(F6:F406)/400.

12 Make a graph showing these various signals, such as Fig. 8.4-1.

13 Vary the phase angle ϕ, the offset ε, and the noise amplitude na, and record the resulting changes in the average, in cell E2.

You will notice that the average is maintained reasonably well even for a signal-to-noise ratio of 1. When the averaging is done over a far larger number of cycles, synchronous detection can pull otherwise invisible signals out of noise. The average is directly proportional to the cosine of the phase angle ϕ between the signal and the reference.

Typical analytical applications of lock-in amplification occur in atomic absorption spectrometry, where it lets us discriminate between emission and absorption at the very same wavelength, and in classical infrared spectrometry, where the light sources are of low intensity, the detectors have low sensitivity, and all surrounding materials radiate as well. By mechanically **chopping** (i.e., interrupting) the beam from the light source at, e.g., 13 Hz, and by using a reference signal tied to the rotating chopper blades, an

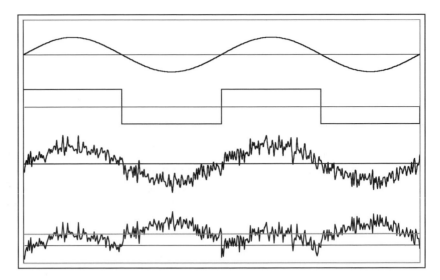

Fig. 8.4-1, from top to bottom: A reference sine wave, the same converted into a square wave, a noisy sine wave signal, and the same as multiplied by the square wave. In this example, the average value of the bottom curve (displayed in color) is within 3% of its value in the absence of noise.

infrared spectrometer ignores the heat emitted by its human operator, unless that operator can nimbly dance back and forth at 13 Hz, while also staying in phase with the chopper blades.

8.5   Data smoothing

'Noisy' experimental data sometimes need to be smoothed. In this context, smoothing is *not* meant to be the drawing of a continuous curve, with continuous derivatives, through all available points, as can be done in Excel simply by double-clicking on a curve and using the command sequence Format Data Series ⇨ Patterns ⇨ Smoothed Line. That method works well for presenting an inherently smooth *theoretical* curve based on relatively few data points, see Figs. 1.3-1 and 1.3-2, but is of little use for noisy *experimental* data, because the curve goes through all points, and thereby tends to emphasize the noise.

Instead, we mean here the use of experimental data that can be expected to lie on a smooth curve but fail to do so as the result of measurement uncertainties. Whenever the data are equidistant (i.e., taken at constant increments of the independent variable) *and* the errors are random and follow a single Gaussian distribution, the least-squares method is appropriate, convenient, and readily implemented on a spreadsheet. In section 3.3 we already encountered this procedure, which is based on least-squares fitting of the data to a polynomial, and uses so-called convoluting integers. This method is, in fact, quite old, and goes back to work by Sheppard (*Proc. 5th*

Congress of Math., Cambridge (1912) II p. 348; *Proc. London Math. Soc.* (2) 13 (1914) 97) and Sherriff (*Proc. Roy. Soc. Edinburgh* 40 (1920) 112), which soon thereafter found its way into the well-known textbook on numerical analysis by E. Whittaker and G. Robinson, *The Calculus of Observations, a Treatise on Numerical Mathematics*, Blackie & Sons, 2nd ed. (1924) p. 290. By and large, however, the community of analytical chemists only became aware of this method more than half a century later, through a paper by Savitzky & Golay (*Anal. Chem.* 36 (1964) 1627).

Even though least-squares methods are sometimes considered to be 'objective', there are still two subjective choices to be made in this application of data fitting to a moving polynomial: what length of polynomial to use, and what polynomial order. Typically, the longer the polynomial, and the lower its order, the more smoothing is obtained, but the larger is the risk of introducing systematic distortion. For example, the Savitzky–Golay tables for smoothing allow data sets up to 25 points long (only odd numbers of data points are used, so that the smoothed data can replace the noisy ones) and offer several choices for polynomial order, such as quadratic & cubic or quartic & quintic. Sherriff already presented convoluting integers for longer data sets as well as for higher-order polynomials.

The tables of convoluting integers listed by Savitzky and Golay contained a large number of errors, and were subsequently corrected by Steinier *et al.* (*Anal. Chem.* 44 (1972) 1906). Subsequently, Madden (*Anal. Chem.* 50 (1978) 1383) gave simple formulas for them, so that these numbers are now readily calculated on a computer. Recently, Barak (*Anal. Chem.* 67 (1995) 2758) extended the method by letting the program self-optimize the polynomial order, so that the user only needs to select the length (i.e., the number of data points) of the moving polynomial, and specify the upper limit of the polynomial order. This program, kindly provided by Prof. Barak, is described in section 10.9.

Below we will use a simple example to explain the principle of the standard, non-self-optimizing method. Say that we have five data pairs x,y such that the x-values are equidistant, with a nearest-neighbor distance δ. For any odd-numbered set of equidistant data (such as the five considered here), the x-value in the middle of the set is the average $\bar{x}$ of the x-values in the set. We now start by subtracting $\bar{x}$ from all five x-values, so that the new x-values will be $-2\delta, -\delta, 0, \delta$, and 2δ.

Now we are ready to use the least-squares analysis. In our example we will fit the five data pairs (shifted as just explained in the x-direction by the amount $-\bar{x}$) to a parabola, i.e., to $y = a_0 + a_1 x + a_2 x^2$. The general formulas for doing that were given in section 3.2, and are here repeated as:

$$a_0 = \begin{vmatrix} \sum x^4 & \sum x^3 & \sum x^2 y \\ \sum x^3 & \sum x^2 & \sum xy \\ \sum x^2 & \sum x & \sum y \end{vmatrix} \Big/ D \qquad (8.5\text{-}1)$$

$$a_1 = \begin{vmatrix} \Sigma x^4 & \Sigma x^2 y & \Sigma x^2 \\ \Sigma x^3 & \Sigma xy & \Sigma x \\ \Sigma x^2 & \Sigma y & N \end{vmatrix} \Bigg/ D \tag{8.5-2}$$

$$a_2 = \begin{vmatrix} \Sigma x^2 y & \Sigma x^3 & \Sigma x^2 \\ \Sigma xy & \Sigma x^2 & \Sigma x \\ \Sigma xy & \Sigma x & N \end{vmatrix} \Bigg/ D \tag{8.5-3}$$

$$D = \begin{vmatrix} \Sigma x^4 & \Sigma x^3 & \Sigma x^2 \\ \Sigma x^3 & \Sigma x^2 & \Sigma x \\ \Sigma x^2 & \Sigma x & N \end{vmatrix} \tag{8.5-4}$$

While these equations look rather formidable, they are readily simplified for our set of equidistant data centered around $x = 0$ (because we subtracted $\bar{x}$). First we consider the expression for D in (8.5-4). It only contains terms in x, and several of these are zero. Specifically, since the data are equidistant in x, all sums of odd powers of x must be zero, i.e., $\Sigma x = -2\delta - \delta + 0 + \delta + 2\delta = 0$ and $\Sigma x^3 = (-2\delta)^3 + (-\delta)^3 + (\delta)^3 + (2\delta)^3 = 0$. Furthermore, since we know the x-values (they are -2δ, $-\delta$, 0, δ, and 2δ), we can evaluate the remaining sums as $\Sigma x^4 = (-2\delta)^4 + (-\delta)^4 + (\delta)^4 + (2\delta)^4 = (16 + 1 + 1 + 16)\,\delta^4 = 34\delta^4$, $\Sigma x^2 = (-2\delta)^2 + (-\delta)^2 + (\delta)^2 + (2\delta)^2 = (4 + 1 + 1 + 4)\delta^2 = 10\delta^2$, and $N = 5$. The entire expression for D therefore reduces to $(34\delta^4) \times (10\delta^2) \times (5) - (10\delta^2) \times (10\delta^2) \times (10\delta^2) = (1700 - 1000)\,\delta^6 = 700\delta^6$:

$$D = \begin{vmatrix} 34\delta^4 & 0 & 10\delta^2 \\ 0 & 10\delta^2 & 0 \\ 10\delta^2 & 0 & 5 \end{vmatrix} = 700\delta^6 \tag{8.5-5}$$

Now look at the expression for a_0 in (8.5-1). It has the same type of terms as D, plus some terms with y, but the latter contain y only to unit power. As before, $\Sigma x = 0$, $\Sigma x^3 = 0$, $\Sigma x^4 = 34\delta^4$, and $\Sigma x^2 = 10\delta^2$. That leaves three terms, in the right-most column of (8.5-1), which we evaluate as follows: $\Sigma y = y_{-2} + y_{-1} + y_0 + y_1 + y_2$, $\Sigma xy = (-2y_{-2} - y_{-1} + y_1 + 2y_2) \times \delta$, and $\Sigma x^2 y = (4y_{-2} + y_{-1} + y_1 + 4y_2) \times \delta^2$. Thus, (8.5-1) becomes

$$a_0 = \begin{vmatrix} 34\delta^4 & 0 & (4y_{-2} + y_{-1} + y_1 + 4y_2)\delta^2 \\ 0 & 10\delta^2 & (-y_{-2} - y_{-1} + y_1 + y_2)\delta \\ 10\delta^2 & 0 & y_{-2} + y_{-1} + y_0 + y_1 + y_2 \end{vmatrix} \Bigg/ 700\delta^6$$

$$= \frac{(34\delta^4)\,(10\delta^2)\,(y_{-2}+y_{-1}+y_0+y_1+y_2) - ((4y_{-2}+y_{-1}+y_1+4y_2)\delta^2)(10\delta^2)(10\delta^2)}{700\delta^2}$$

$$= \frac{(-60y_{-2}+240y_{-1}+340y_0+240y_1-60y_2)\delta^6}{700\delta^6}$$

$$= \frac{-3y_{-2}+12y_{-1}+17y_0+12y_1-3y_2}{35} \tag{8.5-6}$$

Consequently we have simplified the entire expression for a_0 to a set of integer coefficients with which to multiply the various y-values. Now, when we need to smooth the data set, this is all we need. We simply replace the y-value at the central; point, y_0, by the value calculated at $x = x_0$ for the parabola, i.e., by $a_{\text{calc}} = a_0 + a_1x_0 + a_2x_0^2 = a_0$ since $x_0 = 0$.

The same method that reduces (8.5-1) to (8.5-6) can be used to identify a_1 and a_2. For example, we can find the value of the first derivative of the function as $(dy_{\text{calc}}/dx)_{x=0} = (a_1 + a_2x)_{x=0} = a_1$, and its second derivative as $(d^2y_{\text{calc}}/dx^2)_{x=0} = 2a_2$.

Consequently there are separate tables of convoluting integers for smoothing to a quadratic, for finding the first derivative of a quadratic, and for finding its second derivative, all with different entries for data sets containing 5, 7, 9, 11, etc. points. Once the required value is computed, we drop the first point of the data set, add a new one (i.e., we slide the five-point sample past the original data set) and start again, using the same coefficients. This makes the method ideally suited for spreadsheet use. Note that, in all these applications, we do not need the value of $\bar{x}$. However, there are also some cases where $\bar{x}$ is required, as when we use this method for interpolation, see section 10.2.

Now that we understand how the method works, at least in principle, we will take a sine wave, add Gaussian noise, and explore the result of least-squares smoothing.

Instructions for exercise 8.5

1 Open a new spreadsheet.

2 In row 1 enter labels for the increment Δx, and the noise amplitude na.

3 In row 2 enter corresponding numbers, e.g., 0.02*Pi() and 0.2.

4 In row 4 deposit labels for x, sin(x), noise, and the sum s + n of the sine wave and na times the noise.

5 Starting in cell A6 calculate x = 0 (Δx) 7. For $\Delta x = 0.02\pi$ the column will then extend to A117.

6 In column B compute sin(x), in C deposit 'normal' noise of unit standard deviation and zero mean, and in D calculate sin(x) plus the product of the noise amplitude (in B2) and the noise (in C6:C117).

7 First we will use a five-point quadratic as our moving polynomial. For this, the tables list the convoluting integers as $-3, 12, 17, 12$, and -3, and the normalizing factor as 35, see (8.5-6). In cell E8 therefore deposit the instruction $= (-3*D6 + 12*D7 + 17*D8 + 12*D9 - 3*D10)/35$, and copy this down to E115. We start in cell E8, and stop in cell E115, because a five-point polynomial needs two points before, and two points past, the midpoint where it computes its result.

8 In column F we will use a 15-point quadratic, for which the convoluting integers are $-78, -13, 42, 87, 122, 147, 162, 167, 162, 147, 122, 87, 42, -13, -78$, with a normalizing factor of 143. We therefore deposit the instruction $= (-78*D6 - 13*D7 + 42*D8 + 87*D9 + 122*D10 + 147*D11 + 162*D12 + 167*D13 + 162*D14 + 147*D15 + 122*D16 + 87*D17 + 42*D18 - 13*D19 - 78*D20)/143$ in cell F13. Copy this down to cell I10. Note that, for a 15-point polynomial, we now must leave 7 points free on either side.

9 Likewise, in column G, use a 25-point quadratic, for which the convoluting integers are $-253, -138, -33, 62, 147, 222, 287, 342, 387, 422, 447, 462, 467, 462, 447, 422, 387, 342, 287, 222, 147, 62, -33, -138, -253$, while the corresponding normalizing factor is 5175. Place the corresponding instruction $= (-253*D6 - 138*D7 - \cdots - 253*D30)/5175$ in G18, then copy it down to cell I05, since we must leave 12 points free on either side for a 25-point polynomial.

10 Plot graphs of B, D, E, F, and G versus A, or combine them in one graph by giving the data in columns E, F and G an offset of, say, $+1$, $+2$, and $+3$ respectively, together with reference curves which reproduce the data in B with the same offsets.

11 Figure 8.5-1 illustrates what you might see using separate plots. As it happens in this example, the initial noise is mostly positive. Consequently, the smoothed curves start too high. Remember: even perfectly random noise only averages out to zero for a sufficiently large set. As you can see here, even a 25-point data set is clearly not large enough. Of course, your noise data will be different.

Of the various methods described here, signal averaging does not introduce distortion (as long as the instrumental settings don't drift) and requires no assumptions as to the nature of the noise. On the other hand, it is a rather time-consuming way to remove noise, since each point is averaged individually. Least-squares smoothing of equidistant data assumes a relation between neighboring data, and can therefore be much faster than signal averaging, at the risk of distorting the underlying signal.

As described in chapter 7, Fourier transformation can provide another way to smooth these equidistant data, by first transforming the data, then setting the predominantly noise-related frequencies to zero, followed by

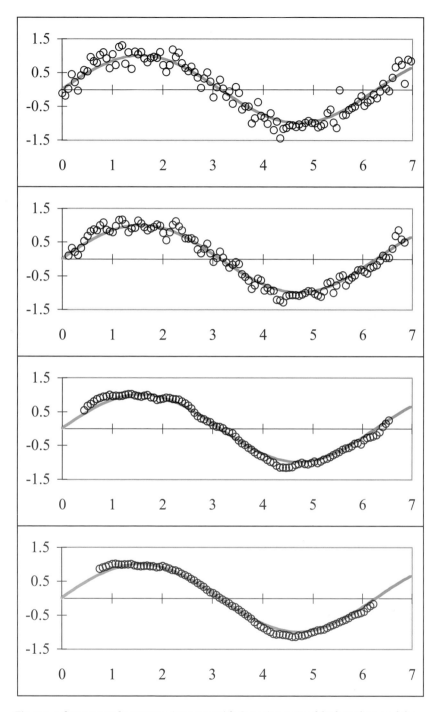

Fig. 8.5-1, from top to bottom: A sine wave with Gaussian noise (black circles), and the same after filtering with a 5-, 15-, and 25-point quadratic polynomial. The noise-free sine wave is shown in color.

inverse transformation. In the above example, Fourier transformation would be very easy indeed, because the noise-free signal is a single sine wave. An advantage of this approach is that no points are lost at the edges; a disadvantage (at least with the FFT macro provided here) is that the number of data points must be an integer power of 2, e.g., 128.

8.6 Peak fitting

Many spectrometric peaks can be described reasonably well in terms of one of two model peak shapes, those of a Gaussian and of a Lorentzian (or Cauchy) peak. Leaving out the normalizing factors as immaterial in the present context, the Gaussian peak can be described as

$$y = A \exp\left[\frac{-(x-B)^2}{C^2}\right] \tag{8.6-1}$$

where A is the amplitude, B specifies the location of the peak center, and C controls the peak width. The full width at half height of the peak is given by $w_{\frac{1}{2}} = 2C\sqrt{\ln 2} \approx 1.665_1\, C$, and the peak area by $AC\sqrt{\pi} \approx 1.772_5\, AC$.

Likewise we can write the Lorentzian peak shape as

$$y = \frac{A}{1 + (x-B)^2/C^2} \tag{8.6-2}$$

for which we find the peak height A, the center of the peak at $x = B$, the full peak width at half height as $w_{\frac{1}{2}} = 2C$, and the peak area as πAC.

Lorentzians are much wider at their base than Gaussian peaks of the same area and height. Mixtures of them can be used to obtain a single peak with variable profile, such as the weighted sum of a Gaussian and a Lorentzian,

$$y = fA \exp\left[\frac{-(x-B)^2}{C^2}\right] + \frac{(1-f)A}{1 + (x-B)^2/C^2} \tag{8.6-3}$$

where the weighting factor f is an additional, adjustable variable with a value ranging from 0 to 1. Alternatively, one can use a weighted product of a Gaussian and a Lorentzian. Other combinations are also possible; e.g., in the theory of absorption of a Doppler-broadened line one encounters the convolution of a Gaussian and a Lorentzian.

Instructions for exercise 8.6

1 Open a new spreadsheet.

2 Enter labels for A, B, C and area in row 1, and enter corresponding values in row 2, such as A = 1, B = 5, and C = 1.

3 In row 4 deposit the column headings x and Gaussian.

4 In A6:A106 place the numbers 0 (0.1) 10.

5 In column B calculate a Gaussian according to (8.6-1).

6 In cell D2 calculate the area as $= 0.1{*}SUM(B6:B106)$, and compare your answer with the theoretical result $AC\sqrt{\pi} \approx 1.772_5\,AC$.

7 Use two cells, A6 and B6, to determine the half-width $w_{1/2}$ using Goal Seek. First deposit the value 4.5 in A6, call Tools ⇨ Goal Seek, then Set cell B6 To value 0.5 By changing cell A6. You will find x = 4.167445 (had you started elsewhere, you might have found the other value, $x = 5.832555$) from which the half-width follows as $2 \times (5 - 4.167445)$ or $2 \times (5.832555 - 5) = 1.6651$.

8 In C6:C106 calculate a Lorentzian according to (8.6-2).

9 It is easy to verify that $w_{1/2} = 2C$ by inspection, since that implies that the function has the value 0.5 at both $B - C$ and $B + C$.

10 It is not so easy to determine its area: merely determining $0.1 \times SUM$ (C6:C106) comes up far short of $\pi = 3.1416$.

11 Make a graph of B6:C106 vs. A6:A106. This immediately shows why the area is not calculated correctly: the Lorentzian peak has such wide tails that summation over merely five half-widths is not enough.

12 Therefore extend the table to row 956, and then determine *half* the area as $0.1 \times SUM(C56:C1056)$. This time the answer is closer, 1.5608, but it is still not the 1.5708 one expects. Even going out to 100 half-widths is not quite enough for the numerical integration of a Lorentzian!

Another expression with variable band shape is

$$y = \frac{A}{\{1 + (2^a - 1)(x - B)^2 / C^2\}^{1/a}} \tag{8.6-4}$$

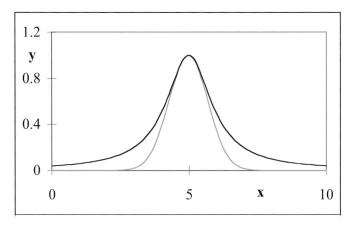

Fig. 8.6-1: A Gaussian curve (colored) and a Lorentzian curve (black), each with $A = 1$, $B = 5$, and $C = 1$.

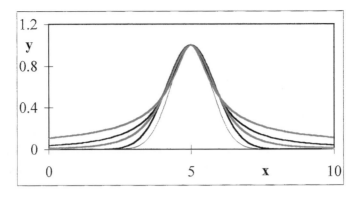

Fig. 8.6-2: Curves with adjustable band shape according to 8.6-4, computed with $A = C = 1$, $B = 5$, and $a = 1 \times 10^{-6}$ (black curve), 0.5 (colored), 1.0 (black), and 1.9 (colored) respectively. For reference, a Gaussian is shown as a thin black curve inside the others (close to the curve for $a = 1 \times 10^{-6}$).

where $0 < a < 2$. This curve is Lorentzian for $a = 1$, and is even wider at the base than a Lorentzian for $1 < a < 2$. It has a half-width of $2C$, and an area of $AC\sqrt{\pi}\,\Gamma(1/a - 1/2)\,/\{\sqrt{2^a - 1}\,\Gamma(1/a)\}$. For $a \geq 2$ the area under the curve is infinite. Figure 8.6-2 shows this function for various values of a.

Not all experimental peaks are symmetrical around their center, and there are various schemes to generate skewed curves to fit such asymmetric experimental data. For example, Losev (*Applied Spectrosc.* 48 (1994) 1289) used a function of the form

$$y = \frac{A}{\exp[-a(x-c)] + \exp[b(x-c)]} \qquad (8.6\text{-}5)$$

where A, a, b, and c are positive numbers. The parameters A and c determine the height and x-position of the peak respectively, while a and b control its shape. The peak has a maximum of height $a^{a/(a+b)}\,b^{b/(a+b)}/(a+b)$, at a distance $[\ln(a/b)]/(a+b)$ from c, and an area $\pi A/\{(a+b)\,\cos[\tfrac{1}{2}\pi\,(a-b)/(a+b)]\}$. Figure 8.6-3 shows what it can look like. By exchanging the values of a and b the curves in Fig. 8.6-3 can be skewed towards the other side.

Figure 8.6-4 illustrates a curve calculated with another expression,

$$y = A\exp\left[-\left\{\frac{\ln[1 + s(x-B)/C]}{s}\right\}^2\right] \qquad \text{for } 1 + s(x-B)/C > 0$$

$$ \qquad (8.6\text{-}6)$$

$$y = 0 \qquad \text{for } 1 + s(x-B)/C \leq 0$$

which has a skewness parameter s, where the sign of s determines in which direction the curve is stretched. When s tends to 0, the curve approaches a Gaussian. (Setting s equal to 0 will lead to the well-known problems

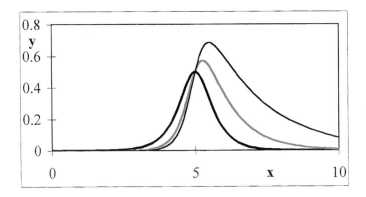

Fig. 8.6-3: Curves with adjustable asymmetry according to 8.6-5, computed with $A = 1$, $c = 5$, and $a = b = 2$ (thick black curve), $a = 3$, $b = 1$ (colored curve), and $a = 3.5$, $b = 0.5$ (thin black curve).

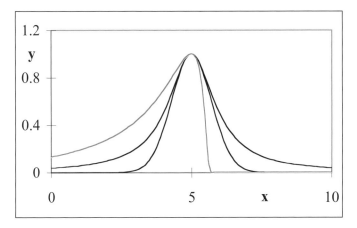

Fig. 8.6-4: A skewed 'Gaussian' curve with $s = -1.5$ (colored) together with the symmetrical Gaussian and Lorentzian curves (black), all with $A = 1$, $B = 5$, $C = 1$.

associated with dividing zero by zero; we therefore use the series expansion $\ln [1 + s(x - B)/C] \approx s(x - B)/C - \frac{1}{2} [s(x - B)/C]^2 + \cdots$ to obtain the limit for $s \to 0$ as $y = A \exp[(x - B)^2/C^2]$, which is indeed Gaussian.) For $-2 < s < 2$ the area under this curve is finite, and is given by $(AC\sqrt{\pi}) \exp[s^2/(4 \ln 2)]$.

For more information on fitting model expressions to peak-shaped experimental curves you might want to consult the review by Fraser & Suzuki in J. A. Blackburn, ed., *Spectral Analysis: Methods and Techniques*, Dekker, 1970, from which some of the above discussion was abstracted.

8.7 Integration

Integration of peak areas is often used to obtain precise values for the concentrations of sample components. Many elegant methods have been devised to integrate analytical functions, but few of these are readily transferable to discrete, experimental data points. Fortunately, integration by summation of the areas of trapezoids of height $(y_i + y_{i+1})/2$ times width $(x_{i+1} - x_i)$ often suffices, and is readily implemented on a spreadsheet. When the signal is available in Fourier-transformed form, integration can be implemented by multiplying all components of the Fourier transform by $j\omega$ before inverse transformation.

We will now use the last part of spreadsheet exercise 6.7 to illustrate some of the practical problems of integration. In exercise 6.7 we computed the area A, retention time t_r, and standard deviation σ_r of a simulated chromatographic peak, based on (6.7-1) through (6.7-3). For $\mu = 0.2$, the simulated peak had its maximum value at a retention time of about 120, and its long-time tail (where the response still made a significant contribution to the integral) reached well beyond $t = 200$. Since the calculation did not extend beyond $t = 200$, some of that tail was missed, and the integral was too low. As a consequence, we found $t_r = 119.90$ instead of 120. Once we understand the origin of this problem, its possible remedy is clear: extend the computation to larger values of t. Indeed, by extending the spreadsheet of Fig. 6.5-1 to row 306 (i.e., to $t = 200$) we obtain $A = 1.00$, $t_r = 120.00$, $\mu t_r = 24.00$, $\sigma_r = 21.90$, and $N_p = 23.99$. Even longer times t would be needed for, say, $\mu = 0.1$, where the peak maximum (at $t_r = 240$) already fell outside the earlier-used time range.

For large values of μ we encountered another difficulty: the number of simulated points was too small to use a simple trapezoidal integration. For example, at $\mu = 0.9$, we had only about ten points that make a significant contribution to the peak. Even though this was a simulation, we could not increase the point density, because $t = 1$ is the smallest step size of the model; in many practical integrations, the step size may be limited by instrumental factors and likewise be fixed. In computing the standard deviation this difficulty was exacerbated because we calculated σ_r^2 as the difference between two larger numbers.

When we have too few points to justify linearizing the function between adjacent points (as the trapezoidal integration does) we can use an algorithm based on a higher-order polynomial, which thereby can more faithfully represent the curvature of the function between adjacent measurement points. The Newton–Cotes method does just that for equidistant points, and is a moving polynomial method with fixed coefficients, just as the Savitzky–Golay method used for smoothing and differentation discussed in sections 8.5 and 8.8. For example, the formula for the area under the curve between x_1 and x_{n+1} is

Table 8.7-1: *The coefficients for Newton–Cotes integration of equidistant points.*

Order n	Coefficients c_i	Denominator
1	1, 1	$\Sigma\, c_i = 2$
2	1, 4, 1	6
3	1, 3, 3, 1	8
4	7, 32, 12, 32, 7	90
5	19, 75, 50, 50, 75, 19	288
6	41, 216, 27, 272, 27, 216, 41	840
7	751, 3577, 1323, 2989, 2989, 1323, 3577, 751	17 280
8	989, 5888, -928, 10496, -4540, 10496, -928, 5888, 989	28 350

$$\int_1^{n+1} f(x)\,dx = n\delta \sum_{i=1}^{n+1} c_i f(x_i) \Big/ \sum_{i=1}^{n+1} c_i \qquad (8.7\text{-}1)$$

where the function $f(x)$ is defined at $n+1$ equidistant x-values in terms of the coefficients c_i, and δ is the spacing of adjacent points on the x-axis. For example, the integral for a five-point fit (so that $n = 5 - 1 = 4$) of data with an x-spacing of 0.1 is

$$\int_1^5 f(x)\,dx = 4 \times 0.1 \times [7f(x_1) + 32f(x_2) + 12f(x_3) + 32f(x_4) + 7f(x_5)]/90 \quad (8.7\text{-}2)$$

where the coefficients are 7, 32, 12, 32, and 7, with a sum of $7 + 32 + 12 + 32 + 7 = 90$. This is a very convenient form for spreadsheet use. Other Newton–Cotes coefficients are listed in Table 8.7-1. Exercise 8.7 illustrates its application.

Note that the trapezoidal rule is the first-order member of this method. In the above example, fourth-order Newton–Cotes integration for $\mu = 0.9$ yields $A = 1.00$, $t_r = 26.67$, $\mu t_r = 24.00$, $\sigma_r = 1.72$, and $N_p = 24.00$. In fact, almost equally accurate results can already be obtained with a second-order Newton–Cotes fit.

Instructions for exercise 8.7

1 Open a new spreadsheet.

2 In A1 enter the label k=, in B1 a value such as 0.3, in C1 the label δ=, and in D1 a corresponding numerical value, e.g., 1.

3 In row 3 enter the labels t, exp, integral, n = 1, and n = 4 respectively.

4 In A5 deposit a zero, in A6 the instruction =A5 + $D$1, in B5 the instruction =EXP($B$1*A5), and in B6 the instruction =EXP($B$1*A6).

5 In C6 compute the integral, =(1/$B$1)*(B6-$B$5).

6 In D6 integrate by the trapezoidal rule as =D5 + (A6 − A5)*(B6 + B5)/2.

7 Copy the instructions to cells in A6:D6 down to row 25.

8 Now go to E9 and deposit there the instruction =4*$D$1*(7*B5 + 32*B6 + 12*B7 + 32*B8 + 7*B9)/90.

9 In cell E13 deposit the instruction =E9 + 4*$D$1*(7*B9 + 32*B10 + 12*B11 + 32*B12 + 7*B13)/90, then copy this instruction to cells E17, E21, and E25.

10 Compare the results of the integration in D25 and E25 with the exact result in C25.

11 Change the value of δ in D1 to 0.1, and verify that result of the integration remains correct upon changing the x-spacing.

12 Change the value of δ to 10, in order to see the limitations of approximating an exponential in terms of a fourth-order polynomial $y = a_0 + a_1 x + a_2 x^2 + a_3 x^3 + a_4 x^4$.

A third common source of difficulties in numerical integrations is truncation error, observed when parameters are carried through the computation to an insufficient number of digits. Fortunately, the automatic double precision of the spreadsheet greatly reduces the importance of truncation errors, so that they only seldom need to be considered. When the integration is performed off-screen, in a function or in a macro, the computation should be specified to use double precision.

Integration is relatively insensitive to noise, but is very sensitive to bias or offset, such as may result from an incorrect zero setting of the measuring instrument, or from some other phenomenon affecting the baseline. That integration is relatively insensitive to noise is readily seen when we consider noise in terms of its Fourier-analyzed components, i.e.,

$$\int [F(t) + \text{noise} + \text{offset}] dt = \int [F(t) + \sum_\omega \sin(\omega t + \varphi) + \sum_\omega \cos(\omega t + \varphi) + C] dt$$

$$= \int [F(t) dt - \frac{1}{\omega} \sum_\omega \cos(\omega t + \varphi) + \frac{1}{\omega} \sum_\omega \sin(\omega t + \varphi) + C \int dt \qquad (8.7\text{-}3)$$

where we have replaced the offset by a constant, C, and neglected the integration constants of the noise components $\sin(\omega t + \varphi)$ and $\cos (\omega t + \varphi)$, assuming that the noise averages out to zero. We see from (8.7-3) that the effect of noise on the integral is attenuated by a factor ω, the angular frequency of that particular noise component. Consequently, the higher the noise frequency with respect to the dominant frequencies in the function $F(t)$, the less it will affect the integral. For most experimental data encountered in chemical analysis, the noise is mostly at frequencies *higher* than the dominant frequencies of the signal, in which case integration reduces their relative importance. However, we have no such luck with baseline correc- tions, instrumental offset and baseline drift, and so-called "pink noise", all of which typically occur at frequencies *below* those of the signal. These often contribute to the integral in direct proportion to the integration interval, even when parts of that interval may not contain much signal. Therefore the integration interval is often taken as small as possible, although this might introduce systematic errors in functions with a 'wide footprint', such as Lorentzian peaks, see the top panel in Fig. 3.3-2.

Even more severe problems arise when we need to integrate areas under overlapping peaks. In that case, it is often not the mechanics of integration that limit the reliability of the results, but the applicability of models that describe the underlying peak shapes, see section 8.6. On the other hand, separating adjacent integration domains by using as domain boundary the minimum in a saddle point between two partially overlapping peaks, can lead to serious systematic errors in both integrals.

8.8 Differentiation

Differentiation is the counterpoint of integration: it is less sensitive to drift and offset at frequencies below those of the signal, but is more strongly affected by noise at frequencies above those of the signal itself. This follows directly from the model used in section 8.6, because

$$\frac{d}{dt}[F(t) + \text{noise} + \text{offset}] = \frac{d}{dt}\left[F(t) + \sum_{\omega} \sin(\omega t + \varphi) + \sum_{\omega} \cos(\omega t + \varphi) + C\right]$$

$$= \frac{dF(t)}{dt} + \omega \sum_{\omega} \cos(\omega t + \varphi) - \omega \sum_{\omega} \sin(\omega t + \varphi) \quad \text{(8.8-1)}$$

where the effect of the constant offset has disappeared (because $dC/dt = 0$) but that of the various noise components is magnified by the multipliers ω. As a consequence, differentiation of experimental data must usually be combined with noise filtering, and then suffers from the signal distortion resulting from such filtering.

The Savitzky–Golay method combines filtering with single or multiple differentiation *in one operation*. Moreover, as we have already seen in section 8.5, it is very convenient for spreadsheet use. In the spreadsheet exercise we will differentiate a noisy sine wave and compare the result with its analytical derivative, a cosine. Then we will compute the second deriva- tive, and again compare the result with the theoretical second derivative, an inverted sine wave. We could also compute that second derivative stepwise, as the derivative of the derivative, but the present route is simpler and loses fewer points at the edges.

Instructions for exercise 8.8

1 Open a new spreadsheet.

2 In row 1 enter labels for the increment Δx, and for the noise amplitude na.

3 In row 2 enter corresponding numbers, e.g., 0.02π and 0.

4 In row 4 deposit labels for x, sin(x), cos(x), − sin(x), noise, the sum s + n of the sine wave and the noise, the first derivative, and the second derivative. (When you try to enter − sin(x), Excel will interpret this as a number. In order to make it a label, highlight the cell, and use Format ⇨ Cells ⇨ Number ⇨ Text, or precede the instruction by an apostrophe.)

5 In cells A6:A117 calculate x = 0 (increment) 6.974.

6 In columns B through D compute sin(x), cos(x) and − sin(x), in column E deposit 'normal' noise of unit standard deviation and zero mean, and in F calculate sin(x) plus the product of the noise amplitude (in B2) and the noise (in E6:E117).

7 For the first derivative, using a five-point quadratic as our moving polynomial, the tables list the convoluting integers as − 2, − 1, 0, 1, and 2, and the normalizing factor as 10 times the increment Δx. In cell G8 therefore deposit the instruction = (− 2*F6 − F7 + F9 + 2*F10)/(10*$B$2), and copy this down to G115.

8 For a second derivative, calculated using a five-point quadratic, we have the convoluting integers 2, − 1, − 2, − 1, 2, with a normalizing factor 7(Δx)^2. Therefore, in cell H8, use = (2*F6 − F7 − 2*F8 − F9 + 2*F10)/(7*$B$2*$B$2) and copy this instruction down to H115.

9 Plot rows D through H versus A.

10 Figure 8.8-1 illustrates that the second derivative will be much noisier than the first, which in turn is much noisier than the data set in column D.

11 The noise can be reduced by using a longer polynomial, such as a 15-point or even a 25-point polynomial. The filtering is stronger, but artificial waves appear, with a period of the polynomial length, as illustrated in Fig. 8.8-2.

As can be seen in Fig. 8.8-1, a hardly noticeable amount of noise in a function can lead to quite dramatic fluctuations in its second derivative.

When the experimental data are not equidistant, a moving polynomial fit can still be used, but the convenience of the Savitzky–Golay method is lost. In that case you may have to write a macro to fit a given data set to a polynomial, and then change the cell addresses to make the polynomial move along the curve. Consult chapter 10 in case you want to write your own macros.

In chapter 7 we saw that Fourier transformation can also be used to differentiate data. Just as the Savitzky–Golay method, Fourier transformation

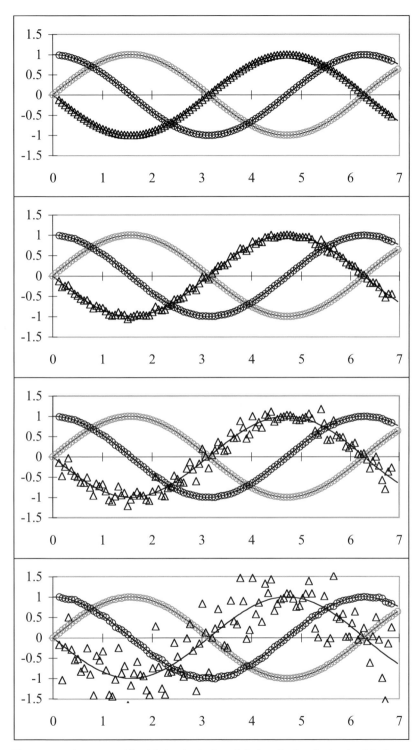

Fig. 8.8-1: A sine wave with added noise (colored circles), its first derivative (black circles) and its second derivative (black triangles), calculated using a five-point quadratic. Noise amplitude used, from top to bottom: 0, 0.0003, 0.001, and 0.003.

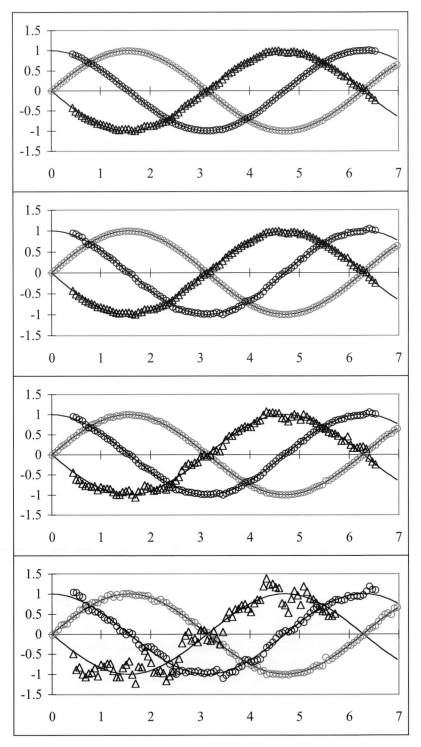

Fig. 8.8-2: A sine wave with added noise (colored circles), its first derivative (black circles) and its second derivative (black triangles), calculated using a 15-point quadratic. Noise amplitude used, from top to bottom: 0.003, 0.01, 0.03, and 0.10.

needs equidistant data. Moreover, the FFT routines provided require that the number of data points be an integer power of 2, and that the two ends match in their values and derivatives. When those additional requirements are met, FFT is a very powerful and efficient method, which provides complete control over the noise filtering while differentiating. Moreover, with FTT one does not lose any points near the edges of the data set.

However, no matter what method you will use, when you anticipate having to differentiate experimental data, make sure that they have minimal noise to start with, because differentiation always enhances noise, and all noise reduction methods introduce distortion, which you will want to keep to a minimum.

8.9 Semi-integration and semi-differentiation

As our final example we will here describe a rather specialized method useful in problems involving signal, heat, and mass transport, insofar as these can be described by a partial differential equation of the type $\partial y/\partial t = a\,\partial^2 y/\partial x^2$ where t is time, x is distance, and a is a constant. Depending on the field, such equations are known as the telegraphers equation in communication theory (where y is voltage or current, and a is the product of the distributed resistance and the distributed capacitance along a transmission line), the equation for heat conduction in physics (where y and a represent temperature and thermal conductivity respectively), or Fick's law in chemistry and biology (where y stands for concentration, and a for diffusion coefficient). Here we will specifically apply this method to planar diffusion of electroactive species in solution.

In many electroanalytical experiments, diffusion is essentially planar and semi-infinite, i.e., the equi-concentration planes are planar rather than curved, and the concentration c at a sufficiently large distance from the electrode, $c_{x\to\infty}$, remains essentially constant during the experiment. If, moreover, the experiment starts at a well-defined time $t = 0$ with a uniform concentration throughout the cell, then the interfacial concentration difference $u = c - c_{x\to\infty}$ is the convolution of the diffusional flux J (the amount of material passing per unit time through a unit cross-sectional area) and $1/\sqrt{\pi Dt}$, i.e.,

$$u = J * \frac{1}{\sqrt{\pi Dt}} \tag{8.9-1}$$

where the asterisk denotes a convolution (not to be mistaken for the programmer's use of the same symbol for multiplication). The faradaic current i_f is directly proportional to the flux J of the electroactive species at the electrode/solution interface, the proportionality constant being nFA, where n is the number of electrons involved in the electrode reaction, F is the Faraday, and A is the electrode area. An equivalent way to describe (8.9-1) is in terms of a semi-integral,

$$u = \frac{1}{\sqrt{D}} \frac{\mathrm{d}^{-\frac{1}{2}} J}{\mathrm{d} t^{-\frac{1}{2}}} \tag{8.9-2}$$

where the notation $\mathrm{d}^{-\frac{1}{2}} y / \mathrm{d} t^{-\frac{1}{2}}$ of **semi-integration** is the counterpart of **semi-differentiation**, $\mathrm{d}^{\frac{1}{2}} y / \mathrm{d} t^{\frac{1}{2}}$, hence the name.

In their book *Fundamentals of electrochemical science* (Academic Press 1994) Oldham and Myland list two different algorithms for semi-integration, of which we have used the more efficient one in section 10.10.

Semi-integration can be used to transform a linear sweep voltammogram into the corresponding stationary current–voltage curve. Likewise, semi-differentiation can be used for the inverse process, i.e., to convert a stationary current–voltage curve into the corresponding linear sweep voltammogram. Consequently, the extensive existing theory for the shapes of polarographic waves for a variety of reaction mechanisms is readily applicable to cyclic voltammetry.

8.10 Interpolation

Interpolating in a data set can often be done by fitting the existing data points around the sought value to an appropriate function, and by then using the parameters of that function to calculate the desired value. For example, when the interpolation is to data within a segment that can be described approximately by a parabola, you can fit the data to the parabola $y = a_0 + a_1 x + a_2 x^2$, and then interpolate the value at the desired x-value x_1 as $y = a_0 + a_1 x_1 + a_2 x_1^2$. It is convenient to use a polynomial, since it will allow you to use a least-squares routine.

Alternatively, you can use Solver to fit data to any analytical function of your fancy, then use the fitting parameters to compute the function at any given point within the range covered.

When you have many data to interpolate, it is usually most convenient to fit the entire data set in the region of interest to an appropriate analytical expression, from which you can then calculate all required data. If you have no idea what would be an appropriate function, or you cannot obtain a reasonably close fit to the data, a small sliding polynomial crawling over the experimental data set may have to be used instead. The Interpolate macro of section 10.2 allows you to fit a moving polynomial of $2n + 1$ equidistant data to a parabola.

In either case, make sure that you use more experimental data points than adjustable parameters, and many more when the data are very noisy. Locating the position of a peak maximum uses a similar approach, and only differs in that one then does not prescribe the x-value but, instead, determines from the calculated parameters where the maximum occurs. For the parabola $y = a_0 + a_1 x + a_2 x^2$, the extremum (maximum or minimum) occurs at $dy/dx = a_1 + 2a_2 x = 0$ or $x = -a_1/2a_2$.

Fourier transformation can also be used for interpolation, provided that the data are equidistant, and the sought values are located at fractions of Δx that are integer multiples of $1/2^n$ where n is, again, an integer, see Fig. 7.2-4.

8.11 Matrix manipulation

In this section we will briefly review the most salient aspects of matrix algebra, insofar as these are used in solving sets of simultaneous equations with linear coefficients. We already encountered the power and convenience of this method in section 6.2, and we will use matrices again in section 10.7, where we will see how they form the backbone of least squares analysis. Here we merely provide a short review. If you are not already somewhat familiar with matrices, the discussion to follow is most likely too short, and you may have to consult a mathematics book for a more detailed explanation. For the sake of simplicity, we will restrict ourselves here to two-dimensional matrices.

A two-dimensional matrix is a *rectangular array* of symbols, called **coefficients**, which are given double indices to specify their position in the array, as in this example of an m-by-n matrix

$$\mathbf{A} = \begin{bmatrix} a_{11} & a_{12} & a_{13} & a_{14} & \cdots & a_{1n} \\ a_{21} & a_{22} & a_{23} & a_{24} & \cdots & a_{2n} \\ a_{31} & a_{32} & a_{33} & a_{34} & \cdots & a_{3n} \\ \vdots & \vdots & \vdots & \vdots & \ddots & \vdots \\ a_{m1} & a_{m2} & a_{m3} & a_{m4} & \cdots & a_{mn} \end{bmatrix} \qquad (8.11\text{-}1)$$

where m denotes the number of rows, and n the number of columns. (You should *memorize the order*: first rows, then columns. As a mnemonic device, just think of the basic electrical RC circuit of a resistor and a capacitor.) We usually denote the entire array by a single, bold-printed capital, such as $\mathbf{A}$, and then refer to its individual coefficients as a_{ij}. In general, a matrix coefficient can be a number, an equation, a mathematical operator, a text string, whatever.

With each matrix we can associate a determinant, which contains only corresponding numbers. Determinants are typically shown within vertical lines rather than brackets. A determinant is a matrix, but not necessarily the other way around, just as a cow is a mammal, but not all mammals are cows. In the type of numerical problems encountered in chemical analysis we often deal with determinants rather than matrices (and this certainly applies to the examples in this book), but we will use the more general matrix notation here.

The size of the above matrix $\mathbf{A}$ is specified as $m \times n$, which is pronounced as m-by-n. Matrices can be added and subtracted as long as they have the same size (by which we mean that they have identical m- and n-values, not just identical products $m \times n$), in which case we merely add or subtract all coefficients with identical indices, i.e., $\mathbf{C} = \mathbf{A} + \mathbf{B}$ implies that all c_{ij} are given by $c_{ij} = a_{ij} + b_{ij}$. Since $a_{ij} + b_{ij} = b_{ij} + a_{ij}$ it follows that $\mathbf{A} + \mathbf{B} = \mathbf{B} + \mathbf{A}$, and the same holds for the difference, $\mathbf{A} - \mathbf{B} = \mathbf{B} - \mathbf{A}$, i.e., matrix addition and subtraction are commutative.

The rules for matrix multiplication are more complicated, because matrix multiplication is *not* commutative, that is, $\mathbf{A}\,\mathbf{B} \neq \mathbf{B}\,\mathbf{A}$. The product $\mathbf{C} = \mathbf{A}\,\mathbf{B}$ of the $m \times n$ matrix $\mathbf{A}$ and the $p \times q$ matrix $\mathbf{B}$ is defined only when the number n of columns in $\mathbf{A}$ is equal to the number p of rows in $\mathbf{B}$, i.e., $n = p$. In that case the coefficients of the product matrix are given by

$$c_{ij} = \sum_{k=1}^{n} a_{in} b_{nj} \qquad\qquad (8.11\text{-}2)$$

Similarly, the product $\mathbf{B}\,\mathbf{A}$ is defined only when the number q of columns in $\mathbf{B}$ equals the number m of rows in $\mathbf{A}$, i.e., $q = m$, in which case the coefficients of the product matrix $\mathbf{D} = \mathbf{B}\,\mathbf{A}$ are

$$d_{ij} = \sum_{k=1}^{m} b_{im} a_{mj} \qquad\qquad (8.11\text{-}3)$$

Consequently there are two products, $\mathbf{C} = \mathbf{A}\,\mathbf{B}$ and $\mathbf{D} = \mathbf{B}\,\mathbf{A}$ which, in general, are quite different.

An important special matrix is the **unit matrix**, which is a *square* matrix (i.e., it has equal numbers of rows and columns) with coefficients that are all 0, except that they are 1 on its main (top-left to bottom-right) diagonal (where the coefficients have equal indices, i.e., where $i = j$). For the appropriately chosen unit matrix (i.e., of such size that the product is defined) we have the property

$$\mathbf{A}\,\mathbf{I} = \mathbf{I}\,\mathbf{A} = \mathbf{A} \qquad\qquad (8.11\text{-}4)$$

Moreover, for any *square* matrix $\mathbf{A}$ with non-zero determinant we can also define a unique **inverse matrix $\mathbf{A}^{-1}$** such that

$$\mathbf{A}^{-1}\mathbf{A} = \mathbf{A}\mathbf{A}^{-1} = \mathbf{I} \qquad\qquad (8.11\text{-}5)$$

Determining the inverse of a matrix by hand is a fairly complicated matter. Fortunately, Excel has a built-in function, MINVERSE, that will perform the inversion. It also has a matrix multiplication function, MMULT, that will calculate the product of two matrices. In order to let the spreadsheet know that your instructions concern an entire block or array rather than an individual cell, these two functions require that you first highlight the entire block to which the instruction applies, and then enter the instruction while *simultaneously* depressing Ctrl, Shift, and Enter.

Somewhat inconsistently, a third matrix operation, TRANSPOSE, is a standard Excel operation. Transposition of a matrix is simply the exchange of rows and columns, i.e., when $\mathbf{A}$ has the coefficients a_{ij}, then its **transpose** $\mathbf{A}^\mathrm{T}$ has the coefficients a_{ji}. To transpose a matrix you highlight it, copy it to the clipboard with Ctrl + c, and select the top left corner of the block where you want the transpose to appear. Only then select Edit ⇨ Paste Special, in the Paste Special dialog box (on its bottom row) select Transpose, and click OK. That will do it.

The values of m and n in matrix $\mathbf{A}$ of (8.11-1) must be positive integers, i.e., members of the set 1, 2, 3, …. When $m=1$ and $n>1$ we have a horizontal **vector**, as in

$$a_1 \quad a_2 \quad a_3 \quad a_4 \dots a_n \qquad (8.11\text{-}6)$$

where the index m has been dropped as unnecessary, while $m>1$ and $n=1$ defines a vertical vector,

$$
\begin{matrix}
a_1 \\
a_2 \\
a_3 \\
\vdots \\
a_m
\end{matrix}
\qquad (8.11\text{-}7)
$$

in which case the index n has been deleted. All rules for matrix operations apply to vectors, so that we need not consider them as special. For $m=1$ and $n=1$ we have a 1×1 matrix, which is simply a **scalar**. In Excel, vectors are represented as matrices with one index 1, such as M$(1,n)$ or M$(m,1)$.

So far we have described the mechanics of matrix operations, but we have yet to demonstrate its power. We already illustrated this in section 6.2, where we took a set of simultaneous equations of the form

$$y_i = a_{i1}x_1 + a_{i2}x_2 + a_{i3}x_3 + \dots + a_{im}x_m \qquad (8.11\text{-}8)$$

which we rewrote in matrix notation as

$$\mathbf{Y} = \mathbf{A}\mathbf{X} \qquad (8.11\text{-}9)$$

and then solved for $\mathbf{X}$ through left-multiplication by $\mathbf{A}^{-1}$,

$$\mathbf{A}^{-1}\mathbf{Y} = \mathbf{A}^{-1}\mathbf{A}\mathbf{X} = \mathbf{I}\mathbf{X} = \mathbf{X} \qquad (8.11\text{-}10)$$

where we have used (8.11-5) and (8.11-4) respectively.

Instructions for exercise 8.11

1 Open a new spreadsheet.

2 In column A, starting with cell A1, enter the numbers 1, 3, 2, and 1.

3 Likewise, in column B enter the numbers 6, 16, 24, 15; in column C the sequence 4, 10, 13, 4; and in column D the numbers 4, 14, 17, and 8.

4 The resulting array should occupy the block A1:D4, and look like

1	6	4	4
3	16	10	14
2	24	13	17
1	15	4	8

This will be our test array, a 4×4 matrix.

5 **Matrix inversion.** In order to invert this matrix, we first select a location for its inverse, by selecting the top left corner cell of where you want the inverse to appear, say F1.

6 Highlight the space required for the result. The inverse of a 4×4 matrix is also a 4×4 matrix, so in this example you need to highlight the block F1:I4.

7 Type = MINVERSE(A1:D4) but *don't* press the Enter key yet.

8 Now hold down the Control and Shift keys, and with both of these down press the Enter key. That will do it: you should see the inverse in block F1:I4.

9 **Matrix multiplication.** As our second exercise we will now multiply the two matrices we already have on the spreadsheet. Since one is the inverse of the other, their product should of course be the unit matrix. The procedure is analogous to that of steps (5) through (8).

10 Go to the left top corner of where you want the product to appear, say A6. Highlight the required area, say A6:D9. Type = MMULT(A1:D4,F1:I4), and hold down Ctrl + Shift while depressing Enter.

11 You should indeed obtain the unit matrix, although the zeros may have some round-off errors, though typically less than $\pm 10^{-15}$.

12 Verify that the product F1:I4 times A1:D4 also yields the unit matrix by calculating that product in F6:I9.

13 Note that the spreadsheet deals with arrays as entire blocks rather than with individual cells. This is perhaps best illustrated by trying to erase *part* of an array. Highlight A6:D8 and press Delete. Nothing will be deleted; instead you will get an error message, "Cannot change part of an array". Acknowledge the message box and highlight A6:D9. Now you will have no problem erasing it.

14 **The determinant.** This is an easy one, because the determinant of a matrix is a scalar (a single number) so you need not use the Ctrl + Shift + Enter trick. Just go to any empty cell, deposit the instruction = MDETERM(A1:D4) and (yes) press Enter. The answer, −196, will appear.

15 Try the same for the determinant of F1:I4. Then multiply the two determinants: their product should be 1.

   In summary, then, only for matrix inversion and multiplication do you need to high-light the entire block where the result should appear, and then enter the instruction while holding down Ctrl + Shift.

8.12

Overflow

So far we have relied on the double precision of the spreadsheet to hold our computed quantities. However, we occasionally encounter situations in which we seem to be limited by the finite number of digits (corresponding to about 15 decimal figures) the spreadsheet uses to represent any measure. Here we will use factorials to illustrate this problem, and a way around it.

The factorial $n!$ of a positive integer n is defined as $n \times (n-1) \times (n-2) \times (n-3) \times \cdots \times 2 \times 1$. Factorials often occur in statistics. For example, given the probability p of observing a given result in a single trial, the binomial probability of $P_{N,p}(n)$ of obtaining that same result n times in a set of N experiments is $p^n (1-p)^{N-n} N! / \{(N-n)! \, n!\}$. Many scientific derivations involve statistics, not just for data analysis, but at a much more fundamental level, as demonstrated, e.g., by the chromatographic plate theory described in sections 6.5 through 6.8.

As illustrated in exercise 8.11, it is very simple to compute factorials of positive integers. The factorial $N!$ of an integer N is, by definition, an integer. But factorials grow so quickly with N that, beyond $N=20$, $N!$ can no longer be represented as an integer. Of course, the spreadsheet can still represents factorials of numbers larger than 20 in scientific notation, to 15 significant digits, which is usually more than enough.

Instructions for exercise 8.12

1 Open a new spreadsheet.

2 In cells A1 and B1 enter labels N and N! respectively.

3 In cell A4 deposit the number 1, in cell A5 the instruction = A4 + 1, and copy this instruction down to row 203.

4 In cell B4 deposit the value 1, in cell B5 the instruction = A5*B4, and also copy this instruction down to row 203.

Take a look at row 174. From there on, $N!$ is no longer computed, because it is just too big to be represented by the software. The spreadsheet fails us because 172! is of the order of 10^{309}, which exceeds the ability of the spreadsheet to represent the number at all: the limit lies at $2^{1024} - 1$, or just above 1.797×10^{308}). The spreadsheet uses the error message #NUM! to warn us that it cannot represent such a large number, a situation called *overflow*. (The same error message is also used in case of *underflow*, when a number is smaller than 2^{-1024}.) Here, then, is a rather common problem: many calculations involving factorials require $N! / \{(N-n)! \, n!\}$, which is a much smaller number than $N!$. But how can we evaluate $N! / \{(N-n)!\, n!\}$ when $N!$ itself is beyond the reach of the spreadsheet? Or when both $N!$ and $(N-n)!\, n!$ exceed the capacity of the spreadsheet?

While we will seldom encounter numbers greater that 1.797×10^{308} in our *measurements*, a number such as $180!$ in a *calculation* is not at all uncommon. We therefore investigate alternative ways to compute $N!$.

The standard approximation for $N!$ is the Stirling formula,

$$N! = \sqrt{2\pi}\, N^{N+\frac{1}{2}} \exp\left[-N + \frac{1}{12N} - \frac{1}{360N^3} + \frac{1}{1260N^5} - \frac{1}{1680N^7} + \cdots \right] \quad (8.12\text{-}1)$$

which is often truncated after the first term in the exponential, i.e., to

$$N! \approx \sqrt{2\pi}\, N^{N+\frac{1}{2}} \exp[-N] \qquad (8.12\text{-}2)$$

5 In cells C4:C203 compute N! using (8.12-2).

The result is not encouraging: now the calculation already stops at $N = 143$ (rather than at $N = 171$) because of numerical overflow in the term $N^{N+\frac{1}{2}}$ of the Stirling formula. While we might want to improve on the numerical accuracy of $N!$ by using additional terms in the exponent of (8.12-1), we will still be limited to $N \leq 142$. Obviously, the Stirling approximation runs into overflow problems before the exact formula does.

Fortunately there is a simpler method that does not let us down so quickly. It merely requires that we go back to the definition of $N!$ as a product of terms, and therefore compute $\ln(N!)$ as

$$\ln(N!) = \ln(N) + \ln(N-1) + \ln(N-2) + \ln(N-3) + \cdots + \ln(2) + \ln(1)$$
$$(8.12\text{-}3)$$

6 In cell D4 deposit the instruction $= \ln(\text{A4})$.

7 In cell D5 deposit $= \text{D4} + \ln(\text{A5})$, and copy this instruction all the way down to row 203.

Notice how easy it is to use (8.11-3) on the spreadsheet. Similarly, when we want to calculate the binomial coefficient $N!/\{(N-n)!\, n!\}$, we might as well compute its logarithm first, because even when $N!$ is too large to be calculated on the spreadsheet, the coefficient $N!/\{(N-n)!\, n!\}$ may not be so large.

To demonstrate that this approach indeed works, pick a value for n, say $n = 24$, and calculate the binomial coefficient $N! / \{(N-n)!\, n!\}$.

8 In cell E27 deposit the instruction $= \text{EXP}(\text{D27} - \text{D3} - \$\text{D}\$27)$, and copy it down all the way to row 203.

We see that the binomial coefficient $N!/\{(N-24)!\,24!\}$ can be computed as far as the column will go in Excel, certainly far beyond where $N!$ itself would be too large for the spreadsheet. This, then, is the approach we have used in section 6.6.

8.13 Summary

In this chapter we have gathered some of the standard methods of numerical analysis, especially those that are useful for computations on experimental data, and for numerical simulations of differential equations. A short chapter obviously cannot do justice to them, and for more details the interested reader might want to consult books on numerical analysis and chemometrics.

In the first sections we have illustrated a number of mathematical methods that are often used in chemical analysis, or are commonly implemented in chemical instrumentation: signal averaging and lock-in amplification (synchronous detection); data smoothing, peak fitting, interpolation, and root finding; non-linear least-squares fitting; integration, and differentiation. Several of these methods were already introduced in earlier chapters, and therefore need only a brief discussion. By now it will be obvious that some methods, such as Savitzky–Golay smoothing and differentiation, lend themselves admirably to even the simplest of spreadsheets, while other techniques, such as Fourier transformation and general least-squares fitting, require a *macro*. Excel not only includes a number of useful macros, but also makes it relatively easy to insert one's own, as will be shown in chapter 10. In chapter 9 we will encounter an intermediate situation, where the spreadsheet can be augmented by user-defined *functions*. These act as small-scale macros, because they operate on the contents of a single spreadsheet cell.

In section 8.11 we reviewed some of the rules of matrix algebra, which will become more and more important as analytical methods become more sophisticated. We already encountered some applications of matrix algebra in sections 6.2 and 6.3, and there are more in chapter 10, since least-squares methods also rely on matrix algebra, although this may not be visible to the user. In general, matrix algebra is the method of choice whenever multiple simultaneous equations must be solved, as in almost all problems of mixture analysis. Moreover, many problems in physics and chemistry involve the solution of simultaneous equations, which is why matrix methods are ubiquitous in modern science. Spreadsheets make their use fast and convenient, by automating their manipulations. All you need to do is understand what matrices are, and how to use them to your advantage.

There is an analogy here with the use of a pocket calculator, although the dimensionality is different. Once you understand what the square root of a number is, or its logarithm, you can relegate the actual computation to your pocket calculator, because it has built-in programs to calculate such functions of *single* numbers. In a similar sense, the two-dimensional spreadsheet allows you to manipulate entire *blocks* of data, using mathematics appropriate for such arrays, namely matrix algebra.

Finally, in section 8.12, we focused briefly on some of the consequence of representing mathematical quantities in a finite number of digits, and saw how we can often get around some of the resulting difficulties.

CHAPTER 9

NUMERICAL SIMULATION OF CHEMICAL KINETICS

9.1 Introduction

In most of the examples used in the preceding chapters, a closed-form solution was available to describe the phenomenon considered, and the spreadsheet was used either to visualize the resulting behavior, or to manipulate and analyze consequent, experimental data. However, more often than we care to admit, we can formulate a problem in mathematical terms, but cannot find a corresponding, mathematical solution. That usually leaves us with two options: either simplify the mathematical model until it becomes solvable, or use a computer to find a *numerical* solution. Neither option is ideal. In simplifying a mathematical model, there is always the risk that one or more important aspects of the problem will be missed. On the other hand, numerical solutions apply only for the specific parameter values used, and the results may therefore be difficult to generalize. Still, more and more problems in science and technology are solved by numerical simulation, including the design of camera lenses, cars, and jet airplanes. In this short chapter we will use the rate expressions of chemical kinetics to illustrate how one can use spreadsheets for the numerical simulation of differential equations.

Much of chemistry involves kinetics. The equilibrium behavior we discussed in chapters 4 and 5 can only be observed because the underlying chemical reactions are sufficiently fast. In other cases, however, the reactions involved may be relatively slow, and their rates can be observed and measured. In practice, large numbers of chemical analyses are based on kinetic measurements, especially the many routine analyses performed by multichannel autoanalyzers in clinical and industrial laboratories. Such kinetic methods of analysis often derive their selectivity from the use of enzymes or other catalysts. They can be relatively fast because one need not wait for the establishment of equilibrium.

For the simplest of kinetic schemes, such as for unidirectional reactions, straightforward mathematical expressions exist that describe the rates at which the reagents are consumed in the process. In order to illustrate how digital simulation works, we will first tackle such simple examples, since they allow us to calibrate the method, and to discuss its constraints and limitations.

Among the various possible approaches we will illustrate the explicit method in section 9.2, and the implicit method in section 9.3. The explicit method is conceptually simpler, and is easier to implement; the implicit method is more accurate. Both methods are, therefore, useful: the former if one requires only a rough idea of the kinetics, the latter for more quantitative considerations.

9.2 The explicit method

9.2a First-order kinetics

We start with the first-order reaction

$$A \xrightarrow{k} \qquad\qquad\qquad (9.2\text{-}1)$$

In what follows we will assume that there are a sufficiently large number of reacting particles to make the kinetics a deterministic process. We can then describe the course of reaction (9.2-1) by the simple first-order differential equation

$$da/dt = -ka \qquad\qquad\qquad (9.2\text{-}2)$$

where a denotes the concentration of the chemical species A. The mathematical solution of (9.2-2) consistent with the initial condition $a_{t=0} = a_0$ is

$$a = a_0 \exp[-kt] \qquad\qquad\qquad (9.2\text{-}3)$$

where $t = 0$ is usually defined as the time at the beginning of the experiment, or at the initiation of the reaction.

We now consider (9.2-2) while pretending not to know its closed-form mathematical solution, (9.2-3). How can we simulate the resulting kinetics without solving the differential equation? Recalling that the differential quotient is really just the limit of a difference equation,

$$\frac{da}{dt} = \lim_{\Delta t \to 0} \frac{\Delta a}{\Delta t} \qquad\qquad\qquad (9.2\text{-}4)$$

where $\Delta a = a_{t+\Delta t} - a_t$, we replace the *differential* equation (9.2-2) by the corresponding *difference* equation,

$$da/dt \approx \Delta a/\Delta t \qquad\qquad\qquad (9.2\text{-}5)$$

which should be a close approximation as long as the increment Δt is sufficiently small, a constraint we will define more precisely below. We therefore rewrite (9.2-2) as

$$\Delta a / \Delta t = -ka \qquad (9.2\text{-}6)$$

so that

$$\Delta a = -ka\Delta t \qquad (9.2\text{-}7)$$

$$a_1 = a_0 + \Delta a = a_0 - ka_0\Delta t = a_0(1 - k\Delta t) \qquad \text{at } t = \Delta t \qquad (9.2\text{-}8)$$

$$a_2 = a_1 - ka_1\Delta t = a_1(1 - k\Delta t) = a_0(1 - k\Delta t)^2 \qquad \text{at } t = 2\Delta t \qquad (9.2\text{-}9)$$

$$a_n = a_{n-1} - ka_{n-1}\Delta t = a_{n-1}(1 - k\Delta t) = a_0(1 - k\Delta t)^n \quad \text{at } t = n\Delta t \qquad (9.2\text{-}10)$$

In other words, we start with $a = a_0$ at $t = 0$, which sets the *initial condition*. We calculate $a_1 = a_0 + \Delta a = a_0 - ka_0\Delta t = a_0(1 - k\Delta t)$ at $t = \Delta t$, then continue by computing $a_2 = a_1 + \Delta a = a_1(1 - k\Delta t)$ at $t = 2\Delta t$, and so on, thereby simulating the entire time course of the concentration a, step-by-step, one interval Δt at a time, simply by multiplying the preceding concentration by $(1 - k\Delta t)$. This is the so-called explicit or Euler method.

Instructions for exercise 9.2.1

1 Open a new spreadsheet.

2 Reserve the top 8 rows for thumbnail sketches.

3 In cell A10 place the label a0 =, in C10 place the label k =, and in E10 the label Δt =. To make a Δ, type a D, highlight it, and select the Symbol font.

4 In cells B10, D10, and F10 enter numerical values for a_0, k, and the time increment Δt, such as 1, 1, and 0.1 respectively.

5 In row 12 deposit the labels time, a simul, a exact, and diff.

6 In column A compute time from 0 to 10 with increments Δt.

7 In B14 place the value of a_0 as = \$B\$10.

8 In B15 calculate the next value of a_1 according to (9.2-8) as = B14*(1–\$D\$10*\$F\$10), and copy this down the entire column (i.e., to the same length as column A).

9 In column C, starting with C14, calculate a according to (9.2-3).

10 In column D compute the difference $a_{exact} - a_{simul}$.

11 In A1:C8 plot a_{simul} and a_{exact} versus time t, and in D1:F8 display a plot of diff = $a_{exact} - a_{simul}$ versus t.

12 The top of your spreadsheet might now look like Fig. 9.2-1.

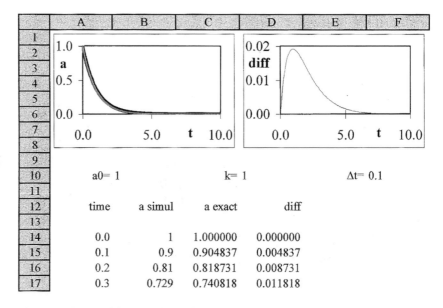

Fig. 9.2-1: The top of the spreadsheet for exercise 9.2-1.

9.2b Numerical accuracy

How good is this simulation? Since we have the exact result available, a comparison would appear to be straightforward. The difference between a_{exact} and a_{simul} follows from the series expansions of (9.2-3) and (9.2-10). We have

$$a_{\text{exact}} = a_0\, e^{-kt} = a_0\, \{1 - (k\,n\,\Delta t) + (k\,n\,\Delta t)^2/2 - (k\,n\,\Delta t)^3/6 + (k\,n\,\Delta t)^4/24 - \cdots\}$$

$$a_{\text{simul}} = a_0\, (1 - k\,\Delta t)^n = a_0\, \{1 - n\,(k\,\Delta t) + (k\,\Delta t)^2 n(n-1)/2$$
$$- (k\,\Delta t)^3 n(n-1)(n-2)/6 + (k\,\Delta t)^4 n(n-1)(n-2)(n-3)/24 - \cdots\}$$

so that

$$a_{\text{exact}} - a_{\text{simul}} = a_0\, \{(k\,\Delta t)^2\, n/2 - (k\,\Delta t)^3\, n(3n-2)/6 + \cdots\} \qquad (9.2\text{-}11)$$

which suggests that, as long as $k\,\Delta t \ll 1$, the error is of the order of $(k\,\Delta t)^2$. However, this is a misleading result, because the above comparison is made at a fixed value of n. When we change Δt at a given k, we must also change the value of n in order to cover the same total time interval τ. In other words, the product $n \times \Delta t = \tau$ should remain constant. We therefore write

$$a_{\text{simul}} = a_0\, (1 - k\,\Delta t)^n = a_0\, (1 - k\tau/n)^n$$

$$= a_0\, \{1 - n\,(k\tau/n) + (k\tau/n)^2 n(n-1)/2 - (k\tau/n)^3 n(n-1)(n-2)/6$$
$$+ (k\tau/n)^4 n(n-1)(n-2)(n-3)/24 - \cdots\}$$

$$= a_0\, \{1 - (k\tau) + (k\tau)^2/2 - (k\tau)^3/6 + (k\tau)^4/24 - \cdots$$
$$- (1/n)\, [(k\tau)^2/2 - (k\tau)^3/2 + (k\tau)^4/4 - \cdots]$$
$$- (1/n^2)\, [(k\tau)^3/3 - 11(k\tau)^4/24 + \cdots] - (1/n^3)\, [(k\tau)^4/4 - \cdots] - \cdots\}$$

$$= a_0\, e^{-kt} - (a_0/n)\, [(k\tau)^2/2 - (k\tau)^3/2 + (k\tau)^4/4 - \cdots]$$
$$- (a_0/n^2)\, [(k\tau)^3/3 - 11(k\tau)^4/24 + \cdots]$$
$$- (a_0/n^3)\, [(k\tau)^4/4 - \cdots] - \cdots$$

so that

$$a_{\text{exact}} - a_{\text{simul}} = (a_0/n)\, [(k\tau)^2/2 - (k\tau)^3/2 + (k\tau)^4/4 - \cdots]$$
$$+ (a_0/n^2)\, [(k\tau)^3/3 - 11(k\tau)^4/24 + \cdots]$$
$$+ (a_0/n^3)\, [(k\tau)^4/4 - \cdots] + \cdots$$

$$= (a_0\, \Delta t/\tau)\, [(k\tau)^2/2 - (k\tau)^3/2 + (k\tau)^4/4 - \cdots]$$
$$+ (a_0\, \Delta t^2/\tau^2)\, [(k\tau)^3/3 - 11(k\tau)^4/24 + \cdots]$$
$$+ (a_0\, \Delta t^3/\tau^3)\, [(k\tau)^4/4 - \cdots] + \cdots \tag{9.2-12}$$

which has a leading term in Δt rather than in $(\Delta t)^2$.

We now return to the deferred question: how small is small enough for the time increments Δt? We find the answer in the series expansion: the relevant combination is always $(k\Delta t)$, in other words, Δt scales as $1/k$. As we have just seen, increments of $1/(10\ k)$ yield errors of the order of a few percent. To be more precise, for $k\, \Delta t = 0.1$, the maximum error of almost $0.02\ a_0$ occurs at $kn\Delta t \approx 1$, where $a \approx 0.36\ a_0$, so that the maximum relative error is about 6%. This may be good enough for many practical purposes.

If we need a more accurate result, we must use smaller increments Δt. In order to reduce the relative error by a factor f, Δt must be reduced by the same factor, which implies that we must simulate f times as many data. That approach, readily implemented in higher-level languages, unfortunately is rather limited on a spreadsheet, because the use of long columns is not only unwieldy but also slows down the computation. Moreover, Excel cannot plot very large data arrays.

Fortunately there is a simple solution, at least in the present example. It is clear that we need more resolution in the computation, but not necessarily in the plots. Equation (9.2-10) suggests how we can leapfrog the computation, by computing only the nth terms. By replacing the instruction for $a_1 = a_0\, (1 - k\, \Delta t)$ by, say, $a_{10} = a_0\, (1 - k\, \Delta t/10)^{10}$, that for $a_2 = a_0\, (1 - k\, \Delta t)^2$ by $a_{20} = a_0\, (1 - k\, \Delta t/10)^{20}$, etc., we achieve the same computational result as we would have obtained by lengthening the column by a factor of 10 but plotting only data from every tenth row, because we effectively compute a at each step $\Delta t = 0.1$ as if we were using ten steps of size 0.01.

Similarly, replacing a_1 by $a_{1000} = a_0\, (1 - k\Delta t/1000)^{1000}$, a_2 by $a_{2000} = a_0\, (1 - k\Delta t/1000)^{2000} = a_{1000}\, (1 - k\Delta t/1000)^{1000}$, and so on, will simulate the effect of a step size $\Delta t/1000$ but without increasing the actual column length.

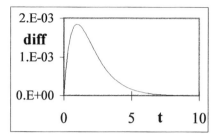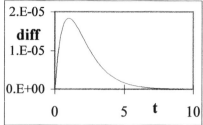

Fig. 9.2-2: By combining the effect of ten (left) or one thousand (right) iteration steps in each row, the computational error can be reduced tenfold or one-thousand-fold respectively. In the latter case, experimental errors will most likely exceed any simulation errors.

13 Make two new columns, one each for a simul and diff.

14 In the new column for a simul calculate a_1 as $a_0 (1 - k\Delta t/10)^{10}$, a_2 as $a_1 (1 - k\Delta t/10)^{10}$, etc.

15 In the new column labeled diff, calculate the difference between a_{exact} and the newly calculated a_{simul}.

16 Plot the difference, and compare it with the values in column D.

17 Repeat the above using for a_1 the relation $a_0 (1 - k\Delta t/1000)^{1000}$, for a_2 the expression $a_1 (1 - k\Delta t/1000)^{1000}$, and so on. Incidentally, you need not worry about such large powers: they will not slow down the computer, because it uses logarithms to calculate the results.

18 Again compute and plot the difference $a_{exact} - a_{simul}$.

19 Your results might look like those in Fig. 9.2-2.

The above example shows that we can use the explicit method to simulate the kinetic transient to any required accuracy by selecting small enough time increments.

9.2c Dimerization kinetics

Next we consider some simple higher-order kinetic schemes, specifically the second- and third-order unidirectional reactions 2A → and 3A → for which, again, closed-form solutions are available to validate our approach. (Since we omit any back reactions, we need not specify the products.) First we consider the unidirectional dimerization reaction

$$A + A \xrightarrow{k} \qquad (9.2\text{-}13)$$

with the corresponding rate equation

$$\mathrm{d}a/\mathrm{d}t = -\, k\, a^2 \qquad\qquad\qquad\qquad\qquad (9.2\text{-}14)$$

Proceeding as in (9.2-5) through (9.2-10), the explicit method now yields

$$\Delta a\,/\,\Delta t = -\, k\, a^2 \qquad\qquad\qquad\qquad\qquad (9.2\text{-}15)$$

$$a_1 = a_0 + \Delta a = a_0 - a_0^2 k\,\Delta t = a_0\,(1 - a_0\, k\,\Delta t) \qquad\qquad (9.2\text{-}16)$$

$$a_2 = a_1\,(1 - a_1\, k\,\Delta t) \qquad\qquad\qquad\qquad\qquad (9.2\text{-}17)$$

$$a_n = a_{n-1}\,(1 - k\, a_{n-1}\,\Delta t) \qquad\qquad\qquad\qquad (9.2\text{-}18)$$

20 Expand the spreadsheet, or open a new, similarly organized one.

21 Compute a_{simul} with the recursive formula $a_n = a_{n-1}\,(1 - k\,a_{n-1}\,\Delta t)$.

22 For a_{exact} use $a(t) = a_0\,/\,(1 + k\,a_0\,t)$.

23 Plot the simulated concentration a_{simul} and the exact result a_{exact} vs. time t.

24 Also plot the algorithmic error diff $= a_{\mathrm{exact}} - a_{\mathrm{simul}}$ against time t.

9.2d A user-defined function to make the spreadsheet more efficient

In the above example, we do not have an explicit expression for a_n in terms of a_0 as in (9.2-10) but, rather, a recursion formula, (9.2-18). Therefore we cannot use the convenient trick that led to the improved numerical accuracy in Fig. 9.2-2. If a higher accuracy is needed, we can (1) use smaller time increments Δt and correspondingly extend the column length, (2) perform the computation off-screen, or (3) look for a more efficient algorithm. In section 9.3 we will explore the third option, the use of a more efficient algorithm. Here we will briefly consider the second option, in which we use a so-called **user-defined function** to do our bidding.

What is needed is a function that will take several **arguments**, in our case the values of a, k, and Δt, apply (9.2-18) n times (where n is a positive integer), and then deposit the result. We therefore create a function to do this; unfortunately, some of the mechanics of entering such a function are different (and slightly more complicated) for Excel 97 and subsequent versions. Below we will describe the two different methods separately. Note that only the method to enter and store the function is different; but that this affects neither the function itself nor its operation. Also notice that these are the same instructions as needed for creating a macro, as described in the next chapter.

If you use either Excel 5 or Excel 95, use the following steps. With the spreadsheet open, select Insert $\rightarrow$ Macro $\rightarrow$ Module. The monitor screen will now show an empty sheet, on which you can now type the function. When you are done typing, click on the tab for the spreadsheet and, bingo, you are back in the spreadsheet. From now on you can go back and forth between the spreadsheet and the function by clicking on the appropriate tab. You need this only when you want to modify the function, or add another.

If you use Excel 97, 98, or 2000, with the spreadsheet open, type Alt + F11 or select Tools → Macro → Visual Basic Editor. Then, select Insert → Module. The monitor screen will now show an empty sheet, on which you can now type the function. When you are done typing, type Alt + F11 to get back to the spreadsheet. From then on, whenever you need to do so, you can go back and forth between the spreadsheet and the function with Alt + F11.

On the blank module sheet, type the text of the function:

```
Function Smallstep(a, k, delt, n)
For i = 1 To n
    a = a * (1 - a * k * delt / n)
Next i
Smallstep = a
End Function
```

Before we use it, we will first explain the working of this function, line by line. The first line defines the function by its name, Smallstep, and specifies what input data the function requires. The last line specifies the end of the function. In between, the value of a is recomputed n times, using a For ... next loop in lines 2 through 4, and the result is specified in line 5. These instructions are written in Visual BASIC, the language that Excel uses for its functions and macros. (Visual BASIC is a modern form of compiled BASIC, with additions to make it suitable for spreadsheet use. Section 10.12 will review some features of Visual BASIC.)

Note that the third line of the function is equivalent to (9.2-18) except that Δt has been replaced by $\Delta t/n$, which is then repeated n times in the For ... next loop. And if you wonder where the labels n and $n-1$ on a have gone, the equal sign in BASIC actually means an **assignment**, i.e., the computer first evaluates the value of a*(1-a*k*delt/n), then assigns that value to the variable a on the left-hand side of the equal sign.

Now that we have defined this function, how do we use it? Just like any other function, such as SIN or LOG. For the sake of this example, let us assume that the calculation of a on the spreadsheet is performed in column E, starting with cell E14. Furthermore we will assume that rows 1 through 8 have been reserved for graphs, that cells A10, C10, and E10 contain labels for a_0, k, and Δt respectively, and that B10, D10 and F10 contain the corresponding numerical values. Finally, let row 12 contain the column headings: cell E12 the label for n, and cell F12 its value.

In cell E14 we would then use the instruction = $B$10 to refer to the initial concentration a_0, in cell E15 we would use = Smallstep(E14,$D$10,$F$10, $F$12), and we would copy this instruction down till, say, cell E115. That's it. The only difference with what we did earlier, at the end of section 9.2b, is that execution of this function may be quite slow when we select a large value for n, because the computer must now perform n calculations for every row of

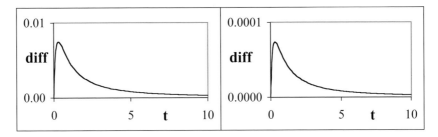

Fig. 9.2-3: By combining ten (left) or one thousand (right) iteration steps in each row through the function Smallstep, we can reduce the computational error in the simulation of dimerization kinetics tenfold or thousand-fold respectively, all without increasing the column length of 101 rows.

column E. The spreadsheet may then appear quite sluggish, or even non-responsive. In that case, look at the bottom left corner of the spreadsheet, where you may see the message Calculating cells: followed by a percentage.

Figure 9.2-3 illustrates some results obtained with this approach: the numerical error is reduced tenfold for $n = 10$, thousand-fold for $n = 1000$.

9.2e Trimerization kinetics

For a unidirectional trimerization reaction

$$3A \xrightarrow{k}$$ (9.2-19)

the rate equation is

$$da/dt = -k_1 a^3$$ (9.2-20)

so that

$$\Delta a / \Delta t = -k a^3$$ (9.2-21)

$$a_1 = a_0 + \Delta a = a_0 - a_0^3 k \Delta t = a_0 (1 - a_0^2 k \Delta t)$$ (9.2-22)

$$a_2 = a_1 (1 - a_1^2 k \Delta t)$$ (9.2-23)

$$a_n = a_{n-1} \{1 - (a_{n-1})^2 k \Delta t\}$$ (9.2-24)

Use the spreadsheet to compare the results of the digital simulation with the exact solution $a = a_0 / \sqrt{1 + 2a_0^2 k t}$.

25 Modify or expand the spreadsheet, or open a new one, similarly organized.

26 Compute a_{simul} from (9.2-24).

27 For a_{exact} use $a(t) = a_0 / \sqrt{(1 + 2 a_0^2 k t)}$.

28 Plot a_{simul} and a_{exact} vs. time t and, in a different graph, diff = $a_{exact} - a_{simul}$ vs. t.

29 Compare your result with Fig. 9.2-4.

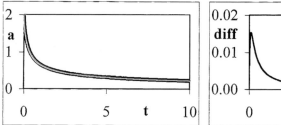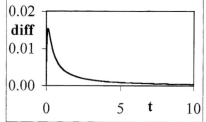

Fig. 9.2-4: The simulation with the explicit method of the unidirectional trimerization reaction (9.2-19) with $a_0 = 2$ and $k = 1$.

Again we can again use a function such as Smallstep, which this time should have as its third line the instruction a = a*(1-a*a*k*delt/n). However, in section 9.3 we will see that it is more effective first to use a more efficient algorithm.

9.2f Monomer–dimer kinetics

We now consider the somewhat more complex kinetic scheme of the establishment of a monomer/dimer equilibrium

$$2A \underset{k'}{\overset{k}{\rightleftharpoons}} B \qquad (9.2\text{-}25)$$

with the associated differential equations

$$da/dt = -ka^2 + 2k'b \qquad (9.2\text{-}26)$$

$$db/dt = +\tfrac{1}{2}ka^2 - k'b \qquad (9.2\text{-}27)$$

where a and b represent the concentrations of the monomer A and the dimer B respectively. Note that these two coupled equations in a and b must obey the mass balance requirement $(a + 2b) = $ constant or $d(a + 2b)/dt = 0$, which leads to the coefficients in (9.2-26) and (9.2-27). Other coefficient combinations are possible as well. For example, we could have selected $da/dt = -2ka^2 + 2k'b$ and $db/dt = +ka^2 - k'b$ instead. The numerical values of the rate constants will of course depend on the convention used, which therefore should be specified.

By considering the mass balance at the beginning of the simulation, $t = 0$, we find that $a + 2b = a_0 + 2b_0$ so that $2b = a_0 + 2b_0 - a$, which can be used to eliminate b from (9.2-26). We then obtain

$$da/dt = -ka^2 - k'a + k'(a_0 + 2b_0) = -ka^2 - ka + k'' \qquad (9.2\text{-}28)$$

where $k'' = k'(a_0 + 2b_0)$.

Conversion of (9.2-28) into the corresponding explicit difference equation yields

$$\Delta a/\Delta t = -ka^2 - k'a + k'' \qquad (9.2\text{-}29)$$

so that

$$a_1 = a_0 + \Delta a = a_0 + (-ka_0^2 - k'\, a_0 + k'')\, \Delta t \qquad (9.2\text{-}30)$$

$$a_2 = a_1 + \Delta a = a_1 + (-ka_1^2 - k'\, a_1 + k'')\, \Delta t \qquad (9.2\text{-}31)$$

$$a_n = a_{n-1} + \Delta a = a_{n-1} + (-k(a_{n-1})^2 - k'\, a_{n-1} + k'')\, \Delta t \qquad (9.2\text{-}32)$$

which will gradually approach equilibrium. In principle, equilibrium is never reached; in practice, it is obtained once all concentration changes have become imperceptibly small. Likewise, in a simulation, we will consider that equilibrium has been reached once the changes in Δa and Δb are smaller than whatever numerical criterion we set.

Even in this case a closed-form mathematical solution exists, so that we can calibrate the accuracy of our simulation. The mathematical solution is

$$a(t) = \frac{-(k'+q)(2ka_0 + k' - q) + (k'-q)(2ka_0 + k' + q)e^{qt}}{2k[(2ka_0 + k' - q) - (2ka_0 + k' + q)e^{qt}]} \qquad (9.2\text{-}33)$$

$$q = \sqrt{(k')^2 + 4k\, k''} \qquad (9.2\text{-}34)$$

Instructions for exercise 9.2-2

1 Open a new spreadsheet.

2 Place labels and numerical values in rows 9 and 10 for a_0, b_0, k, k', and delt.

3 In row 12 deposit labels for time t, for a_{simul}, and for b_{simul}.

4 In column A calculate time t, starting with $t = 0$ and with increments delt.

5 In columns B and C compute values for a_{simul} and b_{simul} respectively, based on (9.2-30) through (9.2-32).

6 Plot a_{simul} and b_{simul} versus t. Figures 9.2-5 and 9.2-6 show such graphs.

7 Compare the result of your simulation with the mathematical solution.

9.2g Polymerization kinetics

Consider the formation of a so-called "living polymer", in which a polymer is built or dissolved through a series of reversible chemical reactions. A notorious example of such a polymer is hemoglobin S, the mutant form of hemoglobin that causes sickle-cell anemia. The term "living" has nothing to do with its occurrence in living tissue, but merely identifies the fact that, depending on the experimental conditions, the polymer can either grow or shrink. And even when its length remains constant, the polymer may grow at

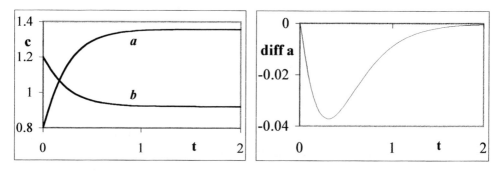

Fig. 9.2-5: Left: explicit simulation of the establishment of the monomer–dimer equilibrium (9.2-25) with $a_0 = 0.8$, $b_0 = 1.2$, $k = 1$, and $k' = 1$. Right: the simulation error $a_{simul} - a_{exact}$.

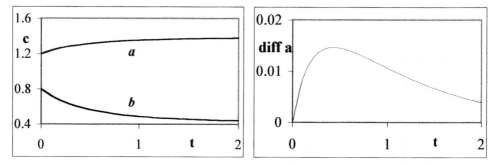

Fig. 9.2-6: Left: explicit simulation of the establishment of the monomer–dimer equilibrium (9.2-25) with $a_0 = 0.8$, $b_0 = 1.2$, $k = 1.5$, and $k' = 0.1$. Right: the simulation error $a_{simul} - a_{exact}$.

one end while dissolving at the other, a process called treadmilling. Specifically, we will use the sequence

$$A_1 + A_1 \underset{k_{11}'}{\overset{k_{11}}{\rightleftharpoons}} A_2 \tag{9.2-35}$$

$$A_1 + A_2 \underset{k_{12}'}{\overset{k_{12}}{\rightleftharpoons}} A_3 \tag{9.2-36}$$

$$A_1 + A_3 \underset{k_{13}'}{\overset{k_{13}}{\rightleftharpoons}} A_4 \tag{9.2-37}$$

$$A_2 + A_2 \underset{k_{22}'}{\overset{k_{22}}{\rightleftharpoons}} A_4 \tag{9.2-38}$$

which is sufficient to illustrate the approach. With this reaction scheme we associate the differential equations

$$da_1/dt = -k_{11} a_1{}^2 - k_{12} a_1 a_2 - k_{13} a_1 a_3 + 2 k_{11}' a_2 + k_{12}' a_3 + k_{13}' a_4 \quad (9.2\text{-}39)$$

$$da_2/dt = -k_{11}' a_2 - k_{12} a_1 a_2 - k_{22} a_2{}^2 + \tfrac{1}{2} k_{11} a_1{}^2 + k_{12}' a_3 + 2 k_{22}' a_4 \quad (9.2\text{-}40)$$

$$da_3/dt = -k_{12}' a_3 - k_{13} a_1 a_3 + k_{12} a_1 a_2 + k_{13}' a_4 \quad (9.2\text{-}41)$$

$$da_4/dt = -k_{13}' a_4 - k_{22}' a_4 + k_{13} a_1 a_3 + \tfrac{1}{2} k_{22} a_2{}^2 \quad (9.2\text{-}42)$$

Proceeding as before, we replace the terms da_i/dt by $\Delta a_i/\Delta t$ in order to obtain explicit expressions for the concentration changes Δa_i, and with these we compute the new concentrations $a_i + \Delta a_i$.

Instructions for exercise 9.2-3

1 Open a new spreadsheet.

2 Reserve the area A1:H12 for the graph (which will contain four species).

3 In rows 13 through 16 place labels and initial values for the concentrations a_1, a_2, a_3, a_4, the rate constants k_{11}, k_{12}, k_{13}, k_{22}, k_{11}', k_{12}', k_{13}', k_{22}', and Δt.

4 In row 18 deposit labels for time t and for the concentrations a_1 through a_4.

5 In column A calculate t, starting with $t = 0$ in cell A20, and using the increment Δt.

6 Starting in row 20 of columns B through E, compute a_1, a_2, a_3, and a_4 based on (9.2-39) through (9.2-42) after substitution of $\Delta a /\Delta t$ for da/dt. For example, the value of a_1 in cell B21 would be calculated as $a_1 + \Delta a_1 = a_1 + (-k_{11} a_1{}^2 - k_{12} a_1 a_2 - k_{13} a_1 a_3 + 2 k_{11}' a_2 + k_{12}' a_3 + k_{13}' a_4) \Delta t$, where a_1 through a_4 are found in B20 through E20 respectively.

7 In column F check that the concentrations calculated obey the mass balance relation $a_1 + 2 a_2 + 3 a_3 + 4 a_4 = $ constant.

8 Plot the various concentrations as a function of time t. Figure 9.2-7 illustrates the top of such a spreadsheet.

The above illustrates that the explicit method provides a simple and readily implemented approach to simulate the course of reaction kinetics, even for rather complicated reaction mechanisms, as long as the initial concentrations of the reactants and products, and the rate constants, are known. The problem need not have a known, closed-form mathematical solution. Non-linear relations are no impediment to the simulation, because the equations are linearized. Such linearization is an acceptable approximation as long as the time increments Δt are sufficiently small. In section 9.2d we have seen how we can exploit a user-defined function to make the increments Δt smaller, by moving some of the computations off the spreadsheet, without having to change either the time range covered or the column length used. In principle, the same method can of course be

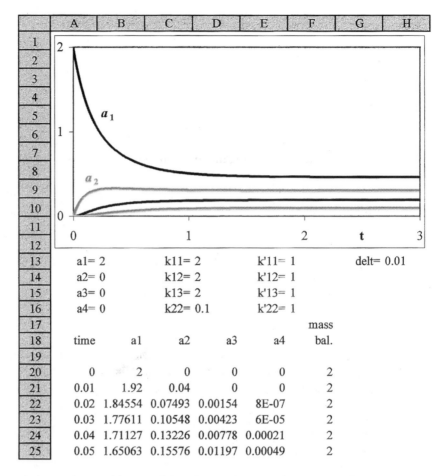

	A	B	C	D	E	F	G	H
13	a1= 2		k11= 2		k'11= 1		delt= 0.01	
14	a2= 0		k12= 2		k'12= 1			
15	a3= 0		k13= 2		k'13= 1			
16	a4= 0		k22= 0.1		k'22= 1			
17						mass		
18	time	a1	a2	a3	a4	bal.		
19								
20	0	2	0	0	0	2		
21	0.01	1.92	0.04	0	0	2		
22	0.02	1.84554	0.07493	0.00154	8E-07	2		
23	0.03	1.77611	0.10548	0.00423	6E-05	2		
24	0.04	1.71127	0.13226	0.00778	0.00021	2		
25	0.05	1.65063	0.15576	0.01197	0.00049	2		

Fig. 9.2-7: The top of the spreadsheet for exercise 9.2-3. The curves are, from top to bottom, for a_1, a_2, a_3, and a_4.

used here, by using different functions for the various species. However, this can slow down the computation considerably. In section 9.3 we will encounter more efficient ways to achieve a similar result.

In kinetic studies one typically encounters the opposite problem: how to extract a mechanism and the corresponding rate parameters from the experimental data. Numerical simulation cannot do that directly. What it can do is test whether an assumed mechanism plus assumed rate parameters will generate concentration profiles that agree with the experimental observations.

9.3

Implicit numerical simulation

The explicit method described in the previous section is very straight-forward, since it only uses already available data. But that is also its Achilles heel, because it can introduce a *systematic bias* as it uses, in each step Δt, the concentration(s) computed for the end of the previous time increment. While the resulting errors can be made vanishingly small in principle, by sufficiently reducing the time interval Δt used, it is usually more effective to use a more efficient algorithm.

When the concentration of a reagent is a continuously decreasing function of time, and the steps Δt are not vanishingly small, the systematic bias will lead to a small but consistent overestimate of that concentration. Over many steps, even a rather small overestimate can accumulate to generate a sizable composite error. Similarly, when the concentration of a product increases, its concentration change in every interval Δt will be computed on the basis of its previous concentration, which again will lead to a small but systematic (and often accumulating) error.

An improved simulation might therefore be obtained by using an estimate of the *average* concentration value during the period Δt. This is done in the implicit method, which considers the previous data point as well as the next, yet to be determined point in the computation. In fact, there are many different implicit methods. Here we only illustrate the simplest of them, which assumes that all variables change linearly over a sufficiently small interval Δt.

Let the dependent variable a at $t = 0$ have the value a_0, and let its value at $t = \Delta t$ be $a_1 = a_0 + \Delta a$. Assuming that a varies in a linear fashion with t in the small interval Δt, we now use the *average* value $(a_0 + a_1) / 2 = (2a_0 + \Delta a) / 2 = a_0 + \frac{1}{2} \Delta a$ as an improved estimate of a during the interval Δt, and therefore substitute it in the right-hand side of the differential equation (9.2.2). This should be more realistic than the implication of (9.2-6) that, over the interval Δt, the variable a retains the value it had at the beginning of that interval.

9.3a First-order kinetics

For the first-order unidirectional reaction (9.2-1) we therefore write the difference equation as

$$\Delta a / \Delta t = - k \, (a + \frac{1}{2} \Delta a) \tag{9.3-1}$$

Equation (9.3-1) contains terms in Δa on both sides of the equal sign. We solve it for Δa as

$$\Delta a = - a \, k \, \Delta t \, / \, (1 + \frac{1}{2} k \, \Delta t) \tag{9.3-2}$$

so that

$$a_1 = a_0 + \Delta a = a_0 \, (1 - \tfrac{1}{2} \, k \, \Delta t) \, / \, (1 + \tfrac{1}{2} \, k \, \Delta t) \qquad (9.3\text{-}3)$$

$$
\begin{aligned}
a_2 = a_1 + \Delta a &= a_1 \, (1 - \tfrac{1}{2} \, k \, \Delta t) \, / \, (1 + \tfrac{1}{2} \, k \, \Delta t) \\
&= a_0 \, \{(1 - \tfrac{1}{2} \, k \, \Delta t) \, / \, (1 + \tfrac{1}{2} \, k \, \Delta t)\}^2 \qquad (9.3\text{-}4)
\end{aligned}
$$

$$a_n = a_0 \, \{(1 - \tfrac{1}{2} \, k \, \Delta t)/(1 + \tfrac{1}{2} \, k \, \Delta t)\}^n \qquad (9.3\text{-}5)$$

which can be compared with (9.2-8) through (9.2-10) respectively.

Instructions for exercise 9.3-1

1 Modify the spreadsheet used for exercise 9.2-1, or open a new spreadsheet and model it after Fig. 9.2-1.

2 Calculate a using (9.3-3) instead of (9.2-8).

3 Again calculate the algorithmic error $a_{\text{simul}} - a_{\text{exact}}$ by comparison with the exact result.

Comparison of the results for $a_{\text{simul}} - a_{\text{exact}}$ obtained here with those of the explicit method of section 9.2 indicates that the implicit method significantly improves the accuracy of the computation. Moreover, by using (9.3-5) to squeeze a number of small steps in one row, we can further reduce the error, which now goes as $(\Delta t)^2$. This is illustrated in Fig. 9.3-1, which compares the results obtained for the first-order reaction A $\rightarrow$ with $a_0 = 1$, $k = 1$, and a column length of 101 rows, i.e., for $t = 0$ (0.01) 10. The message is clear: while the implicit method takes more programming effort, it is much more precise. Consequently, when we need only a qualitative ("quick-and-dirty") estimate, the explicit method will often do, while for quantitative analysis the implicit method is the way to go.

9.3b Dimerization kinetics

For the reaction 2A $\xrightarrow{k}$ the implicit method leads to

$$\Delta a \, / \, \Delta t = - \, k \, (a + \tfrac{1}{2} \, \Delta a)^2 = - \, k \, \{a^2 + a \, \Delta a + \tfrac{1}{4} \, (\Delta a)^2\} \approx - \, k \, (a^2 + a \, \Delta a) \qquad (9.3\text{-}6)$$

where we neglect the higher-order correction term $\tfrac{1}{4} \, (\Delta a)^2$, thereby linearizing the expression. Once this is done, the remainder is straightforward:

$$\Delta a = - \, k \, a^2 \, \Delta t \, / \, (1 + a \, k \, \Delta t) \qquad (9.3\text{-}7)$$

$$a_1 = a_0 \, / \, (1 + a_0 \, k \, \Delta t) \qquad (9.3\text{-}8)$$

$$a_2 = a_1 \, / \, (1 + a_1 \, k \, \Delta t) = a_0 \, / \, (1 + 2 \, a_0 \, k \, \Delta t) \qquad (9.3\text{-}9)$$

$$a_n = a_0 \, / \, (1 + n \, a_0 \, k \, \Delta t) \qquad (9.3\text{-}10)$$

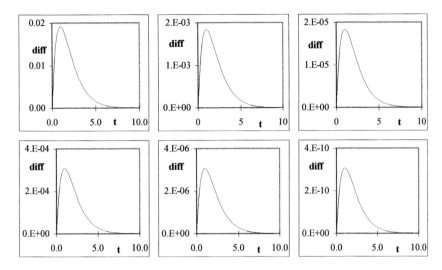

Fig. 9.3-1: The difference $a_{simul} - a_{exact}$ for the explicit (top) and implicit (bottom) method, from left to right using 1, 10, and 1000 steps per row.

4 Calculate a using (9.3-8) instead of (9.2-16).

Note that (9.3-10) is equivalent to the *exact* solution $a = a_0 / (1 + a_0 \, k \, t)$ of equation (9.2-14) because $t = n \, \Delta t$, so that, in this particular case, the implicit method does not generate any error at all.

9.3c Trimerization kinetics

For the reaction $3A \xrightarrow{\;k\;}$ the implicit approach yields

$$\Delta a / \Delta t = - k (a + \tfrac{1}{2} \Delta a)^3 = - k \{a^3 + \tfrac{3}{2} a^2 \Delta a + \tfrac{3}{4} a (\Delta a)^2 + \tfrac{1}{8} (\Delta a)^3\}$$
$$\approx - k \, a^2 (1 + \tfrac{3}{2} \Delta a) \tag{9.3-11}$$

$$\Delta a = - a^3 \, k \Delta t / (1 + \tfrac{3}{2} a^2 k \Delta t) \tag{9.3-12}$$

$$a_1 = a_0 (1 + \tfrac{1}{2} a_0{}^2 k \Delta t) / (1 + \tfrac{3}{2} a_0{}^2 k \Delta t) \tag{9.3-13}$$

$$a_2 = a_1 (1 + \tfrac{1}{2} a_1{}^2 k \Delta t) / (1 + \tfrac{3}{2} a_1{}^2 k \Delta t) \tag{9.3-14}$$

$$a_n = a_{n-1} \{1 + \tfrac{1}{2} (a_{n-1})^2 k \Delta t\} / \{1 + \tfrac{3}{2} (a_{n-1})^2 k \Delta t\} \tag{9.3-15}$$

5 Calculate a using (9.3-13) instead of (9.2-22).

6 Also compute the inherent error of the simulation by calculating the difference with the theoretical result, $a = a_0 / \sqrt{(1 + 2 \, a_0{}^2 \, k \, t)}$.

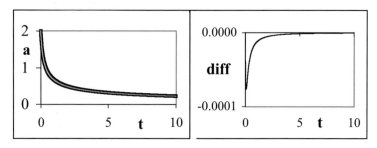

Fig. 9.3-2: The simulation with the implicit method of the unidirectional trimerization reaction (9.2-19) with $a_0 = 2$ and $k = 1$. Comparison with Fig. 9.2-4 shows that the absolute algorithmic error is now more than two orders of magnitude smaller.

The above results (Fig. 9.3-2) again indicate that the implicit method yields considerably better results than the explicit method; on the other hand, it requires more effort to formulate the appropriate equations. This is further illustrated in the two remaining examples.

9.3d Monomer–dimer kinetics

For the monomer–dimer reaction scheme (9.2-25) and the associated differential equation (9.2-28) the implicit method yields the difference equation

$$\Delta a / \Delta t = - k (a + \tfrac{1}{2}\Delta a)^2 - k' (a + \tfrac{1}{2}\Delta a) + k''$$
$$\approx - k a (a + \Delta a) - k' (a + \tfrac{1}{2}\Delta a) + k'' \qquad (9.3\text{-}16)$$

where we have neglected higher-order terms such as in $(\Delta t)^2$. Consequently

$$\Delta a = (- k a^2 - k' a + k'') / \{1/\Delta t + k a + \tfrac{1}{2} k'\} \qquad (9.3\text{-}17)$$

$$a_1 = a_0 + (- k a_0^2 - k' a_0 + k'') / (1/\Delta t + k a_0 + \tfrac{1}{2} k') \qquad (9.3\text{-}18)$$

$$a_2 = a_1 + (- k a_1^2 - k' a_1 + k'') / (1/\Delta t + k a_1 + \tfrac{1}{2} k') \qquad (9.3\text{-}19)$$

$$a_n = a_{n-1} + (- k (a_{n-1})^2 - k' a_{n-1} + k'') / (1/\Delta t + k a_{n-1} + \tfrac{1}{2} k') \qquad (9.3\text{-}20)$$

Instructions for exercise 9.3-2

1 Modify the spreadsheet used for exercise 9.2-2, or open a new spreadsheet and model it after it.

2 Calculate a using (9.3-20) instead of (9.2-32).

3 Calculate b from $b = b_0 + \tfrac{1}{2}(a_0 - a)$.

4 Plot the reagent concentrations.

5 Compare the result obtained with the exact solution, as given by (9.2-33) and (9.2-34).

6 Also compare the magnitude of the errors of the implicit and explicit method in this case.

9.3e Polymerization kinetics

In the preceding case, with only two concentrations, a and b, the changes Δa and Δb can readily be found because, for two concentrations, we could use the mass balance to express the change in one concentration in terms of that of the other. However, when there are more than two reacting species, the equations become more complicated, in which case it is usually simpler (and often necessary) to use matrix algebra. Below we will illustrate that approach for the reaction scheme (9.2-35) through (9.2-38).

Instead of (9.2-39) through (9.2-42) we now have

$$
\begin{aligned}
\Delta a_1/\Delta t = & -k_{11}\, a_1\,(a_1+\Delta a_1) - k_{12}\,(a_1\,a_2 + \tfrac{1}{2}\,a_1\,\Delta a_2 + \tfrac{1}{2}\,a_2\,\Delta a_1) \\
& -k_{13}\,(a_1\,a_3 + \tfrac{1}{2}\,a_1\Delta a_3 + \tfrac{1}{2}\,a_3\,\Delta a_1) + 2\,k_{11}'\,(a_2+\tfrac{1}{2}\Delta a_2) \\
& + k_{12}'\,(a_3+\tfrac{1}{2}\Delta a_3) + k_{13}'\,(a_4+\tfrac{1}{2}\Delta a_4)
\end{aligned}
\tag{9.3-21}
$$

$$
\begin{aligned}
\Delta a_2/\Delta t = & -k_{11}'\,(a_2+\tfrac{1}{2}\Delta a_2) - k_{12}\,(a_1\,a_2 + \tfrac{1}{2}\,a_1\,\Delta a_2 + \tfrac{1}{2}\,a_2\,\Delta a_1) \\
& -k_{22}\,a_2\,(a_2+\Delta a_2) + \tfrac{1}{2}\,k_{11}\,a_1\,(a_1+\Delta a_1) + k_{12}'\,(a_3+\tfrac{1}{2}\Delta a_3) \\
& + 2\,k_{22}'\,(a_4+\tfrac{1}{2}\Delta a_4)
\end{aligned}
\tag{9.3-22}
$$

$$
\begin{aligned}
\Delta a_3/\Delta t = & -k_{12}'\,(a_3+\tfrac{1}{2}\Delta a_3) - k_{13}\,(a_1\,a_3 + \tfrac{1}{2}\,a_1\,\Delta a_3 + \tfrac{1}{2}\,a_3\,\Delta a_1) \\
& + k_{12}\,(a_1\,a_2 + \tfrac{1}{2}\,a_1\,\Delta a_2 + \tfrac{1}{2}\,a_2\,\Delta a_1) + k_{13}'\,(a_4+\tfrac{1}{2}\Delta a_4)
\end{aligned}
\tag{9.3-23}
$$

$$
\begin{aligned}
\Delta a_4/\Delta t = & -(k_{13}'+k_{22}')\,(a_4+\tfrac{1}{2}\Delta a_4) + k_{13}\,(a_1\,a_3 + \tfrac{1}{2}\,a_1\,\Delta a_3 + \tfrac{1}{2}\,a_3\,\Delta a_1) \\
& + \tfrac{1}{2}\,k_{22}\,a_2\,(a_2+\Delta a_2)
\end{aligned}
\tag{9.3-24}
$$

where we have replaced $(a_i+\tfrac{1}{2}\Delta a_i)^2 = a_i^2 + a_i\Delta a_i + \tfrac{1}{4}(\Delta a_i)^2$ by $a_i\,(a_i+\Delta a_i)$, and $(a_i+\tfrac{1}{2}\Delta a_i)(a_j+\tfrac{1}{2}\Delta a_j) = a_i a_j + \tfrac{1}{2}\,a_i\Delta a_j + \tfrac{1}{2}\,a_j\Delta a_i + \tfrac{1}{4}\,\Delta a_i\,\Delta a_j$ by $a_i a_j + \tfrac{1}{2}\,a_i\Delta a_j + \tfrac{1}{2}\,a_j\Delta a_i$, thereby removing all higher-order terms $(\Delta a_i)^2$ and $\Delta a_i\Delta a_j$.

After sorting (9.3-21) through (9.3-24) in terms of the various Δa_i's we obtain

$$
\begin{aligned}
(1/\Delta t + k_{11}\,a_1 + &\tfrac{1}{2}\,k_{12}\,a_2 + \tfrac{1}{2}\,k_{13}\,a_3)\,\Delta a_1 + (\tfrac{1}{2}\,k_{12}\,a_1 - k_{11}')\,\Delta a_2 \\
& + (\tfrac{1}{2}\,k_{13}\,a_1 - \tfrac{1}{2}\,k_{12}')\,\Delta a_3 + (-\tfrac{1}{2}\,k_{13}')\,\Delta a_4 \\
& = -k_{11}\,a_1^2 - k_{12}\,a_1\,a_2 - k_{13}\,a_1\,a_3 + 2\,k_{11}'\,a_2 + k_{12}'\,a_3 + k_{13}'\,a_4
\end{aligned}
\tag{9.3-25}
$$

$$
\begin{aligned}
(\tfrac{1}{2}\,k_{12}\,a_2 - \tfrac{1}{2}\,k_{11}\,a_1)\,\Delta a_1 + &(1/\Delta t + \tfrac{1}{2}\,k_{11}' + \tfrac{1}{2}\,k_{12}\,a_1 + k_{22}\,a_2)\,\Delta a_2 \\
& + (-\tfrac{1}{2}\,k_{12}')\,\Delta a_3 + (-k_{22}')\,\Delta a_4 = -k_{11}'\,a_2 - k_{12}\,a_1\,a_2 \\
& - k_{22}\,a_2^2 + \tfrac{1}{2}\,k_{11}\,a_1^2 + k_{12}'\,a_3 + 2\,k_{22}'\,a_4
\end{aligned}
\tag{9.3-26}
$$

$$
\begin{aligned}
(-\tfrac{1}{2}\,a_2\,k_{12} + \tfrac{1}{2}\,a_3\,k_{13})\,\Delta a_1 + &(-\tfrac{1}{2}\,a_1\,k_{12})\,\Delta a_2 + (1/\Delta t + \tfrac{1}{2}\,k_{12}' + \tfrac{1}{2}\,a_1\,k_{13})\,\Delta a_3 \\
& + (-\tfrac{1}{2}\,k_{13}')\,\Delta a_4 = -k_{12}'\,a_3 - k_{13}\,a_1\,a_3 + k_{12}\,a_1\,a_2 + k_{13}'\,a_4
\end{aligned}
\tag{9.3-27}
$$

$$
\begin{aligned}
(-\tfrac{1}{2}\,a_3\,k_{13})\,\Delta a_1 + &(-\tfrac{1}{2}\,k_{22}\,a_2)\,\Delta a_2 + (-\tfrac{1}{2}\,a_1\,k_{13})\,\Delta a_3 + (1/\Delta t + \tfrac{1}{2}\,k_{13}' + \tfrac{1}{2}\,k_{22}') \\
& \Delta a_4 = -(k_{13}'+k_{22}')\,a_4 + k_{13}\,a_1\,a_3 + \tfrac{1}{2}\,k_{22}\,a_2^2
\end{aligned}
\tag{9.3-28}
$$

which comprise four equations in four unknowns, Δa_1, Δa_2, Δa_3, and Δa_4. We write these in the symbolic form $\mathbf{C\,D} = \mathbf{B}$, where $\mathbf{C}$ is the 4×4 matrix of the coefficients of the terms Δa_i,

$$\mathbf{C} = \begin{vmatrix} (1/\Delta t + k_{11}a_1 + \frac{1}{2}k_{12}a_2 + \frac{1}{2}k_{13}a_3) & (\frac{1}{2}\,k_{12}\,a_1 - k_{11}') & \cdots & -\frac{1}{2}\,k_{13}' \\ \frac{1}{2}\,k_{12}\,a_2 - \frac{1}{2}\,k_{11}\,a_1 & \cdots & & \cdots - k_{22}' \\ -\frac{1}{2}k_{12}\,a_2 + \frac{1}{2}\,k_{13}\,a_3 & \cdots & & \cdots - \frac{1}{2}\,k_{13}' \\ -\frac{1}{2}k_{13}\,a_3 & \cdots & & \cdots \; 1/\Delta t + \frac{1}{2}\,k_{13}' + \frac{1}{2}\,k_{22}' \end{vmatrix}$$

$$(9.3\text{-}29)$$

while $\mathbf{D}$ and $\mathbf{B}$ are vectors,

$$\mathbf{D} = \begin{vmatrix} \Delta a_1 \\ \Delta a_2 \\ \Delta a_3 \\ \Delta a_4 \end{vmatrix}, \quad \mathbf{B} = \begin{vmatrix} -k_{11}\,a_1{}^2 - k_{12}\,a_1\,a_2 - k_{13}\,a_1\,a_3 + 2\,k_{11}'\,a_2 + k_{12}'\,a_3 + k_{13}'\,a_4 \\ -k_{11}'\,a_2 - k_{12}\,a_1\,a_2 - k_{22}\,a_2{}^2 + \frac{1}{2}\,k_{11}\,a_1{}^2 + k_{12}'\,a_3 + 2\,k_{22}'\,a_4 \\ -k_{12}'\,a_3 - k_{13}\,a_1\,a_3 + k_{12}\,a_1\,a_2 + k_{13}'\,a_4 \\ -(k_{13}' + k_{22}')\,a_4 + k_{13}\,a_1\,a_3 + \frac{1}{2}\,k_{22}\,a_2{}^2 \end{vmatrix}$$

$$(9.3\text{-}30)$$

The concentration changes $\mathbf{D}$ are then found as $\mathbf{D} = \mathbf{C}^{-1}\mathbf{C}\,\mathbf{D} = \mathbf{C}^{-1}\mathbf{B}$, i.e., by inverting the matrix $\mathbf{C}$, and by subsequently calculating the product $\mathbf{C}^{-1}\mathbf{B}$. Note that both $\mathbf{B}$ and $\mathbf{C}$ contain the concentrations a_1 through a_4, which are the concentrations at the start of the interval Δt. At the beginning of the computation these are the initial concentrations. Thereafter, these concentrations must be updated to their most recently computed values before the concentration changes during the next interval can be computed. Clearly, such a calculation is far too tedious and time-consuming to do on the spreadsheet itself, where matrix inversion and matrix multiplication must be initiated manually for every time step Δt. Instead, this is the type of problem that, on a spreadsheet, is best done with a function specifically written for that purpose.

As an exercise, after you have familiarized yourself with the material in the next chapter, you might want to try to write such a function. It should have as its input the previous concentrations, the rate parameters, and the time increment. It should then calculate (for this specific reaction scheme) the concentrations at time $t + \Delta t$, and write these concentrations back onto the spreadsheet. Use the matrix inversion subroutine for the hard work.

The examples in this section give you a taste of the implicit method. In fact, they only illustrate its simplest form, in which we assume a linear dependence of all concentrations during each interval Δt, yielding results that exhibit an algorithmic inaccuracy proportional to $(\Delta t)^2$. More sophisticated indirect methods are available, most prominently among them the various higher-order Runge–Kutta schemes (the above examples illustrate the second-order Runge–Kutta approach), and the Adams–Bashforth and Adams–Moulton formulas that also take earlier points into account. Obviously, numerical integration of differential equations is an area of specialized knowledge, of which we have given here only a few simple examples.

For more details the reader should consult one of the many specialized books on that topic.

In section 9.2 we illustrated one explicit method, Euler's forward method. In the present section, we likewise used only one type of implicit method, based on the trapezoidal or midpoint rule. All our examples have used constant increments Δt; higher computational efficiency can often be obtained by making the step size dependent on the magnitudes of the changes in the dependent variables. Still, these examples illustrate that, upon comparing equivalent implicit and explicit methods, the former usually allow larger step sizes for a given accuracy, or yield more accurate results for the same step size. On the other hand, implicit methods typically require considerably more initial effort to implement.

While the examples given here have dealt only with chemical reaction kinetics, the method illustrates how one can, in general, solve single as well as coupled differential equations. Euler's explicit method is useful as a qualitative tool: it is easily implemented, and can provide a reasonably close result when Δt is sufficiently small. The latter requirement, however, may make the explicit method impractical on a spreadsheet. For quantitative work, an implicit method is usually required, as it provides a better approximation given the limited number of iterations practical on a spreadsheet.

9.4 Some applications

This section will illustrate how one can use the implicit method to simulate several interesting kinetic schemes, such as an autocatalytic reaction, and heterogeneous catalysis. Then we will see the ramifications of an often used (and sometimes misused) simplification, the steady-state assumption. Finally, we will simulate a prototypical oscillating reaction.

9.4a Autocatalysis

In an autocatalytic reaction, a reaction product catalyzes the reaction, i.e., enhances its net rate. We will here assume a simple model sequence in which the reaction A $\rightarrow$ C starts with the pure starting material, A. The reaction product, C, catalyzes a second, faster reaction pathway, A + C $\rightarrow$ 2C, so that the reaction will speed up after some catalyst C has been formed:

$$\text{A} \xrightarrow[k_1]{} \text{C} \tag{9.4-1}$$

$$\text{A} + \text{C} \xrightarrow[k_2]{} 2\text{C} \tag{9.4-2}$$

The corresponding rate equations for the concentrations a and c of A and C respectively are

$$\frac{da}{dt} = - k_1 a - k_2 ac \tag{9.4-3}$$

$$\frac{dc}{dt} = k_1 a + k_2 ac \tag{9.4-4}$$

with the mass balance $a + c = a_0 + c_0 = a_0$ since the initial concentration c_0 of C is assumed to be zero. We can therefore use $c = a_0 - a$ to convert (9.4-3) into

$$\frac{da}{dt} = - (k_1 + k_2 a_0) a + k_2 a^2 \tag{9.4-5}$$

which can be formulated for the implicit method as

$$\begin{aligned}\frac{\Delta a}{\Delta t} &= - (k_1 + k_2 a_0)(a + \Delta a/2) + k_2 (a + \Delta a/2)^2 \\ &\approx - (k_1 + k_2 a_0)(a + \Delta a/2) + k_2 a(a + \Delta a) \\ &\approx - k'(a + \Delta a/2) + k_2 a(a + \Delta a)\end{aligned} \tag{9.4-6}$$

where $k' = a_0 k_1 + k_2 a_0$, so that

$$\left(\frac{1}{\Delta t} + (k_1 + k_2 a_0)/2 - k_2 a\right)\Delta a = - (k_1 + k_2 a_0) a + k_2 a^2 \tag{9.4-7}$$

$$a_1 = a_0 + \Delta a = a_0 \frac{1/\Delta t - k'/2}{1/\Delta t + k'/2 - k_2 a_0} \tag{9.4-8}$$

$$a_n = a_{n-1} \frac{1/\Delta t - k'/2}{1/\Delta t + k'/2 - k_2 a_{n-1}} \tag{9.4-9}$$

which is in a form suitable for the simulation.

Instructions for exercise 9.4-1

1 Open a spreadsheet.

2 In its top row, specify values for a_0, Δt, k_1, and k_2, and calculate the resulting value of $k' = k_1 + k_2 a_0$.

3 Calculate a using (9.4-8) and (9.4-9).

4 Plot a as a function of time t.

5 Vary the values of k_1 and k_2 (and thereby of k') and observe that the reaction can exhibit an induction period when $k_1 \ll k_2 a_0$.

Figure 9.4-1 illustrates the distinct time course of the concentration a in this case, with a decomposition rate that starts slowly, then accelerates as more catalyst C is formed. Higher-order autocatalytic reactions exhibit an

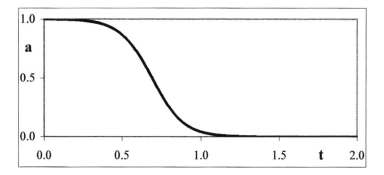

Fig. 9.4-1: An autocatalytic reaction following (9.4-1) and (9.4-2), simulated with $a_0 = 1$, $k_1 = 0.01$, and $k_2 = 10$.

even more sudden transition, such as the Landolt oxidation of sulfite with iodate, which can be represented approximately by (9.4-1) plus $A + 2C \xrightarrow{k_2}$ 3C. Such reactions are often called *clock reactions*.

9.4b Heterogeneous catalysis

In heterogeneous catalysis, the catalytic action is provided on or by an interface. A simple example is the decomposition of ammonia vapor on a tungsten surface: after the ammonia gets adsorbed, it decomposes into hydrogen and nitrogen molecules. Although the actual kinetics of this process are rather more complicated, we will model it here in terms of a simple sequence of fast adsorption followed by a rate-determining interfacial decomposition. Such a model differs from that of a simple first-order reaction in that adsorption is a non-linear process: in general, doubling the ammonia vapor pressure does not lead to a doubling of the amount of ammonia adsorbed. Below we will use the Langmuir adsorption model, which relates the amount adsorbed, Γ, to the concentration (or vapor pressure) c as

$$bc = \frac{\Gamma}{\Gamma_{max} - \Gamma} \tag{9.4-10}$$

or

$$\Gamma = \frac{bc}{1 - bc} \Gamma_{max} \tag{9.4-11}$$

where $b = k_{des}/k_{ads}$ is the equilibrium adsorption constant, and Γ_{max} is the maximum amount of adsorbate that the interface can accommodate. The Langmuir model is based on adsorption equilibrium for an interface with a fixed number of equivalent adsorption sites. At equilibrium, we require that the adsorption rate $v_{ads} = k_{ads} c (1 - \Gamma/\Gamma_{max})$ be equal to the desorption rate

$v_{des} = k_{des} \Gamma / \Gamma_{max}$, where Γ / Γ_{max} is the fraction of adsorption sites that are occupied, and $1 - \Gamma / \Gamma_{max}$ the corresponding fraction of unoccupied sites. We will assume here that the reagent is the only adsorbing species.

For the reaction sequence $A \rightarrow A_{ads} \rightarrow B$, where the second step is rate-determining, we now write the rate expression for the concentration of species A as $da/dt = -k\Gamma$, where Γ denotes the amount of A adsorbed per unit interfacial area, or

$$\frac{da}{dt} = -k\Gamma = \frac{-kba\Gamma_{max}}{1 - ba} \qquad (9.4\text{-}12)$$

so that

$$\frac{\Delta a}{\Delta t} = \frac{-kb\Gamma_{max}(a + \Delta a/2)}{1 + b(a + \Delta a/2)} \qquad (9.4\text{-}13)$$

or

$$(1 + b(a + \Delta a/2))\Delta a = (-kb\Gamma_{max}(a + \Delta a/2))\Delta t \qquad (9.4\text{-}14)$$

from which we obtain (by neglecting the quadratic term in Δa)

$$\Delta a = \frac{-kb\Gamma_{max}a\,\Delta t}{1 + ba + kb\Gamma_{max}\Delta t/2} \qquad (9.4\text{-}15)$$

$$a_n = a_{n-1} + \Delta a = a_{n-1}\frac{1 + ba_{n-1} - kb\Gamma_{max}\Delta t/2}{1 + ba_{n-1} + kb\Gamma_{max}\Delta t/2} \qquad (9.4\text{-}16)$$

For $b \rightarrow \infty$ the above result approaches a linear decay, $a_n \rightarrow a_{n-1} - k\Gamma_{max}\Delta t$ or, equivalently, $a \rightarrow a_0 - k\Gamma_{max}\,t$, see Fig. 9.4-2. This would correspond with a zeroth-order reaction, for which $da/dt = -k\,\Gamma_{max}$, a constant, the same result as would have been obtained by changing Γ in (9.4-12) into Γ_{max}.

Instructions for exercise 9.4-2

1 Open a spreadsheet.

2 In its top row, specify values for a_0, Δt, b, k, and Γ_{max}.

3 Starting with $t = 0$, compute and plot a as a function of time t.

4 Also plot the function $a_0 - k\Gamma_{max}\,t$.

5 Vary the values of b and observe that, when b tends to infinity, da/dt approaches a constant value, and a approaches $a_0 - k\Gamma_{max}\,t$, which corresponds to a zeroth order reaction.

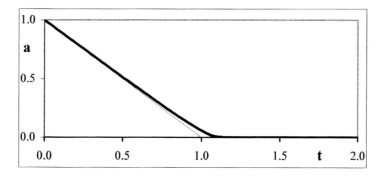

Fig. 9.4-2: A heterogeneous reaction according to (9.4-12), simulated with $a_0 = 1$, $k = 0.1$, $b = 50$, and $\Gamma_{max} = 10$. The thin straight line represents the response to the 0th order rate expression $da/dt = - k\Gamma_{max}$.

9.4c The steady-state approximation

A steady-state approximation is often used in order to simplify the mathematical description of complicated reaction mechanisms. Below we will use simulation to illustrate when such a simplifying assumption is appropriate, and when it is not. We will use the reaction sequence $A + B \rightleftharpoons C \rightarrow$ products. C can either be a chemically identifiable species, or a presumed or hypothetical intermediate, such as a transition state or activated complex. The rate equations are

$$\frac{da}{dt} = \frac{db}{dt} = - k_1 ab + k_{-1} c \tag{9.4-17}$$

$$\frac{dc}{dt} = k_1 ab - (k_{-1} + k_2)c = k_1 ab - k' c \tag{9.4-18}$$

where the rate constant k_1 applies to $A + B \rightarrow C$, k_{-1} to $C \rightarrow A + B$, and k_2 to $C \rightarrow$ products, and where we have used the abbreviation $k' = k_{-1} + k_2$. In order to avoid the need for matrix inversion, we will slightly simplify the problem by assuming $a_0 = b_0$ so that, during the entire process, $a = b$, in which case (9.4-17) and (9.4-18) reduce to

$$\frac{da}{dt} = - k_1 a^2 + k_{-1} c \tag{9.4-19}$$

$$\frac{dc}{dt} = k_1 a^2 - k' c \tag{9.4-20}$$

which can be written in a form suitable for implicit numerical simulation as

$$\frac{\Delta a}{\Delta t} = - k_1 (a + \Delta a/2)^2 + k_{-1}(c + \Delta c/2) \approx - k_1 a(a + \Delta a) + k_{-1}(c + \Delta c/2) \tag{9.4-21}$$

$$\frac{\Delta c}{\Delta t} = k_1(a + \Delta a/2)^2 - k'(c + \Delta c/2) \approx k_1 a(a + \Delta a) - k'(c + \Delta c/2) \qquad (9.4\text{-}22)$$

Equations (9.4-21) and (9.4-22) can be written as

$$\left(\frac{1}{\Delta t} + k_1 a\right)\Delta a + (k_{-1}/2)\,\Delta c = -k_1 a^2 + k_{-1} c \qquad (9.4\text{-}23)$$

$$(-k_1 a)\,\Delta a + \left(\frac{1}{\Delta t} + k'/2\right)\Delta c = k_1 a^2 - k' c \qquad (9.4\text{-}24)$$

so that

$$\Delta a = \frac{\begin{vmatrix} -k_1 a^2 + k_{-1} c & -k_{-1}/2 \\ k_1 a^2 - k' c & 1/\Delta t + k'/2 \end{vmatrix}}{\begin{vmatrix} 1/\Delta t + k_1 a & -k_{-1}/2 \\ -k_1 a & 1/\Delta t + k'/2 \end{vmatrix}}$$

$$= \frac{(-k_1 a^2 + k_{-1} c)(1/\Delta t + k'/2) - (-k_{-1}/2)(k_1 a^2 - k' c)}{(1/\Delta t + k_1 a)(1/\Delta t + k'/2) - (-k_{-1}/2)(-k_1 a)} \qquad (9.4\text{-}25)$$

$$\Delta c = \frac{\begin{vmatrix} 1/\Delta t + k_1 a & -k_1 a^2 + k_{-1} c \\ -k_1 a & k_1 a^2 - k' c \end{vmatrix}}{\begin{vmatrix} 1/\Delta t + k_1 a & -k_{-1}/2 \\ -k_1 a & 1/\Delta t + k'/2 \end{vmatrix}}$$

$$= \frac{(1/\Delta t + k_1 a)(k_1 a^2 - k' c) - (-k_1 a^2 + k_{-1} c)(-k_1 a)}{(1/\Delta t + k_1 a)(1/\Delta t + k'/2) - (-k_{-1}/2)(-k_1 a)} \qquad (9.4\text{-}26)$$

from which we can compute a_n as $a_{n-1} + \Delta a$ and c_n as $c_{n-1} + \Delta c$. And, yes, it is possible to clean these expressions up a bit, but why not let the spreadsheet do the work instead of us.

Instructions for exercise 9.4-3

1 Open a spreadsheet, leaving room at its top for a graph.

2 Label and enter values for a_0, c_0, Δt, $1/\Delta t$, k_1, k_{-1}, and $k' = k_{-1} + k_2$.

3 Name k_1, k_{-1}, k', and $1/\Delta t$; in the example below the names kk (k1 cannot be used since it is a valid cell address), km (m for minus), kd (d for dash) and dt will be used. Note that you cannot use names that you have already used elsewhere in the same workbook, i.e., in the same collection of spreadsheets.

4 Starting with $t = 0$, compute a and c as a function of time t. For example, with the

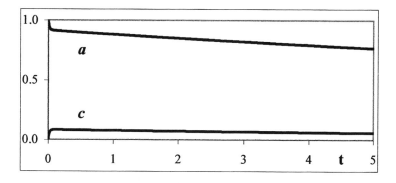

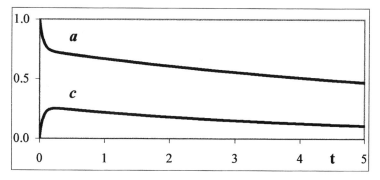

Fig. 9.4-3: The concentrations a and c of the reagent and reaction intermediate respectively, for (top): $a_0 = 1$, $c_0 = 0$, $k_1 = 5$, $k_{-1} = 50$, $k_2 = 0.5$, and (bottom): the same except that $k_{-1} = 10$.

columns for t, a and c starting on row 18, the expression for $a_n = a_{n-1} + \Delta a$ in cell B19 can be based on (9.4-24), and should read = B18 + ((− kk*B18* B18 + km*C18) * (dt + kd/2) − (− km/2) * (kk*B18*B18 − kd*C18)) / ((dt + kk*B18) * (dt + kd/2) − (− km/2) * (− kk*B18)).

5 Plot a and c versus t. Figure 9.4-3a gives two examples.

While the concentration c of the intermediate is never exactly constant, for some parameter combinations it is reasonably so, at least after an initial transient. In that case we can approximate (9.4-18) to $dc/dt = k_1ab − k'c \approx 0$ so that $c \approx k_1ab/k'$, in which case (9.4-17) becomes $da/dt = db/dt = − k_1k_2ab/k'$. This approach, based on the assumption that the concentration of the intermediate is constant, is called the steady-state approximation. In the above example, the steady-state approximation will be valid whenever $k_2 \ll k \ll k_{-1}$.

9.4d Oscillating reactions: the Lotka model

The majority of chemical reactions exhibit a monotonic time course, but it is not unusual for concentrations of intermediates in a series of coupled reactions, such as the concentration of B in the sequence A → B → C, to rise and fall. Less often, but still in quite a number of well-documented cases, concentrations go up and down more than once, and reactions that exhibit this behavior are called oscillatory. Typically, such oscillations will eventually die out once some or all of the starting material has been consumed. However, some reactions can be kept to oscillate indefinitely by keeping their initial concentrations constant, i.e., by replenishing any reagent lost.

Oscillating chemical reactions have been known for almost two centuries, i.e., they are as old as chemistry itself. Some involve homogeneous kinetics, i.e., with the participating species either all in solution or all in the vapor phase. More commonly they involve heterogeneous kinetics, such as electrochemical oscillators, of which some were known already to Faraday. Moreover, it is now known that many biological systems incorporate oscillating reactions as clocks, which maintain biologically important rhythms, such as the circadian (approximately daily) cycle.

In our example we will consider the earliest model of an oscillating homogeneous chemical reaction (A. J. Lotka, *J. Am. Chem. Soc.* 42 (1920) 1595; *Proc. Natl. Acad. Sci. USA* 6 (1920) 410), which is based on the reaction sequence

$$A + B \xrightarrow{k_1} 2B \tag{9.4-27}$$

$$B + C \xrightarrow{k_2} 2C \tag{9.4-28}$$

$$C \xrightarrow{k_3} \text{products} \tag{9.4-29}$$

where we will assume that the concentration a of A is kept constant, so that $da/dt = 0$. The corresponding rate expressions for b and c are

$$\frac{db}{dt} = k_1 a b - k_2 b c \tag{9.4-30}$$

$$\frac{dc}{dt} = k_2 b c - k_3 c \tag{9.4-31}$$

so that

$$\frac{\Delta b}{\Delta t} = k_1 a (b + \Delta b/2) - k_2 (b + \Delta b/2)(c + \Delta c/2) \tag{9.4-32}$$

$$\frac{\Delta c}{\Delta t} = k_2 (b + \Delta b/2)(c + \Delta c/2) - k_3 (c + \Delta c/2) \tag{9.4-33}$$

$$\left(\frac{1}{\Delta t} - k_1 a/2 + k_2 c/2 \right) \Delta b + (k_2 b/2) \Delta c = k_1 a b - k_2 b c \tag{9.4-34}$$

$$(-k_2 c/2)\,\Delta b + \left(\frac{1}{\Delta t} - k_2\,b/2 + k_3/2\right)\Delta c = k_2\,bc - k_3\,c \tag{9.4-35}$$

$$\Delta b = \frac{\begin{vmatrix} k_1 ab - k_2 bc & k_2 bc/2 \\ k_2 bc - k_3 c & 1/\Delta t - k_2 b/2 + k_3/2 \end{vmatrix}}{\begin{vmatrix} 1/\Delta t - k_1 a/2 + k_2 c/2 & k_2 b/2 \\ -k_2 c/2 & 1/\Delta t - k_2 b/2 + k_3/2 \end{vmatrix}}$$

$$= \frac{(k_1 ab - k_2 bc)(1/\Delta t - k_2 b/2 + k_3/2) - (k_2 b/2)(k_2 bc - k_3 c)}{(1/\Delta t - k_1 a/2 + k_2 c/2)(1/\Delta t - k_2 b/2 + k_3/2) - (k_2 b/2)(-k_2 c/2)} \tag{9.4-36}$$

$$\Delta c = \frac{\begin{vmatrix} 1/\Delta t - k_1 a/2 + k_2 c/2 & k_1 ab - k_2 bc \\ -k_2 c/2 & k_2 bc - k_3 c \end{vmatrix}}{\begin{vmatrix} 1/\Delta t - k_1 a/2 + k_2 c/2 & k_2 b/2 \\ -k_2 c/2 & 1/\Delta t - k_2 b/2 + k_3/2 \end{vmatrix}}$$

$$= \frac{(1/\Delta t - k_1 a/2 + k_2 c/2)(k_2 bc - k_3 c) - (k_1 ab - k_2 bc)(-k_2 c/2)}{(1/\Delta t - k_1 a/2 + k_2 c/2)(1/\Delta t - k_2 b/2 + k_3/2) - (k_2 b/2)(-k_2 c/2)} \tag{9.4-37}$$

from which we can compute b_n as $b_{n-1} + \Delta b$ and c_n as $c_{n-1} + \Delta c$.

Instructions for exercise 9.4-4

1 Open a spreadsheet, and at its top reserve space for a graph.

2 Label and enter values for a, b_0, c_0, Δt, k_1, k_2, k_3, and $1/\Delta t$.

3 Name a, k_1, k_2, k_3, and $1/\Delta t$; in the example below the names a, kk1, kk2, kk3, and tt will be used. (We use kk1 as the *name* for k_1 because k1 denotes a cell address, and therefore cannot be used as a name.)

4 Starting with $t = 0$, compute b and c as a function of time t. For example, with the columns for t, b and c starting on row 18, the expression for $b_n = b_{n-1} + \Delta b$ in cell B19, based on (9.4-35), might read = B18 + ((kk1*a*B18–kk2*B18* C18)*(tt − kk2*B18/2 + kk3/2) − (kk2*B18/2)*(kk2*B18*C18 − kk3*C18))/((tt − kk1 *a/2 + kk2*C18/2)*(tt − kk2*B18/2 + kk3/2) − (kk2*B18/2)*(− kk2*C18/2)).

5 Plot b and c versus t. For some parameter combinations you should observe oscillating concentrations, see Fig. 9.4-4. Play with the concentrations a, b_0, and c_0, and the rate constants k_1, k_2, and k_3, to find different patterns. Both frequency, amplitude, and waveforms vary with these parameters.

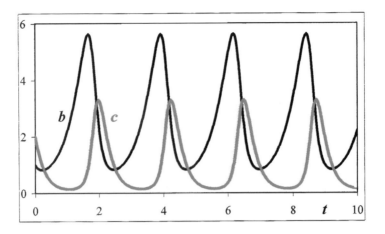

Fig. 9.4-4: The oscillating concentrations b and c in the Lotka oscillator, computed for $a = 1$, $b_0 = 1$, $c_0 = 2$, $k_1 = 2$, $k_2 = 2$, $k_3 = 5$, and $\Delta t = 0.01$.

9.5 Summary

This chapter is devoted to numerical integration, and more specifically to the integration of rate expressions encountered in chemical kinetics. For simple cases, integration yields closed-form rate equations, while more complex reaction mechanisms can often be solved only by numerical means. Here we first use some simple reactions to develop and calibrate general numerical integration schemes that are readily applicable to a spreadsheet. We then illustrate several non-trivial applications, including catalytic reactions and the Lotka oscillator.

The approach is, of course, not restricted to the rate expressions of chemical kinetics, and can be applied to a great variety of differential equations. We have used chemical kinetics here because these are important in chemical analysis, and provide a specific topic to illustrate the method. If one can write down the differential equation of a problem, one can also solve it numerically on the spreadsheet. It can be done the quick (but somewhat crude) explicit way illustrated in section 9.2, or more precisely (but with more initial effort) through the implicit method of sections 9.3 and 9.4.

In this chapter we have briefly introduced user-defined functions. These allow us to extend the range of available spreadsheet functions. They work very efficiently at the level of single-cell instructions. Macros operate in a similar way, but are more effective in dealing with entire blocks of data. The next chapter shows in fair detail how to write macros, and illustrates this with many worked-out examples. Once the material in chapter 10 has been digested, writing more complicated user-defined functions (such as for section 9.3e) should not present any problems. With the facility to make your own functions and macros, there is virtually no limit to what you can do on a spreadsheet.

CHAPTER **10**

SOME USEFUL
MACROS

10.1
What is a macro?

A spreadsheet macro is a computer program that can be called from inside a spreadsheet. The simplest macros merely allow the user to record a sequence of spreadsheet operations, which the computer then memorizes, and can repeat upon demand. All modern spreadsheets have facilities for recording and using macros to repeat a given series of instructions. While these can be quite useful for routine, repetitive computational tasks, their benefit in the present context is small, although we may occasionally use them to find out how Excel encodes visually entered information.

The major focus of this chapter will be on macros that *extend* the already considerable power of the spreadsheet, by incorporating external program instructions. Starting with Excel 5, the **macro language** (i.e., the computer language used to encode the macro) of Excel is VBA, which is sufficiently flexible and powerful to allow the spreadsheet user to introduce additional mathematical operations of his or her own choice, operations that are not already part of the usual spreadsheet repertoire. Earlier versions of Excel used a less transparent and certainly much less powerful macro language, called XLM, which will not be discussed here.

The letters VBA stand for **Visual BASIC for Applications**, an extension of BASIC, which itself is an acronym for Beginners' All-purpose Symbolic Instruction Code. BASIC was developed in the early 1960s as an easily-learned interpreter-based computer language. It has since been updated by Borland and Microsoft to a fully compiled language, and has become very similar to early versions of FORTRAN, the original FORmula TRANslation language of IBM. This is useful because FORTRAN remains the dominant computer language in the physical sciences, and program conversion from earlier FORTRAN code into modern BASIC is very easy. The extension of BASIC into VBA allows it to interact directly with the spreadsheet.

Anything that can be coded in modern, compiled BASIC can be

incorporated as a macro into any Excel spreadsheet more recent than version 4, and can then be executed by the computer whenever that macro is called. This is especially useful because, by themselves, spreadsheets are poorly suited for some often-used types of computations, such as those based on **iterations**, in which computational sequences are repeated until some internal criterion is satisfied. Many well-documented general-purpose higher-language programs are described lucidly in, e.g., W. H. Press, B. P. Flannery, S. A. Teukolsky & W. T. Vetterling, *Numerical Recipes, the Art of Scientific Computing*, Cambridge University Press (1986). This excellent text contains a number of well-tested computer programs in FORTRAN. More recent editions have provided the same programs in other computer languages, such as C and Pascal. Of direct interest to us is a small book by J. C. Sprott, *Numerical Recipes: Routines and Examples in BASIC, a companion manual to Numerical Recipes: the Art of Scientific Computing*, Cambridge University Press (1991), which contains machine translations into BASIC of all the programs of the *Numerical Recipes*. The two should be consulted together: the *Numerical Recipes* for its explanatory text, the BASIC companion manual for its specific BASIC programs.

Note that the version of VBA included with recent versions of Excel is a dialect of Visual BASIC. It is, at the same time, a subset and an extension, because it is specialized for use with a spreadsheet. Therefore, if you want to learn more about VBA, consult books on VBA for Excel rather than those on Visual BASIC, because quite a few of the commands of Visual BASIC do not exist in VBA, and vice versa. Of the many books available, useful early introductions are E. Boonin, *Using Excel Visual Basic for Applications, the fast and easy way to learn*, Que (1996), and R. Jacobson, *Microsoft Excel / Visual Basic, Step by Step*, Microsoft Press (1995). As a general reference book use the *Microsoft Excel / Visual Basic Programmer's Guide*, Microsoft Press (1995). An extensive recent manual is J. Green, *Excel 2000 VBA*, Wrox Press (1999). Section 10.12 briefly summarizes the main syntactic features of VBA.

10.1a The macro module

A spreadsheet macro is a computer program, and in the spreadsheet is listed on a separate "sheet" called a **module**. It differs from a normal worksheet in that it does not contain any cells, but is blank except for the text printed on it. In Excel 5 and Excel 95, modules are conveniently stored as sheets of the workbook to which they belong. In fact, when you select Insert ⇨ Macro, you will find three choices: Module, Dialog, and MS Excel 4.0 Macro, of which you select Module. Likewise, when you right-click on the sheet tab, and select Insert, you will get a choice of Worksheet, Chart, Module, Dialog, and MS Excel 4.0 Macro, of which you again select Module. When you start a new project, use either method to generate a new module sheet, with the default name Module1. Once a module exists, you can effortlessly switch between spreadsheet and module, simply by clicking on their tabs. To summarize the procedure for Excel 5 and Excel 95:

1 Open a new spreadsheet.
2 Make a new Module sheet with the regular menu sequence Insert ⇨ Macro ⇨ Module. An Insert dialog box will appear.
3 Alternatively, right-click on the tab at the bottom of your spreadsheet (most likely labeled Sheet1) and in the resulting sub-menu click on Insert, on Module, then on OK.
4 In either case you will see a blank sheet labeled Module1.
5 Verify that you can readily switch between Module1 and Sheet1 by clicking on their tabs.

Starting with Excel 97, the modules are also stored together with the workbook, but they are not as readily visible. To generate a new module in Excel 97, Excel 98, or Excel 2000 requires the command sequence Tools ⇨ Macro ⇨ Visual Basic Editor, followed by Insert ⇨ Module. From then on, you can switch between spreadsheet and module with Alt + F11 if your keyboard has an F11 key. Otherwise, use File ⇨ Close and Return to Microsoft Excel (or Alt + Q) to move from module to spreadsheet, and Tools ⇨ Macro ⇨ Visual Basic Editor to get from spreadsheet to module. Unfortunately, you can no longer use the spreadsheet tab to create or switch to a module. In summary, starting with Excel 97 the corresponding procedure is as follows:

1 Open a new spreadsheet.
2 Select the VBA Editor with Tools ⇨ Macro ⇨ Visual Basic Editor, which will show you a screen with its left-hand side not just empty but dark, and with its own menu bar.
3 Using that menu bar, select Insert ⇨ Module. The dark space will become white, and will be your module.
4 If your keyboard has an F11 key, verify that you can readily switch between the module and the spreadsheet by clicking Alt + F11.
5 Also verify that you can switch from module1 to spreadsheet with File⇨Close and Return to Microsoft Excel (or Alt + Q).
6 Once you have written a macro, you can also switch from the spreadsheet to the macro module with Tools ⇨ Macro ⇨ Macros (or Alt + F8), selecting the macro by clicking on its name, and then depressing the Edit key.
7 While testing and debugging a macro, it is convenient to use the function key F5 to switch directly to the Macro dialog box, without leaving the VBA editor.]

Writing macros usually involves making and correcting mistakes. The VBA editor will catch quite a few **syntax errors** as you write or modify the macro. However, there are many errors the VBA editor cannot foresee, because they depend on particular parameter values that are yet to be computed, such as sqr(a) when a becomes negative. Such **run-time errors** typically will throw you back to the macro, where you can correct them if you can guess what the (often rather cryptic) error message is trying to tell you.

But that is not all. Often, the corrected macro will refuse to run, and even the macro icon will look funny. In that case, you may first need to **reset** the macro. You do this, in the VBA editor, with the command sequence Run ⇒ Reset or its shortcut, Alt + r, Alt + r. In order to simplify matters and save time, it is often useful during testing and debugging to reset the macro routinely with the sequence Alt + r, Alt + r, F5.

This chapter will deal only with modules. Dialog sheets allows you to design your own dialog boxes, which can give your macros a slick, professional-looking user interface. However, learning to construct functional dialog boxes requires a substantial investment of time and effort. Fortunately, almost the same functionality can be obtained with simple message and input boxes, which are much easier to program, and do not require the use of dialog sheets. In the physical sciences, the emphasis is usually on what a macro can do in terms of its mathematical prowess rather than on its convenience of data entry, or on controlling data access. In this book we will therefore leave dialog sheets alone. You, my reader, will judge for yourself whether the worked-out macros described in this chapter would have benefited from a more elaborate user interface. And if you feel strongly about it, by all means learn how to design dialog boxes, and incorporate them in your macros.

10.1b Reading and modifying the contents of a single cell

As our introduction to using a macro we will write several macros and see how they operate. As our first example we will take a highlighted cell and *read* the *value* it contains. Open a new spreadsheet, open the Visual Basic editor, insert a new module, and in it type:

```
'Read a cell value
Sub read()
myvalue = Selection.Value
MsgBox "The cell value is " & myvalue
End Sub
```

1 First we will explain, line-by-line, what these various instructions mean. The first line (starting with an apostrophe) is a **comment** line; it is ignored by the computer, and is purely for human consumption. In this case it summarizes what the macro will do. The apostrophe always identifies that what follows on that line (i.e., to the right of the apostrophe) is merely comment, and should not be executed by the computer.

VBA will show comments *in dark green*, which on some monitor screens may not show as significantly different from black. In that case you may want to select bright green instead. Do this in the VBA editor with Tools ⇒

Options, in the Options dialog box select the Editor Format tab, Code Colors, Comment Test, then select as Foreground the bright green strip. Also select bright blue for the Keyword Text, then exit with OK.

2 The second line indicates the actual beginning of the macro. It always starts with the word Sub, short for **subroutine** (a stand-alone program that can be used as a module), and then gives it a name (here: "read") to call it by. The brackets are to specify any information that must be passed into and out of the subroutine. A macro, which is a special type of subroutine, has no such **arguments**, so the brackets enclose nothing, but they are nonetheless required, as is the name.

3 The next line defines that myvalue (or any other name you might want to give it) should contain the value shown in whatever cell you will select in the spreadsheet. (This is merely a name, just as an unknown parameter might be given the symbol x in an algebra problem, or a concentration the symbol c or the name conc. In VBA one can use long, self-explanatory names, algebra-like single-letter symbols, or anything in-between, as long as the name starts with a letter, and contains neither empty spaces nor any of the special characters ., !, #, $, %, or &. Long, composite names can be made more easily readable with interspersed capitals, as in myValue, or with underscores, as in my_value.

When the program encounters the line myvalue = Selection.Value, it will assign to the parameter called myvalue the value it finds in the high-lighted cell. You can read the line as: assign to myvalue the Value property from the Selection. Value and Selection are terms that VBA recognizes, and as such will be shown *in dark blue* on the monitor screen. They will also be capitalized automatically, even if you don't.

4 Next comes a check: the message box (which VBA spells as MsgBox) will display the value to verify that it has been read correctly. Message boxes are very useful for this purpose, especially during the debugging stage; they can be removed (by deleting the entire line or, simpler, by preceding them with an apostrophe) after testing has shown that that particular program segment works.

5 The final line lets the computer know that this is the end of the subroutine.

6 Now switch to the spreadsheet, fill several cells with numbers, or with for-mulas that produce numerical values, and highlight one of them.

7 Select Tools ⇨ Macro

8 In the resulting Macro dialog box you should see the macro name in the large window. When you select it by clicking on it, the name of the macro will also appear in the top window labeled Macro Name/Reference. Then click on the Run button.

9 You should now see a message box that contains the text "The cell value is" followed by the value read. For example, if you highlight a cell that con-tains the value 3.456, then the message should read "The cell value is 3.456". If you select a cell containing the formula = SQRT(9) + 1.7 you should see "The cell value is 4.7", and so on.

10 The message box insists on being acknowledged, i.e., you *must* click on its OK button before you can do anything else in the spreadsheet.

11 Try several cells, of varying content, and verify that the program works.

12 By assigning a function key to the macro you can avoid the rather laborious procedure of steps 7 and 8 for calling it. In order to do so, again select Tools ⇨ Macro, and in the Macro dialog box click on "read" to select this particular macro. (If a macro has already been selected, and you want another macro, click on that one.) Now click on Options; in the resulting dialog box click in the small window for the Shortcut key, which should now show a tick mark, then click in the window for Ctrl + and enter the letter z. (Any letter will do. However, in assigning letters to macros, avoid those Ctrl-letter combinations that you may want to use for editing purposes, such as Ctrl + x for cut, Ctrl + c for copy, and Ctrl + v for paste.) Return to the main dialog box by clicking on the OK button, and execute the macro by clicking on Run. From now on, you can call the macro merely by depressing the Ctrl key and, simultaneously, the z key (or whatever other letter you assigned to the macro). Try it.

13 A macro can not only read information, it can also manipulate it. We will demonstrate this by changing the value of the cell we have just read. To do so, return to Module1, and modify the macro as shown below. For your convenience, all new or modified lines are printed here in bold. (Such boldfacing should not be entered in the macro; fortunately, it can't.)

```
'Read and change a cell value
Sub read()
myvalue = Selection.Value
MsgBox "The cell value is " & myvalue
myvalue = myvalue + 4
Selection.Value = myvalue
End Sub
```

14 The line myvalue = myvalue + 4 should not be taken literally, as in an algebraic equation. Instead, it should be read as an **assignment**, in which the value of the right-hand side is assigned to the parameter specified in the left-hand part.

15 It clearly would have been preferable to write this line as myvalue ⇐ myvalue + 4, i.e., with ⇐ instead of an equal sign, but no corresponding character was available on the teletypewriter keyboards for which the early computer languages, including BASIC, were developed. A few recently created computer languages indeed use the symbol : = which somewhat resembles the back arrow ⇐ and explicitly shows a directionality. A few recent VBA instructions also use this symbol (which you will encounter in, e.g., the syntax of input boxes), but most still employ the traditional BASIC assignment symbol =.

16 The next line again illustrates the directionality of the assignment. Unlike an equality, which can be read from either side, Selection.Value = myvalue writes the value of myvalue over the earlier value, quite the

opposite from the line `myvalue = Selection.Value`, which reads that value but does not change anything. Again, the line should be read as if it were printed as `Selection.Value ⇐ myvalue`, and the earlier line as `myvalue ⇐ Selection.Value`.

17 Try the modified macro. When you select an empty cell, the message box will report that it is empty, and thereafter the macro will add 4 to that value, to give the cell the value 4; when you repeat the process with the same cell, the macro will up the ante to 8, then to 12, and so on.

18 Modify the macro to perform another operation, such as a multiplication, or whatever. Experiment with it.

19 It is often undesirable to overwrite an existing answer. We need not do that, but instead can write the modified result somewhere else in the spread-sheet, say immediately to the right of the highlighted cell, or just below it. (Of course, if the latter cell already contains something, that will now be overwritten, so be careful where you deposit the output of your macro.) Modify the macro as follows, then test it.

```
'Read and change a cell value
Sub read()
myvalue = Selection.Value
MsgBox "The cell value is " & myvalue
myvalue = myvalue + 4
Selection.Offset(0,1).Select
Selection.Value = myvalue
End Sub
```

20 The first number in the offset specifies the number of rows to be shifted (down), the second the number of columns (to the right). We can move two cells to the left with, say, `Offset(0,-2)`, one cell down and two to the left with `Offset(1,-2)`, and so on, provided that there is space to go to. Try it out.

21 When the offset directs the result to a cell outside the spreadsheet, an error message appears, warning you of a **run-time error**, i.e., of an error that could not have been foreseen by the computer when it checked the program (because, in this example, the program could not know at that time what cell you were going to select) but that only occurred during the *execution* of the macro. The message box that you will see actually describes the error, albeit somewhat cryptically, as "Offset method of Range class failed". So now you know! Pressing <u>G</u>oto will get you back to Module1, and will highlight the offending instruction. Modify it to move to the right and down, and verify that it is now working smoothly again. As you can see from this example, program crashes in VBA are almost as gentle as those in Excel, and the error messages only barely more informa-tive.

Before you leave the VBA editor, you must 'reset' it. You can do this with <u>R</u>un ⇒ <u>R</u>eset, or with Alt + R, Alt + R. Only after that is done should you return to the spreadsheet with Alt + F11. If you forget to reset the editor, you can return to the spreadsheet but you cannot rerun the macro; in that case, go back to the VBA editor with Alt + F11, reset it, and return to the spreadsheet.

22 We will now return to our rather simple first macro, and extend its usefulness by letting it read and display several self-explanatory properties of a highlighted cell.

```
'Read the cell address, value and formula
Sub read()
myaddress = Selection.Address
MsgBox "The cell address is " & myaddress
myvalue = Selection.Value
MsgBox "The value is " & myvalue
myformula = Selection.Formula
MsgBox "The formula is " & myformula
End Sub
```

Return to the spreadsheet, select a cell, enter a formula in it, such as = 3.4 + 2*LOG(7) or whatever else suits your fancy, and test it.

23 This brings up a point worth mentioning here. Excel and Visual BASIC grew up separately, and have only recently been linked. Some aspects of the marriage still need to be ironed out. This is evident in the VBA math functions. While most of these are the same in Excel and VBA, there are a number of differences.

All VBA math functions are coded with three letters, while Excel functions have no such constraints. Consequently, the sign of x is given by $\text{sgn}(x)$ in VBA, but by $\text{sign}(x)$ in Excel; $\sqrt{x}$ is $\text{sqr}(x)$ in VBA, $\text{sqrt}(x)$ in Excel; a random number is generated in VBA with $\text{rnd}(x)$, in Excel with $\text{rand}()$; and $\arctan(x)$ is specified by $\text{atn}(x)$ in VBA but in Excel by $\text{atan}(x)$. It is annoying that the macro editor of Excel does not (also) accept the Excel functions, or at least alerts the user to these spelling differences. It is outright confusing when the meaning is changed: $\text{fix}(x)$ in VBA does not always give the same result as $\text{fixed}(x,0)$ in Excel. The worst offender is $\log(x)$ which, in VBA, represents the *natural*, e-based logarithm, whereas in Excel (and in almost everyone else's nomenclature) it is written as $\ln(x)$. VBA does not even have a symbol for the 10-based logarithm, so that the 10-based $\log(x) = \ln(x)/\ln(10)$ in Excel must be written in VBA as $\log(x)/\log(10)$. Sorry, even Bill Gates naps sometimes.

24 Try to fool the macro by entering a letter, word or sentence. The message box will return the correct address, but will not be able to distinguish between a value and a formula. Instead, in both cases it will merely repeat the text it finds.

25 Note that, in reading a cell address, the macro editor automatically adds dollar signs for absolute addressing.

10.1c Reading and modifying the contents of a block of cells

We now extend the macro to read not just the contents of a single cell, but of an entire, rectangular block of cells. Such a block constitutes an **array**, which can contain many different values and formulas. To test it we therefore specify a particular cell in that array, as shown here.

```
'Read an array
Sub ReadArray1()
Dim myaddress As Variant, myvalue As Variant, myformula As Variant
myaddress = Selection.Address
MsgBox "The array covers the range " & myaddress
myvalue = Selection.Value
MsgBox "The value of cell (1,1) is " & myvalue(1,1)
MsgBox "The value of cell (3,2) is " & myvalue(3,2)
myformula = Selection.Formula
MsgBox "The formula in cell (1,1) is " & myformula(1,1)
MsgBox "The formula in cell (3,2) is " & myformula(3,2)
End Sub
```

1 This latest extension illustrates how a macro handles arrays. The array must be **dimensioned**, i.e., the computer must be told to expect a *multi-valued parameter* that is organized in rows and/or columns. In VBA it is easiest to dimension an array As Variant, which leaves its precise size and nature unspecified. (The only disadvantage of always using As Variant is that it is inefficient in terms of computer memory and execution speed. Fortunately, with modern computers,we only need to consider those factors in programs that involve rather large data arrays.)

2 Test this macro on an array of formulas, numbers, or text (i.e., one or more letters, words, or sentences). Observe that it will return the address range whether the cells involved are empty or not. It will return no information for the value or formula of an empty cell (but that it comes up empty-handed will stop neither the program nor the computer), and it obviously will not be able to distinguish between a value and a formula when a cell merely contains a number, or text. Still, where there is valid information to be had, the macro will read it properly, and the message boxes will report it.

3 An array is always assumed to have two dimensions: rows, and columns. Therefore, when you select a column, and want to read, say, the value in its second cell, you must specify this cell as cell (2,1), even though there is only one column. Likewise, in a single row, the fifth cell is identified as cell (1,5) because this cell has the row index 1, and the column index 5.

The following macro demonstrates an even more extensive use of message boxes, and a simplification in that only one Array is used, which can then be examined Item by Item. (Note again that VBA considers everything to the right of an apostrophe as a comment. For short instructions and comments, writing the comments to the right of the instruction leads to compact yet very readable code.) Enter this macro and test it on a small sample array.

```
Sub ReadArray2()                        ' defines macro and names it
Dim Array As Variant                    ' dimensions the array
nr = Selection.Rows.Count               ' counts the number of rows
MsgBox "The number of rows is " & nr    ' displays nr
```

```
nc = Selection.Columns.Count          ' counts the number of columns
MsgBox "The number of columns is " & nc ' displays nc
Array = Selection.Value               ' selects highlighted array
MsgBox "The highlighted range is " & Selection.Address
MsgBox "The top left cell has the address " & _
  Selection.Item(1, 1).Address
MsgBox "The top left cell contains the value " & _
  Selection.Item(1, 1).Value
MsgBox "The top left cell contains the formula " & _
      Selection.Item(1, 1).Formula
MsgBox "The bottom left cell has the address " & _
      Selection.Item(nr, 1).Address
MsgBox "The bottom left cell contains the value " & _
      Selection.Item(nr, 1).Value
MsgBox "The bottom left cell contains the formula " & _
      Selection.Item(nr, 1).Formula
MsgBox "The bottom right cell has the address " & _
      Selection.Item(nr, nc).Address
MsgBox "The bottom right cell contains the value " & _
      Selection.Item(nr, nc).Value
MsgBox "The bottom right cell contains the formula " & _
      Selection.Item(nr, nc).Formula
End Sub
```

In VBA, a space followed by an underscore indicates that the line is to be continued. VBA can accommodate lines with up to 256 characters (including spaces), but such long lines are difficult to print, or to read on a monitor screen.

10.1d Two different approaches to modifying a block of cells

As our last example in this section we will take a highlighted block of data, and raise all of its elements to some power, for which we will here select 3. (With an integer power it is somewhat easier to verify by simple inspection that the macro works properly. However, you are free to pick any other number, not necessarily either positive or integer.) Type the following macro, give it a shortcut key, and test it.

```
'Cube all elements of an array
Sub Power()
For Each cell In Selection.Cells
    cell.Value = cell.Value ^ 3
Next cell
End Sub
```

This example illustrates how easy it is to use a **For Each ... Next loop** to address all cells in a given, highlighted block, even without ever specifying it as an array. The name 'cell' specifies the individual cells in the highlighted block, and the loop addresses each of them in turn. (Any other name instead of 'cell' would have worked just as well. You can see that 'cell' is not recognized by the VBA compiler because it is not capitalized. (But "Cells" in

"Selection.Cells" at the end of the third line is both required and recognized by the compiler.)

For readers familiar with the way arrays are addressed in more traditional languages such as FORTRAN or BASIC, the following, more conventional macro would have achieved the same objective:

```
'Cube all elements of an array
Sub Power1()
Dim myArray As Variant
Dim rn As Integer, cn As Integer
myArray = Selection.Value
For rn = LBound(myArray, 1) To UBound(myArray, 1)
  For cn = LBound(myArray, 2) To UBound(myArray, 2)
    myArray(rn, cn) = myArray(rn, cn) ^ 3
  Next cn
Next rn
Selection.Value = myArray
End Sub
```

1 The second dimension statement, "Dim rn As Integer, cn As Integer", specifies that the parameters called rn and cn (for **row n**umber and **c**olumn **n**umber respectively) are integers. It is useful to dimension them as such, because VBA treats integers quite differently from general numbers. When no dimension is specified, VBA assumes by default that the parameter is As Variant, its most flexible dimension. Note that a statement such as "Dim rn, cn As Integer" might be misleading, because it is interpreted by VBA as "Dim rn" followed by "Dim cn As Integer", so that it dimensions rn simply by name (and therefore assigns it by default the dimension As Variant), whereas it dimensions cn as an integer.

    A parameter involved in specifying the cell coordinates in an array is called an index; if there is any ambiguity about their nature, indices should be declared as integers. (Special VBA functions such as LBound and UBound, for lower bound and upper bound respectively, always specify integers, and therefore need not be dimensioned as such.)

2 Below the assignment "Array = Selection.Value" we encounter two nested For Next loops. In each loop a calculation is repeated a specified number of times. In this example, each numerical value in the array is in turn raised to its cube power (or to any other power we might have specified instead of 3). Incidentally, note that the indentation used facilitates our reading of the macro, by showing what lines are involved in the two different loops.

3 In order to keep track of the various numbers to be cubed in the highlighted block, the inner loop (the three lines starting with "For cn = " and ending with "Next cn") calculates the cube of all numbers on a given row of the highlighted area. The program determines the first and last column of the array as LBound(Array, 2) and UBound(Array, 2) respectively, where the 2 denotes that these are the Lower and Upper bounds of the columns. (Array notation specifies rows first, then columns, so that rows are the first array dimension, and columns its second dimension.)

4 The outer loop (starting with "For rn = " and ending with "Next rn") makes sure that the calculation is then repeated for all elements of a given column in the highlighted area, their row indices being specified by LBound(Array, 1) To UBound(Array, 1), where 1 denotes the rows. Again, the macro automatically determines the lower and upper bounds of the highlighted area, so that you need not supply that information.

5 The next line, between "Next rn" and "End Sub", assigns the (now altered) values of Array to the Values of the Selection. In other words, this line writes the results of the macro calculation back into the highlighted area, thereby overwriting the original data.

6 It may be useful to dissect the macro Power1, because it contains five common, distinctive parts of a well-written Excel macro:

   a The first and last lines define the macro as such, and specify its name.
   b Lines 2 and 3 specify the dimensions used.
   c The fourth line assigns the values in the highlighted block to the array called Array, i.e., it *reads* the selected data.
   d The next five lines constitute the heart of the macro, because they are the lines that specify the actual calculation. In fact, only the third one of these five does the actual calculation; the others are needed to specify which data should be cubed. Note that there is nothing special about the particular mathematical operation (cubing) used here: any other operation that can be coded in BASIC could have been used instead.
   e Finally, the penultimate line *writes* the results, in this case back into the active block.

7 On the other hand, the macro Power dispenses with much of this, and simply condenses steps (b) through (e) into a simple, single For Each … Next loop.

8 Why discuss these two macros here, when they achieve the same purpose, while one is so much simpler than the other? The reason is that they are not quite equivalent. For an array of numbers, defined and stored as their values, macros such as Power and Root are our first choice, since they are indeed simpler to read as well as faster to execute. However, they have an Achilles heel: since they modify one cell at the time, in an unspecified order, they are not reliable when the cell values are mutually dependent.

9 For example, use two columns, A and B. Enter the value 1 in both cells A1 and B1. Enter the number 2 in cell A2, highlight cells A1 and A2, and drag the cells down by their common handle to, say A10. On the other hand, in cell B2 we deposit the instruction = B1 + 1, and then copy and paste this into cells B3:B10. In both columns we will now see the sequence 1, 2, 3, …, 9, 10.

10 When we use Power on A1:A10 we will obtain 1, 8, 27, 64, 125, 216, 343, 512, 729, and 1000. However, when we use Power on B1:B10, we will instead get the sequence 1, 8, 729, 3.89E + 08, 5.89E + 25, 2.04E + 77, 8.49E + 231 or ########, and from there on nonsense. Clearly, the macro calculates B2 as 2^3 before it starts to compute the cube of B3, so that B3 at that time reads

$B2 + 1 = 8 + 1 = 9$, hence we find the result $9^3 = 729$ instead. In B4 the macro then computes $730^3 = 3.87E + 08$, and from there the size of the results quickly escalates to exceed the numerical capacity of the spreadsheet (approximately 9.9×10^{307}) in B8, which is why we see the overflow sign ########. Therefore one should use For Each … Next loops only on cells containing values that are independent of all other data in the selected block.

11 Power1 first reads the numerical values of all cells, and stores them in an array. The mathematical operations are then performed on these stored data, so that there is no ambiguity involved. Consequently, Power1 yields correct results for either column. However, Power1 may fail when the selected block is too large (which, depending on the amount of available memory, and the version of Excel used, typically will happen only when the array has more than a few thousand elements). In that case, use Special Copy Values to duplicate the numerical values of the array elements in another block, then use a macro such as Power that does not suffer from size limitations.

12 Verify that the macros Power and/or Power1 work with one- or two-dimensional arrays, i.e., with blocks regardless of their size, as long as they contain two or more cells in the array. Note that the macro will not accept a single cell as valid input: an array must contain more than one cell.

10.1e Numerical precision

In order to test how well macros such as Power or Power1 work, we now make a complementary macro to counteract the effects of the first. To do so, switch back to the Module1 page, highlight one of the macros (either Power or Power1) in its entirety, and use Ctrl + c and Ctrl + v to deposit a copy of it lower on the same Module1 page. Then modify the copy to read (if you use Power)

```
Sub Root ()
For Each Cell In Selection.Cells
  Cell.Value = Cell.Value ^ 1/3
Next Cell
End Sub
```

where we have again boldfaced the two changes to be made, one in the macro name (you are again welcome to select a different name here), the other in the mathematical operation, which now calculates the inverse operation, i.e., the cube root if you earlier raised to the cube power. If you use Power1, make the corresponding two changes in that macro instead.

1 Assign the new macros separate shortcut keycodes. (Such keycodes are case sensitive, i.e., they distinguish between lower case and capitals. You can therefore select Ctrl + Z, which you obtain by simultaneously depressing

the Ctrl, Shift and z keys, which is easily done with three fingers.) Use a pair of corresponding macros (Power & Root, or Power1 & Root1) to test their performance, by first raising numbers to some power, and by then taking their corresponding roots. This should undo the effect of the first macro, and recover the original values.

2 Let the spreadsheet compute the differences between the original and final data. First make an array containing, say, some simple integers, and copy it. On the copy, now test the operation of the two macros, e.g., by first using the Power macro three times in succession on the same data, followed by three uses of the Root macro. In principle, this should return the original data. In practice, some round-off errors may creep in, typically of the order of 10^{-6} or so, which are easily visualized by computing in a third array the differences between the original data and the copy. The small differences shown may be perfectly satisfactory for most purposes, but since in Excel all standard spreadsheet operations are performed at a much higher precision, it is useful to fix the macro which, apparently, worked in the so-called **single-precision** data mode.

3 The cause of the problem lies in the numbers 3 and 1/3 used to raise the individual terms of the array to their cube power, and to take their cube root. When we raised the individual cells in the block (or the individual elements in Array) to their cube power, we reduced the results to single precision. To demonstrate this and, at the same time, fix the problem, add a dimension statement to *both* macros, reading "Dim p As Double", and add a line specifying the value of p (e.g., either "p = 3" or "p = 1/3"). Finally, in the For Each … Next (or in the nested For … Next loop) refer to p rather than to 3 (or 1/3). The macros now should read like the example shown below, where we have also added a title and some comment lines and, as before, bold-faced the most recent changes.

```
' ROOT
'
' An example showing how to capture data from a high-
' lighted block of cells, and raise all its terms to
' some power, in double precision. Note that the input
' data are overwritten by the output data.
'
Sub Root()
Dim p As Double
p = 1 / 3
For Each Cell In Selection.Cells
   Cell.Value = Cell.Value ^ p
Next Cell
End Sub
```

4 Verify that repeated use of the Power and Root (or Power1 and Root1) macros now does lead to computational errors with relative values smaller than 10^{-15}. This final example illustrates that, while the spreadsheet always (and automatically) treats data as double precision, the mathematical operations in VBA must be told specifically to do so!

In general, VBA allows us to leave the parameter dimensions unspecified. However, we have already seen that dimensioning is *necessary* for arrays, and for using double precision. Moreover, dimensioning is *advisable* for integers, because it saves computer memory and speeds up macro execution. Dimensioning is even advisable in general, because it can help us catch typos, an aspect to which we will return later. Regardless of the reasons for it, a parameter dimension must always be specified *before* that parameter is used in the macro. During macro writing, it is often convenient to place the dimension statement at the top of the block of instructions where the parameter is first used. At the final clean up you may then want to collect and combine all dimension statements, and place them at the top of the macro. This makes it easier to find them, guards against dimensioning the same parameter twice, and reduces the line count of the macro.

10.1f Communication via boxes

We have already seen in some of the sample macros of the previous section that a **message box** is readily incorporated in a macro, and can then help to verify that a macro works properly. It can also deliver a warning or other message to the user, and it can even carry a limited amount of information back to the macro. There are two other types of boxes that can be used if more information needs to be exchanged between the user and the program. An **input box** is essentially a message box *with a window* in which the user can place specific information (such as numerical values, formulas, addresses, text) that can then be read by the macro. A **dialog box** is the most versatile type of communication box, since it can contain any number of input windows and buttons, of all the types encountered in Excel dialog boxes. Dialog boxes can give macros a professional appearance. However, their construction often takes too long to make them worthwhile for the simple programs with which we deal here, and we will therefore not discuss them. Instead, we will use the much simpler input boxes to achieve the same goal. Below, then, we restrict the discussion to message and input boxes, in that order.

All message boxes carry at least one button, usually just the OK button that must be pressed to make the box disappear and to continue the macro, thereby making sure that the user has acknowledged the message. The use of message boxes can be extended so that they carry a limited amount of information from the user back to the macro during the execution of that macro. For example, the algorithms for forward and inverse Fourier transformation are very similar, and one might want to use a single macro instead, combined with a message box that provides the necessary binary information so that the user can choose one or the other. Imagine that, in this case, the needed information is stored in the value of an integer iSign, which can have the value of either $+1$ or -1. (How to use such information in the macro will be illustrated in section 10.5.) Likewise, the weighted least squares program of section 10.7 has two forms, depending on whether the intercept is variable or is set equal to zero. In this case, a single program with a message box can again be used instead of two separate macros. Below we show two examples of message boxes that could be used to provide information to the Fourier transformation program. Try them out.

```
Sub Box1()
Ans = MsgBox("Forward Fourier transform?", vbYesNo)
If Ans = vbYes Then iSign = 1
If Ans = vbNo Then iSign = -1
MsgBox "The value of iSign is " & iSign        'for testing
End Sub
```

```
Sub Box2()
Msg = Msg & "Press YES for a forward transform, "
Msg = Msg & Chr(13)
Msg = Msg & "press NO for an inverse transform."
Config = vbYesNo + vbQuestion
Title = "Forward or inverse FT?"
Ans = MsgBox(Msg, Config, Title)
If Ans = vbYes Then iSign = 1
If Ans = vbNo Then iSign = -1
MsgBox "The value of iSign is " & iSign        'for testing
End Sub
```

You will recognize that both message boxes accomplish the same purpose, the only difference being that the second is somewhat more elaborate, with a title and a question mark. Chr(13) is the character code for a linefeed, so that the message appears on two lines. The general syntax of a message box can be either

```
MsgBox "message", configuration, "title"
```

or

```
Answer = MsgBox ("message", configuration, "title")
```

while the message, configuration and/or title can be specified either directly following MsgBox or in separate lines. Here *message* denotes the (obligatory) message, which can consist of text and/or numbers: text must be within quotation marks ("), while text and specific output information must be separated by an ampersand, &. The optional *configuration* refers to buttons and/or icons. When configuration is left blank, only an OK button will show, the status of which can be determined by calling vbOK. Possible two-button combinations are designated with the self-explanatory names vbOKCancel, vbYesNo, and vbRetryCancel, and there are also two three-button combinations, vbYesNoCancel and vbAbortRetryIgnore. There are four possible icons, addressable as vbInformation, vbQuestion, vbExclamation and vbCritical respectively. (Excel 97 and later versions have a few more, including three with specific default buttons.) When both a button (other than the default vbOK) and an icon is required, they are both specified, separated by a + sign, as in Box2(). Finally, the optional *title* defines the text in the blue space at the very top of the message box. If unspecified, the default title is Microsoft Excel. Below, several of these combinations are shown. Play with them to see what they do!

```
MsgBox "Nicely done! The answer was indeed " & 79
MsgBox "Nicely done! The answer was indeed " & 79, vbExclamation
MsgBox "Correctly answered!",, "Congratulations"
  'Note the two commas needed to make "Congratulations" a title
MsgBox "Correctly answered!", vbExclamation, "Congratulations"
MsgBox "Correctly answered! Do you want" & Chr(13) & _
  "to try your luck one more time?", vbYesNo + vbExclamation, _
  "Congratulations"
  'Note the plus sign to specify buttons as well as an icon,
  'and a more compact way to write a multi-line message.
```

Despite these embellishments, a message box works mostly as a one-way messenger: it can give you information *from* the macro, but it can only carry limited, binary information (of the yes/no type) back to the macro. The simplest way to provide general (i.e., numerical or textual) information *to* the macro is through an **input box**. As our first example of an input box consider Sub Box3(). The optional items in parentheses are, in this order, the message and the box title. We use a message box on the next line to verify that the program has indeed read the information. Check it out.

```
Sub Box3()
myValue = InputBox("Enter your favorite number here", "Number")
MsgBox "You chose the number " & myValue   'optional verification
End Sub
```

The following example shows that the entered number has really been read properly, because the macro does not merely accept it as some kind of text but considers it as a number, and rejects incorrect offerings. Here, just for the sake of variety, the (obligatory) message and the (optional) title are specified in separate lines, and an (optional) default value is specified as well. Again, if we want to specify a default value but no title, two commas need to separate the message from the default to let the program know what is what. Here the arguments (such as Message, Title, and Default) of the input box are specified in advance, which makes the code longer but perhaps somewhat easier to read.

```
Sub Box4()
Message = "Enter an integer between 1 and 100 here:"
Title = "Integer"
Default = "1"
myValue = InputBox(Message, Title, Default)
If myValue < 1 Then
  MsgBox "The selected number is too small."
  End
End If
If myValue > 100 Then
  MsgBox "The selected number is larger than 100."
  End
End If
```

```
If myValue - Int(myValue) <> 0 Then
  MsgBox "The selected number is not an integer."
  End
End If
MsgBox "You chose the number " & myValue
End Sub
```

As our next example we use the input box to specify a single address, then verify that the address was read correctly by displaying it in a message box. When an input box is used to read either a single address or an address range of a multi-cell array (a block), preface the InputBox instruction with the command `Set`, and use `Application.` before `InputBox`. Then use that address to select the range, as in

```
Sub Box5()
Dim myAddress As Range
Set myAddress = Application.InputBox _
  (Prompt:="The address is:", Type:=8)
myAddress.Select
MsgBox "The address is " & myAddress.Item(1, 1).Address
End Sub
```

The address can be entered either as text, or by highlighting the cell involved with the mouse pointer. In the latter case it may be shown in the message box in so-called row-column notation, as in R3C7 for row 3 column 7, i.e., G3. Note that Type: $= 0$ will read a formula, Type: $= 1$ a number, Type: $= 2$ a text string, Type: $= 4$ will yield a logical value (True when the address contains a number, False when it is empty or contains a zero), Type: $= 8$ will read a cell reference as a Range, and Type: $= 64$ will read an array.

In order to read other information at that address, such as a value, you can proceed as in

```
Sub Box6()
Dim myAddress As Range
Set myAddress = Application.InputBox _
  (Prompt:="The address is:", Type:=8)
myAddress.Select
myValue = Selection.Value
MsgBox "Address " & myAddress.Item(1, 1).Address & _
  " contains the value " & myValue
End Sub
```

although it would be simpler to turn things around, i.e., to use the input box to read the value and, as an afterthought, obtain the associated address.

Our final example shows how to read the address range of a multi-cell array (a block), and to verify that you have indeed read that information correctly:

```
Sub Box7()

Dim myRange As Range
Set myRange = Application.InputBox _
  (Prompt:="The range is:", Type:=8)
```

```
myRange.Select

' Verify that we have read the range values

Dim NR As Integer, NC As Integer
NR = myRange.Rows.Count
NC = myRange.Columns.Count
MsgBox "The range is " & myRange.Item(1, 1).Address & _
   ":" & myRange.Item(NR, NC).Address
Dim myValue As Variant
myValue = Selection.Value
For j = 1 To NC
  For i = 1 To NR
    MsgBox "Address number " & i & "," & j & " is " & _
    Selection.Item(i, j).Address & _
    " and contains the value " & myValue(i, j)
  Next i
Next j

End Sub
```

Because a message box cannot display a multi-address range by simply calling it as `MsgBox "The range is " & myRange`, we have circumvented this problem here by specifying the first and last address in the array, and by then displaying that information in the spreadsheet format, with a colon as separator. The second message box illustrates how to pack a lot of information into a single, small message box.

Finally, note that one can readily trip up these last three boxes by entering a range into Box5 or Box6, or a single address in Box7. This may be acceptable if you use the macro only yourself. However, if you intend the macro to be shared with others, you should add some lines of code to catch such likely errors. In the worked-out macro programs of the next sections you will encounter several examples of such input-checking code.

10.1g Subroutines

When the same task is performed at different places in a macro, or is needed in several macros, we can simply repeat the code for it every time it is needed. However, when that task is nontrivial, it is often more convenient to isolate the corresponding code, and to place it in an independent **subroutine**, that can be invoked ('called') whenever needed, from anywhere inside a macro, or from inside any other macro. For example, in section 10.5 we will use a Fourier transform, and we will also need to perform several Fourier transformations in section 10.6. Rather than repeat the same code several times, it then makes more sense to place the code for computing the Fourier transformation in a separate subroutine, callable from anywhere. Moreover, since a subroutine is a free-standing program, it is easily tested as such.

The subroutine call must specify all parameters that the subroutine must have to do its job, and those it must communicate back to the calling program. This is where the brackets are used: they specify the data that must go into and/or come out of the subroutine.

Below we will illustrate this with an example so simple that you normally would not write a separate subroutine for it.

```
Sub myMacro()
a = 1
b = 2
c = 0
MsgBox "(1): Before the subroutine call, c = " & c
Call mySum(a, b, c)
MsgBox "(4): After the subroutine call, c = " & c
End Sub

Sub mySum(a, b, c)
MsgBox "(2): a = " & a & ", b = " & b & ", c = " & c
c = a + b
MsgBox "(3): a = " & a & ", b = " & b & ", c = " & c
End Sub
```

Note that the brackets must contain the parameters that go into the sub-routine (here: *a* and *b*) as well as those that it must report out (in this example: *c*). The number of these parameters must be the same, and their data type must also match. However, their names may differ, as illustrated here:

```
Sub mySum(q, r, s)
MsgBox "(2): q = " & q & ", r = " & r & ", s = " & s
s = q + r
MsgBox "(3): q = " & q & ", r = " & r & ", s = " & s
End Sub
```

This subroutine, when called by myMacro( ), will translate *a* into *q*, and *b* into *r*, will then compute *s*, and return it to myMacro( ), which will interpret the latter as *c*. The translations are done purely on the basis of the order in which the parameters appear inside the brackets, so you must be careful with these. In fact, you could write the following subroutine.

```
Sub mySum (b, c, a)
a = b + c
End Sub
```

and get away with it, although it might confuse a reader (though not the computer) that the symbols used in the subroutine are a permutation of those in the calling program. Try it out. It will show you that the subroutine assigns parameter values exclusively on the basis of their positions in the subroutine call, not on the basis of their names. This makes it easy to call the subroutine from different macros, but initially may be confusing. In that case, use the same names throughout.

Note that the output parameters will be computed but will not be trans-ferred back to the calling macro when they are not specified in the subroutine call. For example, try the following combination, which calculates c all right, but fails to make it available to the calling macro:

```
Sub myMacro ()
a = 1
b = 2
c = 0
MsgBox "Before the subroutine call, c = " & c
Call mySum (a, b)
MsgBox "After the subroutine call, c = " & c
End Sub

Sub mySum (a, b)
MsgBox "a = " & a & ", b = " & b & ", c = " & c
c = a + b
MsgBox "a = " & a & ", b = " & b & ", c = " & c
End Sub
```

Such problems with data transfer can be avoided by declaring the parameters involved as Public, in which case they are accessible throughout the entire VBA module. Public parameters must be declared at the very beginning of the module, in the format of a dimension statement, but with `Public` instead of `Dim`, as illustrated below. Note that, in this case, we can merely invoke the subroutine by stating its name, leaving out its brackets and, even, the instruction `Call`.

```
Public a, b, c

Sub myMacro ()
a = 1
b = 2
c = 0
MsgBox "Before the subroutine call, c = " & c
mySum
MsgBox "After the subroutine call, c = " & c
End Sub

Sub mySum
MsgBox "a = " & a & ", b = " & b & ", c = " & c
c = a + b
MsgBox "a = " & a & ", b = " & b & ", c = " & c
End Sub
```

The program shown in section 10.9 illustrates this format.

10.2 A case study: interpolating in a set of equidistant data

As an example of writing a macro we will here illustrate its progress, including its validation at various stages. As our example we will use the Savitzky–Golay method to interpolate in a set of equidistant data points. In other words, given a set of data x,y for evenly spaced x-values, we want to estimate a value for y at an arbitrary value for x inside the range covered. In this example we will first consider interpolating in a set of seven adjacent data points, and we will assume that, over the limited range of those seven adjacent data, their y-values can be approximated satisfactorily by the quadratic expression $y = a + bx + cx^2$.

The macro must of course be based on an algorithm, for which we will use the Savitzky–Golay method, which we briefly encountered in section 3.3, described in somewhat more detail in section 8.5, and further apply in section 10.9. The macro should do the following:

1 read the x,y data
2 read the specific x-value XX where the function must be interpolated
3 compute and display the corresponding, interpolated y-value YY

In our example we will assume that the x- and y-values are stored in columns on the spreadsheet. We will read them separately, so that those columns need not be adjacent to each other. We will therefore use three input boxes, one each to enter the x-values, the y-values, and the specific x-value for which we want to interpolate the data. We will use a message box to display the resulting y-value.

You may often want to start writing a macro by first testing its algorithm. Once the central logic of the macro is working smoothly, you can then add the segments dealing with data input and output. Alternatively, when the algorithm is straightforward and well-understood, it may be easier to start at the data input stage (which often takes the most effort), and to work from there in a linear fashion. In the present example we will use the latter approach.

10.2a Step-by-step

No matter how we approach the task of writing a macro, we have to do it step-by-step. It is easiest to validate every small part as we write it, by including performance tests at every stage, and by progressing to the next segment of code only after the earlier parts have been checked and verified. The performance testing is best done by including message boxes that validate the code, and by repeatedly going back and forth between the macro and the spreadsheet to verify the code at every step. After validation, 'comment-out' the test lines by preceding them with, say, a triple quote, ''', or '/, in order to distinguish them from those comment lines (starting with a single') you want to keep in the final version. Then, when the macro is complete, remove all test lines starting with ''' or '/. (In order to save printed space, we have removed the test lines after their first use in the examples given below.)

First we write and validate code to read the x-values. Start with a new spreadsheet, open a module (Insert ⇨ Macro ⇨ Module in Excel 5 and Excel

95, Tools ⇨ Macro ⇨ Visual Basic Editor, followed by Insert ⇨ Module, in Excel 97 and later versions) and in it type

```
Sub Interpolate()

' Read the range of X-values

Dim myXRange As Range
Set myXRange = Application.InputBox _
   (Prompt:="The X-range is:", Type:=8)
myXRange.Select

' Verify that we have read the X-values

Dim NX As Integer
NX = myXRange.Rows.Count
MsgBox "The range is " & myXRange.Item(1, 1).Address & _
   ":" & myXRange.Item(NX, 1).Address
Dim XValue As Variant
XValue = Selection.Value
For i = 1 To NX
  MsgBox "Address number " & i & " is " & _
     Selection.Item(i, 1).Address & _
     " and contains the value " & XValue(i, 1)

Next i

End Sub
```

Now switch back to the spreadsheet, deposit a few numbers in a column, call the macro (with Tools ⇨ Macro, followed by clicking on the name of the macro and, then, on Run) and verify that the message boxes indeed display the correct addresses and their contents. The actual reading takes only three lines of code (or four, when we count as two the continued line following the space plus underscore); the other lines are there merely to make sure that the input box works properly.

After we have satisfied ourselves that the input box works (either by outlining the range on the spreadsheet, or by typing it in the window of the input box as, say, B3:B7) we add a few lines to make sure that only a single column is selected. We accomplish this with an End statement inside the If... Then loop; the End statement is reached only when the number of columns, NC, differs from 1. The message box just before that End statement specifies the nature of the problem. Make sure that this part also works, and only then delete the verification lines to obtain

```
Sub Interpolate()

' Read the range of X-values

Dim XValue As Variant
Dim NX As Integer, NC As Integer
Dim myXRange As Range
Set myXRange = Application.InputBox _
   (Prompt:="The X-range is:", Type:=8) _
```

```
         myXRange.Select
         NX = myXRange.Rows.Count
         NC = myXRange.Columns.Count
         If NC <>1 Then
            MsgBox "Enter only ONE column."
            End
         End If
         Xvalue = Selection.Value

      End Sub
```

Since we need to read in both *x*- and *y*-values, we can just copy this fragment for the *y*-data, and make the few necessary adjustments, such as defining NC as integer only once, and replacing NX, XValue and myXRange by NY, YValue and myYRange. Moreover, we include a check that both columns have the same length; if they don't, the macro should again terminate after informing the user what went wrong.

```
Sub Interpolate()

' Read the range of X-values

Dim XValue As Variant
Dim NX As Integer, NC As Integer
Dim myXRange As Range
Set myXRange = Application.InputBox _
   (Prompt:="The X-range is:", Type:=8) _
myXRange.Select
NX = myXRange.Rows.Count
NC = myXRange.Columns.Count
If NC <>1 Then
   MsgBox "Enter only ONE column."
   End
End If
XValue = Selection.Value
' Read the range of Y-values

Dim YValue As Variant
Dim NY As Integer
Dim myYRange As Range
Set myYRange = Application.InputBox(Prompt:="The Y-range is:", Type:=8) _
   myYRange.Select
NY = myYRange.Rows.Count
NC = myYRange.Columns.Count
If NC <>1 Then
   MsgBox "Enter only ONE column."
   End
End If
If NY <> NX Then
   MsgBox "The sets of X and Y values must have equal lengths."
   End
End If
YValue = Selection.Value

End Sub
```

We still need to read the specific *x*-value, here called XX, for which we seek an interpolated *y*-value, YY. Although we could specify XX in the same way, by placing it on the spreadsheet and reading it there, it is perhaps easier to just type its value in the window. This is accomplished by adding three lines of code to the end of the macro, together with a comment heading, just before the End Sub line,

```
' Read XX, the value of X where the function is to be interpolated

Dim XX As Double
XX = Application.InputBox(Prompt:="Interpolate for X =", Type:=1)
MsgBox "The specified X-value is " & XX
```

This completes the data input stage, and we now consider the actual interpolation algorithm. First we determine the *x*-value in the table with a value closest to that of XX, and find its index (i.e., its rank number in the data set) NClosest. We then make sure that there are at least three points on either side of NClosest, otherwise we cannot use our seven-point *interpolation* routine reliably. Again, insert the following code fragment at the end of the macro, just before the End Sub line:

```
' Find the index NClosest of the point that is closest to XX

Delta = XValue(2, 1) - XValue(1, 1)
NClosest = Int((XX - XValue(1, 1) + Delta) / Delta)
If XX < XValue(1, 1) Or XX > XValue(NX, 1) Then
   MsgBox "The value lies outside the range of the data set."
   End
End If
If NClosest - 3 < 1 Or NClosest + 3 > NX Then
   MsgBox "The value lies too close to the edge of the data set."
   End
End If
MsgBox "The index NClosest is " & NClosest
```

Once more, verify that this code works, including those situations where the macro shuts down. After you are satisfied that everything works, remove the message boxes that display the values of XX and NClosest; the other message boxes should of course stay.

We are now ready to implement the central part of the algorithm. To this end, we read the seven *y*-values from three places before NClosest to three places beyond it, i.e., for

```
For i = 1 To 7
   Y(i) = Yvalue(NClosest - 4 + i, 1)
Next i
```

and compute the value for YY using the Savitzky–Golay coefficients for a quadratic seven-point smoothing (A), first derivative (B), and second derivative (twice the value in the Savitzky–Golay tables, since $[d^2(a + bx + cx^2) / dx^2]_{x=0} = 2c$),

```
Dim A As Double, B As Double, C As Double
Dim X As Double, YY As Double
A = (5 * Y(1) - 3 * Y(3) - 4 * Y(4) - 3 * Y(5) + 5 * Y(7)) / 42
B = (-3 * Y(1) - 2 * Y(2) - Y(3) + Y(5) + 2 * Y(6) + 3 * Y(7)) / 28
C = (-2 * Y(1) + 3 * Y(2) + 6 * Y(3) + 7 * Y(4) + 6 * Y(5) _
    + 3 * Y(6) - 2 * Y(7)) / 21
X = (XX - XValue(NClosest, 1)) / Delta
YY = (A * X ^ 2) / 2 + B * X + C
MsgBox "The value of Y is " & YY
```

Putting all this together, with a few comment lines, and a message box for the final result, yields the following working macro:

```
Sub Interpolate()

' This macro finds the X-value nearest to a specified value
' XX in a series of equidistant X-values, selects the three
' closest X-values on either side of it, fits a quadratic
' through the seven corresponding Y-values, and evaluates
' this quadratic at the desired value XX.

' The user specifies the columns containing the X- and Y-values,
' and the specific value XX for which the corresponding Y-value
' YY is sought. The output is displayed in an output box.

' Read the range of X-values

Dim XValue As Variant
Dim NX As Integer, NC As Integer
Dim myXRange As Range
Set myXRange = Application.InputBox _
   (Prompt:="The X-range is:", Type:=8)
myXRange.Select
NX = myXRange.Rows.Count
NC = myXRange.Columns.Count
If NC <>1 Then
   MsgBox "Enter only ONE column."
   End
End If
XValue = Selection.Value

' Read the range of Y-values

Dim YValue As Variant
Dim NY As Integer
Dim myYRange As Range
Set myYRange = Application.InputBox _
   (Prompt:="The Y-range is:", Type:-8)
myYRange.Select
NY = myYRange.Rows.Count
NC = myXRange.Columns.Count
If NC <>1 Then
   MsgBox "Enter only ONE column."
   End
End If
```

```
If NY <> NX Then
  MsgBox "The sets of X and Y values have unequal lengths."
  End
End If
YValue = Selection.Value

' Read XX, the value of X where the function is to be interpolated

Dim XX As Double
XX = Application.InputBox(Prompt:="Interpolate for X =", Type:=1)

' Find the index NClosest of the point that is closest to XX

Delta = XValue(2, 1) - XValue(1, 1)
NClosest = Int((XX - XValue(1, 1) + Delta) / Delta)
If XX < XValue(1, 1) Or XX > XValue(NX, 1) Then
  MsgBox "The value lies outside the range of the data set."
  End
End If
If NClosest - 3 < 1 Or NClosest + 3 > NX Then
  MsgBox "The value lies too close to the edge of the data set."
  End
End If

' Read the relevant Y-values

Dim Y(7) As Double
For i = 1 To 7
  Y(i) = YValue(NClosest - 4 + i, 1)
Next i

' Compute the interpolated value YY

Dim A As Double, B As Double, C As Double
Dim X As Double, YY As Double
A = (5 * Y(1) - 3 * Y(3) - 4 * Y(4) - 3 * Y(5) + 5 * Y(7)) / 42
B = (-3 * Y(1) - 2 * Y(2) - Y(3) + Y(5) + 2 * Y(6) + 3 * Y(7)) / 28
C = (-2 * Y(1) + 3 * Y(2) + 6 * Y(3) + 7 * Y(4) + 6 * Y(5) _
   + 3 * Y(6) - 2 * Y(7)) / 21
X = (XX - XValue(NClosest, 1)) / Delta
YY = (A * X ^ 2) / 2 + B * X + C
MsgBox "The value of Y is " & YY

End Sub
```

10.2b The finished product

A somewhat more elaborate version of this macro is shown below. It allows the user to select the length L of the moving polynomial, with L an odd integer between 5 and 45. (The upper limit to L is set to avoid overflow in the computation. Large L-values should be avoided anyway except when data are closely-spaced yet extremely noisy.) The macro computes the coefficients in the expressions for *A*, *B*, and *C* from formulas given by Madden in *Anal. Chem.* 50 (1978) 1383. The result will be accurate only insofar as the data fit a parabola; alternative formulas for fitting to higher-order expressions are also given by Madden. The same approach can be used for other applications of the Savitzky–Golay method, such as smoothing and differentiation. A sophisticated program for such applications, written by Prof. Barak and described in *Anal. Chem.* 67 (1995) 2758 automatically selects the optimal polynomial degree. An adaptation of Barak's program is shown in section 10.9; it contains a subroutine that can readily be used to make the interpolation macro more general. A version using special dialog boxes can be found on the web site of this book.

In this chapter we stay away from dialog boxes because their construction is rather complicated, and we use the much simpler message and input boxes instead. However, there is a disadvantage associated with the use of multiple input boxes: a user making a data entry mistake in a later box may have to start all over again, with the first box. Below we illustrate how to avoid this complication, in the rather elaborate checking of the entered value of L, which uses GoTo statements that bring the user back to the beginning of that input box rather than abort the entire process. Incidentally, general use of GoTo instructions in computations is not advisable, because it can make the code difficult to read and, consequently, error-prone. In the present case, however, the GoTo instruction is used only for the data input stage, where its use is both transparent and appropriate.

```
'''''''''''''''''''''''''''''''''''''''''''''''''''''''''''''''''''''''''''''''
'''''''''''''''''''''''''''''''''''''''''''''''''''''''''''''''''''''''''''''''
''''''''''''''''''''''''''^^^^^^^^^^^^^^^^^^^^^^^^^^''''''''''''''''''''''''''''
''''''''''''''''''''''''^                          ^''''''''''''''''''''''''''
'''''''''''''''''''''''^      INTERPOLATION        ^''''''''''''''''''''''''''
'''''''''''''''''''''''^                          ^''''''''''''''''''''''''''
''''''''''''''''''''''''^^^^^^^^^^^^^^^^^^^^^^^^^^''''''''''''''''''''''''''''
'''''''''''''''''''''''''''''''''''''''''''''''''''''''''''''''© R. de Levie
'''''''''''''''''''''''''''''''''''''''''''''''''''''''''''''''''''''''''''''''
```

```
Sub Interpolate()
' This macro finds the X-value nearest to a specified value XX
' in a series of EQUIDISTANT X-values by fitting a QUADRATIC
' through the L nearest Y-values, and evaluating this quadratic
' at the desired value XX. The number L of data points is user-
' selectable; L should be an odd integer between 5 and 45.

' The user specifies the columns containing the X- and Y-values,
' and the specific value XX for which the corresponding Y-value
' YY is sought. The output is displayed in an output box.


' Read the range of X-values

Dim XValue As Variant, YValue As Variant
Dim NC As Integer, NX As Integer, NY As Integer
```

```
Dim i As Integer, L As Single, LL As Integer
Dim m As Integer, mm As Single, s As Integer
Dim A As Double, B As Double, C As Double
Dim AA As Double, BB As Double, CC As Double
Dim AAA1 As Double, AAA2 As Double, BBB As Double, CCC As Double
Dim X As Double, XX As Double, YY As Double,
Dim Delta As Double, NClosest As Double
Dim myXRange As Range, myYRange As Range
Dim Message As String, Title as String, Default As String
Set myXRange = Application.InputBox _
   (Prompt:="The X-range is:", Type:=8)
myXRange.Select
NX = myXRange.Rows.Count
NC = myXRange.Columns.Count
If NC <> 1 Then
   MsgBox "Enter only ONE column."
   End
End If
XValue = Selection.Value
' Read the range of Y-values
Set myYRange = Application.InputBox _
   (Prompt:="The Y-range is:", Type:=8)
myYRange.Select
NY = myYRange.Rows.Count
NC = myXRange.Columns.Count
If NC <> 1 Then
   MsgBox "Enter only ONE column."
   End
End If
If NY <> NX Then
   MsgBox "The sets of X and Y values" & Chr(13) & _
   "must have the same lengths."
   End
End If
YValue = Selection.Value

' Read XX, the value of X where the function is to be interpolated

Dim XX As Double
XX = Application.InputBox(Prompt:="Interpolate for X =", Type:=1)

' Read L, the length of the data string to be used.
' The default value is 7.

Line1:
Message = "Define the string length L to be used;" & Chr(13) & _
   "select an odd integer between 5 and 45." & Chr(13) & _
   " " & Chr(13) & _
   " " & Chr(13) & _
   "Use a string of length L = "
Title = "String length"
Default = "7"
L = InputBox(Message, Title, Default)
LL = CInt(L)
```

```
If LL - L <> 0 Then
  MsgBox "The value selected for the string" & Chr(13) & _
  "length L must be an INTEGER."
  GoTo Line1
Else
  GoTo Line2
End If

Line2:
m = (LL - 1) / 2
mm = (L - 1) / 2
If m - mm <> 0 Then
  MsgBox "The value selected for the string" & Chr(13) & _
  "length L must be an ODD integer."
  GoTo Line1
Else
  GoTo Line3
End If

Line3:
If L < 5 Then
  MsgBox "The value selected for the" & Chr(13) & _
  "string length L is TOO SMALL."
  GoTo Line1
Else
  GoTo Line4
End If

Line4:
If L > 45 Then
  MsgBox "The value selected for the" & Chr(13) & _
  "string length L is TOO LARGE."
  GoTo Line1
Else
  GoTo Line5
End If

' Find the index NClosest of the point that is closest to XX

Line5:
Dim Delta As Double, NClosest As Double
Delta = XValue(2, 1) - XValue(1, 1)
NClosest = Int(Int(2 * ((XX - XValue(1, 1) + Delta) / Delta) + 1) / 2)
If XX < XValue(1, 1) Or XX > XValue(NX, 1) Then
  MsgBox "The value lies outside the range of the data set."
  End
End If
If NClosest - m < 1 Or NClosest + m > NX Then
  MsgBox "The value lies too close to the edge of the data set."
  End
End If

' Read the relevant Y-values

ReDim Y(LL) As Double
Dim i As Integer, s As Integer
```

```
For i = 1 To LL
   Y(i) = YValue(NClosest - m + i - 1, 1)
Next i

' Compute the interpolated value YY

A = 0
B = 0
C = 0
For i = 1 To LL
   s = -m - 1 + i
   AA = (30 / (2 * m + 3)) * (3 * s * s - m * (m + 1)) / (2 * m + 1) / _
   (2 * m - 1) / (m + 1) / m
   A = A + Y(i) * AA
   B = B + Y(i) * 3 * s / (2 * m + 1) / (m + 1) / m
   CC = 3 * (3 * m * m + 3 * m - 1 - 5 * s * s) / (2 * m + 3) / _
   (2 * m + 1) / (2 * m - 1)
   C = C + Y(i) * CC
Next i

X = (XX - XValue(NClosest, 1)) / Delta
YY = (A * X ^ 2) / 2 + B * X + C
MsgBox "The value of Y is " & YY

End Sub
```

10.3 Propagation of imprecision

Sections 2.3 and 2.4 describe how one can use a spreadsheet to propagate single or multiple standard deviations without using calculus. Once we realize that this can be done, it is only a small further step to automate the process as a macro.

By far the largest part of such a macro is devoted to reading in the data. Here, as in the preceding section, we use successive input boxes for the independent parameters x_i, the associated standard deviations σ_i, and the function F through which these standard deviations must propagate. It is convenient to separate the cases of one and multiple variables ($N=1$ and $N>1$ respectively, where N is the number of independent parameters x_i), because VBA cannot assign an array to a single cell. At the end of the section we test whether there are valuable data in the cell to the right of F.

You may wonder why arrays such as newXiValue and SFiValue are assigned in the first block of instructions. The answer is that it serves to specify the actual sizes of these Variant arrays. A dimension statement such `Dim newXiValue As Variant` does not do this, because it leaves the actual array size unspecified. (Alternatively, after the macro has determined the value of N, we could have dimensioned newXiValue with `ReDim newXiValue (1 To N) As Double`, in which case it should subsequently be treated as a unidimensional vector rather than as a two-dimensional array. The same goes for SFiValue.)

The propagation section starts with suppressing screen updating, which speeds up the computation (because the monitor screen need not be redrawn every time a parameter is changed) and keeps the screen serene. When input boxes are used, as in the present example, screen updating in the macro should only be suppressed *after* the input data have been read, otherwise it would not be possible to enter information into the input boxes by highlighting the corresponding cells.

The computational "trick" is to vary, one at a time, the variables x_i, and to record the resulting changes in F. This is conveniently done with the VBA instruction `Selection.Replace A, B` which allows us to substitute instruction B for instruction A. We then read the result, and change things back again with `Selection.Replace B, A`. It is best to inspect the case $N = 1$ first; once that is grasped, the case $N > 1$ becomes more readily intelligible since it follows the same logic.

The data output section is rather straightforward, except that we use the information whether there are data next to F. In that case, the answer is presented in a message box; otherwise it will be written next to F.

Here is the macro:

```
''''''''''''''''''''''''''''''''''''''''''''''''''''''''''''''''''''''''''''''''''''
''''''''''''''''''''''''''''^^^^^^^^^^^^^^^^^^^^^^^^^^^'''''''''''''''''''''''''''''''
''''''''''''''''''''''^                               ^''''''''''''''''''''''''''''
''''''''''''''''''''''^         PROPAGATION           ^''''''''''''''''''''''''''''
''''''''''''''''''''''^                               ^''''''''''''''''''''''''''''
''''''''''''''''''''''''''''''^^^^^^^^^^^^^^^^^^^^^^^^^^^''''''''''''''''''''''''''''
'''''''''''''''''''''''''''''''''''''''''''''''''''''''''''''© R. de Levie
''''''''''''''''''''''''''''''''''''''''''''''''''''''''''''''''''''''''''''''''''''''
```

```
Sub Propagation()

' This macro takes an arbitrary function F of N mutually
' independent parameters, and the corresponding standard
' deviations of those parameters, and computes the resulting,
' propagated standard deviation in a single function F.

' Requirements: for the sake of convenient data input, the N in-
' dependent input parameters must be placed either in a contiguous
' row or in a contiguous column. Their N standard deviations must
' follow the same format (again either in a contiguous row or column,
' consistent with the format of the N input parameters).

' When the input data are in columns (or rows), the standard deviation
' of the single function F will be placed directly to the right of (or
' below) that function, in italics, provided that this cell is either
' unoccupied or its contents can be overwritten. Otherwise, the result
' will be displayed in a message box.

' Notation used:
' N:  the number of input parameters
```

```
' X:  single input parameter (for N=1)
' S:  the corresponding, single standard deviation of X
' Xi: multiple input parameters (for N>1)
'     NOTE: THESE MUST BE IN A SINGLE, CONTIGUOUS ROW OR COLUMN
' Si: standard deviations of the multiple input parameters
'     NOTE: THESE MUST BE IN A SINGLE, CONTIGUOUS ROW OR COLUMN
' F:  the single function through which the imprecision propagates
' VF: the propagated variance of the function F
' SF: the propagated standard deviation of the function F
' FR: cell immediately to the right of cell containing F

' Select the input parameters of the function

Dim myRange1 As Range
Dim N As Integer, NR As Integer, NC As Integer
Dim XiValue As Variant, newXiValue As Variant, SFiValue As Variant
Set myRange1 = Application.InputBox(prompt:="The input parameters of" & _
   Chr(13) & "the function are located in:", _
   Title:="Propagation InputBox 1: Input Parameters", Type:=8)
myRange1.Select
NR = Selection.Rows.Count
NC = Selection.Columns.Count
If NR = 0 Then End
If NR <> 1 And NC <> 1 Then
   MsgBox "The input parameters should be placed either in a" & Chr(13) & _
   "single, contiguous row or in a single, contiguous column."
   End
ElseIf NR = 1 And NC = 1 Then
   N = 1
   Dim XValue As Double
   XValue = Selection.Value
ElseIf NR = 1 And NC > 1 Then
   N = NC
   XiValue = Selection.Value
   newXiValue = Selection.Value   '' dimensioning the Variant array
   SFiValue = Selection.Value      ' dimensioning the Variant array
ElseIf NR > 1 And NC = 1 Then
   N = NR
   XiValue = Selection.Value
   newXiValue = Selection.Value    ' dimensioning the Variant array
   SFiValue = Selection.Value      ' dimensioning the Variant array
End If

' Select the standard deviations of the input parameters

Dim myRange2 As Range
Dim NRS As Integer, NCS As Integer
Set myRange2 = Application.InputBox(prompt:="The standard deviations of " _
   & "the" _ & Chr(13) & "input parameters are located in:", _
   Title:="Propagation InputBox 2: Standard Deviations", Type:=8)
myRange2.Select
NRS = Selection.Rows.Count
NCS = Selection.Columns.Count
If NRS = 0 Then End
```

```
If NRS <> NR Or NCS <> NC Then
   MsgBox "The number of standard deviations must match the" & Chr(13) & _
          "the number of input parameters, and both must be" & Chr(13) & _
          " in the same format, either as row or as column."
   End
End If
If N = 1 Then
   Dim SValue As Double
   SValue = Selection.Value
ElseIf N > 1 Then
   Dim SiValue As Variant
   SiValue = Selection.Value
End If

' Select the function

Dim myRange3 As Range
Dim NNR As Integer, NNC As Integer
Dim FValue As Double
Set myRange3 = Application.InputBox(prompt:="The function is located " _
   & "in:", _ Title:="Propagation InputBox 2: Function", Type:=8)
myRange3.Select
NNR = Selection.Rows.Count
NNC = Selection.Columns.Count
If NNR = 0 Then End
If NNR = 1 And NNC = 1 Then FValue = Selection.Value
If NNR <> 1 Or NNC <> 1 Then
     MsgBox "The propagation is computed only for a SINGLE function."
   End
End If

' Test and prepare the default output cell
' for the standard deviation of the function F

Dim m As Integer
Dim FFValue As Double
If N = 1 Or NC = 1 Then
   Selection.Offset(0, 1).Select
   m = 0
   FFValue = Selection.Value
   If FFValue = 0 Or IsEmpty(FFValue) Then
     m = m
   Else
     m = m + 1
   End If
   Selection.Offset(0, -1).Select
End If

If N > 1 And NR = 1 Then
   Selection.Offset(1, 0).Select
   m = 0
   FFValue = Selection.Value
   If FFValue = 0 Or IsEmpty(FFValue) Then
     m = m
   Else
     m = m + 1
```

```
  End If
  Selection.Offset(-1, 0).Select
End If

Application.ScreenUpdating = False
Dim SFValue As Double, VFValue As Double, newFiValue As Double
Dim i As Integer, j As Integer
VFValue = 0
myRange1.Select

' Compute the standard deviation SF in F for N=1

If N = 1 Then
  Dim newXValue As Double, VFiValue As Double
  newXValue = XValue * 1.000001
  Selection.Replace XValue, newXValue
  myRange3.Select
  Dim newFValue As Double
  newFValue = Selection.Value
  myRange1.Select
  Selection.Replace newXValue, XValue
  SFValue = Abs(1000000 * (newFValue - FValue) * SValue / XValue)
End If

' Compute the standard deviation SF in F for column input

If N > 1 And NC = 1 Then
  For i = 1 To N
    newXiValue(i, 1) = XiValue(i, 1) * 1.000001
    Selection.Replace XiValue(i, 1), newXiValue(i, 1)
    myRange3.Select
    newFiValue = Selection.Value
    myRange1.Select
    Selection.Replace newXiValue(i, 1), XiValue(i, 1)
    VFiValue = 1000000 * (newFiValue - FValue) * SiValue(i, 1) / _
      XiValue(i, 1)
    VFValue = VFValue + VFiValue ^ 2
    SFValue = Sqr(VFValue)     ' Note: VBA uses Sqr, Excel uses SQRT
  Next i
End If

' Compute the standard deviation SF in F for row input

If N > 1 And NR = 1 Then
  For j = 1 To N
    newXiValue(1, j) = XiValue(1, j) * 1.000001
    Selection.Replace XiValue(1, j), newXiValue(1, j)
    myRange3.Select
    newFiValue = Selection.Value
    myRange1.Select
    Selection.Replace newXiValue(1, j), XiValue(1, j)
    VFiValue = 1000000 * (newFiValue - FValue) * SiValue(1, j) / _
      XiValue(1, j)
    VFValue = VFValue + VFiValue ^ 2
    SFValue = Sqr(VFValue)     ' Note: VBA uses Sqr, Excel uses SQRT
  Next j
End If
```

```
Dim Answer As String

If N > 1 And NC = 1 Then
  myRange3.Select
  If m > 0 Then
    Answer = MsgBox("The cell to the right of the function is not empty." & _
    Chr(13) & "Can its contents be overwritten?", vbYesNo, "Overwrite?")
    If Answer = vbYes Then m = 0
  End If
  If m = 0 Then
    Selection.Offset(0, 1).Select
    Selection.Font.Italic = True
    Selection.Value = SFValue
  Else
  MsgBox "The standard deviation of the function is " & SFValue
  End If
End If
If N > 1 And NR = 1 Then
  myRange3.Select
  If m > 0 Then
    Answer = MsgBox("The cell below the function is not empty." & _
    Chr(13) & "Can its contents be overwritten?", vbYesNo, "Overwrite?")
    If Answer = vbYes Then m = 0
  End If
  If m = 0 Then
    Selection.Offset(1, 0).Select
    Selection.Font.Italic = True
    Selection.Value = SFValue
  Else
    MsgBox "The standard deviation of the function is " & SFValue
  End If
End If

End Sub
```

10.4 Installing and customizing a macro

In this section we will describe how you can install a macro from an outside source, and how you can customize your macros. The specific macros described in sections 10.2 through 10.8 of this chapter need not be typed in (which would be a tedious and error-prone task) but can be downloaded conveniently from the web site http://uk.cambridge.org/chemistry/resources/delevie Store these on your computer in a file called, say, ExcelAnalysisMacros. The short practice macros of section 10.1 must of course be typed in, which is also how you would enter any other macros you have written yourself.

10.4a Installing external macros

In order to install one or more macros from a file such as ExcelAnalysisMacros, highlight the macro(s) (e.g., with Edit ⇒ Select All) and copy them with Ctrl + c to the clipboard. For incorporation into an Excel 5 or Excel 95 workbook, start (or, if already open, switch to) Excel, and open (or go to) the spreadsheet where you want the macros installed. Use Insert ⇨ Macro ⇨ Module to open a new module, and deposit the selected macro(s) into the module with Ctrl + v. Click on a spreadsheet tab to return to that spreadsheet. You should now have access to the macro(s) through Tools ⇨ Macro, which should list them.

In Excel 97, Excel 98 or Excel 2000, again after first having copied the macro(s) to the clipboard, open the spreadsheet, then use F11 or Tools ⇨ Macro ⇨ Visual Basic Editor followed by Insert ⇨ Module to open the module. (When the new module already contains the instruction Option Explicit, erase that line, or convert it into a comment by placing a hyphen in front of it. Option Explicit enforces the good programming practice of dimensioning all variables, but can give a beginner a lot of unnecessary grief.) Then deposit the file in the module with Ctrl + v. Switch back to the spreadsheet with Alt + F11, and verify that you have installed the macro(s) with Tools ⇨ Macro ⇨ Macros, which should show them.

Now look at the imported macros in their module. Much of their text should appear on the monitor screen in black, but their headings and comments (i.e., lines or line fragments that start with ') should automatically show in dark green, while specific instructions recognized as such by VBA should appear on the monitor screen in dark blue. (Whether these colors are easily discernable on your screen depends on the monitor used.) The colors show that the VBA editor recognizes the file properly, and is doing its job checking the programs, line by line.

Any macro can be called as follows. Start with Tools ⇨ Macro (in Excel 5 or Excel 95) or with Tools ⇨ Macro ⇨ Macros (in Excel 97 and more recent versions), in order to get the Macro dialog box. Double-click on the name of the macro, which thereupon will appear in the top window of that box. Then click on Run. This sequence always works, but it requires four mouse clicks, and therefore can get rather tedious when you need to repeat these steps many times. Incidentally, if the macro name does not appear in the Macro dialog box, the computer cannot find it. In that case the macro is perhaps not named properly, as Sub *macroname*( ), or is stored in another workbook.

Below we describe several methods to make a macro more easily callable. You might want to follow the first (and easier) of these procedures also when writing your own macros, because you will need to call a macro repeatedly during its debugging stage, and anything that facilitates debugging is fair game.

10.4b Assigning a shortcut key

Excel macros can be called most conveniently with a simple shortcut key, for which you have to remember the specific letter used for the particular macro. Say that you want to call the Fourier transform routine of section 10.5, with its two drivers, the macros Forward( ) and Inverse( ). Note that you can *only* call *macros*; you cannot call subroutines, since they require additional input information. Macros are subroutines without input information, which is why they sport empty brackets. A shortcut key assignment cannot be made in VBA, but must be made from the spreadsheet that contains the macro.

Starting either from an active spreadsheet, select Tools ⇨ Macro or, in Excel 97 and beyond, Tools ⇨ Macro ⇨ Macros. In the large window of the resulting Macro dialog box you will now see all macros active in this workbook. Select (i.e., single-click on) the one to which you want to assign a shortcut key, e.g., the macro Forward; it will then appear in the top window of that dialog box. (Do not double-click on it, because then you will have selected to run it.) Now click on Options. In the resulting Macro Options dialog box, click on the small, square window in front of the Shortcut Key, then click on the window to the right of Ctrl + , enter the letter of your choice (e.g., f), and click OK. This will get you back to the Macro dialog box. Click Run, and off you go. Or, if you do not want to run the macro at this time, but merely assign the shortcut key, click Cancel. The next time, all you need to do to run Forward( ) is to depress the control key Ctrl and the letter key f simultaneously (i.e., depress the control key just a bit longer, both before and after the letter key is depressed).

Now repeat the same procedure for Inverse( ), to which you must of course assign a different letter, such as F, or g. Note that the standard shortcut commands in Excel use the lower case (so that you could even assign Ctrl + C (i.e., Ctrl + Shift + c) to a macro without interfering with the copy command Ctrl + c). Incidentally, if you want to use the same macro repeatedly (as you might during debugging), you can avoid the above procedure by re-invoking the macro with Ctrl + y. This command recalls the last macro you have used during the present session. But beware: if you use Ctrl + y without having used a macro before during the same session, it will close your spreadsheet!

10.4c Embedding in a menu

While you can make all macros work by selecting them with Tools ⇨ Macros or, more conveniently, by using an assigned shortcut key combination, Ctrl + letter, it may be preferable to incorporate often-used macros into the menu structure of Excel. This obviates the need to remember the specific keystroke combination, and it certainly looks slick to have your own macro on call in your spreadsheet menu. At any rate, by the time you have incorporated all the macros from the diskette in your spreadsheet, you may prefer to have explicit labels rather than having to memorize ten or more different shortcut letters. Fortunately, the two methods tolerate each other, so that you can use them both.

In Excel 97 and later versions, embedding a button in a toolbar can be done quite easily, by adding two small macros that operate automatically whenever a workbook is opened or closed. The following macros, when added to the macro Interpolate( ), will insert a button labeled Interpolation in the Tools menu when a workbook is opened that contains this macro, and will remove the button again upon closing that workbook.

```
Sub Auto_Open()
Set Item = CommandBars("Tools").Controls.Add(Type:=msoControlButton)
With Item
  .Caption = "Interpolation"   'Inserts the button in the toolbar
  .OnAction = "Interpolate"    'Calls the macro
End With
End Sub

Sub Auto_Close()
  CommandBars("Tools").Controls("Interpolation").Delete
End Sub
```

Unfortunately, the above method does not work in Excel 5 and Excel 95. Below we will describe how to include the forward and inverse Fourier transform macros (to be described in section 10.5) into the Excel menu bar under the heading Tools.

Go to the module where the macro is located, and select Tools ⇨ Menu Editor. (As already mentioned in section 1.12, the menu items displayed by Excel sometimes change depending on where you are in the program. Here is such a case: when you are not in an active module, you will not find the Menu Editor under Tools.) In the resulting Menu Editor dialog box, in the large window with the heading Menu, highlight &Tools. Then, in the adjacent Menu Items window, highlight the name of the menu item *above which* you want to insert the new menu item. For our example, click on &Goal Seek. If instead you want to place your macro at the very bottom, highlight (End of Menu).

Now click on the Insert button. In the Caption box type the name you want to use in referring to the macro; for our example type &Fourier transform. The ampersand, &, indicates that the next letter will be underlined in the menu, and can then be used to invoke the macro by keystroke. We will use F since that letter has not yet been used in the standard menu under Tools. For Convolution and Weighted Least Squares you may want to use u and W respectively.

Since there are two Fourier transform macros, we need to install both. Therefore highlight (End of Submenu) in the S̲ubmenu Items window. Then click again on Insert, in the space above (End of Submenu) type &Forward, highlight the M̲acro window, and in it type the name of the macro, in this case also Forward (without ampersand, and without empty brackets). Subsequently, again highlight (End of Submenu), click on Insert, type &Inverse, highlight the M̲acro box, and type Inverse. Finally, click on OK.

This should do it. Check it by switching to a spreadsheet, and selecting T̲ools. You should now see, between A̲utoCorrect and G̲oal Seek, the label F̲ourier transform, followed by an arrow pointing to the right. Click on F̲ourier transform, and the two choices will appear, F̲orward, and I̲nverse. Then verify that they indeed work. If not, try again; you can D̲elete what does not work from the Menu Editor dialog box.

A final warning: when workbooks have customized macros embedded in their menus, make sure that only one workbook is open at the time. With several open workbooks, each with special additions to their toolbars, Excel tries to combine them, but may not always succeed, especially when the various workbooks contain non-identical macros that use the same name.

10.4d Miscellany

It is usually most convenient to store macros with the particular workbook where they will be used. If you later switch to another workbook, just copy those macros to the new workbook if you need them there as well. However, if you anticipate using a macro often, and you use your own personal computer, you may want to store it in the Personal Macro Workbook, which will keep it as a Personal.xls file in the XLStart folder. All files stored in that folder are opened automatically when Excel is started, and will then be available to any spreadsheet. To keep them out of view, hide them with W̲indows ⇨ H̲ide. To modify these files, first use W̲indows ⇨ U̲nhide, which will show them in the Unhide dialog box. Select the particular hidden file you want to edit, in this case Personal, and click on the OK button.

Of course, when you install a macro, you might also want to verify that it performs its task properly. Typically, when you write a macro yourself, count on spending at least as much time debugging it as you have already spent on writing the macro in the first place, unless you have verified every step as you went along. Ready-made macros from, say, a diskette or from an Internet source should not give you such problems, but that may not always be the case. For example, you may encounter difficulties when trying to run an Excel 97 macro in Excel 5, since software is seldom forward-compatible.

By now you are used to the fact that Excel allows at least three different ways of accomplishing any particular task. And, indeed, there are yet other ways to install a macro in Excel. The most professional method is to generate an add-in, in which case the macro will appear to the user as a fully integrated, tamper-proof part of Excel, somewhat like Solver. The disadvantages are that preparing such an add-in is a rather elaborate procedure, and that the user will not have access to its code. Since the primary purpose of this chapter is not so much to provide you with specific add-ins but, instead, to show you how to write and incorporate your *own* macros, we take the alternative approach of storing the macros in or with the spreadsheet in such a way that you can later edit and modify them. That, of course, also has its downside, because it makes them vulnerable to tampering and accidental erasure.

In order to simplify macro writing as much as possible, we have not used dialog boxes, but have instead used only message and input boxes or, as in the examples of sections 10.5 through 10.7, done without input boxes altogether. In the latter case, the input data must fit in a single array, and must conform to a specific format; the output then appears in a fixed place relative to the input.

The macros presented here have some rudimentary checks, but are by no means foolproof: they can be tripped up by anyone setting his or her mind to it. However, protection against all possible exigencies would double their length, and make them much more difficult to read, and therefore has not even been attempted.

Finally, before we return to more sample macros, some advice. When you want to write your first substantive macro, look at the sample macros of this chapter, see whether they contain parts that you can use, and then do so by copying those parts and incorporating them into your own macro. Often you can patch together the skeleton of a new macro from snippets of pre-existing macros, much faster than if you were to start from scratch. (You will still need to verify that these fragments do the job you expect of them in their new context.) While the copyright notice in the macro headings should discourage unauthorized commercial use, readers are specifically allowed and encouraged to copy and use these macros or parts thereof as templates for their own, non-commercial use. As you gain more experience in macro writing, and become a more confident macro scribe, you will not need such crutches anymore, and will develop your own writing style.

10.5 Fourier transformation

Below we show a macro for the forward Fourier transform, and then describe how it can be modified to perform an inverse Fourier transform. The macro is readily incorporated in the spreadsheet. It uses no input boxes, but instead requires that the input data are organized as an array of n data pairs, where n is an integer power of 2. The data should be entered in three adjacent columns, the second column containing the real parts of the input data, and the third column the associated imaginary parts, while the first column should contain the associated *equidistant* times (or, for the inverse transform: frequencies). If the input data are real, the third column should be left blank and/or be filled with zeros. Likewise, if the input data are imaginary, the second column should contain blanks and/or zeros.

The output will appear in the three columns immediately to the right of the input data, thereby overwriting any prior data in that location. The three output columns of the forward transform will display the frequency, the real part of the output, and its imaginary part; the output of the inverse transform will show time instead of frequency.

The Analysis Toolpak of Excel already includes a Fourier transform function, available through Tools ⇨ <u>D</u>ata Analysis ⇨ Fourier Analysis. Unfortunately that function suffers from three serious limitations: (1) it only accepts real inputs, (2) it does not properly scale its output, and (3) it generates its output in the form of labels, which need to be extracted using the = IMREAL( ) and = IMAGINARY( ) functions before they can be plotted or otherwise used in subsequent calculations. Although it is possible to work around those limitations, it is far easier to avoid them by starting afresh, and to include frequency and time scales at the same time. This is what we have done here.

10.5a Forward Fourier transformation

```
' Forward Fourier transformation

' Forward Fourier transformation of an array of complex data,
' arranged in a single block of three adjacent columns, the
' first column containing time, the second column the corres-
' ponding real components of the input, and the third column
' the imaginary components. With only real input data, either
' leave the third column blank, or fill it with zeros; likewise
' for purely imaginary data, the second column should only con-
' tain zeros and/or blanks. The times should be equidistant,
' and be centered around zero.

' The output is written in the three columns to the right
' of the input data block, thereby overwriting any prior
' data in that region. The output columns contain (from left
' to right) the frequency, the real components of the output
' data, and their imaginary components.

' The frequency scale of the output has as its units the
' inverse of the units of the input time scale. In other
' words, when the input times are given in seconds, the
' output frequencies are in Hertz; when the input times
' are in ms, the output frequencies are in kHz; etc. Only
' the numerical values of time and frequency are provided,
' not their units. To initiate the transform, highlight the
' input array, then call the macro.
```

```
Sub ForwardFourier()

' Check the array length n, which must be at least 2,
' and must be a power of two

Dim rnmax As Integer, cnmax As Integer, length As Integer
Dim rn As Integer, cn As Integer, n As Integer
n = 0
rnmax = Selection.Rows.Count
length = rnmax

If length < 2 Then
  MsgBox "There must be at least two rows."
  End
End If

Do While length > 1
  length = length / 2
Loop
If length <> 1 Then
  MsgBox "The number of rows must be a power of two."
  End
End If

' Check that there are three input columns

cnmax = Selection.Columns.Count
If cnmax <> 3 Then
  MsgBox "There must be three input columns, one for time, and" & _
  Chr(13) & "one each for the real and imaginary parts of the input."
  End
End If

' Read the input data

Dim dataArray As Variant
dataArray = Selection.Value

' Check that the first column has its first two elements

Dim check1 As Double, check2 As Double
check1 = VarType(dataArray(1,1))
If check1 = 0 Then
  MsgBox "Enter the top left value."
  End
End If
check2 = VarType(dataArray(2,1))
If check2 = 0 Then
  MsgBox "Enter a value in row 2 of the first column."
  End
End If

' Rearrange the input data

Dim nn As Integer
nn = 2 * rnmax
ReDim Term(nn) As Double
For rn = 1 To rnmax / 2
  Term(2 * rn - 1) = dataArray(rnmax / 2 + rn, 2)
```

```
   Term(2 * rn) = dataArray(rnmax / 2 + rn, 3)
   Term(rnmax + 2 * rn - 1) = dataArray(rn, 2)
   Term(rnmax + 2 * rn) = dataArray(rn, 3)
Next rn

' Check that the output does not overwrite valuable data

Selection.Offset(0, 3).Select
Dim outputArray As Variant
outputArray = Selection.Value
Dim z As Double
For rn = 1 To rnmax
   For cn = 1 To cnmax
      z = outputArray(rn, cn)
      If IsEmpty(z) Or z = 0 Then
         n = n
      Else
         n = n + 1
      End If
   Next cn
Next rn

Dim Ans As String
If n > 0 Then
   Ans = MsgBox("There are data in the space where the output will be" & -
   Chr(13) & " written. Proceed anyway and overwrite those data?", -
   vbYesNo)
   If Ans = vbNo Then
      Selection.Offset(0, -3).Select
      End
   End If
End If

' Calculate the frequency scale

Dim interval As Double
interval = (dataArray(2, 1) - dataArray(1, 1)) * rnmax
For rn = 1 To rnmax
   dataArray(rn, 1) = (-(rnmax / 2) + rn - 1) / interval
Next rn

' The following is the forward Fourier transform routine FOUR1
' from J. C. Sprott, "Numerical Recipes: Routines and Examples
' in BASIC", Cambridge University Press, Copyright (C) 1991 by
' Numerical Recipes Software. Used by permission. Use of this
' routine other than as an integral part of the present book
' requires an additional license from Numerical Recipes Software.
' Further distribution is prohibited. The routine has been modi-
' fied to yield double-precision results, and to conform to the
' standard mathematical convention for Fourier transformation.

Dim tr As Double, ti As Double, theta As Double, wtemp As Double
Dim wi As Double, wr As Double, wpi As Double, wpr As Double
Dim i As Integer, istep As Integer, j As Integer
Dim m As Integer, mmax As Integer
j = 1
```

```
For i = 1 To nn Step 2
  If j > i Then
    tr = Term(j)
    ti = Term(j + 1)
    Term(j) = Term(i)
    Term(j + 1) = Term(i + 1)
    Term(i) = tr
    Term(i + 1) = ti
  End If
  m = Int(nn / 2)
  While m >= 2 And j > m
    j = j - m
    m = Int(m / 2)
  Wend
  j = j + m
Next i

mmax = 2
While nn > mmax
  istep = 2 * mmax
  theta = -2 * [Pi()] / mmax
  wpr = -2 * Sin(0.5 * theta) ^ 2
  wpi = Sin(theta)
  wr = 1
  wi = 0
  For m = 1 To mmax Step 2
    For i = m To nn Step istep
      j = i + mmax
      tr = wr * Term(j) - wi * Term(j + 1)
      ti = wr * Term(j + 1) + wi * Term(j)
      Term(j) = Term(i) - tr
      Term(j + 1) = Term(i + 1) - ti
      Term(i) = Term(i) + tr
      Term(i + 1) = Term(i + 1) + ti
    Next i
    wtemp = wr
    wr = wr * wpr - wi * wpi + wr
    wi = wi * wpr + wtemp * wpi + wi
  Next m
  mmax = istep
Wend

For rn = 1 To rnmax / 2
  dataArray(rn, 2) = Term(rnmax + 2 * rn - 1) / rnmax
  dataArray(rn, 3) = Term(rnmax + 2 * rn) / rnmax
  dataArray((rnmax / 2) + rn, 2) = Term(2 * rn - 1) / rnmax
  dataArray((rnmax / 2) + rn, 3) = Term(2 * rn) / rnmax
Next rn

' Write the output data

Application.ScreenUpdating = False
Selection.Value = dataArray

End Sub
```

10.5b Descriptive notes

We will first discuss some aspects of the forward Fourier transform macro that the writer of an Excel macro must provide in order to make the macro convenient to use, and somewhat foolproof. We will not deal with the heart of the macro itself, the part that performs the fast Fourier transformation, which is simply lifted from the *Numerical Recipes*. That routine required only two minor changes: (1) we modified it to compute in double precision, and (2) we changed the sign of the parameter theta in order to bring the definition of Fourier transformation in line with standard mathematical usage, see equation 7.1-1. For more detailed information on Fourier transformation the reader is referred to books dealing specifically with that topic, such as E. O. Brigham, *The Fast Fourier Transform*, Prentice Hall, 1974, 2nd ed. 1988. General programs are available to perform Fourier transformation on data arrays of arbitrary length, but these are more complicated and are not really needed in the present, primarily didactic context, which is why we have settled on the customary restriction to 2^N complex data.

The macro performs a forward Fourier transformation on the highlighted data, which must be organized in a rectangular array of 2^N rows by 3 columns, and are then read by the statement `dataArray   = Selection.Value`. At the end of the macro, the complementary instruction `Selection.Value = dataArray` writes the results of the calculation in the adjacent three columns.

The lines `rnmax  =  Selection.Rows.Count` and `cnmax  = Selection.Columns.Count` determine the size of the array. These parameters are then used to check that the array has appropriate dimensions. If the dimensions are incorrect, the user is warned by a message box, and the program is aborted (through the instruction End following the line specifying the message box). Note that the fixed input format is only required because in this example we use neither input boxes nor a more general dialog box. Either of these could accommodate less rigidly formatted input arrays.

The program tests whether any of the cells that will be used for the output are free of contents. If not, the user is asked whether or not to proceed with the transform. If the answer is no, the program is halted, to give the user time to free the output space while saving its contents. Note that the linefeed character `Chr(13)` is used to divide the text in the message box into two parts of near-equal length.

When graphs are displayed on the spreadsheet, updating the latter can be very slow, since each new point written to the screen typically requires that all graphs be redrawn that display that point, or any other point depending on the just computed value. The speed at which the monitor redraws the graphs can then become rate-limiting. In order to avoid this, screen updating is suppressed during the data output stage, using the instruction `Application.ScreenUpdating = False`. At the end of the macro, at End Sub, `Application.ScreenUpdating` is automatically reset to its default value, True, so that screen updating is restored, all newly computed data are entered, and (finally) all graphs redrawn.

Dimensioning arrays is necessary; dimensioning them as Variant avoids many unnecessary complications. The other parameters can be dimensioned in more standard ways. A convenience of VBA is that dimensions can be defined "on the fly", i.e., during the program rather than at its beginning, so that parameters to be determined by the input array (such as the number of input data) can first be read, and subsequently be used to define an array size. This is done, e.g., in ReDim Term(nn) As Double.

Instead of specifying the value of π we have below used the spreadsheet function Pi( ). This can be done by placing that function inside straight brackets, as in TwoPi = 2 * [Pi()].

10.5c A bidirectional Fourier transformation macro

The forward and inverse Fourier transform routines differ only in the following, relatively minor respects:

a The forward transform calculates a frequency scale, the inverse transform a time scale.
b The forward transform defines theta as $2\pi/rn_{max}$, the inverse transform as $-2\pi/rn_{max}$.
c The forward transform scales the output through division by the maximum number of rows, rn_{max}, while the inverse transform does not.

Since the above are only very small differences, the two procedures can easily be packaged as a *single* routine provided that one can specify which of the two choices (forward or inverse) is to be used. There are several ways to achieve this. For example, one can use a message box as described in section 10.1f, which constitutes a simple, quick way to get the program up and running. Here we will use a somewhat more efficient way (because it does not require the user to answer a query) to achieve the same goal, by using two small macros that set an integer, iSign, to either $+1$ or -1 for the forward and inverse transform respectively. This is followed by the main program which is now in the form of a subroutine, Fourier(iSign), that reads the value of iSign and performs most of the functions of the earlier macro. Finally, in the next section (on convolution) we will also use the Fourier transform routine from the Numerical Recipes. Therefore, in order to avoid needless repetition, we place the FT operation itself in a separate subroutine. So this will be the final structure of the program: we have two short input macros, Forward( ) and Inverse( ), which call the main subroutine Fourier(iSign), which in turn calls FFT(Term, nn, iSign).

Finally, in order to make the macro more convenient to the user, we have made it to accept two input formats, i.e., with the time or frequency scales either centered around zero or starting at zero, as specified by the additional integer variable jSign.

```
' ' ' ' ' ' ' ' ' ' ' ' ' ' ' ' ' ' ' ' ' ' ' ' ' ' ' ' ' ' ' ' ' ' ' ' ' ' ' ' ' ' ' ' ' ' ' ' ' '
' ' ' ' ' ' ' ' ' ' ' ' ' ' ' ' ' ' ' ' ' ' ' ' ' ' ' ' ' ' ' ' ' ' ' ' ' ' ' ' ' ' ' ' ' ' ' ' ' '
' ' ' ' ' ' ' ' ' ' ' ' ' ' ' ' ' ' ' ' ' ' ^^^^^^^^^^^^^^^^^^^^^^^^^ ' ' ' ' ' ' ' ' ' ' ' ' ' '
' ' ' ' ' ' ' ' ' ' ' ' ' ' ' ' ' ' ' ^                              ^ ' ' ' ' ' ' ' ' ' ' ' ' ' '
' ' ' ' ' ' ' ' ' ' ' ' ' ' ' ' ' ' ' ^   FOURIER  TRANSFORMATION   ^ ' ' ' ' ' ' ' ' ' ' ' ' ' '
' ' ' ' ' ' ' ' ' ' ' ' ' ' ' ' ' ' ' ^                              ^ ' ' ' ' ' ' ' ' ' ' ' ' ' '
' ' ' ' ' ' ' ' ' ' ' ' ' ' ' ' ' ' ' ' ^^^^^^^^^^^^^^^^^^^^^^^^^^^^ ' ' ' ' ' ' ' ' ' ' ' ' ' '
' ' ' ' ' ' ' ' ' ' ' ' ' ' ' ' ' ' ' ' ' ' ' ' ' ' ' ' ' ' ' ' ' ' '© R. de Levie
' ' ' ' ' ' ' ' ' ' ' ' ' ' ' ' ' ' ' ' ' ' ' ' ' ' ' ' ' ' ' ' ' ' ' ' ' ' ' ' ' ' ' ' ' ' ' ' ' '
```

' Forward or inverse Fourier transformation

' Fourier transformation of an array of complex data, arranged in a single
' block of three adjacent columns, the first column containing the variable
' (e.g., time, or frequency), the second column the real components of the
' input data, the third column their imaginary components. For purely real
' input data, either leave the second column blank, or fill it with zeros.
' Likewise, for purely imaginary input data, the second column should only
' contain zeros, blanks, or combinations thereof.

' The macro accepts two input formats: (1) data with time or frequency
' values centered around zero, and (2) data with time or frequency values
' starting at zero. The output format is commensurate with the input format.

' The output is written in the three columns to the right of the input
' data block, thereby overwriting any prior data in that region. The
' output columns contain (from left to right) frequency or time, the
' real components of the output data, and their imaginary components.

' To initiate the transform, highlight the three columns of the input
' array, and call one of the two macros, Forward() or Inverse().

' The macro uses two drivers to set iSign to +1 (for a forward transform)
' or to -1 (for inverse transformation). The main subroutine Fourier reads
' the input information, makes a number of checks, and computes the output.
' For the actual Fourier transformation it calls the subroutine FFT.

```
Sub Forward()
Dim iSign As Integer
iSign = 1
Call Fourier(iSign)
End Sub
' ' ' ' ' ' ' ' ' ' ' ' ' ' ' ' ' ' ' ' ' ' ' ' ' ' ' ' ' ' ' ' ' ' ' ' ' ' ' ' ' ' ' ' ' ' ' ' ' '

Sub Inverse()
Dim iSign As Integer
iSign = -1
Call Fourier(iSign)
End Sub
' ' ' ' ' ' ' ' ' ' ' ' ' ' ' ' ' ' ' ' ' ' ' ' ' ' ' ' ' ' ' ' ' ' ' ' ' ' ' ' ' ' ' ' ' ' ' ' ' '

Sub Fourier(iSign)

' Check the array length n, which must be a power of 2,
' and be at least 2

Dim cn As Integer, cnmax As Integer, rn As Integer, rnmax As Integer
Dim n As Integer, nn As Integer, jSign As Integer
```

```
Dim dataArray As Variant, outputArray As Variant, z As Variant
Dim Ans As String
Dim length as Single
Dim check1 As Double, check2 As Double, interval As Double, denom As Double
n = 0
rnmax = Selection.Rows.Count
length = CSng(rnmax)

If length < 2 Then
  MsgBox "There must be at least two rows."
  End
End If
Do While length > 1
  length = length / 2
Loop
If length < 0.9999 Or length > 1.0001 Then
  MsgBox "The number of rows must be a power of two."
  End
End If

' Check that there are three input columns

cnmax = Selection.Columns.Count
If cnmax <> 3 Then
  MsgBox "There must be three input columns, one for" & _
  Chr(13) & "time, frequency, etc., the next two for the" & _
  Chr(13) & "real and imaginary parts of the input data."
  End
End If

' Read the input data

dataArray = Selection.Value

' Check that the first column has its first two elements

check1 = VarType(dataArray(1, 1))
If check1 = 0 Then
  MsgBox "Enter the top left value."
  End
End If
check2 = VarType(dataArray(2, 1))
If check2 = 0 Then
  MsgBox "Enter a value in row 2 of the first column."
  End
End If

' Determine what input convention is used:
' jSign = -1 for data centered around zero;
' jSign = +1 for data starting at zero.

interval = (dataArray(2, 1) - dataArray(1, 1)) * rnmax
If dataArray(1, 1) > (-0.5 * interval / rnmax) And _
  dataArray(1, 1) < (0.5 * interval / rnmax) Then jSign = 1
If dataArray(1, 1) < (-0.5 * interval / rnmax) And _
  dataArray(rnmax / 2 + 1, 1) > (-0.5 * interval / rnmax) And _
  dataArray(rnmax / 2 + 1, 1) < (0.5 * interval / rnmax) _
  Then jSign = -1
```

```
If jSign = 0 Then
  MsgBox "The input format is incorrect. " & _
  Chr(13) & "It should either be centered" & _
  Chr(13) & "around zero, or start at zero."
  End
End If

' Read and rearrange the input data

nn = 2 * rnmax
ReDim Term(nn) As Double, Re(rnmax) As Double, Im(rnmax) As Double

If jSign = -1 Then
  For rn = 1 To rnmax / 2
    Term(2 * rn - 1) = dataArray(rnmax / 2 + rn, 2)
    Term(2 * rn) = dataArray(rnmax / 2 + rn, 3)
    Term(rnmax + 2 * rn - 1) = dataArray(rn, 2)
    Term(rnmax + 2 * rn) = dataArray(rn, 3)
  Next rn
End If

If jSign = 1 Then
  For rn = 1 To rnmax
  Term(2 * rn - 1) = dataArray(rn, 2)
  Term(2 * rn) = dataArray(rn, 3)
  Next rn
End If
' Check that the output does not overwrite valuable data

Selection.Offset(0, 3).Select
outputArray = Selection.Value
Dim z As Variant
For rn = 1 To rnmax
  For cn = 1 To cnmax
    z = outputArray(rn, cn)
    If IsEmpty(z) Or z = 0 Then
      n = n
    Else
      n = n + 1
    End If
  Next cn
Next rn
Dim Ans As String
If n > 0 Then
  Ans = MsgBox("There are data in the space where the output will be" & _
  Chr(13) & " written. Proceed anyway and overwrite those data?", vbYesNo)
  If Ans = vbNo Then
    Selection.Offset(0, -3).Select
    End
  End If
End If

' Calculate and write the frequency or time scale

If jSign = -1 Then
  For rn = 1 To rnmax
    dataArray(rn, 1) = (-(rnmax / 2) + rn - 1) / interval
```

```
    Next rn
End If

If jSign = 1 Then
  For rn = 1 To rnmax / 2
    dataArray(rn, 1) = (rn - 1) / interval
  Next rn
  For rn = (rnmax / 2 + 1) To rnmax
    If iSign > 0 Then
      dataArray(rn, 1) = (rn - rnmax - 1) / interval
    Else
      dataArray(rn, 1) = (rn - 1) / interval
    End If
  Next rn
End If

' Calculate the Fourier transform

Call FFT(Term, nn, iSign)

' Arrange and write the output data

denom = (rnmax + 1 + iSign * (rnmax - 1)) / 2

If jSign = -1 Then
  For rn = 1 To rnmax / 2
    dataArray(rn, 2) = Term(rnmax + 2 * rn - 1) / denom
    dataArray(rn, 3) = Term(rnmax + 2 * rn) / denom
    dataArray((rnmax / 2) + rn, 2) = Term(2 * rn - 1) / denom
    dataArray((rnmax / 2) + rn, 3) = Term(2 * rn) / denom
  Next rn
End If

If jSign = 1 Then
  For rn = 1 To rnmax
    dataArray(rn, 2) = Term(2 * rn - 1) / denom
    dataArray(rn, 3) = Term(2 * rn) / denom
  Next rn
End If

Application.ScreenUpdating = False

Selection.Value = dataArray

End Sub
'''''''''''''''''''''''''''''''''''''''''''''''''''''''''''''''''''''

Sub FFT(Term, nn, iSign)

' The following is the forward Fourier transform routine FOUR1
' from J. C. Sprott, "Numerical Recipes: Routines and Examples
' in BASIC", Cambridge University Press, Copyright (C) 1991 by
' Numerical Recipes Software. Used by permission. Use of this
' routine other than as an integral part of the present book
' requires an additional license from Numerical Recipes Software.
' Further distribution is prohibited. The routine has been modi-
' fied to yield double-precision results, and to conform to the
' standard mathematical convention for Fourier transformation.

Dim tr As Double, ti As Double, theta As Double, wtemp As Double
```

```
Dim wi As Double, wr As Double, wpi As Double, wpr As Double
Dim i As Integer, istep As Integer, j As Integer
Dim m As Integer, mmax As Integer
j = 1
For i = 1 To nn Step 2
  If j > i Then
     tr = Term(j)
     ti = Term(j + 1)
     Term(j) = Term(i)
     Term(j + 1) = Term(i + 1)
     Term(i) = tr
     Term(i + 1) = ti
  End If
  m = Int(nn / 2)
  While m >= 2 And j > m
     j = j - m
     m = Int(m / 2)
  Wend
  j = j + m
Next i
mmax = 2
While nn > mmax
  istep = 2 * mmax
  theta = 2 * [Pi()] / (-iSign * mmax)
  wpr = -2 * Sin(0.5 * theta) ^ 2
  wpi = Sin(theta)
  wr = 1
  wi = 0
  For m = 1 To mmax Step 2
    For i = m To nn Step istep
       j = i + mmax
       tr = wr * Term(j) - wi * Term(j + 1)
       ti = wr * Term(j + 1) + wi * Term(j)
       Term(j) = Term(i) - tr
       Term(j + 1) = Term(i + 1) - ti
       Term(i) = Term(i) + tr
       Term(i + 1) = Term(i + 1) + ti
    Next i
    wtemp = wr
    wr = wr * wpr - wi * wpi + wr
    wi = wi * wpr + wtemp * wpi + wi
  Next m
  mmax = istep
Wend

End Sub
```

10.6
Convolution and deconvolution

The macros for convolution and deconvolution are variations on the Fourier transform macros of section 10.5, and have a very similar structure. There are two macros, Convolve( ) and Deconvolve( ), that are merely drivers to distinguish between convolution (kSign $=+1$) and deconvolution (kSign $=-1$) in the main subroutine, Convolution(kSign). The latter in turn repeatedly calls the fast Fourier transformation subroutine FFT(Term, nn, iSign). Note that the input has the same format as that for Fourier transformation (three input columns, of which the first is used for the independent variable, such as time t), but that the second and third input columns now should contain two different, *real* functions. There is only one output column, which again contains only real numbers. For convolution, the input columns are t, x, and y, and the output column is x*y. For deconvolution, the input columns are t, x*y, and y (note the order), while the output column is x.

High-frequency noise can interfere with deconvolution. We therefore include and adjustable Hanning window. The adjustable parameter w selects the window function $\{0.5*[1 + \cos(\pi x / x_{\max})]\}^{w}$. The window has no effect for $w = 0$. For $w = 1$ it provides the standard Hanning window. Stronger noise reduction (with a correspondingly larger signal distortion) is obtainable with $w > 1$.

In deconvolution we divide XY by Y, where X is the Fourier transform of x, and Y that of y. Whenever $Y = 0$ this will lead to a problem, which is avoided by substituting a small but non-zero value, *min*, for Y. However, the magnitude of *min* can be crucial: if it is too small, a particular frequency can be magnified out of proportion, giving rise to an oscillatory solution. If, on the other hand, *min* is taken too large, it may distort the final result. When the relative uncertainty in the data is of the order of ε, the optimal value of *min* appears to lie between ε and ε^2, but we are unaware of an automatic solution to this problem. You may therefore have to experiment with the value of *min* when problems occur, in order to fine-tune the solution. Initially take *min* as ε^2; if that first guess leads to oscillations (i.e., regular sinusoidal fluctuations of constant amplitude, as distinct from noise), increase the value of *min* in steps of an order of magnitude until a non-oscillatory result is obtained. In that case it may be wise to include a second input box for the value of *min*.

```
'''''''''''''''''''''''''''''''''''''''''''''''''''''''''''''''''''''''
'''''''''''''''''''''''''''''''''''''''''''''''''''''''''''''''''''''''
''''''''''''''''''''''''''''^^^^^^^^^^^^^^^^^^^^^^^^^^^^^^^''''''''''''''''''''''
'''''''''''''''''''''''''^                        ^'''''''''''''''''''''''
'''''''''''''''''''''''''^         CONVOLUTION        ^'''''''''''''''''''''''
'''''''''''''''''''''''''^                        ^'''''''''''''''''''''''
'''''''''''''''''''''''''''''''^^^^^^^^^^^^^^^^^^^^^^^^^^^^^''''''''''''''''''''''
''''''''''''''''''''''''''''''''''''''''''''''''''''''''''''''''''''© R. de Levie
'''''''''''''''''''''''''''''''''''''''''''''''''''''''''''''''''''''''

' Convolution or deconvolution of two columns of real data, arranged in
' a single block of three adjacent columns, the first column containing
' the variable (time), the second column the real components of the first
' data set, the third column the real components of the second data set.
'
```

```
' The macro accepts any input format in which the t-scale contains 2^N
' equidistant values (where N is a positive integer >1), regardless of
' its starting value. While the output is independent of the starting
' value of the t-scale, it does depend on the phase relationship between
' the two input signals. In order to avoid phase shifts, the window
' function (in the third column) should be centered.

' For deconvolution, the macro incorporates a Hanning window of
' adjustable width; the default setting, w = 0, corresponds to a
' rectangular window (i.e., the absence of filtering); w = 1
' yields the standard Hanning window; while w > 1 gives an
' extra-narrow window (more noise rejection, and more distortion).

' Upon selection of w = 0, a second filtering option appears, based on
' a rectangular window. It allows the user to set the highest n frequen-
' cies equal to zero. The default setting, 0, is again for no filtering.

' The output is written in one column, immediately to the right of the
' input data block, thereby overwriting any prior data in that region.
' After writing the output, the activated area returns to its original
' position and contents.

' To initiate the transform, highlight the three columns of the input
' array, and call the macro.

' The following are the two drivers for the convolution subroutine. They
' set kSign either to +1 (for convolution) or to -1 (for deconvolution).
''''''''''''''''''''''''''''''''''''''''''''''''''''''''''''''''''''''''

Sub Convolve()
Dim kSign As Integer
kSign = 1
Call Convolution(kSign)
End Sub

''''''''''''''''''''''''''''''''''''''''''''''''''''''''''''''''''''''''

Sub Deconvolve()
Dim kSign As Integer
kSign = -1
Call Convolution(kSign)
End Sub

''''''''''''''''''''''''''''''''''''''''''''''''''''''''''''''''''''''''

Sub Convolution(kSign)

' Check the array length n, which must be a power of 2,
' and be at least 2

Dim rn As Integer, cn As Integer
Dim rnmax As Integer, length As Integer
Dim cnmax As Integer, N As Integer, nn As Integer
N = 0
Dim z As Double
z = 0
rnmax = Selection.Rows.Count
nn = 2 * rnmax
```

```
cnmax = 3
length = rnmax

If length < 2 Then
  MsgBox "There must be at least two rows."
  End
End If
Do While length > 1
  length = length / 2
Loop
If length < 0.9999 Or length > 1.0001 Then
  MsgBox "The number of rows must be a power of two."
  End
End If

' Check that there are three input columns

cnmax = Selection.Columns.Count
If cnmax <> 3 Then
  MsgBox "There must be three input columns, one for the" & _
  Chr(13) & "independent variable (e.g., time), the next two" & _
  Chr(13) & " for the two functions to be convolved or deconvolved."
  End
End If

' Read the input data

Dim inputArray As Variant, dataArry As Variant
inputArray = Seleciton.Formula
dataArray = Selection.Value

' Check that the first column has its first two elements

Dim check1 As Double, check2 As Double, dataSpacing As Double
check1 = VarType(dataArray(1, 1))
If check1 = 0 Then
  MsgBox "Enter the top left value."
  End
End If
check2 = VarType(dataArray(2, 1))
If check2 = 0 Then
  MsgBox "Enter a value in row 2 of the first column."
  End
End If
dataSpacing = (dataArray(2, 1) - dataArray(1, 1))

' Read and rearrange the input data from the second and third columns

ReDim Term2(nn) As Double, Term3(nn) As Double, Term4(nn) As Double
  For rn = 1 To rnmax
  Term2(2 * rn - 1) = dataArray(rn, 2)
  Term2(2 * rn) = z
  Term3(2 * rn - 1) = dataArray(rn, 3)
  Term3(2 * rn) = z
Next rn

' Check that the output does not overwrite valuable data.
```

```
Selection.Offset(0, 1).Select
Dim outputArray As Variant
Dim q As Single
outputArray = Selection.Value
For rn = 1 To rnmax
  q = outputArray(rn, 3)
  If IsEmpty(q) Or q = 0 Then
    n = n
  Else
    n = n + 1
  End If
Next rn
Selection.Offset(0, -1).Select
Dim Answer As String, vAnswer As String
If n > 0 Then
  Answer = MsgBox("There are data in the column where the output will be" & _
  Chr(13) & " written. Proceed anyway and overwrite those data?", vbYesNo)
  If Answer = vbNo Then End
End If

' Transform the data from the second and third columns

Dim iSign As Integer
iSign = 1
Call FFT(Term2, nn, iSign)
Call FFT(Term3, nn, iSign)

' Multiply in the frequency domain for convolution, or divide
' in the frequency domain for deconvolution. For deconvolution,
' a minimum value "min" (arbitrarily set here to 1E-12) prevents
' division by zero. Moreover, an input box is provided for
' optional noise filtering.

Dim rnm2 As Double, drnmax As Double
drnmax = CDbl (rnmax)
rnm2 = drnmax * drnmax
Dim min As Double, D As Double
If kSign = 1 Then
  For rn = 1 To rnmax
    Term4(2 * rn - 1) = (Term2(2 * rn - 1) * Term3(2 * rn - 1) _
      - Term2(2 * rn) * Term3(2 * rn)) / (average3 * rnm2)
    Term4(2 * rn) = (Term2(2 * rn - 1) * Term3(2 * rn) _
      + Term2(2 * rn) * Term3(2 * rn - 1)) / (average3 * rnm2)
  Next rn
End If

If kSign = -1 Then

' Select the Hanning window parameter w

  Dim w As Double, ww As Integer
  Dim WindowFunction as Double, TwoPi As Double
  min = 0.000001
  ww = 1
  TwoPi =2 * [Pi()]
  Dim Message1 As String, Title1 As String, Default1 As String
```

```
Message1 = "Enter the window parameter, a non-negative number." &
          Chr(13) & _
          "The default is 0, which corresponds to a rectangular" &
          Chr(13) & _
          "window (no filtering); select 1 for a normal Hanning
          filter."
Title1 = "Adjustable Hanning Window"
Default1 = "0"
w = InputBox(Message1, Title1, Default1)
If w = 0 Then ww = 0
If w < min Then w = min
For rn = 1 To rnmax / 2 - 1
  D = Term3(2 * rn - 1) * Term3(2 * rn - 1) + Term3(2 * rn) _
    * Term3(2 * rn)) * rnmax * dataSpacing
  If D < min Then D = min
  Term4(2 * rn - 1) = (Term2(2 * rn - 1) * Term3(2 * rn - 1) _
    + Term2(2 * rn) * Term3(2 * rn)) / D
  Term4(2 * rn) = (Term2(2 * rn) * Term3(2 * rn - 1) _
    - Term2(2 * rn - 1) * Term3(2 * rn)) / D
  WindowFunction = (0.5 * (1 - Cos(TwoPi * (2 * rn + rnmax + 1) _
    / (2 * rnmax)))) ^ w
  Term4(2 * rn - 1) = Term4(2 * rn - 1) * WindowFunction
  Term4(2 * rn) = Term4(2 * rn) * WindowFunction
Next rn
For rn = (rnmax / 2 to rnmax
D = (Term3(2 * rn - 1) * Term3(2 * rn -1) + Term3(2 * rn) _
  * Term3(2 * rn)) * rnmax * dataSpacing
If D < min Then D = min
Term4(2 * rn - 1) = (Term2(2 * rn - 1) * Term3(2 * rn - 1) _
  + Term2(2 * rn) * Term3(2 * rn)) / D
Term4(2 * rn) = (Term2(2 * rn) * Term3(2 * rn - 1) _
  - Term2(2 * rn - 1) * Term3(2 * rn)) / D
  WindowFunction = (0.5 * (1 - Cos(TwoPi * (2 * rn - rnmax + 1) _
  / (2 * rnmax)))) ^ w
  Term4(2 * rn - 1) = Term4(2 * rn - 1) * WindowFunction
  Term4(2 * rn) = Term4(2 * rn) * WindowFunction
Next rn

' Select a high-frequency cut-off in the rectangular filter
' This option is offered only when w = 0 (i.e., a rectangular
' window) is selected in the Adjustable Hanning Filter box.
  If ww = 0 Then
    Dim nfzero As Single
    Dim nfz As Integer
    Dim Message2 As String, Title2 As String, Default2 As String
    Message2 = "Enter the NUMBEr of highest frequencies" & Chr(13) & _
      "you want to filter out. If you don't want" & Chr(13) & _
      "any filtering, just press the OK button."
    Title2 = "Noise Filter"
    Default2 = "0"
    nfzero = InputBox(Message2, Title2, Default2)
    If nfzero < 0 Then nfz = 0
    If nfzero > 0 And nfzero < rnmax / 2 Then nfz = Int(nfzero)
```

```
      If nfzero > rnmax / 2 Then nfz = rnmax / 2
      If nfz > 0 Then
      For rn = (rnmax / 2 + 2 - nfz) To (rnmax / 2 + nfz)
         Term4 (2 * rn - 1) = z
         Term4 (2 * rn - 2) = z
      Next rn
    End If
  End If
End If

' Calculate output data

iSign = -1
Call FFT(Term4, nn, iSign)

' Arrange and write the output data

For rn = 1 To rnmax / 2
   outputArray((rnmax / 2) + rn, 3) = Term4 (2 * rn - 1)
Next rn
For rn = rnmax / 2 + 1 To rnmax
   outputArray(rn - (rnmax / 2), 3) = Term4 (2 * rn - 1)
Next rn

Application.ScreenUpdating = False
Selection.Offset(0, 1).Select
Selection.Value = outputArray
Selection.Offset(0, -1).Select
Selection.Value = inputArray
End Sub
```

10.7 Weighted least squares

While for most readers an initial understanding of the operation of least-squares equations is most easily achieved in terms of specific expressions, such as those given in chapters 2 and 3, a corresponding computer program is more efficiently written in terms of matrix algebra. Moreover, such a program can then accommodate any number of independent variables. Because Excel does not contain a weighted least-squares routine, one is provided here. At the same time, this will serve to illustrate how to incorporate matrix expressions and subroutines in a macro.

10.7a The algorithm

Here we merely summarize the methods used; for details the reader should consult textbooks on statistics, such as N. R. Draper & H. Smith, *Applied Regression Analysis*, 2nd ed., Wiley 1981. The dependent input variables y_i are collected in an input vector **Y**, the corresponding weights w_i in a vector **W**, and the dependent variable(s) x_i in the matrix **X**. In addition, the latter contains a first column of 1's or 0's, depending on whether the fitted curve has a variable intercept or is forced to go through the origin.

Denoting vectors and matrices in bold, and using the shorthand notation $\mathbf{M}'$ for the transpose of matrix $\mathbf{M}$, and $\mathbf{M}''$ for its inverse (as used in the comments in the subroutine; VBA code will not allow the usual superscripts, T for transpose, and -1 for inverse), the sought coefficients are then calculated in a vector $\mathbf{b}$ as

$$\mathbf{b} = (\mathbf{X}'\,\mathbf{W}\,\mathbf{X})''\,\mathbf{X}'\,\mathbf{W}\,\mathbf{Y}$$

which applies for *any* number of independent variables. The corresponding variances are calculated as the diagonal elements of the variance–covariance matrix $\mathbf{V}$

$$\mathbf{V} = (\mathbf{X}'\,\mathbf{W}\,\mathbf{X})''\,(\mathbf{Y}'\,\mathbf{W}\,\mathbf{Y} - \mathbf{b}'\,\mathbf{X}'\,\mathbf{W}\,\mathbf{Y})\,/\,(N - P)$$

where N is the number of data points, and P the number of dependent variables. This formalism yields the results for the unweighted least-squares analysis when all w_i's are set equal to 1, in which case $\mathbf{W}$ is the unit matrix.

The weighted least-squares routine shown below provides the adjustable parameters and their standard deviations. If that is all you need, you may want to use it also as your general least-squares routine, especially after you have incorporated it in a menu or given it a toolbar icon (in which case it is easier to use and more readily accessible than the Regression routine in the Analysis ToolPak). When using it for unweighted least squares, merely leave the second column empty. Alternatively, if you desire the routine to provide more statistical information, you can modify it to do so. Remember, *you* are at the controls here.

10.7b Implementation

As in the Fourier transform example of section 10.2 we use the highlighted area rather than a dialog box. While this is much faster in use, it requires that the reader remember the required format. In the present case the format is as follows:

1 The first column must contain the so-called "dependent" variable, here called y; the variable that is assumed to contain all experimental errors.
2 The second column must contain the corresponding weights w_i. If no weights are assigned, either leave this column blank or fill it with 1's. However, *the column cannot be left out.*
3 The next column(s) must contain the "independent" variable(s) x. There can be as many independent variables as required. Of course, there must be at least one more input datum in the y-column than there are x-columns, otherwise the problem is not properly defined for least-squares analysis, which requires more data points than adjustable parameters.

In practice, least-squares methods are most useful when the problem is strongly over-defined, i.e., when the number of y-values is much in excess of this minimum.

As already mentioned, the program uses an x-matrix that contains one more column than specified by the user. This is because the user specifies x_1, x_2, etc., whereas the program calculates a fit to either $y = a_0 + a_1x_1 + a_2x_2 + \cdots$ or, when the fitted curve is supposed to go through the origin, $y = a_1x_1 + a_2x_2 + \cdots$. These two options correspond to an implied x-variable of value 1 or 0 respectively. Thus there are two options: either the program calculates the constant a_0 in $y = a_0 + a_1x_1 + a_2x_2 + \cdots$, or it sets a_0 equal to zero, thereby forcing the fitted curve through the origin, as in a proportionality. As programmer you could choose to provide these options (for "general" or for "through the origin") via a message box that pops up during the program, or by providing two different program entries. Here we have chosen the latter route, which we find more user-friendly. Consequently the weighted least-squares program is made to be a subroutine to two small programs, WLS0 and WLS1, that merely set a parameter p equal to either zero or one respectively. (Note that the transfer of the value of p makes the main least-squares program a subroutine, since macros have no input parameters.) The weighted least-squares subroutine, in turn, calls on three smaller subroutines to perform the three matrix manipulations: transpose, invert, and multiply. In order to keep the number of such subroutines to a minimum, vectors are treated as matrices with one row or column.

Near the end of the macro the standard deviation is calculated from the variance. When the routine is tested with noise-free data, so that the variance is zero, the program will attempt to take the square root of zero, which it cannot do. As a standard precaution, in order to prevent the resulting error message and associated problems, the variance is first tested, and a minimum value of 10^{-40} inserted when the program computes a smaller value.

The output will appear in two rows *below* the input data, in the order A_0, A_1, A_2, ..., A_n. This format was chosen because the output fits precisely in a block of the same width as the input array; moreover, this arrangement leaves the right-hand side of the input block free for subsequent data manipulations. The output is made to stand out with italic and bold type. Alternatively you could use color for that purpose, with an instruction such as `ActiveCell.Font.ColorIndex = 7`. This option was not exercised here because colors may give problems with some black-and-white printers.

Before the program is ready to write its output, it makes a quick, barely noticeable foray into the region where it wants to write, to make sure that it will not overwrite valuable data. If it finds data, it will then ask the user whether or not to overwrite these.

So here goes:

```
''''''''''''''''''''''''''''''''''''''''''''''''''''''''''''''''''''
''''''''''''''''''''''''''''''''''''''''''''''''''''''''''''''''''''
''''''''''''''''''''''''^^^^^^^^^^^^^^^^^^^^^^^^^^^'''''''''''''''''''
'''''''''''''''''''''''^                          ^''''''''''''''''''
'''''''''''''''''''''''^ WEIGHTED  LEAST  SQUARES  ^''''''''''''''''''
'''''''''''''''''''''''^                          ^''''''''''''''''''
''''''''''''''''''''''''^^^^^^^^^^^^^^^^^^^^^^^^^^^'''''''''''''''''''
''''''''''''''''''''''''''''''''''''''''''''''''''''''© R. de Levie
''''''''''''''''''''''''''''''''''''''''''''''''''''''''''''''''''''
```

```
' The function of the following two drivers is merely to set the
' value of one parameter, p, equal to either one or zero, in order
' to choose between a general weighted least squares fitting (p = 1)
' or one that forces the curve through the origin (p = 0).

Sub WLS0()
Dim p As Double
p = 0
Call WeightedLeastSquares(p)
End Sub
```

```
'''''''''''''''''''''''''''''''''''''''''''''''''''''''''''''''''''''

Sub WLS1()
Dim p As Double
p = 1
Call WeightedLeastSquares(p)
End Sub
```

```
'''''''''''''''''''''''''''''''''''''''''''''''''''''''''''''''''''''

' WEIGHTED LEAST SQUARES

' This subroutine computes the parameters and their standard de-
' viations for a weighted least squares fit to data in 3 or more
' columns. The columns must be arranged as follows: the first
' column must contain the dependent variable Y, the second its
' weights W, and the next column(s) the dependent variable(s) X.
'
' The weights column must be included, but it may be left blank,
' in which case all data will be assigned unit weights.
' Therefore, if an unweighted least-squares fit is desired,
' either leave the second column empty, or fill it with 1's.
'
' The subroutine requires an input parameter p: p = 1 causes
' a general weighted least squares fit to the data, while p = 0
' forces the fit to pass through the origin. The subroutine must
' therefore be called by a small driver which sets the value of p.

Sub WeightedLeastSquares(p)

' Determination of the array size:

Dim rmax As Integer               'rmax = maximum number of rows
rmax = Selection.Rows.Count
Dim m As Integer, n As Integer
Dim cmax As Integer               'cmax = max. number of columns
cmax = Selection.Columns.Count
```

```
Dim ccmax As Integer
ccmax = cmax - 1
Dim i As Integer, j As Integer
Dim u As Double, z As Double
u = 1
z = 0
Dim SRR As Double, varY As Double, sumW As Double

' Check that the number of columns is at least 3:

If cmax < 3 Then
  MsgBox "There must be at least three columns, one" & _
  Chr(13) & "for Y, one for W, and one or more for X."
  End
End If

' Check that rmax > cmax, so that the number of data
' points is sufficient to define the problem:


If rmax < cmax Then
  MsgBox "There must be at least " & cmax & " input" & _
  Chr(13) & " data to define the problem."
  End
End If

' Dimension the arrays:

Dim dataArray As Variant
Dim outputArray As Variant
ReDim yArray(1 To rmax, 1 To 1) As Double
ReDim wArray(1 To rmax, 1 To rmax) As Double
ReDim xArray(1 To rmax, 1 To ccmax) As Double
ReDim ytArray(1 To 1, 1 To rmax) As Double
ReDim ytwArray(1 To 1, 1 To rmax) As Double
ReDim ytwyArray(1 To 1, 1 To 1) As Double
ReDim xtArray(1 To ccmax, 1 To rmax) As Double
ReDim xtwArray(1 To ccmax, 1 To rmax) As Double
ReDim pArray(1 To ccmax, 1 To ccmax) As Double
ReDim piArray(1 To ccmax, 1 To ccmax) As Double
ReDim qArray(1 To ccmax, 1 To 1) As Double
ReDim bArray(1 To ccmax, 1 To 1) As Double
ReDim btArray(1 To 1, 1 To ccmax) As Double
ReDim btqArray(1 To 1, 1 To 1) As Double
ReDim M1(1 To rmax, 1 To rmax) As Double
ReDim M2(1 To rmax, 1 To rmax) As Double
ReDim vArray(1 To ccmax, 1 To ccmax) As Double

  ' Read the dataArray, then Fill the various input arrays: yArray,
  ' wArray, and xArray. The wArray contains zero's except that it
  ' has the individual, normalized weights as its diagonal elements.

dataArray = Selection.Value

For i = 1 To rmax
  yArray(i, 1) = dataArray(i, 1)
Next i
```

```
For i = 1 To rmax
  If IsEmpty(dataArray(i, 1)) Then
    MsgBox "Y-value(s) missing"
    End
  End If
Next i
For i = 1 To rmax
  If IsEmpty(dataArray(i, 2)) Then dataArray(i, 2) = u
Next i
For j = 3 To cmax
  For i = 1 To rmax
    If IsEmpty(dataArray(i, j)) Then
      MsgBox "X-value(s) missing"
      End
    End If
  Next i
Next j

sumW = z
For i = 1 To rmax
  sumW = sumW + dataArray(i, 2)
Next i
For i = 1 To rmax
  For j = 1 To rmax
    wArray(i, j) = z
  Next j
  wArray(i, i) = dataArray(i, 2) * rmax / sumW
Next i
For i = 1 To rmax
  For j = 1 To rmax
  Next j
Next i

For i = 1 To rmax
  xArray(i, 1) = p
Next i
For j = 3 To cmax
  For i = 1 To rmax
    xArray(i, (j - 1)) = dataArray(i, j)
  Next i
Next j

' Compute b = (X' W X)" X' W Y, where ' and t denote
' transposition, and " and i indicate inversion

' The various arrays and their dimensions (rows, columns) are:
' Y         = yArray    ( rmax,    1)
' W         = wArray    ( rmax,  rmax)
' X         = xArray    ( rmax, ccmax)
' X'        = xtArray   ( ccmax, rmax)
' X' W      = xtwArray  ( ccmax, rmax)
' X' W X    = pArray    ( ccmax,ccmax)
' (X' W X)" = piArray   ( ccmax,ccmax)
' X' W Y    = qArray    ( ccmax,    1)
' b         = bArray    ( ccmax,    1)
```

```
Call Transpose(xArray, rmax, ccmax, xtArray)
Call Multiply(xtArray, ccmax, rmax, wArray, rmax, xtwArray)
Call Multiply(xtwArray, ccmax, rmax, xArray, ccmax, pArray)
Call Invert(pArray, ccmax, piArray)
Call Multiply(xtwArray, ccmax, rmax, yArray, 1, qArray)
Call Multiply(piArray, ccmax, ccmax, qArray, 1, bArray)

' Check against overwriting spreadsheet data

Selection.Offset(2, 0).Select
outputArray = Selection.Value
Selection.Offset(-2, 0).Select

m = 0
For i = rmax - 1 To rmax
   For j = 1 To cmax
      If IsEmpty(outputArray(i, j)) Then
         m = m
      Else
         m = m + 1
      End If
   Next j
Next i
Dim Answer As String
If m > 0 Then
   Answer = MsgBox("There are data in the two lines below the " & Chr(13) & _
   "input data array. Can they be overwritten? ", vbYesNo, "Overwrite?")
   If Answer = vbNo Then End
End If
' The additional arrays and their dimensions (rows, columns) are:
' Y'          = ytArray    ( 1,   rmax)
' Y' W        = ytwArray   ( 1,   rmax)
' Y' W Y      = ytwyArray  ( 1,      1)
' b'          = btArray    ( 1, ccmax)
' b' X' W Y   = btqArray   ( 1,      1)

Call Transpose(yArray, rmax, 1, ytArray)
Call Transpose(bArray, ccmax, 1, btArray)
Call Multiply(ytArray, 1, rmax, wArray, rmax, ytwArray)
Call Multiply(ytwArray, 1, rmax, yArray, 1, ytwyArray)
Call Multiply(btArray, 1, ccmax, qArray, 1, btqArray)

' Calculate SRR = Y'WY - b'X'WY, and then varY, the variance of y,
' as varY = SRR/(rmax-ccmax); and vArray, the variance/covariance
' matrix, as V = (X'WX)" times varY, of which we here only use
' the diagonal elements, i.e., the variances.

SRR = ytwyArray(1, 1) - btqArray(1, 1)
varY = SRR / (rmax - ccmax)

For i = 1 To ccmax
   For j = 1 To ccmax
      vArray(i, j) = varY * piArray(i, j)
   Next j
Next i
```

```
ActiveCell.Offset(rmax, 0).Select
For j = 1 To cmax
  ActiveCell.Font.Bold = True
  ActiveCell.Font.Italic = True
  ActiveCell.Offset(0, 1).Select
Next j
ActiveCell.Offset(1, -cmax).Select
For j = 1 To cmax
  ActiveCell.Font.Italic = True
  ActiveCell.Offset(0, 1).Select
Next j
ActiveCell.Offset(-1, -cmax).Select
ActiveCell.Value = "Coeff.:"
ActiveCell.Offset(0, 1).Select
For j = 1 To ccmax
  ActiveCell.Value = bArray(j, 1)
  ActiveCell.Offset(0, 1).Select
Next j

If p = 0 Then
  ActiveCell.Offset(0, -ccmax).Select
  ActiveCell.Value = "zero"
  ActiveCell.Offset(1, -1).Select
Else
  ActiveCell.Offset(1, -cmax).Select
End If
ActiveCell.Value = "St.Dev.:"
ActiveCell.Offset(0, 1).Select

For j = 1 To ccmax
If vArray(j, j) < 1E-40 Then
    ActiveCell.Value = "<1E-20"
  Else
  ActiveCell.Value = Sqr(vArray(j, j))
  End If
  ActiveCell.Offset(0, 1).Select
Next j

If p = 0 Then
  ActiveCell.Offset(0, -ccmax).Select
  ActiveCell.Value = "zero"
  ActiveCell.Offset(1, -1).Select
Else
  ActiveCell.Offset(1, -cmax).Select
End If

End Sub

''''''''''''''''''''''''''''''''''''''''''''''''''''''''''''''''''

Sub Transpose(M1, r1, c1, Mout)

' Computes the transpose Mout of matrix M1
'    r1: number of rows in M1
'    c1: number of columns in M1
' Mout will have c1 rows and r1 columns
```

```
Dim i As Integer, j As Integer
For i = 1 To c1
  For j = 1 To r1
    Mout(i, j) = M1(j, i)
  Next j
Next i
End Sub
```

. . . .

```
Sub Multiply(M1, r1, c1, M2, c2, Mout)
' Computes the product of two matrices: Mout = M1 times M2
'    r1: number of rows in M1
'    c1: number of columns in M1
'    c2: number of columns in M2
' M2 must have c1 rows; Mout will have r1 rows and c2 columns

Dim i As Integer, j As Integer, k As Integer
Dim z As Double
For i = 1 To r1
  For j = 1 To c2
    Mout(i, j) = z
    For k = 1 To c1
      Mout(i, j) = Mout(i, j) + M1(i, k) * M2(k, j)
    Next k
  Next j
Next i
End Sub
```

. . . .

```
Sub Invert(M1, r1, Mout)

' The input matrix is M1, the output matrix is Mout
'    r1: number of rows in M1
' both M1 and Mout are square, with r1 rows and r1 columns

ReDim BB(1 To n, 1 To n) As Double
ReDim ipivot(1 To n) As Double
ReDim Index(1 To n) As Double
ReDim indexr(1 To n) As Double
ReDim indexc(1 To n) As Double
Dim big As Double, dummy As Double, pivinv As Double
Dim irow As Integer, icol As Integer
Dim i As Integer, j As Integer
Dim k As Integer, L As Integer, LL As Integer, n As Integer
n = r1 + 1
Dim u As Double
u = 1
Dim z As Double
z = 0

' Copy the input matrix in order to retain it

For i = 1 To r1
  For j = 1 To r1
    Mout(i, j) = M1(i, j)
  Next j
Next i
```

```
' The following is the Gauss-Jordan elimination routine GAUSSJ
' from J. C. Sprott, "Numerical Recipes: Routines and Examples
' in BASIC", Cambridge University Press, Copyright (C) 1991 by
' Numerical Recipes Software. Used by permission. Use of this
' routine other than as an integral part of the present book
' requires an additional license from Numerical Recipes Software.
' Further distribution is prohibited. The routine has been modi-
' fied slightly, primarily to yield double-precision results.
For j = 1 To r1
  ipivot(j) = z
Next j
For i = 1 To r1
  big = z
  For j = 1 To r1
    If ipivot(j) <> u Then
      For k = 1 To r1
        If ipivot(k) = z Then
          If Abs(Mout(j, k)) >= big Then
            big = Abs(Mout(j, k))
            irow = j
            icol = k
          End If
        ElseIf ipivot(k) > 1 Then Exit Sub
        End If
      Next k
    End If
  Next j
  ipivot(icol) = ipivot(icol) + 1
  If irow <> icol Then
    For L = 1 To r1
      dummy = Mout(irow, L)
      Mout(irow, L) = Mout(icol, L)
      Mout(icol, L) = dummy
    Next L
    For L = 1 To r1
      dummy = BB(irow, L)
      BB(irow, L) = BB(icol, L)
      BB(icol, L) = dummy
    Next L
  End If
  indexr(i) = irow
  indexc(i) = icol
  If Mout(icol, icol) = z Then Exit Sub
  pivinv = u / Mout(icol, icol)
  Mout(icol, icol) = u
  For L = 1 To r1
    Mout(icol, L) = Mout(icol, L) * pivinv
  Next L
  For L = 1 To r1
    BB(icol, L) = BB(icol, L) * pivinv
  Next L
  For LL = 1 To r1
    If LL <> icol Then
```

```
      dummy = Mout(LL, icol)
      Mout(LL, icol) = z
      For L = 1 To r1
         Mout(LL, L) = Mout(LL, L) - Mout(icol, L) * dummy
      Next L
      For L = 1 To r1
         BB(LL, L) = BB(LL, L) - BB(icol, L) * dummy
      Next L
    End If
  Next LL
Next i
For L = r1 To 1 Step -1
  If indexr(L) <> indexc(L) Then
    For k = 1 To r1
       dummy = Mout(k, indexr(L))
       Mout(k, indexr(L)) = Mout(k, indexc(L))
       Mout(k, indexc(L)) = dummy
    Next k
  End If
Next L
Erase indexc, indexr, ipivot
For i = 1 To r1
  For j = 1 To r1
  Next j
Next i
End Sub
```

10.8 More about Solver

Excel's multi-parameter non-linear least-squares routine. Solver, an implementation of the Levenberg–Marquardt algorithm, is a generally useful tool. In section 10.8a we describe an addition that can make it even more useful, and in section 10.8b we briefly indicate how you can call Solver from your macro.

10.8a Adding uncertainty estimates to Solver

Solver, can provide parameters that fit a user-specified function to a set of data, but does not yield estimates of the precision of those parameters. Here we exploit the approach used in section 10.3 to compute the precision of the parameters obtained with Solver. We will make the usual assumptions that, in fitting a function $F(a_i)$ to N experimental data pairs x,y, all indeterminate uncertainties are restricted to y, and follow a single Gaussian distribution. Furthermore we will assume that Solver has already been used to find a solution y_{calc} based on a mathematical model expression of the type $y_{calc} = F(a_i)$, where a_i are the parameters Solver has adjusted. We can then use a second macro, called SolverAid, to estimate the standard deviations of those parameters a_i.

In order to do so, we first evaluate the partial differential quotients $y_{n,i} = \partial y_{n,\text{calc}} / \partial a_i$ where n identifies the particular data pair, i.e., $1 \le n \le N$, while i specifies a given parameter a_i. The numerical differentiation uses the method explained in sections 2.3 and 2.4, and already implemented in the propagation macro of section 10.3. We then compute the matrix

$$P_{ij} = \sum_{n=1}^{N} y_{n,i}\, y_{n,j} \tag{10.8-1}$$

and use this to obtain the standard deviations of the parameters a_i through

$$\sigma_i = \sqrt{\frac{P_{ii}^{-1}\chi^2}{N-P}} \tag{10.8-2}$$

where P is the number of data points (i.e., the maximum value of i), and

$$\chi^2 = \sum_{n=1}^{N} (y_{n,\text{exp}} - y_{n,\text{calc}})^2 \tag{10.8-3}$$

Moreover, the program computes the standard deviation of the over-all fit,

$$\sigma_Y = \sqrt{\frac{\chi^2}{N-P-1}} \tag{10.8-4}$$

with which to compute the standard deviations in those parameters a_i. The numerical manipulations require a matrix inversion, for which we use the subroutine $\text{Invert}(M_1, r_1, M_{out})$ listed in section 10.7, which therefore need not be repeated here. Subroutines can be called from any macro. We change the parameters a_i one at a time, and determine the resulting changes in $y_{n,\text{calc}}$.

Alternatively one can modify the references to these parameters in the computation of Y_{calc} (E. J. Billo, *Excel for Chemists, A Comprehensive Guide*, Wiley-VCH 1997 pp. 297–299). However, when y_{calc} refers to these parameters *indirectly*, through intermediate parameters, this can yield incorrect results, e.g., when we instruct Solver to vary pK_a while we first compute $K_a = 10^{-pK_a}$ in a separate cell, and then refer to that cell in the computation of Y_{calc}. The problem will disappear by designating the cell containing K_a as the variable instead, but the approach taken here simply avoids this unnecessary complication.

SOLVERAID

© R. de Levie

```
Sub SolverAid()

' This macro takes the results of Solver and computes the
' corresponding standard deviations.

' The standard deviations of the fit, and the standard
' deviation(s) of the parameter(s) will be placed directly,
' in italics, to the right of the corresponding parameters,
' provided that those cells are unoccupied. Otherwise, the
' result(s) will be displayed in message boxes. It is,
' therefore, most convenient to leave blank the spreadsheet
' cells to the right of the parameters and of SRR.

' Notation used:
' P1:    single Parameter determined by Solver
' PP:    multiple parameters determined by Solver
'        NOTE: THESE MUST BE IN A SINGLE, CONTIGUOUS COLUMN
' SP:    standard deviations on those parameters
' SRR:   the sum of the residuals squared
'        used to optimize Solver
' SY:    the standard deviation on the function
' YC:    the Y-Values computed with the parameters P
'        NOTE: THESE MUST BE IN A SINGLE, CONTIGUOUS COLUMN
' c:     prefix denoting the number of columns:
'        cP = columns of PP, cY = columns of YC
'        Note: cP and cX should be 1
'r:      prefix denoting the number of rows:
'        rP = rows of PP, rY = rows of YC

' Select the computed Solver parameter P1 or parameters

Dim myRange1 As Range
Set myRange1 = Application.InputBox(prompt:= _
   "The parameters determined by Solver are located in:", _
   Title:="SolverAid InputBox 1: Solver parameters", Type:=8)
myRange1.Select
Dim rP As Integer
If Selection.Columns.Count <> 1 Then End
rP = Selection.Rows.Count
If rP = 0 Then
   End
ElseIf rP = 1 Then
   Dim P1Value As Variant
   P1Value = Selection.Value
ElseIf rP > 1 Then
   Dim PPValue As Variant
   PPValue = Selection.Value
End If

' Test and prepare the default output range for
' the standard deviations of the parameters, SP

Dim n As Integer
n = 0
Selection.Offset(0, 1).Select
If rP = 1 Then
```

```
   SP1Address = Selection.Address
   SP1Value = Selection.Value
   If IsEmpty(SP1Value) Then
      n = 0
   Else
      n = 1
   End If
Else
   Dim SPValue As Variant
   SPValue = Selection.Value
   For i = 1 To rP
      z = SPValue(i, 1)
      If IsEmpty(z) Then
         n = n
      Else
         n = n + 1
      End If
   Next i
End If

' Select the computed chi-squared value, SRR

Dim myRange2 As Range
Set myRange2 = Application.InputBox(prompt:= _
   "The sum of squares of the residuals is located in:", _
   Title:="SolverAid InputBox 2: SRR", Type:=8)
myRange2.Select
Dim cSRR As Integer, rSRR As Integer
cSRR = Selection.Columns.Count
rSRR = Selection.Rows.Count
If cSRR <> 1 Then End
If rSRR <> 1 Then End
Dim SRRValue As Double
SRRValue = Selection.Value

' Test and prepare the default output range for
' the standard deviation of the fit, SYY

Selection.Offset(0, 1).Select
Dim nn As Integer
nn = 0
SYValue = Selection.Value
If IsEmpty(SYValue) Then
   nn = 0
   SYAddress = Selection.Address
Else
   nn = 1
End If

' Select the computed Y-values, YC

Dim myRange3 As Range
Set myRange3 = Application.InputBox(prompt:= _
   "The column containing Ycalc is:", _
   Title:="SolverAid InputBox 3: Ycalc", Type:=8)
myRange3.Select
```

```
Dim rY As Integer
rY = Selection.Rows.Count
If rY <= rP + 1 Then
  MsgBox " The number N of data pairs must be at least" & Chr(13) & _
  "larger by one than the number of parameters P."
  End
End If

Dim YCValue As Variant
YCValue = Selection.Value
Application.ScreenUpdating = False

' Compute the standard deviations SY of the fit

Dim SY As Double
SY = Sqr(SRRValue / (rY - rP))

' Compute the partial differentials and the
' standard deviations for the one-parameter case

If rP = 1 Then
  myRange1.Select
  P1Value = P1Value * 1.000001
  Selection = P1Value
  myRange3.Select
  Dim YYValue1 As Variant
  YYValue1 = Selection.Value
  ReDim D1(1 To rY) As Double, DD1(1 To rY) As Double
  Dim SDD1 As Double, SP1 As Double
  SDD1 = 0
  For j = 1 To rY
    D1(j) = (YYValue1(j, 1) - YCValue(j, 1)) / (0.000001 * P1Value)
    DD1(j) = D1(j) * D1(j)
    SDD1 = SDD1 + DD1(j)
  Next j
  P1Value = P1Value / 1.000001
  myRange1.Select
  Selection.Value = P1Value
  SP1 = SY / Sqr(SDD1)
  If nn = 0 Then
    myRange2.Select
    Selection.Offset(0, 1).Select
    ActiveCell.Font.Italic = True
    Selection.Value = SY
  Else
    MsgBox "The standard deviation of the fit is " & SY
  End If
  If n = 0 Then
  myRange1.Select
    Selection.Offset(0, 1).Select
    ActiveCell.Font.Italic = True
    Selection.Value = SP1
  Else
    MsgBox "The standard deviation of the parameter is " & SP1
  End If
```

```
' Compute the partial differentials for the multi-parameter case

Else
  ReDim D(1 To rP, 1 To rY) As Double
  Dim YYValue As Variant
  For i = 1 To rP
    myRange1.Select
    PPValue(i, 1) = PPValue(i, 1) * 1.000001
    Selection.Value = PPValue
    myRange3.Select
    Dim YYValue As Variant
    YYValue = Seleciton.Value
    For j = 1 To rY
      D(i, j) = (YYValue(j, 1) - YCValue(j, 1)) / (0.000001 *
      PPValue(i, 1))
    Next j
    PPValue(i, 1) = PPValue(i, 1) / 1.000001
  Next i
    myRange1.Select
    Selection.Value = PPValue

  ReDim DD(1 To rP, 1 To rP, 1 To rY) As Double
  For i = 1 To rP
    For ii = 1 To rP
      For j = 1 To rY
        DD(i, ii, j) = D(i, j) * D(ii, j)
      Next j
    Next ii
  Next i

  ReDim SDD(1 To rP, 1 To rP) As Double
  ReDim SDDinv(1 To rP, 1 To rP) As Double
  For i = 1 To rP
    For ii = 1 To rP
      SDD(i, ii) = 0
    Next ii
  Next i
  For i = 1 To rP
    For ii = 1 To rP
      For j = 1 To rY
        SDD(i, ii) = SDD(i, ii) + DD(i, ii, j)
      Next j
    Next ii
  Next i

  Call Invert(SDD, rP, SDDinv)

  For i = 1 To rP
    SPValue(i, 1) = SY * Sqr(SDDinv(i, i))
  Next i
  Dim Answer As String
  If nn > 0 Then
    Answer = MsgBox("There are data in the cells to the" _
    & Chr(13) & "right of the sum of squares of the" _
```

```
           & Chr(13) & "residuals. Can they be overwritten?", _
           vbYesNo, "Overwrite?")
         If Answer = vbYes Then nn = 0
       End If
       If n = 0 Then
         myRange1.Select
         Selection.Offset(0, 1).Select
         Selection.Font.Italic = True
         For i = 1 To rP
           Selection.Value = SPValue
         Next i
       Else
         For i = 1 To rP
           MsgBox "The standard deviation of parameter #" & i & " is " _
           & SPValue(i, 1)
         Next i
       End If
       If n > 0 Then
         Answer = MsgBox("There are data in the cells to the right of the" _
         & Chr(13) & "Solver parameters. Can they be overwritten?", _
         vbYesNo, "Overwrite?")
         If Answer = vbYes Then nn = 0
       End If
       If nn = 0 Then
         myRange2.Select
         Selection.Offset(0, 1).Select
         ActiveCell.Font.Italic = True
         Selection.Value = SYY
       Else
         MsgBox "The standard deviation of the fit is " & SYY
       End If
     End If
     myRange1.Select
     Selection.Offset(0, 1).Select

     End Sub
```

10.8b Incorporating Solver into your macro

Because Solver is such a generally useful tool, it may be desirable to call it from inside a macro. This can be especially helpful in iterative procedures, where Solver must be called repeatedly. In the more recent versions of Excel this is indeed possible. It requires that you open your spreadsheet, select the VBA editor (with Alt + F11), and use Tools ⇒ References. In the resulting References – VBAProject dialog box, under Available References, find and activate SOLVER.xls. To do so, click on Browse, select Files of type: Microsoft Excel Files (*.xls,*.xla) and, in Look in:, find where Solver is located. For example, you may have to double-click on Systemdisk(C:), Program Files, Microsoft Office, Office, Library, and Solver. In the File name window type Solver.xla, and Open.

This will bring you back to the References – VBAProject dialog box, with SOLVER.xls at the bottom of the list. Move it up with the Priority up button, so that all selected items are together, then click OK. Once this is done, VBA will know where to find Solver, and you can address and manipulate it from within your macro.

Say that you want to fit an equation with four adjustable parameters, located in cells B1:B3 and F1 respectively (leaving space for the appropriate labels to the left, and for SolverAid uncertainty estimates to the right), and that your criterion of best fit is to minimize a number in cell F3 where you might, e.g., compute a value with a least-squares function such as SUMXMY2(D7:D23,G7:G23). The following macros shows you how you might do so. The first instruction, `SolverReset`, returns all Solver settings back to their default settings. The instruction `SetCell:="$F$3"` designates cell F3 as the target cell, equivalent to the Solver command Set Target Cell. The Solver line Equal To followed by the options Max, Min, and Value of: is replaced in the macro by `MaxMinVal:=`, where 1 selects Max, 2 picks Min, and 3 specifies Value of. (In the latter case, follow this instruction with `ValueOf:=` with the appropriate numerical value.) The Solver instruction By Changing Cells: is replaced in VBA code by `ByChange:=` followed by the addresses of the cells containing the adjustable parameters. The optional next lines add the constraint, B2 ≥ 3.2, equivalent to the Solver instruction Subject to the Constraints, with the options ≤, =, ≥, integer, and binary as relations 1 through 5 respectively, and the constraint that F1 be an integer. All the SolverOptions can likewise be set with the command `SolverOptions`. Finally, `SolverSolve` instructs Solver to proceed. The code fragment is shown here as an independent macro, so that you can easily test it, but you can of course incorporate these instructions in your own macro, and then invoke Solver as often as you wish. Note the use of the directional assignment := (for ⇐), of commas to separate the various instructions, and of quotes and dollar signs to denote absolute addresses.

```
Sub SolverDriver ()

SolverReset
SolverOk SetCell:="$F$3", MaxMinVal:=2, ByChange:="$B$1:$B$3,$F$1"
SolverAdd CellRef:=Range ("B2"), Relation:=3, FormulaText:=3.2
SolverAdd CellRef:=Range ("F1"), Relation:=4
SolverSolve

End Sub
```

10.9 Smoothing and differentiating equidistant data

In smoothing data with a moving polynomial, we want to remove as much noise as possible, while causing minimal signal distortion. Noise removal is usually optimal with the longest polynomial of lowest order. Selecting the length of the polynomial to be used is a subjective choice, which determines the resolution of features one wants to retain in the processed curve. When the curve contains sharp features that are believed to be significant, the length of the moving polynomial should be short enough to avoid averaging-out these features.

It is usually even less transparent how to select the optimal order of the moving polynomial. Again there is a trade-off: the lower the polynomial order, the more the finer details are filtered out. The lower the order, the less the curve can follow the noise-related excursions around the average. On the other hand, for minimal signal distortion one would select high-order polynomials. Maximal noise reduction and minimal signal distortion are clearly conflicting requirements, each pulling in a different direction. Consequently, a compromise is usually sought to provide maximal smoothing for minimal distortion. Moreover, different parts of a curve (such as broad and narrow peaks in a spectrum) may require different polynomial orders for optimal filtering in different parts of the data set.

Professor Barak has recently extended the algorithm for least-squares smoothing and differentiation of equidistant data points (*Anal. Chem.* 67 (1995) 2758) by letting the program self-optimize the order of the polynomial used. This self-optimization is performed each time the fitted data segment moves. Consequently, the resulting curve has a composite order. The optimization is based on the *F* test using a preset probability level, here 5%. This program takes the guesswork out of the assignment of polynomial order. Especially when the signal contains both broad and narrow features in a single curve, this method often outperforms smoothing filters of fixed order. However, one should keep in mind that polynomials often provide poor approximations for commonly encountered analytical features, such as Gaussian and Lorentzian peaks. In such cases, a polynomial approximation, even when uniformly converging, will only yield satisfactory smoothing when the length of the moving polynomial covers only a relatively small part of the feature. This requires that each feature is represented by a sufficiently large number of measured data.

The program, written by Prof. Barak, is more sophisticated than the examples we have given so far, both in its algorithm and in its use of dialog boxes and help files. The macro is available on the web site. Figure 3.3–2f illustrates how it works for the data of Fig. 3.3–2b.

If operating the program, highlight the data (in the form of a single, contiguous, column), then call the macro SGBSmooth. In the resulting dialog box, using the scrollbars provided, enter the desired filter length (i.e., the number of data points to be contained in the moving polynomial), and the maximum allowable order of the fitting polynomials. The maximum polynomial order must be smaller than the filter length, but should exceed by at least one or two the maximum number of inflection points anticipated in any noise-free curve segment. (The minimum order is best kept at 0. It does *not* set a lower limit to the order used, but only affects the first fit.) Also indicate whether you want to compute the 1st and/or 2nd derivative of the curve, in which case you will be asked to provide the equidistant data spacing Δx.

The output will appear immediately to the right of the input data, in two to four columns. Make sure that this space is empty, because the output will overwrite any data in those columns. The output will show, from left to right, the smoothed data, the polynomial order used, and (if requested) the values of the 1st and/or 2nd derivative respectively.

Below we show a derivative version, called ELS for equidistant least squares, that uses the same algorithm but with easier-to-make input boxes. It provides two options: a fixed or a variable (self-optimizing) order, selected by the macros ELSfixed and ELSoptimized respectively. The main program is the subroutine ELS, which itself calls on several functions and on a subroutine, ConvolutionFactors. The latter need not be a separate subroutine, since it is only called once in the program. However, by placing it in a separate subroutine it becomes available for use in other programs, such as Interpolation, that could benefit from it.

Here is the macro:

```
'''''''''''''''''''''''''''''''''''''''''''''''''''''''''''''''''''''''''''''''
'''''''''''''''''''''''''''''''''''''''''''''''''''''''''''''''''''''''''''''''
''''''''''''''''''''''''^                          ^''''''''''''''''''''''''''''
''''''''''''''''''''''''^           EQUIDISTANT    ^''''''''''''''''''''''''''''
''''''''''''''''''''''''^         LEAST SQUARES    ^''''''''''''''''''''''''''''
''''''''''''''''''''''''^                          ^''''''''''''''''''''''''''''
'''''''''''''''''''''''''''''''''''''''''''''''''''''''''''''''''''''''''''''''
''''''''''''''''''''''''''''''''''''''''''''© P. Barak, pwbarak@facstaff.wisc.edu
'''''''''''''''''''''''''''''''''''''''''''''''''''''''''''''''''''''''''''''''

'Declare calculational variables
Public GramPoly() As Double, Weight() As Double, Y() As Double

Sub ELSfixed()

'selects a fixed polynomial order by setting iOrder equal to 1

Dim iOrder As Integer
iOrder = 1
Call ELS(iOrder)

End Sub

''''''''''''''''''''''''''''''''''''''''''''''''''''''''''''''''''''''''''''''
Sub ELSoptimized()

'selects a variable polynomial order (between 1 and a user-selectable
'maximum value,MaxOrder) by setting iOrder equal to -1
```

```
Dim iOrder As Integer
iOrder = -1
Call ELS(iOrder)

End Sub
```

. . . .

```
Public Sub ELS(iOrder)
```

' This program for a least-squares fit to equidistant data y,x with
' a moving polynomial uses the approach pioneered by Sheppard [Proc.
' London Math. Soc. (2) 13 (1914) 81] and Sherriff [Proc. Royal Soc.
' Edinburgh 40 (1920) 112], and subsequently advocated by Whittaker
' & Robinson [The calculus of observations, Blackie & Son, first
' published in 1924 and reprinted through 1965] and by Savitzky & Golay
' [Anal. Chem. 36 (1964) 1627]. It computes smoothed values of the data
' set, or (if so desired) its first or second derivative. The user
' selects the length of the moving polynomial.

' There are two options: in ELSfixed() the user also selects the
' order of the polynomial, whereas in ELSoptimized() the program
' optimizes the order of the polynomial (between 1 and an upper
' limit set by the user) each time the moving polynomial slides
' one data point along the data set, using an algorithm described
' by Barak [Anal. Chem. 67 (1995) 2758].

' The program compares the ratio of the variances for a given order
' and that for the next-lower order with the corresponding F test
' as its first criterion. Since symmetrical functions often contain
' mostly even powers, a single, final comparison is made between the
' variances of the next-higher and the next-lower order. If the latter
' is not desired, simply comment out the section following the comment
' line "Second test for optimum Order".

' The length of the moving polynomial, PolyLength, must be an odd inte-
' ger between 3 and 31, not exceeding the length of the data set, NPts.
' The selected (maximum) order of the polynomial, MaxOrder, must be a
' positive integer, MaxOrder > 0. Moreover, for ELSfixed, MaxOrder <
' PolyLength, while for ELSoptimized we have MaxOrder < PolyLength – 1.

' The output is written in one or two columns to the right of the
' input data. The first output column contains the smoothed or dif-
' ferentiated data. The second column, which appears only with
' ELSoptimized(), displays the order selected by the program. Make
' sure that the output space is free, or can be overwritten.

' Some of the abbreviations and indices used:
'   NPts: number of points in the data set, computed by the macro
'   from the data range provided in ELS InputBox 1: Input data.
'   PolyLength: length (in number of points) of the moving polynomial,
'   selected in ELS InputBox 2: Length of Moving Polynomial.
'   m: number of points on each side of the central point in the
'   moving polynomial, calculated as m = (PolyLength-1)/2.
'   t: index for the individual points in the moving
'   polynomial; t ranges from -m to +m.
'   k: index for the position of the center of the moving polynomial

```
'          in the data set; k ranges from m+1 to NPts-m.
'       j:  index for the order of the polynomial.
'       OptOrder: array of Order values.
'       s:  index for the order of the derivative; s ranges from 0 to
'          DerivOrder.
'       MaxOrder:  the order of the polynomial (from ELS InputBox 3:
'          Polynomial Order) or its maximum value (from ELS InputBox 3:
'          Maximum Polynomial Order). MaxOrder must be an integer, > 0, and
'          < PolyLength for ELSfixed, or < (PolyLength-1) for ELSoptimized.
'       Order:  working value of the polynomial order. In ELSfixed,
'          Order = MaxOrder; in ELSoptimized, Order starts at 1, and
'          has a maximum value of MaxOrder.
'       DerivOrder: order of the highest-order derivative to be computed,
'          selected by the answer to ELS InputBox 4: Derivative Order
'             DerivOrder = 0 for smoothing,
'             DerivOrder = 1 for the 1st derivative,
'             DerivOrder = 2 for the 2nd derivative.
'       tries:  number of attempts to enter data in input box.
'       Y() and YData():  the input data containing the entire data set.
'       FValueTable:  obtained directly from the Excel function FInv.
'       OutputData:  the final (smoothed or derivative) result.
'       OutputOrder:  the values of OptOrder used in ELSoptimized.

' The main routine, ELS(iOrder) calls the functions GenFact() and Smooth().
'   It also calls the subroutine ConvolutionFactors, which calculates the
'   Gram polynomials and the corresponding convolution weights.

' Preliminaries and data input

Dim a As Integer, b As Integer, DerivOrder As Integer, i As Integer
Dim ii As Integer, j As Integer, jj As Integer, k As Integer
Dim m As Integer, MaxOrder As Integer, n As Integer, NPts As Integer
Dim Order As Integer, PolyLength As Integer, q As Integer, s As Integer
Dim t As Integer, Tries2 As Integer, Tries3 As Integer, Tries4 As Integer

Dim GenFact As Long

Dim AA As Double, BB As Double, DeltaX As Double, FTest1 As Double
Dim FTest2 As Double, GenFact As Double, Length As Double,
Perc As Double Dim Sum As Double Dim SumSq As Double, SumXY As Double,
SumY As Double Dim SumY2 As Double

Dim SumSquares() As Double, SumX2() As Double

Dim OptOrder As Variant, OutputData As Variant, OutputOrder As Variant
Dim TestContents As Variant, YData As Variant

Dim myRange As Range

Dim Ans As String

Dim z

Set myRange = Application.InputBox(prompt:="The input data are " _
  & "located in column:", Title:="ELS InputBox 1: Input data", Type:=8)
myRange.Select
If Selection.Columns.Count <> 1 Then End
NPts = Selection.Rows.Count
```

```
If NPts = 0 Then End
YData = myRange.Value
OutputData = myRange.Value          'in order to define the array size
OutputOrder = myRange.Value         'in order to define the array size
OptOrder = myRange.Value            'in order to define the array size

'   Test and prepare the default output range

n = 0
Selection.Offset(0, 1).Select
TestContents = Selection.Value
For i = 1 To NPts
   z = TestContents(i, 1)
   If IsEmpty(z) Then
      n = n
   Else
      n = n + 1
   End If
Next i

If iOrder = 1 Then
   If n > 0 Then
      Ans = MsgBox(" There are data in the column where the output " _
         & "will" & Chr(13) & "be written. Proceed anyway and overwrite " _
         & "those data?", vbYesNo)
      If Ans = vbNo Then
         MsgBox ("Safeguard the data in the highlighted area by" & _
            Chr(13) & "moving them to another place, then try again.")
      End
      End If
   End If
   Selection.Offset(0, -1).Select
End If

If iOrder = -1 Then
   Selection.Offset(0, 1).Select
   TestContents = Selection.Value
   For i = 1 To NPts
      z = TestContents(i, 1)
      If IsEmpty(z) Then
         n = n
      Else
         n = n + 1
      End If
   Next i

   If n > 0 Then
      Ans = MsgBox("There are data in the TWO columns where the output " _
         & "will" & Chr(13) & "  be written. Proceed anyway and overwrite " _
         & "those data?", vbYesNo)
      If Ans = vbNo Then
         MsgBox ("Safeguard the data in the highlighted area by" & _
            Chr(13) & "moving them to another place, then try again.")
      End
      End If
```

```
    End If
    Selection.Offset(0, -2).Select
  End If

'  Select the length of the moving polynomial, PolyLength

Tries2 = 0
Line2:
Length = InputBox(prompt:="The length of the moving polynomial is:", _
    Title:="ELS InputBox 2: Polynomial Length")
PolyLength = CInt(Length)

' Make sure that PolyLength is an odd integer
' larger than 0 and smaller than NPts

If (Length <= 0 Or Length >= NPts Or PolyLength - Length <> 0 _
  Or CInt((PolyLength - 1) / 2) - ((Length - 1) / 2) <> 0) Then
  MsgBox "The length of the moving polynomial must" & Chr(13) & _
    "      be an odd integer larger than zero and " & Chr(13) & _
    "smaller than the length of the input column."
  Tries2 = Tries2 + 1
  If Tries2 = 2 Then End
  GoTo Line2
End If

' Select the order of the moving polynomial, MaxOrder

Tries3 = 0
Line3:
If iOrder = 1 Then
  MaxOrder = InputBox(prompt:="The order of the moving" _
    & "polynomial is:", Title:="ELS InputBox 3: Polynomial Order")
Else
  MaxOrder = InputBox(prompt:="The maximum order of the moving " _
  & "polynomial is:", Title:="ELS InputBox 3: Polynomial Order")
End If

' Make sure that MaxOrder > 0 and that either MaxOrder < PolyLength
' (for ELSfixed) or MaxOrder < PolyLength - 1 (for ELSoptimized).

If iOrder = 1 Then
  If (MaxOrder <= 0 Or MaxOrder >= PolyLength) Then

    MsgBox "The order of the moving polynomial" & Chr(13) & _
           "must be larger than zero, and smaller" & Chr(13) & _
           "than the length of the moving polynomial."
    Tries3 = Tries3 + 1
    If Tries3 = 2 Then End
    GoTo Line3
  End If
Else
  If (MaxOrder <= 0 Or MaxOrder >= PolyLength - 1) Then
    MsgBox "The maximum order of the moving polynomial" & _
      Chr(13) & "must be larger than zero, and smaller than" & _
      Chr(13) & "the length of the moving polynomial minus 1."
    Tries3 = Tries3 + 1
    If Tries3 = 2 Then End
```

```
      GoTo Line3
   End If
End If
End If

' Select smoothing, first derivative, or second derivative

Tries4 = 0
Line4:
DerivOrder = InputBox(prompt:="Select the order of the derivative" &
     Chr(13) _
   & "(either 1 or 2); for smoothing, select 0." & Chr(13) & Chr(13) _
   & "The order of the derivative is:", Title:="ELS InputBox 4:" _
   & "Derivative Order")

' Make sure that DerivOrder has the value 0, 1, or 2

If DerivOrder = 0 Then
   GoTo Line6
ElseIf DerivOrder = 1 Then
   GoTo Line5
ElseIf DerivOrder = 2 Then
   GoTo Line5
Else
   MsgBox " The order of the moving polynomial must be" & Chr(13) & _
      "either 0 (for smoothing), 1 (for the first" & Chr(13) & _
      "derivative), or 2 (for the second derivative)."
   Tries4 = Tries4 + 1
   If Tries4 = 2 Then End
   GoTo Line4
End If

Line5:
DeltaX = InputBox(prompt:="The data spacing in x is:", _
   Title:="ELS InputBox 4: X Increment")

Line6:
m = (PolyLength - 1) / 2

ReDim Y(1 To NPts), OptOrder(1 To NPts, 1)
For i = 1 To NPts
   Y(i) = YData(i, 1)
   OptOrder(i, 1) = MaxOrder
Next i

Call ConvolutionFactors(PolyLength, MaxOrder, DerivOrder)

' THE FOLLOWING SECTION IS USED ONLY BY ELSoptimized

If iOrder = -1 Then

   ReDim SumX2(1 To MaxOrder)
   For j = 1 To MaxOrder
      Sum = 0
      For i = -m To m
         Sum = Sum + GramPoly(i, j, 0) ^ 2
      Next i
      SumX2(j) = Sum
   Next j
```

```
' Calculate FValueTable(MaxOrder,PolyLength)

  ReDim FValueTable(1 To MaxOrder, 1 To PolyLength)
  jj = 0
  Do
    jj = jj + 1
    For ii = 1 To MaxOrder
      FValueTable(ii, jj) = Application.FInv(0.05, ii, jj)
    Next ii
  Loop Until jj = PolyLength

  For k = m + 1 To NPts - m

    ReDim SumSquares(0 To MaxOrder)
' Calculate SumSquares for Order = 0
    Order = 0
    SumY = 0
    SumY2 = 0
    For t = -m To m
      SumY = SumY + Y(k + t)
      SumY2 = SumY2 + Y(k + t) ^ 2
    Next t
    SumSquares(0) = SumY2 - SumY ^ 2 / (2 * t + 1)
' Calculate SumSquares for Order = 1
    Order = 1
    SumSq = 0
    For t = -m To m
      SumSq = SumSq + (Smooth(PolyLength, k, Order, t, 0) - Y(k + t)) ^ 2
    Next t
    SumSquares(1) = SumSq
' Test whether one-higher order satisfies the criterion
    Do
      Order = Order + 1
      If Order > MaxOrder Then GoTo line10

' Calculate SumSquares for Order > 1
      SumXY = 0
      For t = -m To m
        SumXY = SumXY + Y(k + t) * GramPoly(t, Order, 0)
      Next t
      SumSquares(Order) = SumSquares(Order - 1) - SumXY ^ 2 / SumX2(Order)

' First test for optimum Order

      FTest1 = (SumSquares(Order - 1) - SumSquares(Order)) * _
        (PolyLength - Order) / (SumSquares(Order))
    Loop Until (FTest1 / FValueTable(1, PolyLength - Order - 1)) < 1
' Second test for optimum Order
    If Order < MaxOrder Then
```

```
        Order = Order + 1
        SumXY = 0
        For t = -m To m
           SumXY = SumXY + Y(k + t) * GramPoly(t, Order, 0)
        Next t
        SumSquares(Order) = SumSquares(Order - 1) - SumXY ^ 2 / SumX2(Order)
        FTest2 = (SumSquares(Order - 2) - SumSquares(Order)) * _
           (PolyLength - Order) / (SumSquares(Order))
        If (FTest2 / FValueTable(2, PolyLength - Order - 1)) < 1 Then
           Order = Order - 1
        End If
      End If
line10:
    OptOrder(k, 1) = Order - 1

    Perc = 100 * (k / NPts)
    Percentage = Int(Perc)
    Application.StatusBar = "Calculation   & Percentage & "% done."

  Next k

End If

' THIS ENDS THE PART USED ONLY BY ELSoptimized

' Prepare the output files

For k = m + 1 To NPts - m
  If k = m + 1 Then
    For t = -m To -1
       OutputData(k + t, 1) = Smooth(PolyLength, k, OptOrder(k + t, 1), _
          t, DerivOrder) / (DeltaX ^ DerivOrder)
       OutputOrder(k + t, 1) = OptOrder(k, 1)
    Next t
  End If

  OutputData(k, 1) = Smooth(PolyLength, k, OptOrder(k, 1), 0, DerivOrder) _
     / (DeltaX ^ DerivOrder)
  OutputOrder(k, 1) = OptOrder(k, 1)

  If k = NPts - m Then
    For t = 1 To m
       OutputData(k + t, 1) = Smooth(PolyLength, k, OptOrder(k + t, 1), _
          t, DerivOrder) / (DeltaX ^ DerivOrder)
       OutputOrder(k + t, 1) = OptOrder(k, 1)
    Next t
  End If
Next k

' Write the output files

Selection.Offset(0, 1).Select
Selection.Value = OutputData
If iOrder = -1 Then
  Selection.Offset(0, 1).Select
  Selection.Value = OutputOrder
  Selection.Offset(0, -1).Select
```

```
End If

Application.StatusBar = False

End Sub
'' ' ' ' ' ' ' ' ' ' ' ' ' ' ' ' ' ' ' ' ' ' ' ' ' ' ' ' ' ' ' ' ' ' ' ' ' ' ' ' ' ' ' ' ' ' ' '

Function GenFact(a, b)

' Computes the generalized factorial

Dim gf As Double, j As Integer

gf = 1
For j = (a - b + 1) To a
  gf = gf * j
Next j
GenFact = gf

End Function
'' ' ' ' ' ' ' ' ' ' ' ' ' ' ' ' ' ' ' ' ' ' ' ' ' ' ' ' ' ' ' ' ' ' ' ' ' ' ' ' ' ' ' ' ' ' ' '

Public Function Smooth(PolyLength, k, j, t, s)

' Computes the appropriately weighted sum of the Y-values

Dim i As Integer, m As Integer, Sum As Double
m = (PolyLength - 1) / 2

Sum = 0
For i = -m To m
  Sum = Sum + Weight(i, t, j, s) * Y(k + i)
Next i
Smooth = Sum

End Function
'' ' ' ' ' ' ' ' ' ' ' ' ' ' ' ' ' ' ' ' ' ' ' ' ' ' ' ' ' ' ' ' ' ' ' ' ' ' ' ' ' ' ' ' ' ' ' '

Public Sub ConvolutionFactors(PolyLength, MaxOrder, DerivOrder)

' Calculates tables of GramPoly(i = -m to m, k = -1 to MaxOrder,
'  s = 1 to DerivOrder), and of Weight(i = -m to m, t = -m to m, k = -1 to
'  MaxOrder, s = -1 to DerivOrder)

Dim i As Integer, k As Integer, m As Integer, s As Integer, t As Integer
Dim AA As Double, BB As Double

m = (PolyLength - 1) / 2
ReDim GramPoly(-m To m, -1 To MaxOrder, -1 To DerivOrder)

'Evaluate the Gram polynomials for DerivOrder=0

For i = -m To m
  GramPoly(i, 0, 0) = 1
  GramPoly(i, 1, 0) = 0
Next i
For i = -m To -1
  GramPoly(i, 1, 0) = i / m
Next i
For i = 1 To m
```

```
    GramPoly(i, 1, 0) = i / m
  Next i
  For k = 2 To MaxOrder
    AA = 2 * (2 * k - 1) / (k * (2 * m - k + 1))
    BB = ((k - 1) * (2 * m + k)) / (k * (2 * m - k + 1))
    For i = 0 To m
      GramPoly(i, k, 0) = AA * i * GramPoly(i, k - 1, 0) - _
        BB * GramPoly(i, k - 2, 0)
    Next i
    For i = -m To -1
      If k Mod 2 = 0 Then
        GramPoly(i, k, 0) = GramPoly(-i, k, 0)
      Else
        GramPoly(i, k, 0) = -GramPoly(-i, k, 0)
      End If
    Next i
  Next k

  'Evaluate the Gram polynomials for DerivOrder>0

  If DerivOrder > 0 Then
    For s = 1 To DerivOrder
      For i = -m To m
        GramPoly(i, -1, s) = 0
        GramPoly(i, 0, s) = 0
      Next i
      For k = 1 To MaxOrder
        AA = 2 * (2 * k - 1) / (k * (2 * m - k + 1))
        BB = ((k - 1) * (2 * m + k)) / (k * (2 * m - k + 1))
        For i = -m To m
          GramPoly(i, k, s) = AA * (i * GramPoly(i, k - 1, s) + _
          s * GramPoly(i, k - 1, s - 1)) - BB * GramPoly(i, k - 2, s)
        Next i
      Next k
    Next s
  End If

  'Calculate the convolution weights

  ReDim Weight(-m To m, -m To m, -1 To MaxOrder, -1 To DerivOrder)

  For k = 0 To MaxOrder
    AA = (2 * k + 1) * GenFact(2 * m, k) / GenFact(2 * m + k + 1, k + 1)
    For s = 0 To DerivOrder
      For i = -m To m
        For t = -m To m
          Weight(i, t, k, s) = Weight(i, t, k - 1, s) _
          + AA * GramPoly(i, k, 0) * GramPoly(t, k, s)
        Next t
      Next i
    Next s
  Next k

End Sub
```

For a large data set, ELS may take its sweet time. The progress of the calculation is therefore displayed on the status bar with the instruction
`Application.StatusBar` = *message* below line10 of the main subroutine
ELS(iOrder), and is removed at the end of that subroutine.

Again, the above program can readily be modified. For example, a reader
who wants to use the equidistant least-squares program to compute a third
derivative (for which the convolution integers are not available in the usual
tables) will find that this can be done with a few changes around InputBox 4,
because the limitation to a second-order derivative is not inherent to the
algorithm, but was inserted merely to illustrate how this is done. Similarly,
in order to modify the criterion used in the F-test to, say, 1%, one only
needs to change the value 0.05 in the line `FValueTable(i,j)` =
`Application.FInv(0.05,i,j)` to 0.01. Alternatively, the reader might
want to make that criterion a user-selectable parameter, to be entered
through an additional input box.

10.10 Semi-integration and semi-differentiation

In sections 6.11 and 8.9 we encountered a rather special case of convolution
involving the function $1/\sqrt{t}$. A macro to perform this operation was used in
section 6.11, and is listed below. It uses input boxes for the data spacing (in
time) and the measured currents, then calculates the convolution of that
current with $1/\sqrt{t}$.

```
' In its semi-integration mode, this macro takes the faradaic current of a
' linear sweep voltammogram, and transforms it into the underlying current-
' voltage curve corrected for the time-dependence of planar diffusion. The
' resulting curve is then amenable to further mathematical analysis along
' the lines pioneered by Koutecky, and summarized by, e.g., Heyrovsky &
' Kuta in their book on the Principles of Polarography (Academic Press,
' 1966). In its semi-differentiation mode, the macro instead converts a
' stationary current-voltage curve into the corresponding linear sweep or
' cyclic voltammogram.

Sub SemiIntegrate()
Dim iIntDif As Integer
iIntDif = 1
Call Semi(iIntDif)
End Sub

''''''''''''''''''''''''''''''''''''''''''''''''''''''''''''''''''''''''''''

Sub SemiDifferentiate()
Dim iIntDif As Integer
iIntDif = -1
Call Semi(iIntDif)
End Sub

Sub Semi(iIntDif)
```

```
' ' ' ' ' ' ' ' ' ' ' ' ' ' ' ' ' ' ' ' ' ' ' ' ' ' ' ' ' ' ' ' ' ' ' ' ' ' ' ' ' ' ' ' ' ' ' ' ' ' ' ' ' ' ' ' ' ' '
' ' ' ' ' ' ' ' ' ' ' ' ' ' ' ' ' ' ' ' ' ' ' ' ' ' ' ' ' ' ' ' ' ' ' ' ' ' ' ' ' ' ' ' ' ' ' ' ' ' ' ' ' ' ' ' ' ' '
' ' ' ' ' ' ' ' ' ' ' ' ' ' ' ' ' ' ' ^^^^^^^^^^^^^^^^^^^^^^^^^^ ' ' ' ' ' ' ' ' ' ' ' ' ' ' ' ' ' '
' ' ' ' ' ' ' ' ' ' ' ' ' ' ' ' ' ' ^                              ^ ' ' ' ' ' ' ' ' ' ' ' ' ' ' ' ' '
' ' ' ' ' ' ' ' ' ' ' ' ' ' ' ' ' ^     SEMI-INTEGRATION AND    ^ ' ' ' ' ' ' ' ' ' ' ' ' ' ' ' ' '
' ' ' ' ' ' ' ' ' ' ' ' ' ' ' ' ' ^     SEMI-DIFFERENTIATION    ^ ' ' ' ' ' ' ' ' ' ' ' ' ' ' ' ' '
' ' ' ' ' ' ' ' ' ' ' ' ' ' ' ' ' ^                              ^ ' ' ' ' ' ' ' ' ' ' ' ' ' ' ' ' '
' ' ' ' ' ' ' ' ' ' ' ' ' ' ' ' ' ^^^^^^^^^^^^^^^^^^^^^^^^^^ ' ' ' ' ' ' ' ' ' ' ' ' ' ' ' ' ' '
' ' ' ' ' ' ' ' ' ' ' ' ' ' ' ' ' ' ' ' ' ' ' ' ' ' ' ' ' ' ' ' ' ' ' ' ' ' '© R. de Levie
' ' ' ' ' ' ' ' ' ' ' ' ' ' ' ' ' ' ' ' ' ' ' ' ' ' ' ' ' ' ' ' ' ' ' ' ' ' ' ' ' ' ' ' ' ' ' ' ' ' ' ' ' ' ' ' ' ' '

' The macro uses a simple semi-integration algorithm given by Oldham which
' is adequate to illustrate the approach. The semi-differential is then
' found by taking first differences.

' The input data should be stored in a single, continuous column. The
' resulting transformed output data will be written in the column
' immediately to the right of the input data. That column is therefore
' best left free.

' Parameter names used:
'       cnmax        number of columns in highlighted data block
'       rnmax        number of rows in highlighted data block
'       dataArray    array of all input data
'       rn           index specifying row number in input data array
Dim cnmax As Integer, rnmax As Integer
Dim iIntDif As Integer, j As Integer, k As Integer, n As Integer
Dim a As Double, s As Double, deltaTime As Double
Dim myRange As Range
Dim inputData As Variant, semiInt As Variant, outputData As Variant
Dim testValue As Variant, z As Variant
Dim myAddress As String, Ans As String
n = 0

' Input the time increment and use it to compute the parameter s

If iIntDif = 1 Then
  deltaTime = InputBox("Enter the time increments, in seconds.", _
    "SemiIntegration InputBox 1: Time increments")
Else
  deltaTime = InputBox("Enter the time increments, in seconds.", _
    "SemiDifferentiation InputBox 1:Time increments")
End If
s = Sqr(Abs(deltaTime))

' Enter the input data

If iIntDif = 1 Then
  Set myRange = Application.InputBox(Prompt:= _
    "The current values are located in:", Title.= _
    "SemiIntegration InputBox 2: Input data", Type:=8)
Else
  Set myRange = Application.InputBox(Prompt:= _
    "The current values are located in:", Title:= _
    "SemiDifferentiation InputBox 2: Input data", Type:=8)
End If
```

```
myRange.Select
If Selection.Columns.Count <> 1 Then
   End
Else
   rnmax = Selection.Rows.Count
   inputData = Selection.Value
   semiInt = Selection.Value
   output Data = Selection.Value
   myAddress = Selection.Address
End If

' Check that the output will not overwrite valuable data

Selection.Offset(0, 1).Select
testValue = Selection.Value
For j = 1 To rnmax
   z = testValue(j, 1)
   If IsEmpty(z) Or z = 0 Then
     n = n
   Else
     n = n + 1
   End If
Next j

If n > 0 Then
   Ans = MsgBox("There are data in the space where the output will be" & _
   Chr(13) & " written. Proceed anyway and overwrite those data?", _
   vbYesNo)
   If Ans = vbNo Then
     Selection.Offset(0, -1).Select
     End
   End If
End If

' Compute the semi-integral

ReDim P(1 To rnmax) As Double
P(1) = 1

For j = 2 To rnmax
   P(j) = P(j - 1) * (j - 1.5) / (j - 1)
Next j
For k = 1 To rnmax
   a = 0
   For j = 1 To k
     a = a + P(k - j + 1) * inputData(j, 1)
   Next j
   semiInt(k, 1) = a * s
Next k

If iIntDif = -1 Then
   For k = 2 To rnmax
     outputData(k, 1) = (semiInt(k, 1) - semiInt(k - 1, 1)) _
     / deltaTime
   Next k
   semiInt(1, 1) = ""
```

```
Else
   For k = 1 To rnmax
      outputData(k, 1) = semiInt(k, 1)
   Next k
End If

' Write the result

Selection.Value = outputData

End Sub
```

In order to test such a macro, we can verify that it indeed yields approximately correct results for functions for which the semi-integral is known, such as the constant a or the function $\sqrt{at}$, for which the semi-integrals are $2a\sqrt{t/\pi}$ and $\frac{1}{2} t \sqrt{\pi a}$ respectively. Such tests indicate that the data density is of crucial importance, because the algorithmic errors are roughly inversely proportional to the data spacing.

10.11 Reducing data density

Sometimes we may want to see the effect of data density. On a pre-existing experimental data set we can seldom increase the data density, but we can get an idea of the resulting computational errors by observing the effect of reducing the data density. Below we will show two ways in which to achieve this, by either picking every nth data point, or by averaging over groups of n adjacent data. The structure of the macro is similar to that of SemiIntegrate(), and by now will be self-explanatory.

```
' The Thin macros take a set of equidistant data, and generate a new data set
' that contains only one out of every n data points of the original set. When
' ThinCull() is used, the thinned data are obtained by deleting the other
' data. In ThinToAv() the new data set is obtained by averaging over the
' thinned point and its deleted neighbors, in which case the thinning factor n
' is restricted to odd integers.

Sub ThinCull()
Dim iThin As Integer
iThin = 1
Call DataThinner(iThin)
End Sub

' . . . . . . . . . . . . . . . . . . . . . . . . . . . . . . . . . . . . . . . . . . . . . . . . . . .

Sub ThinToAv()
Dim iThin As Integer
iThin = -1
Call DataThinner(iThin)
End Sub

' . . . . . . . . . . . . . . . . . . . . . . . . . . . . . . . . . . . . . . . . . . . . . . . . . . .

Sub DataThinner(iThin)
```

```
''''''''''''''''''''''''''''''''''''''''''''''''''''''''''''''''''''''''''
''''''''''''''''''''''''''''''''''''''''''''''''''''''''''''''''''''''''''
'''''''''''''''''''''''''''''^^^^^^^^^^^^^^^^^^^^^^^^^^'''''''''''''''''''''''
''''''''''''''''''''''''''''^                      ^''''''''''''''''''''''''''
'''''''''''''''''''''''''''^          DATATHINNER    ^'''''''''''''''''''''''''
''''''''''''''''''''''''''''^                      ^''''''''''''''''''''''''''
'''''''''''''''''''''''''''''^^^^^^^^^^^^^^^^^^^^^^^^^^'''''''''''''''''''''''
''''''''''''''''''''''''''''''''''''''''''''''''''''''''''''''''''''© R. de Levie
''''''''''''''''''''''''''''''''''''''''''''''''''''''''''''''''''''''''''''
```

```vba
' The original data set should be stored in cn adjacent columns, where cn
' denotes the number of columns. The output will be written in the cn columns
' immediately to the right of the original data. Those columns are therefore
' best left free.

' The output array has the same dimension as the input array, but only con-
' tains data in its top 1/n part.

' Parameter names used:
'     cn              the number of columns in the highlighted data block
'     rn              the number of rows in the highlighted data block
'     inputArray      the array of all input data
'     outputArray     the array of all output data
'     n               the thinning factor
Dim cn As Integer, i As Integer, j As Integer, k As Integer
Dim m As Integer, n As Integer, rn As Integer, rnnew As Integer
Dim nn As Single, s As Single, sum As Single
Dim inputData As Variant, outputData As Variant
Dim testValue As Variant, z As Variant
Dim myRange As Range
Dim Ans As String
n = 0

' Input the time increment and use it to compute the parameter s

nn = InputBox("The thinning factor is:", _
   "DataThinner InputBox 1: Thinning factor")
n = Int(nn)
If iThin = 1 Then
  If nn / n <> 1 Then
    MsgBox "The thinning factor must be an INTEGER."
    End
  End If
ElseIf iThin = -1 Then
  s = (nn - 1) / 2
  i = (n - 1) / 2
  If s / i <> 1 Then
    MsgBox "The thinning factor must be an ODD INTEGER."
    End
  End If
End If

' Input the original data set

Set myRange = Application.InputBox(Prompt:= _
  "The current values are located in:", Title:= _
```

```
           "DataThinner InputBox 2: Input data", Type:=8)
myRange.Select
cn = Selection.Columns.Count
rn = Selection.Rows.Count
inputData = Selection.Value
outputData = Selection.Value

' Check that the output will not overwrite valuable data

Selection.Offset(0, cn).Select
testValue = Selection.Value
For i = 1 To rn
  For j = 1 To cn
    z = testValue(i, j)
    If IsEmpty(z) Or z = 0 Then
      m = m
    Else
      m = m + 1
    End If
  Next j
Next i

If m > 0 Then
  Ans = MsgBox("There are data in the space where the output will be" & _
  Chr(13) & " written. Proceed anyway and overwrite those data?", _
  vbYesNo)
  If Ans = vbNo Then
    Selection.Offset(0, -cn).Select
    End
  End If
End If

' Compose the output set

rnnew = Int(rn / n)

For i = 1 To rn
  For j = 1 To cn
    If i <= rnnew Then
      If iThin = 1 Then
        outputData(i, j) = inputData(i * n, j)
      Else
        If i * n + (n - 1) / 2 < rn Then
          sum = 0
          For k = (1 - n) / 2 To (n - 1) / 2
            sum = sum + inputData(i * n + k, j)
          Next k
          outputData(i, j) = sum / nn
        Else
          outputData(i, j) = ""
        End If
      End If
    Else
      outputData(i, j) = ""
    End If
  Next j
```

```
Next i
' Write the result
Selection.Value = outputData
End Sub
```

10.12   An overview of VBA

Earlier in this Appendix we gave a number of examples of VBA, but so far we have not discussed its general structure. We will do so now. VBA is a recent extension of BASIC, a computer language structurally rather similar to FORTRAN, the original FORmula TRANslator program of IBM. VBA was developed in the early nineties as the macro language for Microsoft Windows applications, and differs somewhat depending on the specific application for which it is intended. In this workbook we use the version of VBA provided with Excel. It uses structured programming, and therefore (for those of you who may have learned earlier versions of BASIC) no longer carries line numbers. Moreover, the earlier type declaration suffix (%, !, #, etc.) has been replaced by an (optional) dimension statement. The VBA extensions add capabilities to manipulate not just numbers and text but entire spreadsheet **objects**, such as a worksheet, a block of spreadsheet cells, a chart, or a box.

Apart from its object-related instructions, VBA is very much like any other modern form of compiled BASIC, and it will accept code written in traditional BASIC with relatively minor modifications. Below we will review those aspects of VBA that are most useful in science (most books on VBA stress accounting and data base aspects, because that is where spreadsheets are used even more), and we will illustrate them in simple examples or complete sample macros you can try out and modify.

10.12a   Objects

An object can contain many other objects, e.g., a worksheet can consist of several ranges and charts; the ranges can have many arrays, which in turn can include a multitude of cells; the charts likewise can contain many objects such as axes, curves, labels, etc. Likewise, a group of objects can form a collection, which it itself an object. Consequently there is a whole hierarchy of objects, from the more all-inclusive to the more specific, and VBA describes all specific objects by enumerating the entire hierarchy, from the general to the particular, separated by periods, as in

```
Application.Workbooks("Book1").Worksheets("Sheet1").Range("A1:C5")
```

which denotes the Range A1:C5 in Sheet1 of Book1. (Object names in VBA must start with a letter, may contain up to 40 letters, numbers and/or underscores, but can contain no spaces or punctuation marks. This is why Sheet1 and Book1 are written *without* spaces.) Certain names are automatically interpreted by VBA as object names, and are then capitalized by the VBA editor as shown here, and printed in blue. It is good practice to type them in lower case, and to let the VBA editor do the capitalization because, when the editor does that, you will know that you are using a recognized name. The same applies to properties and methods.

When there is no ambiguity, the more inclusive objects need not be enumerated. For example, when only one sheet is open, we can just use

```
Range("A1:C5")
```

or, when the range was earlier specified by highlighting it, simply as

```
Selection
```

10.12b Properties and methods

Objects can have specific **properties** that can be read or assigned, such as their value:

```
myArray = Range("A1:C5").Value
```

and **methods**, which are actions affecting the object, such as changing its value or its appearance, reading its address or its formula, copying or clearing the contents, as in

```
Selection.ClearContents
```

Just as objects can contain many other objects, a specific object can have several properties and methods. For instance, a cell can have as properties an address, a formula, and a value, and as methods we can select it, clear it, copy it, and so on.

The above illustrate the general structure of object-related VBA instructions: they form a sequence of increasingly more specific objects, separated by periods, followed by either a property or a method. Objects are like nouns, they don't *do* anything. Properties again don't do anything. Just as in normal language, a complete sentence requires a noun, or an implied noun. Among the objects, properties, and methods, only the last takes any action. For that reason, a list of objects with or without a property does not make a line of program code: they are like a series of nouns without a verb, i.e., they make an incomplete sentence. Only the last two examples shown, `myArray = Range("A1:C5").Value` and `Selection.ClearContents`, lead to action by the program, and can stand alone as a line of code, the first because it assigns the object to a name (myArray), the second because it includes a method (ClearContents).

On the whole, VBA commands use self-explanatory English words, and are therefore easy to *read*. While this enumerative quality of VBA tends to make it very readable, it can also lead to rather wordy code. Moreover, you may experience difficulties with VBA when you want to *write* code but don't know the exact term to use. Excel provides several help files to get you past this initial hurdle. The specific examples given in this chapter are meant to familiarize you with the most useful terms.

10.12c Data types

VBA recognizes three different data types: textual, numerical, and logical. There are two types of numbers: general numbers and integers, and both can be represented in single or double precision. Note that the spreadsheet always uses double precision, but that VBA needs to be told specifically to do so, otherwise the macro computes in single precision instead.

Arrays can be specified by size, as in myRow(0 To 2) or myArray(1 To 3, 1 To 10), or can be left As Variant. Avoid ambiguous dimensioning such as myRow(2) which has 3 elements, with indices 0, 1, and 2, unless Option Base 1 is used, which deletes use of the index 0, in which case myRow(2) has only two elements, 1 and 2. Arrays defined As Variant always start with index 1, regardless of the Option Base statement.

Vectors and arrays are best defined first by type (or only by name), then redimensioned just before use, once their sizes are known:

```
Dim Data As Variant, Magnitude(), index() As Single
. . .
Redim Magnitude (1 To N) As Double
. . .
Redim index (nTop to nBottom)
. . .
Redim Data (1 To iRows, 1 To iColumns)
. . .
```

Text can be any combination of letters, numbers, punctuation marks and spaces. Text includes the names of variables used in computations. Here are some examples of text:

```
roadmap
R2D2
4U
Hello!
Please enter a number between 1 and 100
```

Names of variables must begin with a letter, and form an unbroken string of no more than 255 letters, numbers, and/or the underscore, _, so that only the first two items in the above list can serve as such. Names of variables cannot be identical to VBA instructions, such as Value, Address, Range, but modified names such as Xvalue, newAddress, or myRange can be used. Capital letters can be used to make long strings of letters more easily readable, as in myErrorEstimate. Alternatively one can use the underscore as a word separator, as in my_name or X_Y_value.

VBA does not require that the names of variables be dimensioned, although doing so is good practice. Note that a variable name containing a typo will be considered as the name of a valid new variable. Typos can be caught by including at least one capital in each variable name, and by typing all names in lower case. When they have been dimensioned properly, even if only by name, the VBA editor will substitute the capitals, and failure to do so will identify errand variable names. You can also use Option Explicit, which requires you to dimension all variables by name, and then will catch misspelled names. If you use Option Explicit, it must be the first line in the macro module: `Option Explicit`. In that case you must specify the *names* of the variables used in dimension statements, e.g., `Dim Jones`, or `Dim Y, Z`. Specifying their *nature* (e.g., `Dim Jones As String`, or `Dim Y as Single`, `Z As Double`) is good programming practice, but is not required.

Unspecified variables are considered as Variants, which are most flexible, but which also take up most space. Large macros will run faster when, e.g., integers and text strings are specified as such. Text used as a string of symbols (as in a fixed cell address or range, such as B17:C20) or in a message (e.g., Hello!) is often placed in quotation marks.

Logicals are restricted to two values: True and False.

10.12d Expressions

Expressions cannot be used as stand-alone lines, but must be part of another structure in a VBA program. Expressions can yield numerical values, as in

```
72
(3*4)/5

concentration  'assuming that "concentration" has been defined
Sqr(Ca*Ka1)    'VBA uses Sqr where the spreadsheet uses SQRT !
```

or they can return a logical, as in

```
pH < 7              'True when pH < 7, otherwise False
```

10.12e Statements

Statements are complete program instructions, i.e., they are lines of code that can be executed as such. The following are examples of statements, with explanatory comments where needed:

```
x = 1
y = A*exp(-k*t)
Beep           'VBA recognizes this as a complete instruction!
MsgBox "Welcome"
```

```
Cubit          'assuming the existence of a macro of this name
Dim Array As Variant
```

The last example, `Dim`, is a particular type of statement that requires some comment. Dimensioning is not required in VBA; in order to keep matters as simple as possible, we have therefore not used dimensioning in the simple examples of section 10.1. Moreover, if one elects to use dimension statements, these can be placed just before the dimensioned parameter is used; this tends to improve program readability. With all this freedom, why use dimension statements at all? Apart from the usual reasons of efficient memory use, and the occasional need to specify double precision (see section 10.1d), it is a good guard against typos, as mentioned in section 10.12c.

10.12f Active regions

Objects can be defined in VBA as explained above, or they can simply be highlighted (activated, clicked on). In that case there is no need to recite a hierarchy of ever-smaller objects containing the one we mean. Instead, we merely need to identify the selected region as the wanted one. The following objects refer to such highlighted regions:

```
ActiveWorkbook
ActiveWorksheet 'This can be a worksheet,
                'a module sheet, or a dialog sheet
ActiveDialog    'This refers specifically to the topmost
                'dialog sheet, and can therefore differ
                'from the ActiveWorksheet
ActiveCell
Selection       'This can be a range, a graph, a button, etc.
```

For our purposes, cells and ranges are the most useful descriptors, which is why they are discussed below in more detail.

10.12g Cells

We have already seen how cells can be specified, in *absolute* coordinates, e.g. as

```
Worksheets("Sheet1").Range("C11")
Range("C11")     'Provided a Worksheet is active
Cell(11,3)       'Note the index order: first rows, then columns
ActiveCell       'Provided the cell has been highlighted
```

Note that the notations C11 and (11,3) use different conventions: `Range("C11")` states the column first, while `Cell(11,3)` identifies the very same cell but starts with the row number. The `Cell(rows,columns)` notation is convenient for, say, generating the tables of multiplication from 1 to 10:

```
Sub Tables()
For r = 1 To 10
    For c = 1 To 10
        Cells(r,c).Value = r*c
    Next c
Next r
End Sub
```

The instruction `Offset(rows,columns)` is used to specify *relative* addresses. For example, `Range("C11").Offset(3,-1).Select` selects cell B14, which now will become the active cell. The following macro illustrates relative addressing:

```
Sub Compass()
ActiveCell.Value = "Center"
ActiveCell.Offset(-2,0).Value = "N"
ActiveCell.Offset(-2,1).Value = "NE"
ActiveCell.Offset(0,1).Value = "E"
ActiveCell.Offset(2,1).Value = "SE"
ActiveCell.Offset(2,0).Value = "S"
ActiveCell.Offset(2,-1).Value = "SW"
ActiveCell.Offset(0,-1).Value = "W"
ActiveCell.Offset(-2,-1).Value = "NW"
End Sub
```

Note that, in the latter case, the active cell remains at the center. A foolproof macro would of course have to make sure that the active center is located neither in the top two rows nor in the first column of the spreadsheet.

Among the properties of a single cell are its **address**, the numerical **value** it contains, and (if applicable) the **formula** generating that value. Other cell properties are typographic: cell height and width, font type, size, color, etc. All of these properties can be read, set, copied, erased, and so on, by the appropriate methods. Here is a sampling. The first macro reads the value in the active cell, multiplies it by 3, and returns that new value to the cell. It then puts 19, $\sqrt{X} = 4$ and $3 + 2 \times 19 = 41$ in the three cells to its right. The second macro changes the cell appearance by using centered numbers showing three decimal places, presented in bold, underlined blue 10-point Times New Roman characters.

```
Sub ReadnRite()
X = 16
Y = ActiveCell.Value
newY = Y * 3
ActiveCell.Value = newY
ActiveCell.Offset(0,1).Value = 19
ActiveCell.Offset(0,2).Value = Sqr(X)
ActiveCell.Offset(0, 3).Value = 3 + 2 * ActiveCell.Offset(0, 1).Value
End Sub

Sub Appearances()
```

```
Selection.HorizontalAlignment = xlCenter
Selection.NumberFormat = "0.000"
Selection.Font.Bold = True
Selection.Font.Italic = False
Selection.Font.Name = "Times New Roman"
Selection.Font.Size = 10
Selection.Font.Underline = xlSingle
Selection.Font.ColorIndex = 32
End Sub
```

Note that the VBA instruction to take a square root is Sqr, whereas in the spreadsheet the same operation requires four letters: Sqrt.

10.12h Ranges

In VBA, a Range can be any set of cells, from a single cell or a block of cells to a collection of such blocks. The following are examples of Range specifiers:

```
Range("C11:E101")
Range("A1,B3:B8,C11:E101")
Rows(17)
Columns(4)
Range("C11:E101").Rows(17)
Range("C11:E101").Columns(4)
```

In the second example, note that a Range need not be contiguous, and that the quotation marks are around the entire list rather than around individual cells or blocks. In the next two lines, the entire Row17 and the entire Column4 (i.e., column D) respectively are identified. The final examples show the *intersections* of two ranges, by selecting those parts of row 17 or column 4 that lie *inside* block C11:E101. These, then, specify C17:E17 and D11:D101 respectively.

If there is any ambiguity about the worksheet or workbook where the range is to be found, these must be specified, as in

```
Worksheets("Sheet1").Range("C11:E101")
```

Because Range is such a common object in VBA, it can be combined with many different properties and methods. You can Activate a Range, Clear it, Copy it, Cut it, Delete it, and so on. You can also ask for the Address of a Range, for its Areas, Cells, Columns, Count, EntireRow, Formula, Format, etc. You can also refer to entire regions of the worksheet with instructions such as CurrentRegion or UsedRange. For the precise definitions of these, and examples of their use, refer to the Visual Basic Help file, which comes with Excel but may have to be installed if it was left out during the initial installation procedure.

10.12i Subroutines

Subroutines are complete sections of computer code that can calculate parameter values and/or take some spreadsheet actions. Subroutines can be called from other computer code, including from other subroutines. However, only the special subroutines called macros can be called directly from the spreadsheet. Subroutines have the following syntax:

```
Sub name (arguments)
    statements
End Sub
```

As all object names, the subroutine name must start with a letter, and consist of up to 40 letters, numbers and/or underscores. Empty spaces are not allowed, nor are punctuation marks. Separate words are therefore often strung together, as in WeightedLeastSquares, often with caps to indicate where spaces would have been. You can also use the underscore _ as a separator.

The arguments are a list of (zero, one, or more) parameters to be passed *into* or *out of* the subroutine, except when the parameters in question have been declared Public, see section 10.1g. We have already encountered numerous subroutines in sections 10.5, 10.6, and 10.7. For instance, in section 10.7 we used the subroutines WLS0( ) and WLS1( ) to assign a numerical value to p (of either 0 or 1) for use in the subsequent subroutine WeightedLeastSquares(p), while the latter in turn farmed out its matrix operations to the subroutines Transpose(M1,r1,c1,Mout), Multiply(M1,r1,c1,M2,c2, Mout), and Invert(M1,r1,n,Mout). The use of Public parameters was illustrated in section 10.9.

10.12j Macros

Macros are special subroutines that can be called directly from the spreadsheet, either through Tools ⇨ Macro or (when pre-arranged) by direct selection from a menu, a toolbar, or with a shortcut key. We have already encountered numerous macros in this chapter; they all have the syntax:

```
Sub name()
    statements
End Sub
```

While macros cannot have arguments (hence the empty brackets behind their names), they can have subroutines. Sometimes macros are therefore used as "drivers" for more substantial subroutines. For example, the following subroutine specifies the value of the parameter i, then calls the main subroutine to do the work:

```
Sub InverseFT()
Dim iSign As Integer
```

```
iSign = -1
Call Fourier(iSign)
End Sub
```

This allows the main subroutine to be called conveniently by two (or more) different drivers, while responding slightly differently because of the different parameter values passed on to the subroutine. The same approach was also used in section 10.6 to distinguish between convolution and deconvolution, and in section 10.7 to select either the general weighted least squares or that which forces the curve through the origin.

10.12k Functions

Functions are code fragments that return a specific numerical value. The syntax is

```
Function name(arguments)
  statements, including name = expression
End Function
```

As with subroutines, the list of arguments is optional, but the brackets are required even if no argument is passed. One of the statements must define the value to be returned, and assign it to the name of the function. For example, Pi( ) calculates the value of π, Fact(14) computes the factorial $14! = 14 \times 13 \times 12 \times \cdots \times 2 \times 1$, and BesselK(3,1) provides the value of the modified Bessel function of order 1 and argument 3, i.e., $K_1(3)$.

There are two types of functions: those provided with Excel, and those written by the user. Tables 1.6-2 through 1.6-6 list a number of the mathematical functions provided in Excel. In chapter 9 we already encountered user-defined functions, which act like macros, except that they pertain to a single cell, and as output can only produce a numerical value.

10.12l Message boxes

A message box is primarily used to provide textual and/or numerical information to the user; it can also carry logical information back to the program that issued it. Message boxes figured prominently in sections 10.1 and 10.2. In their most basic form they have the syntax

```
MsgBox "message"
```

but they can also carry information, as in

```
MsgBox "The value of x is " & x
```

and their general form is

```
MsgBox "prompt", buttons and/or icons, "title"
```

where the buttons can be OK (the default); OK or Cancel; Yes or No; Yes, No or Cancel; Abort, Retry, or Ignore; and Retry or Cancel. The icons denote a message for your information, a question, an exclamation, or a warning. If you want to use both a button and an icon, use a plus sign between them, as in

```
MsgBox "Do you want to go on?", vbYesNo + vbQuestion, _
"Continue?"
```

The prompt or message will appear inside the box, the title in the blue box top.

A message box can also be used as a function, as in

```
Answer = MsgBox ("Beep?", vbYesNoCancel + vbQuestion)
If Answer = vbYes Then Beep
```

Apart from their obvious uses in finished macros and subroutines, message boxes are convenient during program development for debugging, by displaying intermediate parameter values, and for enforcing a pause in program execution.

10.12m Input boxes

Input boxes are meant to provide textual and/or numerical information to the program. Apart from a prompt, as in the Message Box, they also have a window, in which information can be typed. Several examples were already given in section 10.2. The syntax for the input box *function* is

```
InputBox (prompt, title, default, …, type)
```

of which only the prompt is required. The optional default displays a default value in the window, that will be used unless it is modified. The dots denote several additional, optional specifiers of, e.g., the position of the box on the screen. "Type" specifies, in a numerical code, what kind of information is expected: 0 indicates a formula, 1 a number, 2 a text string, 4 a logical value (i.e., True or False), 8 a cell reference (i.e., an address or address range), 16 an error value, and 64 an array of values. More than one input format can be indicated by specifying several types simultaneously, separated by a plus-sign, as in Type: $= 2 + 4$.

An input box is typically used as a function, as in

```
Password = InputBox ("Please enter your password here",
Entry Sentry)
vAnswer = InputBox ("Enter the multiplier here")
```

or

```
Set myCell = InputBox (Prompt:="Select a cell", Type:=8)
```

which sets the value of myCell to that of the cell reference entered in the window of the Input Box. Sometimes the specific numerical value of a string must be extracted from it with the Val function, as in

```
myVal = Val ("123")
```

The more powerful input box *method* allows one to enter an address either directly or by pointing to it with the mouse, as in

```
Set myCell = Application.InputBox(Prompt:="Select a cell", _
Type:=8)
```

10.12n Dialog boxes

Dialog boxes are the most general of type of boxes in Excel. They provide a large array of input devices, including windows for textual or numerical information, and different types of labels and buttons. In addition to the more than 200 dialog boxes already available in Excel, you can make and use your own custom Dialog Boxes, which will be stored in special Dialog Sheets. They are very flexible, but their construction is too complex to be described here; instead, the reader should consult a book on VBA.

10.12o Collective statements

The With statement relieves the programmer of the need to repeat a common object hierarchy in order to specify multiple methods and/or properties of the same object, and (because the computer does not have to search more than once for the common object either) speeds up program execution as well. The syntax is

```
With common object hierarchy
      .statements
End With
```

where all common statements must start with a period. For example, the macro `Appearance()` in section 10.12g can be replaced by the somewhat longer but more easily readable

```
Sub FormatCell()
Selection.HorizontalAlignment = xlCenter
Selection.NumberFormat = "0.000"
  With Selection.Font
     .Bold = True
     .Italic = False
     .Name = "Times New Roman"
     .Size = 10
     .Underline = xlSingle
     .ColorIndex = 32
  End With
End Sub
```

10.12p For ... Next loops

There are two types of **control loops**. When you know how many times a particular procedure should be repeated, use a **For...Next** loop; when you do not know this in advance, use a **Do** loop instead.

The **For…Next** loop has the syntax

```
For counter = start To end (Step stepsize)
    (Statements)
    (Exit For)
    (Statements)
Next counter
```

as in

```
For j = 4 To 10 (Step 2)
   X(j) = X(j) + 3
Next j
```

which will add 3 to X(4), X(6), X(8) and X(10). Specifying the step size is optional; the default step size is $+1$. Specifying the counter in the closing statement (i.e., the j in Next j) is also optional, but makes the statement easier to read. For … Next statements can include If…Then conditions that may lead to a jump out of the loop (with Exit For), or to end the program. The Fourier transform, convolution, and weighted least-squares programs contain many For … Next loops that can be consulted as examples.

When a For … Next loop contains only a single, short statement, it may be written on one line, with its three parts separated by colons, as in

```
For j = 4 To 10 (Step 2): X(j) = X(j) + 3: Next j
```

For…Next loops may be nested. In that case it is good practice to use separate counters, as in

```
For i = 1 To rowmax
  For j = 1 To colmax
     X(i,j) = X(i,j) + 3
  Next j
Next i
```

where the indentation emphasizes the loop structure, and tends to alert you to incorrectly nested loops. Identation is ignored by the computer.

The **For Each…Next** loop is similar to a For…Next statement except that it will automatically apply the statements to each element of the specified group. The syntax is

```
For Each element In group
    (Statements)
Next element
```

For example,

```
For Each Cell In Selection
   If Cell.Value < 0 Then Cell.Interior.ColorIndex = 3
Next Cell
```

which colors the background red (ColorIndex = 3) of any cell in Selection that has a negative value. You don't have to specify in advance how large Selection will be, or determine it before the loop starts: the computer will figure it out as long as Selection is highlighted. If you prefer to color the numbers themselves, rather than their backgrounds, use

```
For Each Cell In Selection
   If Cell.Value < 0 Then Cell.Font.ColorIndex = 7
Next Cell
```

which displays the numbers in hot pink (`ColorIndex = 7`) instead.

10.12q Do loops

There are three types of **Do loops**: Do…, Do…While, and Do…Until. Of these, the first is somewhat tricky, because it is unconstrained, i.e., it could go on forever, in which case you may have to stop it with Esc or Ctrl + Break. Below we only discuss the constrained Do loops. The **Do…While** loop has the syntax

```
Do While condition
   (Statements)
Loop
```

or

```
Do
   (Statements)
Loop While condition
```

where the location of the While condition makes a difference: Do While …Loop will skip the loop when the condition is not met, whereas Do … Loop While will run the loop at least once. The following example computes the factorial of a number highlighted in the spreadsheet, and displays its result in a message box.

```
Sub Factorial()
N = Selection.Value
c = 1
fact = 1
Do While c < N + 1
   fact = fact * c
   c = c + 1
Loop
MsgBox "factorial(" & N & ") = " & fact
End Sub
```

The syntax of the **Do…Until** loop is similar except that the term 'While' is replaced by 'Until'. In the above example, the Do Until statement would read `Do Until c > N` to yield the same result. As with For…Next loops, Do loops can be nested.

10.12r Conditional statements

The **If...Then** statement is a one-liner statement, with the syntax

```
If condition Then result
```

as in

```
If x <> 0 Then y = 1/x
```

or

```
If b*b-4*a*c < 0 Then MsgBox "The roots are complex."
```

When the condition requires more space than can be accommodated on a single line, the syntax is

```
If condition Then
   results
      .
      .
      .
   results.
End If
```

in order to indicate where the listing of results ends, as in

```
If b*b-4*a*c > 0 Then
   Root1 = (-b = Sqr(b*b-4*a*c))/(2*a)
   Root2 = (-b - Sqr(b*b-4*a*c))/(2*a)
End If
```

The **If...Then, Else** statement has unrestricted length, and uses the syntax

```
If condition Then
   statements
Else
   statements
End If
```

As an example consider solving the quadratic equation $ax^2 + bx + c = 0$ given specific values for a, b, and c:

```
D = b*b-4*a*c                    'D is the discriminant
If D >= 0 Then
   ReRoot1 = (-b+Sqr(D))/(2*a)
   ImRoot1 = 0
   ReRoot2 = (-b-Sqr(D))/(2*a)
   ImRoot2 = 0
Else
   ReRoot1 = -b/(2*a)
   ImRoot1 = Sqr(-D)/(2*a)
   ReRoot2 = ReRoot1
   ImRoot2 = -ImRoot1
End If
```

When more than two choices must be considered in an If…Then, Else statement, these can be handled using **If…Then, ElseIf, Else.** In the above example one might instead write

```
If D > 0 Then
   ReRoot1 = (-b+Sqr(D))/(2*a)
   ImRoot1 = 0
   ReRoot2 = (-b-Sqr(D))/(2*a)
   ImRoot2 = 0
ElseIf D = 0 Then
   ReRoot1 = -b/(2*a)
   ImRoot1 = 0
   ReRoot2 = Root1
   ImRoot2 = 0
Else
   ReRoot1 = -b/(2*a)
   ImRoot1 = Sqr(-D)/(2*a)
   ReRoot2 = ReRoot1
   ImRoot2 = -ImRoot1
End If
```

Note that ElseIf is followed by a condition, followed by Then, whereas Else stands alone. When you have multiple ElseIf statements all evaluating the same expression, the code may be made more efficient as well as easier to read by using **Select Case** statements instead.

10.12s Exit statements

Exiting a program before it has reached its end can be done with the **End** statement, which stops the program. In order to exit a part of the program, such as a loop, but continue the remainder of the program (or, in a set of nested loops, switch to the next-higher loop), use an **Exit For** statement if in a For … Next loop, or an **Exit Do** when in a Do … loop. There are also corresponding **Exit Function** and **Exit Sub** statements which allow you to exit the function or subroutine while not ending the entire program. Such Exit statements usually follow an appropriate If statement.

10.13 Summary

Spreadsheets are very powerful and convenient computational tools to illustrate mathematical relationships, and to solve numerical problems, as demonstrated in this book within the context of analytical chemistry. The original spreadsheets were poorly suited to perform some types of mathematical operations, such as iterations. Fortunately, the open structure of modern Excel allows the user to introduce extra features, by incorporating additional programs that accommodate particular needs. Macros are the most convenient and user-friendly way to give the spreadsheet such additional functionality and power. Moreover, they can interact directly with the spreadsheet through message boxes, input boxes, and dialog boxes, and they can even be used to modify spreadsheet commands.

In this closing chapter we have illustrated a variety of approaches suitable to a primarily didactic environment, by empowering the user to write his or her own programs with a minimum of programming expertise. This is why we have provided a number of easily modifiable examples of macros rather than nicely packaged, tamper-proof but invisible and unalterable add-in routines. The purpose of this chapter, then, is two-fold: (1) to provide a few specific macros which the reader may need for some of the exercises, and perhaps can use subsequently as well, and (2) to give enough specific examples to entice the reader to write macros when and where needed, and to show a variety of ways in which this can be done.

In order to illustrate the various ways in which data can be read from the spreadsheet and the results be returned, the sample macros in sections 10.5 (Fourier transform), 10.6 (convolution), and 10.7 (weighted least squares) read an entire block of input data through the simple command `Array = Selection.Value`, and return an output to the spreadsheet with the complementary write command `Selection.Value = Array`. This is, perhaps, the easiest way to go between the spreadsheet and the macro, although it is limited to one single block of input data. (As illustrated in sections 10.6, input boxes can still be added where needed.) Alternatively, in sections 10.2 (interpolation), 10.3 (propagation of imprecision), 10.8 (Solver imprecision), 10.9 (equidistant least squares), 10.10 (semi-integration & semi-differentiation) and 10.11 (reducing data density) we use input boxes to pass information from spreadsheet to macro. And section 10.9 illustrates the elegant program of Professor Barak for smoothing and differentiating equidistant data sets with a moving polynomial of self-optimized order in two different formats. Use whichever method works for you and for the problem at hand.

Between the input and output stages of a macro, the user has complete control over what to do with the captured data. Because VBA is an extension of BASIC, the programmer does not need to know how to write code for specialized applications, such as a fast Fourier transform, or a matrix inversion, but instead can incorporate well-documented general-purpose programs such as those of the *Numerical Recipes*. Note that these are freely usable only for private use; copyright must be obtained for their commerical use.

The immediacy of access to spreadsheet macros, and the ease of modifying and testing them, invites user experimentation and extension. The macro editor checks the code line-by-line as it is entered, compilation is automatic, and errors seldom lead to computer hang-up. Modern spreadsheets can also serve as gateways to more sophisticated computing; in this respect, macros form the bridge to using higher languages.

Most computing needs of undergraduate (and even graduate) students in chemistry can be met with a spreadsheet such as Excel (beyond version 4) that includes macros accepting higher-language code. Likewise, the spreadsheet can satisfy most graphing needs of students and faculty. This does not mean that spreadsheets are the solution to all computational problems. There is still plenty of room for improvement in Excel macros, especially in the area of error messages, which are often far too cryptic for the novice programmer, and can be outright misleading. Spreadsheet macros may be too slow for complicated numerical simulations, they cannot do formal (rather than numerical) calculations the way programs such as Maple or Mathematica can, and they are also unsuitable for highly specialized applications such as occur in X-ray crystallography, quantum mechanics, or molecular modeling. Writing such programs is so complicated that it is best left to the specialist. But then, not too long ago the same used to be true for much of data analysis based on least-squares or Fourier transformation.

Modern spreadsheets go a long way to make the calculational power of personal computers available to a large scientific audience, in a way that minimizes the psychological and instructional obstacles, and therefore maximizes their potential usefulness.

INDEX